This book is the first comprehensive and coherent introduction to the role of cosmic strings and other topological defects in the universe. This study has been one of the major driving forces in cosmology over the last decade, and lies at the fruitful intersection of particle physics and cosmology. After an introduction to standard cosmological theory and the theory of phase transitions in the early universe, the book then describes, in turn, the properties, formation, and cosmological implications of cosmic strings, monopoles, domain walls and textures. The book concludes with a chapter considering the role of topological defects in inflationary universe models. Ample introductory material is included to make the book accessible to the wide audience of particle physicists, astrophysicists and cosmologists for whom this topic is of immediate interest.

T0224795

CAMBRIDGE MONOGRAPHS ON
MATHEMATICAL PHYSICS

General Editors: P. V. Landshoff, D. R. Nelson, D. W. Sciama, S. Weinberg

COSMIC STRINGS AND
OTHER TOPOLOGICAL DEFECTS

Cambridge Monographs on Mathematical Physics

A. M. Anile *Relativistic Fluids and Magneto-Fluids*
J. Bernstein *Kinetic Theory in the Early Universe*
N. D. Birrell and P. C. W. Davies *Quantum Fields in Curved Space*[†]
D. M. Brink *Semiclassical Methods in Nucleus–Nucleus Scattering*
J. C. Collins *Renormalization*[†]
P. D. B. Collins *An Introduction to Regge Theory and High Energy Physics*[†]
M. Creutz *Quarks, Gluons and Lattices*[†]
F. de Felice and C. J. S. Clarke *Relativity on Curved Manifolds*
B. DeWitt *Supermanifolds, second edition*[†]
P. G. O. Freund *Introduction to Supersymmetry*[†]
F. G. Friedlander *The Wave Equation on a Curved Space-Time*[†]
J. Fuchs *Affine Lie Algebras and Quantum Groups*
J. A. H. Futterman, F. A. Handler and R. A. Matzner *Scattering from Black Holes*
M. Göckeler and T. Schücker *Differential Geometry, Gauge Theories and Gravity*[†]
M. B. Green, J. H. Schwarz and E. Witten *Superstring Theory, volume 1: Introduction*[†]
M. B. Green, J. H. Schwarz and E. Witten *Superstring Theory, volume 2: Loop Amplitudes, Anomalies and Phenomenology*[†]
S. W. Hawking and G. F. R. Ellis *The Large-Scale Structure of Space-Time*[†]
F. Iachello and A. Arima *The Interacting Boson Model*
F. Iachello and P. van Isacker *The Interacting Boson Fermion Model*
C. Itzykson and J.-M. Drouffe *Statistical Field Theory, volume 1: From Brownian Motion to Renormalization and Lattic Gauge Theory*[†]
C. Itzykson and J.-M. Drouffe *Statistical Field Theory, volume 2: Strong Coupling, Monte Carlo Methods, Conformal Field Theory, and Random Systems*[†]
J. I. Kapusta *Finite-Temperature Field Theory*
D. Kramer, H. Stephani, M. A. H. MacCallum and E. Herlt *Exact solutions of Einstein's Field Equations*
N. H. March *Liquid Metals: Concepts and Theory*
L. O'Raifeartaigh *Group Structure of Gauge Theories*[†]
A. Ozorio de Almeida *Hamiltonian Systems: Chaos and Quantization*[†]
R. Penrose and W. Rindler *Spinors and Space-time, volume 1: Two-Spinor Calculus and Relativistic Fields*[†]
R. Penrose and W. Rindler *Spinors and Space-time, volume 2: Spinor and Twistor Methods in Space-Time Geometry*
S. Pokorski *Gauge Field Theories*[†]
V. N. Popov *Functional Integrals and Collective Excitations*[†]
R. Rivers *Path Integral Methods in Quantum Field Theory*[†]
R. G. Roberts *The Structure of the Proton*
W. C. Saslaw *Gravitational Physics of Stellar and Galactic Systems*[†]
J. M. Stewart *Advanced General Relativity*
A. Vilenkin and E. P. S. Shellard *Cosmic Strings and Other Topological Defects*
R. S. Ward and R. O. Wells Jr *Twistor Geometry and Field Theories*[†]
G. F. Bertsch and R. A. Broglia *Oscillations in Finite Quantum Systems*

[†] Issued as a paperback

COSMIC STRINGS
AND OTHER
TOPOLOGICAL DEFECTS

A. VILENKIN

Tufts University

E. P. S. SHELLARD

University of Cambridge

CAMBRIDGE
UNIVERSITY PRESS

CAMBRIDGE UNIVERSITY PRESS
Cambridge, New York, Melbourne, Madrid, Cape Town, Singapore, São Paulo, Delhi

Cambridge University Press
The Edinburgh Building, Cambridge CB2 8RU, UK

Published in the United States of America by Cambridge University Press, New York

www.cambridge.org
Information on this title: www.cambridge.org/9780521654760

First published 1994
First paperback edition 2000
Reprinted 2001

A catalogue record for this publication is available from the British Library

ISBN 978-0-521-39153-5 hardback
ISBN 978-0-521-65476-0 paperback

Transferred to digital printing 2009

To
Inna & Alina
John & Florence

Contents

Preface

The subject of this monograph is the study of the properties, evolution and cosmological implications of topological defects – cosmic strings, domain walls, monopoles and textures. This area of research has expanded tremendously during the last decade and still remains one of the most active fields in modern cosmology. The growing interest in topological defects has at least a threefold motivation: First, defects arise in a wide class of elementary particle models. They are inevitably formed during phase transitions in the early universe, and their subsequent evolution and observational signatures must therefore be understood. Secondly, observational cosmology has posed a number of perplexing enigmas, ranging from the origin of large-scale structure through to the baryon asymmetry of the universe. Credible solutions for some of these cosmic enigmas have invoked topological defects of one kind or another. Finally, apart from their possible astrophysical roles, topological defects are fascinating objects in their own right. Their properties, which are very different from those of more familiar systems, can give rise to a rich variety of unusual physical phenomena.

Our aim in writing this book has been to provide a systematic exposition of the potential role of topological defects in our universe. We hope that it will prove useful for a broad range of readers: We give a pedagogical introduction to the essential concepts for newcomers, while also endeavouring to provide a comprehensive – though not exhaustive – reference text for the specialist. Some background in quantum field theory, general relativity and cosmology will undoubtedly prove helpful, but we have aimed to make the book as self-contained as possible. There are two predominant directions from which researchers move into this area – field theory and astrophysics – and we hope to cater for both with an introduction to the standard cosmology and the essential ingredients of gauge theories and phase transitions in chapters 1 and 2. Following this, in chapter 3, we briefly describe each of the different types of topological defect, along with their classification scheme.

Severe observational constraints restrict the creativity of theorists when dealing with monopoles and domain walls, but the situation is much more favourable for cosmic strings. It is not surprising, therefore, that the literature is dominated by cosmic strings and that the corresponding chapters occupy more than half of this monograph. The reader interested only in astrophysical applications of strings may wish to skip chapters 4–9 and concentrate on 10, 11 and 12. The necessary background from the earlier chapters is reviewed in §10.1 and §12.1.1. The remaining chapters on domain walls, monopoles and textures reflect more recent interest in the cosmology of these defects and their hybrids. Finally, in the last chapter we discuss developments towards reconciling topological defects with inflation, the other major motivating force in theoretical cosmology over the last decade.

It is appropriate to comment on the nature of the growing body of literature on this subject, since this gave added impetus for writing the monograph. Inevitably, it reflects the fact that the understanding of the role of defects has been subject to significant evolution since the early 1980s. To the uninitiated this can appear confusing – sometimes even contradictory – so a major aim has been to provide guidance with a critical review of the literature. Nevertheless, a considerable proportion of the material is new; clarity often dictated that we adopt a different approach to the original and some of the material is entirely new.

Given that rapid development continues and that the observational constraints on defects are steadily improving, we have placed most emphasis on the permanent features of this subject area, while attempting to be as comprehensive and up-to-date as possible. As we have already mentioned, a variety of roles have been suggested for topological defects in a number of cosmological scenarios. Most of these suggestions will probably turn out to be incorrect, and some may even be ruled out before this book is in print. In the sections where we discuss such speculative ideas, however, we have tried to clearly state the underlying assumptions and to point to possible observational tests. Of course, the book unavoidably reflects our own biases and misconceptions, and we invite readers to let us know about any errors or omissions.

We would like to thank a number of colleagues for undertaking to read various chapters, for pointing out a number of errors and for their helpful comments, including Pedro Avelino, Richard Battye, David Bennett, Luis Bettencourt, Robert Brandenberger, Eugene Chudnovsky, Anne Davis, David Garfinkle, Jaume Garriga, Gary Gibbons, Tom Kibble, Robert Leese, Jim Peebles, Martin Rees, Soo-Jong Rey, David Spergel, Albert Stebbins, Chris Thompson and Tanmay Vachaspati. We are especially grateful for critical remarks and useful suggestions from Martin Bucher, Allen Everett and Mark Hindmarsh. We also thank Roger Glisson, Ruth Harris and Sarah Kirkup for TEXing a number of chapters. A.V. would like to acknowledge funding from the National Science Foundation. E.P.S. has been supported by Trinity College and the Science and Engineering Research Council.

Alexander Vilenkin
Paul Shellard   *September 1993*

Paperback preface

Since the publication of the hardback edition in 1994, there have been some significant developments in the cosmology of topological defects. For the most part, these have been quantitative advances or due to observational improvements, rather than fundamental changes in the physical and mathematical framework described in the original text. The role of this extended preface, therefore, is to review the literature over the past six years, drawing the attention of the reader to areas in which there has been the most activity. Highlights include improvements in understanding defect formation, new mechanisms for generating ultrahigh energy cosmic rays and, with a particular emphasis here, tight constraints on large-scale structure formation scenarios.

The limitations of space have made it necessary to be selective in our referencing and, possibly, some serious oversights will have occurred. We are grateful to those readers of the hardback edition who pointed out a number of errors and omissions which we have endeavoured to correct in the main text or in this preface. In this respect we would especially like to thank Thibault Damour. We are also grateful for helpful comments on various topics in the preface from Richard Battye, Luis Bettencourt, Jose Blanco-Pillado, Martin Bucher, Tom Kibble, Carlos Martins, Ken Olum and Tanmay Vachaspati.

0.1 Defect formation

0.1.1 Second-order phase transitions

Defect formation at a second-order phase transition has recently attracted much attention, in large measure due to the possibility of testing various defect formation models in condensed matter experiments. Kibble's [1976] original picture, which we followed in chapter 9, assumed that the initial density of defects is determined by the correlation length ξ of the Higgs field ϕ at the 'freeze-out' temperature T_f. This temperature was identified with the Ginzburg temperature T_G, below which thermal fluctuations are ineffective in restoring the symmetry on the scale of the correlation length. It now appears that this picture needs revision.

Zurek [1985a, 1996] has emphasized the importance of the relaxation timescale $\tau(T)$, which is the time it takes correlations to establish on the length scale $\xi(T)$. As the critical temperature T_c is approached, both $\xi(T)$ and $\tau(T)$ diverge as inverse powers of $(T - T_c)$ with appropriate critical exponents. At some point $\tau(T)$ becomes longer than the characteristic dynamical timescale of the variation of the temperature, $t_D = |T - T_c|/|\dot{T}|$. After this, the growth of the correlation length terminates: correlations do not have enough time to establish on larger scales. Zurek suggested that the freeze-out temperature T_f is determined by the condition $\tau \sim t_D$. A similar suggestion was made by Kibble [1980], although he did not return to it in his later work. We shall refer to this description of defect formation as the Kibble–Zurek picture.

The initial distribution of defects in this picture is given by the distribution of zeros of the Higgs field, smoothed over the correlation lengthscale ξ, in thermal equilibrium at $T = T_f$. It can be determined directly from numerical simulations of field theory thermodynamics and, for strings in an $O(2)$-symmetric $\lambda\phi^4$ model, this has been done by Antunes, Bettencourt & Hindmarsh [1998] and Antunes & Bettencourt [1998]. They studied, in particular, the temperature dependence of the equilibrium fraction of the total string length in infinite strings, $f_\infty(T)$. (To see the importance of this quantity, note that closed loops shrink and disappear soon after the phase transition, and if there were no infinite strings, one would be left with no strings at all.) They found that, as the critical temperature is approached from above, f_∞ decreases slowly from its high-temperature value $f_\infty(\infty) \approx 0.75$ to $f_\infty(T_c) \approx 0.4$, but then drops discontinuously and vanishes at lower temperatures. The continuous decrease in f_∞ is due to the long-range correlations in the Higgs field near T_c [Robinson & Yates, 1996; Scherrer & Vilenkin, 1997]. Correlations decay exponentially on scales greater than ξ, so after smoothing one can expect to recover the value obtained in early lattice simulations with uncorrelated fields, $f_\infty = 0.7 - 0.8$ (see §9.1.2). The precise value of f_∞ is, of course, not required for an order of magnitude estimate of the string density at formation. The important point is that $f_\infty \sim 1$ above T_c.

The Kibble–Zurek picture of defect formation is supported by numerical simulations [Laguna & Zurek, 1997; Yates & Zurek, 1998; Antunes, Bettencourt & Zurek, 1999], by calculations using the methods of nonequilibrium quantum field theory [Stephens, Calzetta, Hu & Ramsey, 1999], and by experiments in liquid ^{3}He [Bauerle *et al.*, 1996; Ruutu *et al.*, 1996,

1998]. However, the most recent ^{4}He experiments do not find defects at the predicted densities [Dodd *et al.*, 1998]. Rivers [1999] attributes the low vortex density in ^{4}He to large thermal fluctuations of the order parameter in the Ginzburg regime, $T_{\mathrm{f}} > T > T_{\mathrm{G}}$. This temperature range is very narrow for ^{3}He, while for ^{4}He the whole experiment takes place within the Ginzburg regime. On the other hand, Bettencourt, Antunes & Zurek [2000] dispute Rivers' conclusions and argue that the defect density is largely unaffected by thermal fluctuations in the Ginzburg regime. Karra & Rivers [1998] have also questioned the assumption that the typical defect separation at formation is given by the correlation length. Their analysis suggests that the two length scales are generally different, although they do coincide under some restricted conditions.

We thus see that, despite significant progress towards understanding defect formation in second-order phase transitions, the picture is still far from complete. This topic continues to inspire collaboration between condensed matter physicists and cosmologists and has been the subject of a series of workshops (see, for example, Davis & Brandenberger [1995]). Parallels between condensed matter physics and early universe cosmology have been explored in Volovik & Vachaspati [1996] and Volovik [1999].

0.1.2 First-order phase transitions

The density of defects formed in a first-order phase transition depends on the efficiency of phase equilibration in bubble collisions. Consider, for example, vortex formation resulting from the collision of three bubbles with different scalar field phases in their interiors. The phase difference between each pair of bubbles can be calculated using the 'geodesic rule', that is, taking the shortest path between θ_1 and θ_2. Naively, one might expect that a vortex will be formed when the three phase differences add up to 2π and that there will be no vortex if they add up to zero. Recent analysis has shown, however, that the physics of defect formation in bubble collisions is more interesting and complicated.

Three bubbles are not likely to collide simultaneously. The collision of the first two bubbles sets up a complicated dynamics of phase oscillation and equilibration. If the phase equilibrates before the third bubble arrives, then no defect is going to be formed. In the case of global symmetry breaking, the phase θ is a massless Goldstone boson, and phase waves from the collision region propagate into the bubble interiors at the speed of light. If the bubble walls expand without damping, then they move with a speed close to the speed of light, and so the circle of intersection of the two bubbles expands superluminally. The phase waves can never catch up with the intersection circle and, therefore, cannot affect the phase at the points of collision with the third bubble. We thus conclude that the geodesic rule should be valid in this case [Melfo & Perivolaropoulos, 1995; Ferrera, 1999].

If there is a substantial damping force acting on the bubble walls, the walls will reach a terminal speed $v \ll 1$. The phase waves will then catch up with the expanding walls, they will get reflected from the walls, and the returning waves will 'flip' the initial phase difference. The resulting phase oscillations are damped due to the expansion of the bubbles, and the phase eventually equilibrates [Ferrera & Melfo, 1996]. As the damping is increased, the arrival of the third bubble is delayed, and vortex formation is less likely. Simple geometric simulations of the nucleation and growth of bubbles, leading to the completion of the phase transition and to the formation of defects have been performed by Borrill, Kibble, Vachaspati, & Vilenkin [1995], Borrill [1996], and by de Laix & Vachaspati [1999]. Field theory simulations in $(2+1)$ dimensions performed by Ferrera [1998] confirm that the density of defects drops as the damping is increased.

In the case of a gauge symmetry breaking, the situation is complicated by the fact that the phase θ is gauge-dependent. One can nonetheless define a gauge-invariant phase difference between two points A and B as

$$\Delta\theta = \int_A^B dx^i(\partial_i\theta - ieA_i). \tag{0.1}$$

The points A and B can be taken at the centres of the colliding bubbles. One then finds that $\Delta\theta$ is constant while the bubbles are well-separated and undergoes damped oscillations after the collision [Kibble & Vilenkin, 1995; Copeland, Saffin & Tornkvist, 1999; Davis & Lilley, 1999]. The oscillation period is determined by the gauge boson mass, and the damping time by the plasma conductivity and the bubble expansion speed. Both timescales are typically much smaller than the bubble radii at collision.

The currents associated with the oscillating phase gradients generate a magnetic field which is concentrated along the intersection circle of the bubble walls, with the total magnetic flux given by $\Phi = \Delta\theta/e$. Once again, in the case of relativistic bubbles the expansion of the circle is superluminal, and the flux cannot escape. In a three-bubble collision, the three magnetic flux tubes combine when the bubbles meet. If the sum of the initial phase differences is $\Delta\theta_{\text{tot}} = 2\pi$, then a vortex is formed carrying a flux quantum $\Phi = 2\pi/e$. On the other hand, if the bubble expansion speed is small, the magnetic field can spread into the exterior region, resulting in a suppression of the defect density.

0.1.3 Formation of semilocal strings

The dynamics of string formation is more complicated in the case of semilocal strings, whose stability is not enforced by topology and depends on dynamical considerations (refer to section 4.2.5). Unlike the 'usual', topologically stable strings, semilocal strings can terminate at Higgs field configurations resembling global monopoles, with the gauge magnetic flux spreading from the end of the string. Shortly after the phase transition, one can expect to have a distribution of short string segments terminating at such monopoles. The segments can contract under the string tension and annihilate. However, the monopoles strongly interact with a force independent of the distance. This force, which for stable strings is greater than the string tension, can pull the monopole and antimonopole at the ends of neighbouring segments together, resulting in a longer string [Hindmarsh, 1993].

Numerical simulations of semilocal string formation in flat spacetime have been performed by Achucarro, Borrill & Liddle [1999]. They found that the process of short segments joining into longer strings is rather efficient and that the resulting length density of semilocal strings can be comparable to that of topologically stable strings obtained from similar initial conditions. The results vary, depending mainly on the parameter $\beta = \lambda/2e^2$ characterising the relative strength of Higgs and gauge couplings, with the string density vanishing at $\beta > 1$, where the semilocal strings are unstable. There remain interesting and outstanding questions such as under what conditions the string joining process is efficient enough to form infinite strings.

For a review of semilocal strings, and of the closely related subject area of electroweak strings, refer to the recent article by Achucarro & Vachaspati [1999].

0.2 Defect evolution

0.2.1 Nambu–Goto string networks

Developing a complete quantitative description of string network evolution remains a challenging problem. However, there has been significant progress in both the numerical and analytic modelling of the properties of a string network.

The analytic 'one-scale' model presented in section 9.3 describes long-string evolution in terms of a variable correlation length L given two free parameters, the loop chopping efficiency c_N and the rms velocity $\langle v_N^2 \rangle^{1/2}$ (see (9.3.16)) where N labels a scale-invariant epoch such as the pure matter or radiation eras. A generalisation known as the 'velocity-dependent one-scale' model [Martins & Shellard, 1996, 2000a] greatly extends its range of applicability by incorporating a variable rms velocity v as well as frictional damping, while reducing the free parameters to a single rescaled chopping efficiency $\tilde{c}$. The coupled equations for L and v

are the following:

$$\frac{dL}{dt} = HL + \mathcal{H}Lv^2 + \tfrac{1}{2}\tilde{c}v, \qquad \frac{dv}{dt} = (1 - v^2)\left(\frac{k}{L} - 2\mathcal{H}v\right), \qquad (0.2)$$

where $\mathcal{H} = H + (2\ell_{\rm f})^{-1}$ is a modification of the Hubble parameter H which includes a friction damping lengthscale $\ell_{\rm f}$ as in (8.1.42). Here, k is a phenomenological parameter which can be related to small-scale structure on the string. It can be found in asymptotic relativistic and non-relativistic regimes by direct comparison with exact solutions and simulations, with a fairly accurate ansatz in a cosmological context being $k = 2\sqrt{2}\pi^{-1}(1 - v^6)/(1 + v^6)$. Having fixed $\tilde{c} = 0.23 \pm 0.04$ by comparison with network simulations in a single scale-invariant regime, this model has been tested and shown to be in good agreement with the large-scale properties of string network simulations in other regimes. In particular, it successfully bridges the matter–radiation transition when both L and v change significantly (see fig. 0.1). A simple modification of (0.2) also covers evolution in open and closed FRW cosmologies at late times [Avelino, Caldwell & Martins, 1997] (see also van de Bruck [1998a] for Λ-models).

In order to explore the small-scale features of an expanding universe string network, an extensive programme of simulations has been performed at ultrahigh resolution using the Allen–Shellard code. Evidence is mounting for a typical loop production scale above the gravitational backreaction cutoff $\ell \sim \Gamma G\mu t$ for GUT-scale strings (refer to §9.3.4). Considerable progress has been made characterizing small-scale structure using multifractal methods and modelling its evolution with Carter's elastic string formalism [Martins & Shellard, 2000b]. (This elastic string and brane description is reviewed in Carter [1997].) For a fuller discussion of other analytic models, particularly the 'three-scale' model of Austin, Copeland & Kibble [1995], please refer to the review article by Hindmarsh & Kibble [1995].

The small-scale properties of Minkowski space networks have also been examined in large parallel simulations using the discrete Smith–Vilenkin algorithm [Vincent, Hindmarsh & Sakellariadou, 1997]. The important dependence of small-scale structure on the numerical loop cutoff was studied and successfully characterized using the 'three-scale' model. The further suggestion that loops will always be produced at the smallest scale physically possible (that is, effectively direct particle emission) may well be a result of the numerical lattice discretization chosen and will be discussed below.

The evolution of a network of cosmic string loops (with no infinite strings) has been analysed using a Boltzmann-type rate equation to describe loop collisions and self-intersections [Copeland, Kibble & Steer, 1998]. The model was able to encompass string thermodynamics results

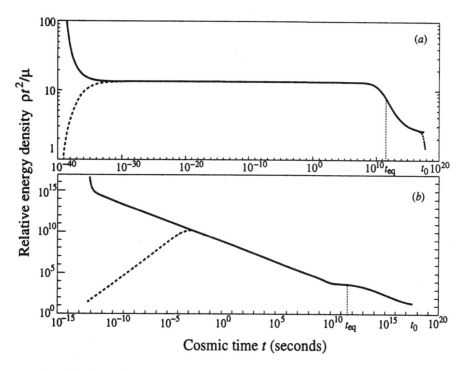

Fig. 0.1. The complete cosmological history of GUT strings (a) and electroweak strings (b) in a flat FRW model (Ω_M=1, h=0.65) [Martins & Shellard, 2000a]. The rescaled energy density $\rho/\mu t^2$ is plotted against cosmic time t for two extreme initial conditions arising from second-order (solid) and first-order (dashed) phase transitions. For GUT strings, note the slow relaxation after equal matter–radiation t_{eq} and, in addition, the late-time evolution (dotted) for a model with a cosmological constant (Ω_Λ=0.8). Electroweak string motion is always damped, only becoming relativistic today, t_0.

in flat space (refer to §6.6) and led to some intriguing possibilities in an expanding universe. Under some circumstances, relative loop densities could grow in the radiation era, but inevitably loops would disappear during the matter era.

0.2.2 Field theory string simulations

Networks of $U(1)$ gauge strings have been studied in abelian-Higgs field theory simulations in flat spacetime [Vincent, Antunes & Hindmarsh, 1998]. Noting a low density of closed loops, the authors argued that long strings lose most of their energy not by loop production, as it is generally believed, but by direct emission of Higgs and gauge bosons. If correct, this claim would dramatically alter the particle content of the background left behind by a string network. However, similar field theory simulations [Moore & Shellard, 1998] showed that the network behaviour could be interpreted using the standard picture of string evolution (0.2), with the lower density of long-lived loops consistent with an expected production size comparable to the string width. Moreover, detailed analysis of oscillating string configurations has found that direct massive particle emission is strongly suppressed when the oscillation frequency is below the mass of emitted particles. In addition, Olum & Blanco-Pillado [1999b] have studied very large amplitude string oscillations showing that the radiation rate is also exponentially suppressed with wavelength.

The evolution of a global string network in an expanding universe has been simulated by Yamaguchi [1999] and Yamaguchi, Yokoyama & Kawasaki [1999]. They found scaling evolution but with string energy densities dramatically lower than those from Nambu network simulations. In the radiation era, they found a relative density $\zeta_r \approx 0.9$ and, in the matter era, $\zeta_m \approx 0.5$, which should be compared with $\zeta_r = 13 \pm 2.5$ and $\zeta_m = 3.5 \pm 1$ from Table 10.1. However, radiation backreaction for global string simulations is extremely high because of their very limited dynamic range. Here, the analogue of $\Gamma G \mu \approx 10^{-4}$ for GUT gauge strings becomes $K^{-1} = \Gamma/(2\pi \ln(t/\delta)) \gtrsim 2$ (see (12.2.2)) which, for cosmological global strings, should be $K^{-1} \approx 0.1$. A radiation backreaction term of the expected magnitude can be added to the one-scale model (0.2) and successfully accommodates the lower densities seen in field theory simulations. Clearly it is vital to incorporate such differences when relying on field theory simulations in a cosmological context.

0.2.3 String radiation backreaction and self-energy

A 'local backreaction approximation' has been proposed for phenomenologically modelling radiative damping by oscillating strings [Battye & Shellard, 1995, 1996a]. This approximation was applied, in the first instance, to global strings using the Kalb–Ramond formalism. Local corrections to the equations of motion were used to separate the divergent self-energy of the string from the radiation field, while also suppressing the 'runaway' solutions ubiquitous in backreaction problems. Comparisons with global string field theory simulations demonstrated that the leading order terms could be used to satisfactorily incorporate radiative damping in Nambu string simulations. This proposal was revisited for strings coupled to axionic, gravitational and dilatonic fields by Buonanno & Damour [1998a]. Using dimensional regularization, they isolated finite local terms in the self-force and found that they lead to antidamping, no damping and damping in the three cases respectively. They suggest that these unphysical results for the axion and gravitational field are due to the failure of the local backreaction approximation for gauge fields, but not for a scalar field like the dilaton. However, since the spectrum and angular distribution of the dilaton, axion and gravitational radiation are very similar, they argue that the dilatonic damping terms can provide a useful model for both axion and gravitational backreaction.

The renormalization of the string self-energy has received detailed attention and a number of erroneous results in the literature have been corrected. In particular, the divergent part of the gravitational self-force of a string, discussed in section 7.7.1, has been shown to vanish [Carter & Battye, 1998]. The nonzero result (7.7.6), which came from varying the connection in the equations of motion, is exactly cancelled by a neglected term arising from varying the d'Alembertian. Similar corrections have been found for strings with axionic and dilatonic couplings [Buonanno & Damour, 1998b] (see also Carter [2000]). For the classical superstring, the divergent axion and dilaton self-interactions cancel as they are equal and opposite.

0.2.4 Superconducting strings

The evolution of string currents has been included in analytic models to describe a superconducting string network [Davis & Dimopoulos, 1998; Martins & Shellard, 1998a]. This is relevant to the origin and evolution of primordial magnetic fields as well as the formation of vortons. Unfortunately, an improved understanding of current loss mechanisms is required before these models acquire a truly quantitative status. The cosmological implications of the formation of vortons have been studied by a number of authors but, given present uncertainties, discrepancies remain in identifying excluded parameter ranges and those which are viable for vorton dark matter [Brandenberger, Carter, Davis & Trodden, 1996; Martins & Shellard, 1998b; Carter & Davis, 1999]. There have been advances in the analytic study of vorton states, with the chiral case proving to be the most tractable [Carter & Peter, 1999].

0.2.5 Nonabelian defect networks

The evolution of nonabelian string networks with several string types is a competition between the intercommuting of like string pairs and the no-crossing behaviour of unlike strings. Whether or not this leads to a scaling solution or a frustrated defect network which dominates the universe has been the subject of numerical study. McGraw [1996, 1998] has studied nonabelian strings in an S^3 model using simple nodal network simulations (like those for Z_3 strings in §9.6), with a new algorithm which tracks the nonabelian fluxes. The freedom to choose different string tensions led to complex new web-like structures, but self-similar evolution was always observed. A discrete lattice model that supports string-like defects with nonabelian windings was used in simulations by Pen & Spergel [1997]. Like McGraw they observed scaling for three string species, but they found strong evidence for network frustration with larger numbers of strings (N=8). There clearly is a need for full nonabelian string simulations to study this evolution more accurately.

On the assumption of string domination, Bucher & Spergel [1999] have studied frustrated network dynamics on scales large compared to the inter-string spacing. Approximating the network as a continuous medium (solid dark matter), they characterized it in terms of the mean density, tension, shear modulus and compressibility. A covariant description in an expanding universe was used to study the response of the frustrated network to linear perturbations and the implications for CMB anisotropy.

These methods have also been applied to frustrated domain wall networks [Battye, Bucher & Spergel, 1999]. Such frustrated networks can potentially reconcile a flat universe with observational evidence for acceleration and a low matter density $\Omega_m \approx 0.3$. At present frustrated walls are difficult to distinguish from quintessence models with an evolving light scalar field, but the situation may improve with future gravitational lensing observations.

0.2.6 Hybrid defect evolution

In models with a symmetry breaking pattern $G \to H \times U(1) \to H \times Z_2$, each monopole formed at the first phase transition gets connected to two strings at the second phase transition. The result is a 'necklace', with monopoles playing the role of beads along the string. The dynamics of necklaces crucially depends on the parameter

$$r = m/\mu d, \qquad (0.3)$$

where m is the monopole mass, μ is the string tension, and d is the average monopole separation along the string. For $r \ll 1$, monopoles are dynamically unimportant, and the behaviour of necklaces on scales $l \gg d$ is essentially the same as that of Nambu–Goto strings. In the opposite case with $r \gg 1$, monopoles dominate the dynamics, resulting in lower velocities and a smaller characteristic lengthscale in a network of necklaces, that is, $v \sim r^{-1/2}$ and $l \sim r^{-1/2}t$. At present, the evolution of necklaces is not well understood. Monopole annihilation tends to decrease r, while frictional damping and gravitational radiation drive r towards larger values. If the latter tendency prevails, the evolution of necklaces may be very different from that of ordinary cosmic strings [Berezinsky & Vilenkin, 1997].

Interactions between different types of defects can play an important role in defect evolution. Consider, for example, a symmetry breaking resulting in the formation of both domain walls and monopoles. This can be arranged in the $SU(5)$ model with an adjoint Higgs field Φ, if the cubic term $\gamma Tr\Phi^3$ in the potential is set to zero. As the walls begin to move because of their tension, they will collide with monopoles. There is no topological obstruction to the unwinding of a monopole when its core is on a domain wall, so one can expect that on collision with walls, monopoles will dissolve into Higgs and gauge boson waves [Dvali, Liu & Vachaspati, 1998]. This expectation has been confirmed in numerical simulations [Pogosian & Vachaspati, 1999b; Alexander, Brandenberger, Easther & Sornborger, 1999]. Monopoles could thus be 'swept away', resolving the cosmic monopole problem (see §14.3). Domain walls can later be removed by introducing a small bias in the Higgs potential (a cubic term with a small γ).

0.3 Large-scale structure and the cosmic microwave sky

Topological defects are one of the few credible alternatives to inflation as a theory for large-scale structure formation. Initial enthusiasm, however, has given way to realism in the face of new observational data and the difficulty in making reliable predictions for these complex nonlinear models. Since the publication of the hardback edition, a number of important developments have taken place. First, there has been a general recognition of the appropriate relativistic methodology for studying defect-induced perturbations. Secondly, considerable computational effort has been expended using these methods to make more precise predictions over a broader range of scales. Finally, observational improvements, particularly from the cosmic microwave background (CMB), have put defect theories under considerable pressure. Indeed, if present calculations are to be believed, then global monopoles and texture theories are ruled out as the primary seeds for large-scale structure formation. However, even in this case, a subsidiary role in structure formation cannot be excluded and distinct defect signatures remain to be tested in future experiments.

What follows is an introduction to the causal compensated perturbation theory necessary for dealing with cosmic defects. The reader more interested in defect predictions confronting observation can skip to §0.3.2.

0.3.1 Relativistic methodology

Perturbation equations for causal sources

In chapter 11 we present methods which are useful for calculating the detailed features of structures generated locally by individual defects. However, a fully relativistic analysis is necessary for longer time periods if we are to deal with perturbations induced on specific lengthscales as they fall within the horizon. Most recent work in the field has been based on the methods first presented in an important paper by Veeraraghavan & Stebbins [1990]. In order to study large-scale structure formation in a specific defect scenario we need to solve the linearized Einstein equations with the defect energy–momentum tensor $T_{\mu\nu}$ explicitly included as a stiff source term. In the simplest flat ($\Omega=1$) FRW model, we can perturb the line element as follows

$$ds^2 = a^2(\tau)[d\tau^2 - (\delta_{ij} - h_{ij})dx^i dx^j]\,, \qquad (0.4)$$

where we have employed the synchronous gauge in which the metric perturbation satisfies $h_{00} = h_{0i} = 0$. The h_{ij} can be decomposed into scalar, divergenceless vector and transverse traceless tensor components. In Fourier space, we can equivalently decompose using rotations about the wavevector direction $\hat{\mathbf{k}}$ to obtain

$$h_{ij} = \frac{1}{3}\delta_{ij}h + (\hat{k}_i\hat{k}_j - \frac{1}{3}\delta_{ij})h^{\mathrm{S}} + (\hat{k}_i h_j^{\mathrm{V}} + \hat{k}_j h_i^{\mathrm{V}}) + h_{ij}^{\mathrm{T}}\,, \qquad (0.5)$$

where $k_i h_i^{\mathrm{V}} = 0$ and $k_i h_{ij}^{\mathrm{T}} = h_{ii}^{\mathrm{T}} = 0$. Here, there are two independent degrees of freedom in each of the vector h_i^{V} and tensor h_{ij}^{T} components, with the other two being the scalar trace $h = h_{ii}$ and the 'anisotropic' scalar $h^{\mathrm{S}} = \frac{3}{2}\hat{k}_i\hat{k}_j h_{ij} - \frac{1}{2}h$. This scalar-vector-tensor (SVT) decomposition similarly decouples the Einstein equations into six equations:

$$\ddot{h} + \frac{\dot{a}}{a}\dot{h} - 3\left(\frac{\dot{a}}{a}\right)^2 \sum_N (3w_N + 1)\Omega_N\delta_N = 8\pi G(T_{00} + T_{ii})\,, \qquad (0.6)$$

$$\ddot{h}^{\mathrm{S}} + 2\frac{\dot{a}}{a}\dot{h}^{\mathrm{S}} + \frac{1}{3}k^2(h - h^{\mathrm{S}}) = -16\pi G T^{\mathrm{S}}\,, \qquad (0.7)$$

$$\ddot{h}_i^{\mathrm{V}} + 2\frac{\dot{a}}{a}\dot{h}_i^{\mathrm{V}} = -16\pi G T_i^{\mathrm{V}}\,, \qquad (0.8)$$

$$\ddot{h}_{ij}^{\mathrm{T}} + 2\frac{\dot{a}}{a}\dot{h}_{ij}^{\mathrm{T}} + k^2 h_{ij}^{\mathrm{T}} = -16\pi G T_{ij}^{\mathrm{T}}\,, \qquad (0.9)$$

where $k = |\mathbf{k}|$ and we have decomposed the defect spatial source term T_{ij} into SVT parts. There are four further constraint equations in which the momenta T_{0i} and the energy density T_{00} appear. Here, Ω_N represents the density parameter for a matter component N, the corresponding density perturbation is $\delta_N \equiv \delta\rho_N/\rho_N$ and the equation of state is $P_N = w_N\rho_N$. If we adopt a frame comoving with a cold dark matter component (labelled 'c'), then the density perturbation δ_c is directly related to any changes in the spatial volume with $\delta_c = \frac{1}{2}h$. With a radiation component Ω_{r}

included, the scalar trace equation (0.6) then becomes simply

$$\ddot{\delta}_c + \frac{\dot{a}}{a}\dot{\delta}_c - \frac{3}{2}\left(\frac{\dot{a}}{a}\right)^2(\Omega_c\delta_c + 2\Omega_r\delta_r) \quad = \quad 4\pi G(T_{00} + T_{ii}), \quad (0.10)$$

$$\ddot{\delta}_r + \frac{1}{3}k^2\delta_r - \frac{4}{3}\ddot{\delta}_c \quad = \quad 0, \quad (0.11)$$

where the second equation arises from radiation stress–energy conservation. The most efficient means by which to solve the sourced equations (0.10) and (0.11) is using Green's functions, that is, using the solutions today $\mathcal{G}_c(k, \tau_0, \tau')$ and $\mathcal{G}_r(k, \tau_0, \tau')$ of the homogeneous equations with a unit source at time τ'. The resulting solution for the matter density perturbation is then

$$\delta_c(\mathbf{k}, \tau_0) = 4\pi \int_{\tau_1}^{\tau_0} \mathcal{G}_c(k, \tau_0, \tau')\, T_+(\mathbf{k}, \tau')\, d\tau', \quad (0.12)$$

where $T_+ = T_{00} + T_{ii}$. This expression sums over all the defect sources, multiplying each by the appropriate transfer function to linearly project its contribution to the present time. Of course, only large scales remain in the linear regime today, so nonlinear evolution must be included at late times on smaller scales.

Compensation

This apparently simple formulation is complicated for practical purposes by the choice of initial conditions. Cosmic defects create isocurvature perturbations within the causal horizon, so throughout cosmic history the defect overdensity must be balanced by a corresponding underdensity (or compensation) in the background matter on superhorizon scales. This can be seen explicitly by constructing a pseudoenergy–momentum tensor $\mathcal{T}_{\mu\nu}$ which incorporates the defect energy–momentum, the density perturbations and the gravitational energy [Weinberg, 1972; Veeraraghavan & Stebbins, 1990]:

$$\mathcal{T}_{00} \quad = \quad T_{00} + \frac{3}{8\pi}\left(\frac{\dot{a}}{a}\right)^2(\Omega_c\delta_c + \Omega_r\delta_r) - \frac{1}{4\pi}\frac{\dot{a}}{a}\dot{\delta}_c, \quad (0.13)$$

$$\mathcal{T}_{0i} \quad = \quad T_{0i} - \frac{1}{2\pi}\left(\frac{\dot{a}}{a}\right)^2\Omega_r v_r^i, \quad (0.14)$$

$$\mathcal{T}_{ij} \quad = \quad T_{ij} + \frac{1}{8\pi}\left(\frac{\dot{a}}{a}\right)^2\Omega_r\delta_r\delta_{ij} - \frac{1}{8\pi}\frac{\dot{a}}{a}\left[\dot{h}_{ij} - \frac{1}{3}\dot{h}\delta_{ij}\right]. \quad (0.15)$$

The pseudoenergy–momentum tensor is constructed in this way so that it obeys an ordinary (not covariant) conservation law,

$$\dot{\mathcal{T}}_{00} = k_i\mathcal{T}_{0i}, \qquad \dot{\mathcal{T}}_{0i} - k_j\mathcal{T}_{ij}. \quad (0.16)$$

Causality in defect theories implies that there can be no correlations on superhorizon scales, so we have

$$\langle T_{ij}(\mathbf{x}', \tau) T_{lm}(\mathbf{x}' - \mathbf{x}, \tau) \rangle = 0, \qquad |\mathbf{x}| > \tau. \tag{0.17}$$

This compact support in real space corresponds at most to white noise in Fourier space for small wavenumbers,

$$\langle |T_{ij}(\mathbf{k}, \tau) T_{lm}^*(\mathbf{k}, \tau)| \rangle \propto k^0, \qquad k\tau \ll 2\pi. \tag{0.18}$$

Combining this with the pseudostress–energy conservation (0.16), it is straightforward to obtain the important result that the power spectrum of the pseudoenergy T_{00} must fall-off asymptotically as

$$\langle |T_{00}^2| \rangle \propto k^4, \qquad k\tau \ll 2\pi. \tag{0.19}$$

Of course, this superhorizon fall-off was well known previously for causal theories, but the cancellation in pseudoenergy shows explicitly the relationship between the defect energy T_{00} and the density perturbations δ_c and δ_r. In defect simulations, it is usual to choose initial conditions in which T_{00} is cancelled directly by an artificial and opposite density perturbation $\delta^I(\mathbf{k}, \tau)$; this is then projected forward and finally added to the subsequent perturbations given by (0.12). However, this k^4 fall-off must also be satisfied throughout the whole evolution because otherwise spurious superhorizon modes will contribute. This is a key issue in numerical simulations of the defect sources and requires accurate energy–momentum conservation.

Estimating the power spectrum

A crucial deficiency in using the expression (0.12) is the inadequacy of the dynamic range of current defect simulations. These are limited to about an order of magnitude in conformal time which gives rise to a twofold problem. First, defects seed structures at any one time over a broad range of lengthscales, so a single simulation is likely to be missing power even at its central wavenumbers. Secondly, only perturbation lengthscales which begin well outside the horizon will be unaffected by the imposition of artificial compensating initial conditions on subhorizon scales. This difficulty can be partially alleviated by a semianalytic approach if we are only interested in the perturbation power spectrum $P(k)$. We can obtain this by squaring the expression (0.12) to find the asymptotic result $(\tau_0/\tau_i \to \infty)$ [Albrecht & Stebbins, 1992a]:

$$P(k) = \langle |\delta_c(\mathbf{k}, \tau_0)|^2 \rangle \approx 16\pi^2 \int_{\tau_i}^{\tau_{s0}} |G_c(k, \tau_0, \tau')|^2 \mathcal{F}(k, \tau') d\tau', \tag{0.20}$$

where the 'structure function' $\mathcal{F}$ is given by the unequal time correlator of the defect source terms

$$(2\pi)^3 \mathcal{F}(k, \tau)\delta^{(3)}(\mathbf{k} - \mathbf{k}') = \int_{-\infty}^{+\infty} \langle T_+(\mathbf{k}, \tau)T_+^*(\mathbf{k}', \tau + \tau')\rangle d\tau'. \quad (0.21)$$

Once the unequal time correlator is reliably calculated, the evaluation of the power spectrum is straightforward over an arbitrary range of length-scales. However, the dynamic range limitations of defect simulations also affect the calculation of the correlators, so typically various asymptotic assumptions are imposed, such as those required by scale-invariance and causality. Unfortunately, the correlator is not scale-invariant during the important matter–radiation transition (see fig. 0.1), so further interpolating approximations must be used.

CMB temperature anisotropies

The Boltzmann equation is needed to accurately evolve photon and neutrino distributions under the influence of gravity and scattering processes. Conceptually, we can describe the generation of CMB anisotropies, like the density perturbation (0.12), as a linear operator acting on the defect source. For a temperature fluctuation at a point $\mathbf{x}$ in the direction $\hat{\mathbf{n}}$ we can write,

$$\frac{\Delta T}{T}(\hat{\mathbf{n}}, \mathbf{x}, \tau_0) = \int_{\tau_1}^{\tau_0} d\tau' \int d^3x' \, \mathcal{G}^{\mu\nu}(\hat{\mathbf{n}}, \mathbf{x} - \mathbf{x}', \tau_0, \tau') \, T_{\mu\nu}(\mathbf{x}', \tau). \quad (0.22)$$

where the $\mathcal{G}^{\mu\nu}$ are the Green's function solutions of the coupled Einstein–Boltzmann equations for specified cosmological parameters. Note that the polarization, through the Stokes' parameters I and Q, can be treated analogously.

It is usual to describe the brightness pattern in spherical harmonics

$$\frac{\Delta T}{T}(\hat{\mathbf{n}}) = \sum_{\ell=0}^{\ell} \sum_{m=-\ell}^{\ell} a_{\ell m} Y_{\ell m}(\hat{\mathbf{n}}), \quad (0.23)$$

with the angular power spectrum for a given multipole ℓ given by the expectation value of the $a_{\ell m}$'s, that is,

$$C_\ell = \sum_{m=-\ell}^{\ell} \frac{|a_{\ell m}|^2}{2\ell + 1}. \quad (0.24)$$

Gaussian theories are described by the C_ℓ's, but for nongaussian theories like defects they provide an incomplete picture. As before, we can decompose the C_ℓ's into scalar, vector and tensor contributions. In contrast to inflationary models, cosmic defects generically produce significant vector contributions, as in the line-like discontinuities generated by strings (discussed in §10.2).

In order to obtain maximal information about the cosmic microwave sky, a full-scale Boltzmann analysis is necessary, that is, solving (0.22) directly with a defect simulation as source (see, for example, Allen, Caldwell, Dodelson, Knox, Shellard & Stebbins [1997]). However, if only the angular power spectrum is required, then it is sufficient to use the unequal time correlators analogously to (0.20) [Albrecht, Coulson, Ferreira & Magueijo, 1996],

$$C_{\mu\nu,\rho\lambda}(k, \tau, \tau') = \langle T_{\mu\nu}(\mathbf{k}, \tau) T_{\rho\lambda}(\mathbf{k}, \tau') \rangle. \qquad (0.25)$$

Exploiting the SVT decomposition and energy–momentum conservation, we need at most the three auto- and cross-correlators of the scalars T_{ii} and T^{S}, and two auto-correlators for the vector and tensor components. As before, these are calculated directly from defect simulations with their accompanying limitations. The correlators, which are symmetric matrices, can be simplified further by diagonalizing and representing them by a limited number of their positive eigenvalues [Pen, Seljak & Turok, 1997]. Once the eigenvalues are obtained, fast Boltzmann codes can then be used to rapidly obtain predictions for defect models in a variety of cosmologies. The methods for calculating CMB temperature anisotropies in cosmic defect theories are described at some length in Stebbins & Dodelson [1997], Albrecht, Battye & Robinson [1999], Uzan, Deruelle & Riazuelo [1999] and Durrer, Kunz & Melchiorri [1999].

0.3.2 Large-scale structure formation

Cosmic string-induced structures

Considerable progress has been made in quantitatively describing the perturbations seeded by cosmic strings and other defects. Typical cold dark matter (CDM) power spectra from expanding universe string network simulations are shown in fig. 0.2 for two flat (Ω=1) cosmologies, one with and one without a substantial cosmological constant [Avelino, Shellard, Wu & Allen, 1998a, 1999]. The string energy scale is normalized to match the COBE-DMR large-angle CMB anisotropy (with $G\mu \approx 1.7 \times 10^{-6}$ [Allen *et al.* 1997]) and, for comparison, a compilation of several large-scale structure surveys is also plotted. The deficiencies of the Ω_{M}=1 CDM model are clear from fig. 0.2(a). Although the density fluctuation on $8h^{-1}$Mpc is acceptable, strings generically produce an excess of small-scale power below this and a severe shortage of large-scale power around $100h^{-1}$Mpc (as with other defects). Substituting hot dark matter can suppress small-scale power, but it does nothing to improve the problem on large scales.

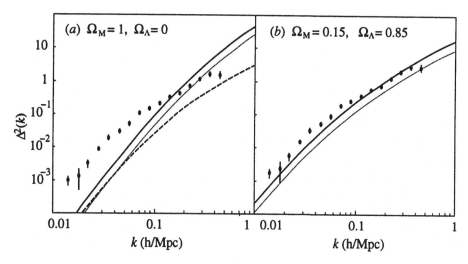

Fig. 0.2. Linear power spectra $\Delta^2(k) = k^3 P(k)/2\pi^3$ obtained from cosmic string simulations with cold dark matter in (a) a flat $\Omega_M=1$ model and (b) with a large cosmological constant $\Omega_\Lambda=0.85$ [Avelino *et al.*, 1998a, 1999]. Correlated small loops (under standard assumptions) significantly boost the infinite string power spectra (dotted). For comparison, the global texture spectrum is shown for an $\Omega_M=1$ model (dashed) [Pen, Seljak & Turok, 1997]. These spectra are all COBE-normalized. Also shown are observational points from a compilation of large-scale structure surveys [Peacock & Dodds, 1994].

More recently, cosmologies with a substantial cosmological constant have become popular for observational reasons and because they shift power from small to large scales in inflationary models. A similar improvement occurs in string models as shown in fig. 0.2(b) where there is a reasonably good fit to the observational points with $\Omega_\Lambda \approx 0.8$. The apparently small bias required ($b \lesssim 1.5$) is almost scale-independent, though considerable uncertainties remain in the amplitude because of the approximations made in the analysis. An important effect underlying these improvements is the departure from scale-invariance of the string network at the matter–radiation transition. The slow relaxation in the matter era from the higher density radiation strings causes a significant boost to the scalar power in the density fluctuations which is enhanced in Λ-cosmologies. Additional power also comes from small loops which were found to be highly correlated with the long strings, as shown by the boosts in fig. 0.2. An improvement in the shape of the string power spectrum is also apparent in open models, but greater bias is required [Ferreira, 1995; Avelino, Caldwell & Martins, 1997; Avelino *et al.*, 1998a].

These results are in fair quantitative agreement with the semianalytic power spectrum of Albrecht & Stebbins [1992a] ($\Omega_M = 1$, model 'X'), but they are only in qualitative agreement with the phenomenological string model of Battye, Albrecht & Robinson [1998]. However, the apparent shortfall in power observed in the latter case can be improved in more realistic 'wiggly' string models [Pogosian & Vachaspati, 1999a].

Some nongaussianity is expected in the large-scale structure distribution induced by cosmic strings. This has been observed by studying the one-point probability density function and genus statistics [Avelino, Shellard, Wu & Allen, 1999]. However, significant nongaussianity is only evident on scales smaller than about $1.5(\Omega_M h^2)^{-1}$Mpc, that is, on scales today which are nonlinear in any case. The wake-like structures created by a single string are submerged amongst the many contributions from other strings, except paradoxically on small scales where early time contributions before equal matter–radiation are suppressed. The mild nongaussianity that does exist on cluster scales, however, alters the X-ray cluster normalization at $8h^{-1}$Mpc (for strings relative to inflation), because this is very sensitive to the tail of the distribution. Consistent with the COBE-normalized power spectrum, it appears that string models only appear to be viable in cosmologies with $\Omega_M \approx 0.2$–0.3 [van de Bruck, 1998b; Avelino, Shellard & Wu, 1999].

The nonlinear hydrodynamic evolution of cosmic string wakes has been studied by Sornborger, Brandenberger, Fryxell & Olson [1997], Abel, Stebbins, Anninos & Norman [1998] and Wandelt [1998]. These simulations demonstrated interesting biasing mechanisms with enhancements in the baryon to cold dark matter ratio. In contrast, in a hot dark matter universe, string wakes were unable to form interesting structures at early times. Nonlinear structures created in a phenomenological string model have also been studied by Hara, Yamamoto, Mahonen & Miyoshi [1999].

Global texture and monopole power spectra

The linear power spectra induced in global monopole and texture theories have been studied by a number of authors (see, for example, Bennett & Rhie [1993], Pen, Spergel & Turok [1994], Durrer, Howard & Zhou [1996], Pen, Seljak & Turok [1997] and Durrer, Kunz & Melchiorri, 1999]). The texture spectrum in a flat CDM ($\Omega_M=1$) model is illustrated in fig. 0.2(a); the global monopole result is very similar. Textures appear to be substantially less attractive than strings because the COBE-normalized spectrum requires significant bias even at $8h^{-1}$Mpc. This missing small-scale power relative to strings can be partly attributed to the small differences between the unequal time correlators in the matter and radiation eras obtained from field theory simulations of global defects. The shape of the texture spectrum is improved for flat models with a substantial cosmological constant [Durrer, Kunz & Melchiorri, 1999], but the power spectrum amplitude is too low and requires an overall bias in the range 4–6. In addition, large-scale bulk velocities are too small by a similar factor.

0.3.3 Cosmic microwave sky

The greatest observational challenge to cosmic defect theories arises from the patterns they create on the cosmic microwave sky. Although a complete picture of defect-induced CMB fluctuations can only be obtained from a full solution of the coupled Einstein–Boltzmann equations, we can approximately discriminate between the physical mechanisms at work using the Sachs–Wolfe formula

$$\frac{\delta T}{T} = \int h_{ij}\hat{n}^i\hat{n}^j dt + \mathbf{v}\cdot\hat{\mathbf{n}} + \frac{4}{3}\frac{\delta\rho_\gamma}{\rho_\gamma}.$$

Here, the spacetime metric perturbation h_{ij} represents the gravitational effects of the cosmic defects, $\hat{n}$ is the direction of photon propagation, while $\mathbf{v}$ is the matter velocity and ρ_γ is the photon density on the surface of last scattering. For large angular scales, the first term dominates (i.e. the integrated Sachs–Wolfe effect), whereas the second and third terms (i.e. Doppler shifts and intrinsic photon fluctuations) become increasingly important at smaller angular scales.

Integrated Sachs–Wolfe effect

In standard inflation without a cosmological constant, the ISW effect merely reduces to the Newtonian potential Φ on the surface of last scattering. Temperature fluctuations are primarily due to photons climbing out of the (scalar) potential wells as they decouple; these are the same wells around which large-scale structure collapses. For cosmic defects, however, large-angle CMB anisotropies are created dynamically along the whole photon trajectory by the gravitational effects of the evolving network. This includes additional vector and tensor contributions (0.8–0.9) which are not directly related to the induced density perturbations. These extra modes imply that defect models, when normalized by large-angle CMB observations (such as the COBE-DMR experiment), generically produce less scalar power for density fluctuations relative to inflation.

Calculations solely of the ISW effect on large angular scales have been performed using direct defect simulations for cosmic strings [Allen, Caldwell, Shellard, Stebbins & Veeraraghavan, 1996; Mahonen, Hara, Voll & Miyoshi, 1997] and global defects [Bennett & Rhie, 1993; Pen, Spergel & Turok, 1994]. One realisation of a full-sky temperature map induced by a string network is shown in fig. 0.3. The near scale-invariance of defect evolution implies that they produce an angular power spectrum on large

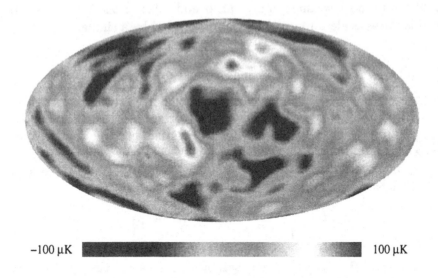

−100 µK ▓▓▓▓▓▓▓▓▓▓▓▓▓▓▓▓▓▓▓▓▓▓▓▓▓▓▓▓▓▓▓▓▓▓▓▓ 100 µK

Fig. 0.3. Full-sky temperature anisotropy map of the cosmic microwave sky induced by a cosmic string network [Allen *et al.*, 1996]. This map is consistent in both power spectrum and statistics with the COBE-DMR experiment.

scales which is nearly flat and consistent with COBE (i.e. for multipoles $\ell < 30$). Despite the distinct CMB signature of individual defects, the superposition of so many contributions between the observer and the surface of last scattering results in Gaussian statistics on large angular scales (a consequence of the central limit theorem). Many of these results were anticipated in a useful analytic model by Perivolaropoulos [1992, 1995].

For much smaller angular scales below the projected inter-defect separation, it should be possible to observe the line-like discontinuities of strings [Bennett, Bouchet & Stebbins, 1990; Gilbert & Perivolaropoulos, 1995] and the 'hot spots' of global monopoles and textures [Coulson, Ferreira, Graham & Turok, 1994; Durrer, Howard & Zhou, 1994] (as discussed, for example, in §10.2). The search for such nongaussian signatures is of great interest for forthcoming high resolution CMB experiments. These may be detectable even if the defect energy scale is below that needed for galaxy formation. A further feature of defect models is that this nongaussian ISW signal dies away at high multipoles ($\ell \gtrsim 1000$) more slowly than the exponentially suppressed inflationary spectrum.

Doppler peaks

The analysis of defect-induced CMB anisotropies on intermediate scales ($100 \lesssim \ell \lesssim 1000$) using Boltzmann codes has revealed angular power spectra quite distinct from inflationary scenarios. Adiabatic photon fluctuations begin to oscillate coherently as soon as they cross the horizon, creating a series of so-called Doppler peaks in the angular power spectrum (see fig. 0.4); the primary peak for inflation is typically at a multipole $\ell \approx 200$. However, photon fluctuations from defects are induced dynamically on scales below the horizon, so a given lengthscale begins to oscillate later. They are at least out of phase by $\pi/2$ with inflationary adiabatic perturbations, so the primary peak is expected at $\ell \approx 300$ or greater [Crittenden & Turok, 1995; Durrer, Gangui & Sakellariadou, 1996; Albrecht *et al.*, 1996]. In fact, however, since the defect sources act incoherently all Doppler peaks tend to be 'washed out' and, if secondary peaks are present at all, they are expected to be weak (see, for example, Magueijo, Albrecht, Coulson & Ferreira [1996]). Such features would be sufficient to distinguish cosmic defects from inflation, but there are actually much more obvious differences when the ISW effect is included.

The extra contributions from the ISW effect, notably the vector modes, tend to swamp out the scalar photon signal from the surface of last scattering (refer to fig. 0.4). Under some circumstances, this can almost entirely eliminate the primary peak as well as any weak secondaries. These effects were first observed using expanding universe defect simulations for strings by Allen *et al.* [1997] and for global defects by Pen, Seljak & Turok [1997].

A full Boltzmann analysis of a scale-invariant matter era string simulation showed a flat vector mode contribution which begins to decline at $\ell \approx 100$ [Allen *et al.*, 1997]. Because the vectors are so dominant, the smaller intrinsic scalar photon contribution did little more than keep the total spectrum flat until $\ell \approx 200$ and there was little or no evidence for significant Doppler peaks. However, the broken scale-invariance of string networks during the matter–radiation transition introduces additional features in the string C_ℓ spectrum in a similar manner to large-scale structure. A phenomenological string model which includes this effect produces a large and significant primary peak which is enhanced in flat models with a cosmological constant (see fig. 0.4) [Battye, Albrecht & Robinson, 1998]. The peak is even stronger when further scalar power is introduced in a modified model which incorporates string 'wiggliness' [Pogosian & Vachaspati, 1999a]. For certain parameter choices, notably $\Omega_\Lambda \gtrsim 0.5$, the primary peak can rise up to amplitudes seen observationally, but the summit is apparently at a relatively high multipole $\ell \approx 400$. In addition, these models do not produce tangible secondary peaks. It remains to be seen whether these effects are confirmed using full string simulations but, in any case, a comparison with the observational data points (also shown in fig. 0.4) is not at present encouraging. Of course, these models do not exhaust the full range of physical string possibilities. Closed Λ-cosmologies, for example, would cause the primary peak to migrate to smaller multipoles. The full role of the gravitational wave

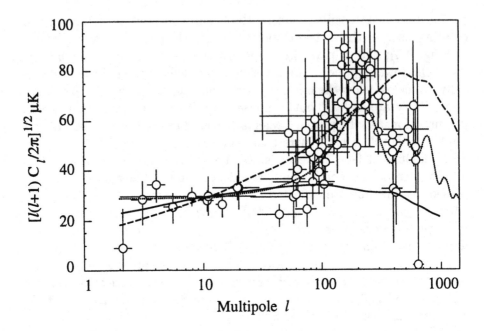

Fig. 0.4.   Angular power spectra for cosmic microwave anisotropy from global texture simulations (solid line) [Durrer, Kunz & Melchiorri, 1999] and from a phenomenological string model (dashed) [Battye, Albrecht & Robinson, 1998], as well as a standard inflation model (dotted). Both cosmic defect models have a significant cosmological constant with $\Omega_\Lambda = 0.8$ for textures and $\Omega_\Lambda = 0.5$ for strings. Recent CMB experimental points are included for comparison.

background produced by the string network remains to be fully explored [Avelino & Caldwell, 1996] and there might be interesting enhancements if the strings decay into particles other than gravitational waves [Contaldi, Hindmarsh & Magueijo, 1999a; Riazuelo, Deruelle & Peter, 1999].

Global defect models were studied using unequal time correlators (0.25) estimated from expanding universe field theory simulations [Pen, Seljak & Turok, 1997; Durrer, Kunz & Melchiorri, 1999]. The angular power spectrum was again dominated by vector modes and showed only a small rise at multipoles near $\ell \approx 100$, before falling away (see fig. 0.4). Texture models showed least evidence of a definable primary peak with a rise of less than 50% in the angular power, whereas many observations indicate rises greater than a factor of four. However, the greater coherence of textures as a source left more wiggles in the declining power spectrum at high multipoles $\ell > 300$. The dependence on the cosmological constant was relatively small and only for $\Omega_\Lambda \approx 0.8$ was a weak secondary peak evident [Durrer, Kunz & Melchiorri, 1999]. The CMB polarization from global defect models has two observational signatures distinguishing them from inflation which might be detectable in future CMB experiments [Seljak, Pen & Turok, 1997]. A significant magnetic-type polarization generated by vector modes was evident on small angular scales and there was only a low cross-correlation between temperature and polarization.

Note that this discussion about cosmic defects and the microwave sky is by no means exhaustive and more comprehensive referencing can be found in Shellard [2000]. There are many other papers on perturbation methods, nongaussian tests, decoherence and causality which, although sometimes quantitatively superseded, have also contributed to key developments.

Future prospects

If present observational indications of a strong primary Doppler peak at $\ell \approx 200$ are confirmed and if clear secondary peaks are discovered, then defect models will be ruled out as the main seeds for large-scale structure formation. However, such a conclusion is subject to a number of caveats. First, the evolution of defect networks and their nonlinear astrophysical effects have not been exhaustively studied, so there cannot be complete confidence in present predictions. Secondly, we have already seen enhancements in the primary peak due to broken scale-invariance in string models and in cosmologies with a cosmological constant. Alternative cosmologies, such as closed models, might push this primary peak in the right direction. However, the options appear limited [Albrecht, Battye & Robinson, 1999] and defect incoherence seems to leave little prospect for creating strong secondary peaks.

Cosmic defects could still play a subsidiary role in structure formation, even if they were not the prime cause. This would require some coincidence between the inflationary and defect energy scales, but there are models, such as hybrid inflation, where this can occur naturally. Such complementary structure formation has been studied in both open and hybrid inflation models [Contaldi, Hindmarsh & Magueijo, 1999b; Avelino, Caldwell & Martins, 1999; Battye & Weller, 2000]. Ultimately, then, an unequivocal test for heavy defects will have to await high resolution CMB experiments searching for their signatures on small angular scales.

0.4 Primordial backgrounds

0.4.1 Gravitational waves

In section 10.4 we discussed the stochastic gravitational radiation background produced by a cosmic string network and the resulting observational constraints from pulsar timing measurements and nucleosynthesis. The analysis of new pulsar timing data was accompanied by suggestions that heavy GUT-scale strings were ruled out [Kaspi, Taylor & Ryba, 1994;

Thorsett & Dewey, 1996]. However, a more sophisticated Bayesian analysis of the same data considerably weakened this conclusion [McHugh, Zalamansky, Vernotte & Lantz, 1996]. Estimates of the gravitational wave background from strings have been reexamined by numerically integrating over loop contributions from both radiation and matter eras [Caldwell, Battye & Shellard, 1996]. The results were relatively insensitive to loop production sizes (if they were near the expected backreaction cutoff, $\ell \sim \Gamma G \mu t$), but the dependence on the loop spectrum was important. For loop radiation dominated by kinks (see §7.5.2), the constraint on the string energy scale was $G\mu \lesssim 5.4(\pm 1.1) \times 10^{-6}$. The constraint assuming loop radiation dominated by cusps was significantly stronger because of the high frequency contribution from loops in the matter era. However, two factors mitigate against the cusp constraint and remain to be fully investigated. Cusp backreaction may introduce a high frequency cutoff and the strongly beamed signature of rare late-time loops does not average so simply into the stochastic background.

Future generations of gravitational wave detectors (such as LIGO II) will have ample sensitivity to probe the predicted spectrum of GUT-scale strings with $G\mu \sim 10^{-6}$. Because the amplitude of the string spectrum is dependent on $\mathcal{N}$, the number of spin degrees of freedom (1.3.25), measurements at different frequencies would provide unprecedented insight into the particle content of the universe at early times.

Violent processes in the early universe, such as strongly first-order phase transitions, can also produce stochastic backgrounds of gravitational waves. Few processes are more violent than the relativistic demise of a hybrid defect network (see, for example, §13.6) and the gravitational radiation spectrum this can produce has been studied by Martin & Vilenkin [1996, 1997]. Strings which disappear when they become attached to domain walls at a time t_w produce a scale-invariant spectrum but only above a cutoff frequency $\nu_w \approx 10^{-11}(G\mu)^{-1}(t_w/1\,\mathrm{s})^{-1/2}\,\mathrm{Hz}$. Monopole–antimonopole pairs which become connected by strings will subsequently begin to oscillate relativistically and radiate. The resulting spectrum will have a lower frequency cutoff $\nu_{\min}$ set by the average monopole separation at this time and an upper limit $\nu_{\max}$ set by the maximum Lorentz factors that can be achieved under string accelerations. Advanced interferometers may be able to detect such a signal even for strings with $G\mu \sim 10^{-8}$, provided $\nu_{\min} \lesssim 100\,\mathrm{Hz} \lesssim \nu_{\max}$.

0.4.2 Dark matter axions

The production of axions through the radiative decay of a network of global strings was discussed in section 12.2.3. As the axion remains a prime cold dark matter candidate this calculation has been studied in some detail by several authors. A more accurate summation over the loop spectra and distributions led to the prediction that a dark matter axion with $\Omega_a \approx 1$ should have a mass of approximately $m_a \sim 100\,\mu\text{eV}$ [Battye & Shellard, 1996b, 1999] (for standard string evolution assumptions and cosmological parameters). An order of magnitude estimate of the additional axion contribution from the decay of the hybrid axion string-wall network demonstrated that this was subdominant, but slightly increased the dark matter axion mass [Lyth, 1992]. These estimates have been questioned on the basis of global string field theory simulations and a lower mass proposed [Hagmann, Chang & Sikivie, 1999]. Unfortunately, however, this work fails to incorporate important modifications to radiative damping when extrapolating to cosmological scales (refer to §0.2.2).

0.4.3 Dilaton constraints

Superstring theories predict the existence of a light scalar field, the dilaton, with gravitational-strength couplings to ordinary matter. In models with cosmic strings, oscillating loops of string will copiously emit dilatons, as long as the characteristic frequency of oscillation is greater than the dilaton mass m_D. The power, spectrum, and angular distribution of the dilaton radiation are very similar to the gravitational case (refer to §7.5).

The dilaton radiation effectively stops at late times, when the loop sizes become much greater than m_D^{-1}, and the dilatons eventually decay with a lifetime $\tau \sim m_{\text{pl}}^2/m_D^3$. The density of any unstable relic particles is subject to a multitude of astrophysical constraints and, in this case, they can be expressed as constraints on the dilaton mass m_D and the string gravitational coupling $G\mu$ [Damour & Vilenkin, 1997]. The resulting constraints are rather stringent: gauge strings with $G\mu \sim 10^{-6}$ are ruled out unless $m_D \gtrsim 100$ TeV, while the currently popular value $m_D \sim 1$ TeV imposes the bound $G\mu \lesssim 10^{-15}$. For global strings, the latter bound is substantially weakened, $G\mu \lesssim 10^{-9}$.

These dilaton constraints can be avoided in models for which the initial scale of the string network is greater than m_D^{-1}, so that dilaton-emitting loops are never formed. If the strings are formed during inflation (see §16.2), the network begins the usual evolution only after its characteristic scale comes within the horizon, which can easily happen at $t > m_D^{-1}$.

0.5 Cosmic rays and high energy phenomena

0.5.1 Ultrahigh energy cosmic rays

Topological defects can produce high energy particles by a variety of mechanisms and can contribute to the observed spectrum of cosmic rays. A particularly intriguing possibility is that ultrahigh energy cosmic rays with $E \gtrsim 10^{11}$ GeV, which are hard to explain by usual astrophysical acceleration mechanisms, may be due to vacuum defects [Sigl, Schramm & Bhattacharjee, 1994]. Defects produce superheavy Higgs and gauge bosons which then decay into light particles. Although defects can naturally give particle energies well in excess of 10^{11} GeV, an explanation of the observed fluxes has proven to be more challenging. Another challenge is to explain the apparent absence of the Greisen–Zatsepin–Kuzmin (GZK) cutoff in the observed cosmic ray spectrum. The cutoff is due to the abrupt drop in the mean free path of protons at $E \approx 6 \times 10^{10}$ GeV, resulting from pion production on the background microwave photons. It is expected in all models with a uniform distribution of sources. Ultrahigh energy photons are also strongly absorbed due to pair production on radio background photons.

Among the specific mechanisms that have been suggested for the production of ultrahigh energy particles is the annihilation of monopole–antimonopole ($M\bar{M}$) bound states, monopolonia [Hill, 1983; Bhattacharjee & Sigl, 1995]. Later work has shown, however, that the required density of monopolonia cannot be achieved without violating observational bounds on the density of free monopoles [Blanco-Pillado & Olum, 1999b].

Another possible source is the annihilation of overlapping string segments near a cusp. This has been numerically simulated by Olum & Blanco-Pillado [1999a], who found that the overlapping segments almost entirely disintegrate into particles (refer to fig. 0.5). Blanco-Pillado & Olum [1999a] classified possible shapes of cusps and showed that, for a generic cusp, a frame of reference can be found where the string velocity at the tip lies in the plane of the cusp. (A cusp whose motion is orthogonal to that plane, like the one illustrated in fig. 7.5, is therefore an exceptional case, not a generic situation.) Taking into account the motion of the cusp and the associated Lorentz contraction of the string core, they showed that the energy released in a generic cusp event is $E \sim \mu(L\delta)^{1/2}$, where L is the characteristic string length and δ is the string thickness. This is much smaller than the earlier estimate $E \sim \mu L^{2/3}\delta^{1/3}$ (see §8.2). The resulting particle flux is below present observational capabilities.

Another class of models involves hybrid monopole–string defects in which each monopole is attached to N strings. In monopole–string networks ($N\geq3$), monopoles are accelerated by the strings and radiate gauge quanta of very high energy (see §9.6 and for a detailed analysis refer to Berezinsky, Martin & Vilenkin [1997]). But once again, the observed flux of cosmic rays cannot be reached without exceeding the observational bound on the γ-ray flux in the energy range 10 MeV–100 GeV [Berezinsky, Blasi & Vilenkin, 1998].

For $N=2$, the hybrid defects have the appearance of necklaces, with

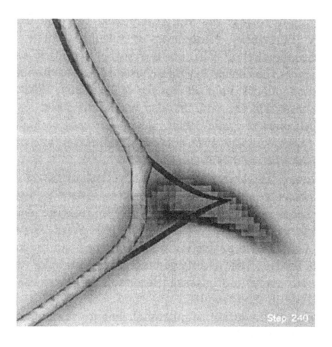

Fig. 0.5. Field theory simulation in the vicinity of a string cusp [Olum & Blanco-Pillado, 1999a]. The light region is the actual core of the string which should be compared with the dark line which follows the Nambu–Goto trajectory. A small section of the string has annihilated and left behind its energy in radiation.

monopoles playing the role of beads along the strings (refer to §0.2.5). Monopoles and antimonopoles trapped in a necklace loop inevitably annihilate. This model can account for the observed flux of ultrahigh energy particles, provided that the value of the parameter r in (0.3) is sufficiently high [Berezinsky & Vilenkin, 1997; Berezinsky, Blasi & Vilenkin, 1998]. Numerical simulations are needed to check whether or not the required large values of r can indeed be reached.

$M\bar{M}$ pairs connected by strings (N=1) typically decay long before the present epoch (see §14.4.2). However, if the monopoles are heavy and the strings are light and, in addition, the monopoles have no magnetic fluxes other than those confined in the strings, then the frictional energy loss is negligible and decay by gravitational radiation can be very slow. The pairs can then live long enough to produce high energy particles at present [Blanco-Pillado & Olum, 1999b]. The observed flux can be explained for monopoles of mass $m \sim 10^{14}$ GeV and electroweak-scale strings, $\mu \sim (100\,\text{GeV})^2$, in which case the lifetime of the pairs is comparable to the present age of the universe. An attractive feature of this model is that the distribution of $M\bar{M}$ pairs follows that of cold dark matter, and the main contribution to the cosmic ray flux is expected to come from the halo of our Galaxy [Berezinsky, Kachelriess & Vilenkin, 1997]. This would explain the absence of the GZK cutoff. Another signature of this model is the anisotropy of the cosmic ray flux caused by the asymmetric position of the Sun in the galactic halo [Dubovsky & Tinyakov, 1998].

Decay of metastable vortons has also been considered as a source of cosmic rays [Masperi & Silva, 1998]. As in the case of $M\bar{M}$ pairs, the distribution of vortons follows that of cold dark matter, and observational signatures of the two models are very similar.

We finally note the interesting possibility that ultrahigh energy particles could themselves be topological defects. Magnetic monopoles can naturally be accelerated to $E \sim 10^{11}$ GeV in galactic magnetic fields [Weiler & Kephart, 1996; Huguet & Peter, 1999; Wick, Kephart, Weiler & Biermann, 2000]. Alternatively, charged vortons can be accelerated in strong electric fields in the vicinity of accreting black holes at the centres of active galactic nuclei and quasars [Bonazzola & Peter, 1997]. No GZK cutoff is expected in these models.

For a review of the origin of ultrahigh energy cosmic rays, refer to Bhattacharjee & Sigl [1999].

0.5.2 Gamma ray bursts

Superconducting cosmic strings moving through magnetized plasma in galaxies and clusters can develop large electric currents and give rise to a variety of high energy phenomena. Oscillating loops of current-carrying string emit short pulses of highly beamed electromagnetic radiation from their cusps. The pulses lose their energy by accelerating particles in the plasma, resulting in ultrarelativistic jets terminating at shocks. Synchrotron radiation of electrons in these jets has been suggested as the possible origin of γ-ray bursts, which still evade more conventional explanation. This model requires strings of symmetry breaking scale $\eta \sim 10^{14}$ GeV [Berezinsky, Hnatyk & Vilenkin, 2000]. An earlier version of this scenario, due to Babul, Paczynski & Spergel [1987] is discussed in section 12.1.4. Particles reflected from an advancing shock can be accelerated to ultrahigh energies, and it is conceivable that the mysteries of γ-ray bursts and of ultrahigh energy cosmic rays have a common resolution.

0.5.3 Baryogenesis

Cosmic defects can naturally provide environments satisfying Sakharov's three criteria for the generation of the baryon (B) asymmetry of the universe: B-violation, CP-violation, and nonequilibrium. The required asymmetry can be produced, for example, in B- and CP-violating decays of superheavy particles emitted by the defects. The particles can either be emitted from string cusps or small loops, or during the decay of hybrid defects, such as monopoles connected by strings or walls bounded by strings. Several scenarios of this sort are mentioned in sections 10.6, 13.6 and 14.4.2. Baryon asymmetry produced by decaying vortons has been discussed by Davis & Perkins [1997] and, in a model involving baryonic condensates on strings, by Brandenberger & Riotto [1998] and Dimopoulos & Davis [2000].

A baryon asymmetry generated in the early universe can be destroyed by non-perturbative electroweak processes. The electroweak vacuum is degenerate, with different minima having different Higgs field windings (i.e. gauge textures). A change in winding is accompanied by a change in the baryon number, due to the anomaly of the baryon current. B-conservation can, therefore, be violated by quantum tunnelling between different vacua [t'Hooft, 1976]. At finite temperatures, B-violating transitions can be caused by thermal fluctuations taking the Higgs and gauge fields over the barrier between two neighbouring vacua through a saddle point configuration, which is called the sphaleron.

Sphaleron processes are in thermal equilibrium above the electroweak phase transition temperature T_{EW}, and any baryon number produced at $T > T_{\mathrm{EW}}$ will be erased, unless it is protected by some conservation law, for example, $B-L$ conservation [Kuzmin, Rubakov & Shaposhnikov, 1985]. The baryon asymmetry can, however, be regenerated at $T \sim T_{\mathrm{EW}}$, when sphaleron processes fall out of equilibrium. This electroweak baryogenesis has been a subject of intense study over the last several years (reviews can be found in Turok [1992], Cohen, Kaplan & Nelson [1993], Rubakov & Shaposhnikov [1996] and Riotto & Trodden [1999]). If the electroweak transition is of first order, then the passage of an expanding bubble wall through some point in space can lead to a rapid variation of the Higgs field at that point, and thus to a significant departure from thermal equilibrium. If in addition the walls have CP-violating interactions, then all three conditions for baryogenesis are satisfied. Analysis shows, however, that in the standard electroweak model the phase transition is too weakly first-order and the CP-violation is too small to explain the observed baryon asymmetry. The minimal supersymmetric extension of the standard model can fulfil these requirements, but the allowed parameter space is small.

An alternative to a first-order phase transition is provided by evolving networks of topological defects, with moving defects playing the same role as the expanding bubble walls. In the case of cosmic strings, this scenario has been worked out by Brandenberger, Davis & Trodden [1994], Trodden, Davis & Brandenberger [1995] and Brandenberger, Davis, Prokopek, & Trodden [1996]. It requires strings whose constituent fields have CP-

violating interactions and are coupled to the electroweak Higgs or gauge fields, so that the electroweak symmetry is restored near the string core. The baryon asymmetry is generated at $T < T_{\text{EW}}$, when B-violating sphaleron processes can be efficient only in the symmetry-restored region around the strings. For the scenario to work, this region has to be wider than the sphaleron radius, $r_{\text{sph}} \sim (e^2 T)^{-1}$. Recent work suggests, however, that this condition is difficult to implement in realistic models [Cline, Espinosa, Moore & Riotto, 1999].

0.6 Defects and inflation

0.6.1 Inflating defects

In inflationary scenarios, an accelerated expansion of the universe is driven by the energy of the false vacuum. All one needs to start inflation is a false vacuum region greater than the horizon. Topological defects have false vacuum in their cores, and if the core radius is greater than the horizon, one can expect that defect cores will inflate [Linde, 1994; Vilenkin, 1994; Linde & Linde, 1994]. This is denoted as 'topological inflation'.

Consider a defect model in which the symmetry breaking field ϕ develops a vacuum expectation value η at large distances from the core, with $V(\eta) = 0$, while the false vacuum energy density inside the core is $V(0) \equiv V_0 > 0$. In flat spacetime, the radius of the core δ_0 is determined by the balance of gradient and potential energies, $(\eta/\delta_0)^2 \sim V_0$. This gives $\delta_0 \sim \eta V_0^{-1/2}$. The horizon radius corresponding to the false vacuum energy density V_0 is given by $H_0^{-1} \sim m_{\text{pl}} V_0^{-1/2}$, and the condition $\delta_0 > H_0^{-1}$ gives $\eta \gtrsim m_{\text{pl}}$. One expects, therefore, that with gravity taken into account, nonsingular defect solutions of fixed thickness exist only when the symmetry breaking scale is below some critical value $\eta_c = C m_{\text{pl}}$ with $C \sim 1$, and that defect cores inflate for $\eta > \eta_c$. Numerical studies of global defect models with quartic potentials support this conjecture and give $C \approx 0.33$ [Sakai, Shinkai, Tachizawa & Maeda, 1996; Sakai, 1996] for domain walls and monopoles and $C \approx 0.23$ for strings [Cho, 1998]. Similar results are obtained for gauge defects, with C weakly dependent on the relative strength of gauge and Higgs couplings. The above values of C are for defects of the lowest topological charge. As the charge is increased, the core radius gets larger and η_c gets smaller [de Laix, Trodden & Vachaspati, 1998].

Once started, topological inflation never ends: the inflating core of the defect cannot disappear for topological reasons. In fact, it can be shown that the volume of the core grows exponentially with time. In a two-dimensional analogy, the core can be pictured as an inflating balloon connected by a throat to the exterior region. The spacetime structure of inflating defects has been studied by Sakai [1996], Cho & Vilenkin [1997] and Cho [1998].

0.6.2 Defects from preheating

A new mechanism of defect formation after inflation has been suggested by Kofman, Linde & Starobinsky [1996] and Tkachev [1996]. As the inflaton field ϕ oscillates about the minimum of its potential, parametric resonance may transfer most of its energy to other boson fields, or to inhomogeneous modes of the field ϕ itself. Field fluctuations excited in this way are much greater than they would be if the inflaton energy were thermalized. Just like thermal fluctuations in equilibrium, these large nonthermal fluctuations can lead to the restoration of broken symmetries. As the fluctuations are redshifted by the subsequent expansion of the universe, the symmetries are broken again and topological defects can be formed. This mechanism has been verified in numerical simulations of a 'chaotic' inflation model with a complex field ϕ and inflaton potential $V(\phi) = \frac{1}{4}\lambda(|\phi|^2 - \eta^2)^2$ [Kasuya & Kawasaki, 1997, 1998; Tkachev, Khlebnikov, Kofman & Linde, 1998]. The formation of cosmic string networks is observed for symmetry breaking scales up to $\eta \sim 10^{16}$ GeV. This mechanism helps alleviate the difficulties in producing superheavy defects in inflationary scenarios discussed in section 16.1.

0.7 Defect gravity and singularities

0.7.1 Nonsingular global strings

The gravitational field of a global string, discussed in section 7.3, has a curvature singularity at a finite distance from the string core. This places global strings in a unique position: the spacetimes of other defects, gauge and global, are all nonsingular. The Cohen–Kaplan metric for a global string (7.3.7) was obtained by solving the combined Einstein and scalar field equations assuming cylindrical symmetry, Lorentz boost invariance along the string, and staticity. The example of domain walls suggests, however, that staticity might be too strong an assumption. And indeed, Gregory [1996] has shown that a nonsingular global string spacetime can be found by allowing expansion along the string axis. The resulting metric has the form

$$ds^2 = e^{A(r)}(dt^2 - e^{2\kappa t}dz^2) - dr^2 - e^{B(r)}d\theta^2. \qquad (0.26)$$

The de Sitter-like expansion along the string is similar to the expansion in the plane of the wall of the domain wall metric (13.3.6). Solutions of this form exist for all values of κ, but it can be shown, using dynamical systems analysis, that there is only a single value for which the metric is nonsingular. This value has been estimated as

$$\kappa \sim \epsilon^{7/4} \exp(-1/\epsilon), \qquad (0.27)$$

where $\epsilon = 8\pi G\eta^2$ and η is the scalar field expectation value outside the string core. For sub-Planckian strings, $\epsilon \ll 1$ and the expansion rate (0.27) is tiny.

The explicit form of $A(r)$ and $B(r)$ in (0.26) is not known, but it can be shown that the metric is closely approximated by the Cohen–Kaplan solution in the range $0 < r \ll r_0$ and has a horizon at $r = r_0$. Here, $r_0 \sim \delta\epsilon^{-1} \exp(1/2\epsilon)$ with δ being the thickness of the string core. For strings with $\eta \lesssim 10^{16}$ GeV, r_0 is well beyond the present cosmological horizon, and the Cohen–Kaplan solution should be an excellent approximation.

0.7.2 Vacuumless defects

Somewhat analogous to 'ordinary' global defects are 'vacuumless' defects which arise in models where the potential $V(\phi)$ has no minima and monotonically decreases from its maximum at $\phi = 0$ to zero at $|\phi| \to \infty$ [Cho & Vilenkin, 1999a; see also D'Hoker & Jackiw, 1982]. For both global and gauge vacuumless strings, the field magnitude $|\phi(r)|$ grows with the distance r from the string core, the energy per unit length $\mu(r)$ within a cylinder of radius r also grows, and the gravitational field becomes singular at a finite value of r [Cho & Vilenkin, 1999b]. For example, in the case of a global vacuumless string with a potential $V(\phi) = \lambda M^{4+n}(M^n + |\phi|^n)^{-1}$, we have $|\phi(r)| \propto r^{2/(n+2)}$, $\mu(r) \propto |\phi(r)|^2$, and the singularity develops at a distance where $|\phi|$ becomes comparable to m_{pl}. Gravitational fields of vacuumless domain walls and global monopoles have horizons and are nonsingular, and the gravitational field of a vacuumless gauge monopole is essentially that of a magnetically charged black hole. As in the case of ordinary global strings, singular vacuumless string spacetimes were obtained assuming staticity, and one can expect that nonsingular solutions may be found when this assumption is relaxed.

0.7.3 Black hole nucleation on strings

In section 14.4 we discussed models of strings ending on magnetic monopoles. In such models strings are metastable and decay through the nucleation of monopole–antimonopole pairs. It was recently realized that, with gravity taken into account, even topologically stable strings may not be immune to this kind of decay, with magnetically charged black holes playing the role of monopoles [Eardley, Horowitz, Kastor & Traschen, 1995; Hawking & Ross, 1995; Gregory & Hindmarsh, 1995].

The existence of magnetically charged black hole solutions has been known for a long time. Like any magnetic monopoles, they should satisfy the Dirac quantization condition, $g = 2\pi n/e$, where g is the magnetic charge. Consider now a black hole with $g = 2\pi/e$ in the context of the simplest $U(1)$ Higgs model. At high temperatures the $U(1)$ symmetry is unbroken, and the magnetic field of the black hole is spherically symmetric. When the symmetry gets broken, the magnetic flux is squeezed into a Nielsen–Olesen string which terminates at the black hole.

The instanton describing the breaking of cosmic strings by nucleation of black hole pairs is given by the Euclideanized C-metric when the string is much thinner than the Schwarzschild radius of the holes and by the Euclideanized Ernst metric in the opposite limit. In both cases, the bounce action is $B \sim m^2/\mu$, where m is the black hole mass and μ is the string tension. A magnetically charged black hole should satisfy $m \geq g m_{pl}$, and the smallest action is obtained for extremal black holes saturating this inequality, $B_{min} \sim (e^2 G\mu)^{-1}$. The nucleation probability, $\mathcal{P} \propto \exp(-B)$, is negligible for sub-Planckian strings with $G\mu \ll 1$.

Unlike the Nielsen–Olesen string, a global $U(1)$ string cannot terminate at a black hole. Otherwise, one would be able to take a closed loop encircling the string with a nontrivial winding number, say, $\Delta\theta = 2\pi$, and continuously shrink it to a point. This argument does not apply to gauge strings, since in the latter case the phase θ is gauge-dependent and a nonsingular gauge potential can only be defined on two (or more) overlapping patches on 2-spheres surrounding the black hole. An explicit construction of the Higgs and gauge fields for a string terminating at a black hole horizon has been given by Achucarro, Gregory & Kuijken [1995].

Another way of seeing that $U(1)$ global strings cannot end on black holes is to note that the global $U(1)$ symmetry can be explicitly broken, as in the case of axionic strings, with the strings becoming boundaries of domain walls. Such strings cannot end because the boundary of a boundary is zero. The same argument applies to any kind of strings which can become boundaries of domain walls, such as Z_2 strings [Preskill, quoted in Hawking & Ross, 1995].

0.8 Supplementary references

Note that most references prior to 1994 can be found in the main references section.

Abel, T., Stebbins, A., Anninos, P., & Norman, M.L. [1998], 'First structure formation: II. Cosmic string + hot dark matter models', *Ap. J.* **508**, 530.

Achucarro, A., Borrill, J., & Liddle, A.R. [1999], 'The formation rate of semilocal strings', *Phys. Rev. Lett.* **82**, 3742.

Achucarro, A., Gregory, R., & Kuijken, K, [1995], 'Abelian Higgs hair for black holes', *Phys. Rev.* **D52**, 5729.

Achucarro, A., & Vachaspati, T. [1999], 'Semilocal and electroweak strings', to appear in *Phys. Rep.* (hep-ph/9904229).

Albrecht, A., Battye, R.A., & Robinson, J. [1999], 'A detailed study of defect models for cosmic structure formation', *Phys. Rev.* **D59**, 023508.

Albrecht, A., Coulson, D., Ferreira, P., & Magueijo, J. [1996], 'Causality, randomness, and the microwave background', *Phys. Rev. Lett.* **76**, 1413.

Alexander, S., Brandenberger, R.H., Easther, R., & Sornborger, A. [1999], 'On the interaction of monopoles and domain walls', preprint (hep-ph/9903254).

Allen, B., Caldwell, R.R., Dodelson, S., Knox, L., Shellard, E.P.S., & Stebbins, A. [1997], 'CMB anisotropy induced by cosmic strings on angular scales $\geq 15''$', *Phys. Rev. Lett.* **79**, 2624.

Allen, B., Caldwell, R.R., Shellard, E.P.S., Stebbins, A., & Veerarghavan, S. [1996], 'Large angular scale CMB anisotropy induced by cosmic strings', *Phys. Rev. Lett.* **77**, 3061.

Antunes, N.D., & Bettencourt, L.M.A. [1998], 'The length distribution of vortex strings in $U(1)$ equilibrium scalar field theory', *Phys. Rev. Lett.* **81**, 3083.

Antunes, N.D., Bettencourt, L.M.A., & Hindmarsh, M. [1998], 'The thermodynamics of cosmic string densities in $U(1)$ scalar field theory', *Phys. Rev. Lett.* **80**, 908.

Antunes, N.D., Bettencourt, L.M.A., & Zurek, W.H. [1999], 'Vortex string formation in a 3D $U(1)$ temperature quench', *Phys. Rev. Lett.* **82**, 2824.

Austin, D., Copeland, E.J., & Kibble, T.W.B. [1995], 'Characteristics of cosmic string scaling configurations', *Phys. Rev.* **D51**, 2499.

Avelino, P.P., & Caldwell, R.R. [1996], 'Entropy perturbations due to cosmic strings', *Phys. Rev.* **D53**, 5339.

Avelino, P.P., Caldwell, R.R., & Martins, C.J.A.P. [1997], 'Cosmic strings in an open universe: quantitative evolution and observational consequences', *Phys. Rev.* **D56**, 4568.

Avelino, P.P., Caldwell, R.R., & Martins, C.J.A.P. [1999], 'Cosmological consequences of string-forming open inflation models', *Phys. Rev.* **D59**, 123509.

Avelino, P.P., Shellard, E.P.S., & Wu, J.H.P. [1999], 'The cluster abundance in cosmic string models for structure formation', to appear in *M.N.R.A.S.* (astro-ph/9906313).

Avelino, P.P., Shellard, E.P.S., Wu, J.H.P., & Allen, B. [1998a], 'Cosmic string-seeded structure formation', *Phys. Rev. Lett.* **81**, 2008.

Avelino, P.P., Shellard, E.P.S., Wu, J.H.P., & Allen, B. [1998b], 'Non-gaussian features of linear cosmic string models', *Ap. J.* **507**, L101.

Avelino, P.P., Shellard, E.P.S., Wu, J.H.P., & Allen, B. [1999], 'Cosmic string loops and large-scale structure', *Phys. Rev.* **D60**, 023511.

Battye, R.A., Albrecht, A., & Robinson, J. [1998], 'Structure formation by cosmic strings with a cosmological constant', *Phys. Rev. Lett.* **80**, 4847.

Battye, R.A., Bucher, M., & Spergel, D. [1999], 'Domain wall dominated universes', DAMTP preprint (astro-ph/9908047).

Battye, R.A., & Shellard, E.P.S. [1995], 'String radiative backreaction', *Phys. Rev. Lett.* **75**, 4354.

Battye, R.A., & Shellard, E.P.S. [1996a], 'Radiative backreaction on global strings', *Phys. Rev.* **D53**, 1811.

Battye, R.A., & Shellard, E.P.S. [1996b], 'Axion string constraints', *Phys. Rev. Lett.* **73**, 2954 (1994), Erratum **76**, 2203 (1996).

Battye, R.A., Shellard, E.P.S. [1999], 'The spectrum of radiation from axion strings', *Nucl. Phys. Proc. Suppl.* **72**, 88.

Battye, R.A., & Weller, J. [2000], 'Cosmic structure formation in hybrid inflation models', *Phys. Rev.* **D61**, 043501.

Bauerle *et al.* [1996], 'Simulated cosmic strings in a "big bang" in superfluid ^{3}He at 100μK', *Nature* **382**, 332.

Berezinsky, V., Blasi, P., & Vilenkin, A. [1998], 'Signatures of topological defects', *Phys. Rev.* **D58**, 103515.

Berezinsky, V., Hnatyk, B., & Vilenkin, A. [2000], 'Superconducting cosmic strings as gamma ray burst engines', preprint (astro-ph/0001213).

Berezinsky, V., Kachelriess, M., & Vilenkin, A. [1997], ' Ultra-high energy cosmic rays without a GZK cutoff', *Phys. Rev. Lett.* **79**, 4302.

Berezinsky, V., Martin, X., & Vilenkin, A. [1997], 'High energy particles from monopoles connected by strings', *Phys. Rev.* **D56**, 2024.

Berezinsky, V., & Vilenkin, A. [1997], 'Cosmic necklaces and ultrahigh energy cosmic rays', *Phys. Rev. Lett.* **79**, 5202.

Bettencourt, L.M.A., Antunes, N.D., & Zurek, W.H. [2000], 'The Ginzburg regime and its effect on topological defect formation', preprint (hep-ph/0001205).

Bhattacharjee, P., & Sigl, G. [1995], 'Monopole annihilation and highest energy cosmic rays', *Phys. Rev.* **D51**, 4079.

Bhattacharjee, P., & Sigl, G. [1999], 'Origin and propagation of extremely high energy cosmic rays', preprint (astro-ph/9811011).

Blanco-Pillado, J.J., & Olum, K.D. [1999a], 'The form of cosmic string cusps', *Phys. Rev.* **D59**, 063508.

Blanco-Pillado, J.J., & Olum, K.D. [1999b], 'Monopole–antimonopole bound states as a source of ultrahigh energy cosmic rays', *Phys. Rev.* **D60**, 083001.

Bonazzola, S., & Peter, P. [1997], 'Can high energy cosmic rays be vortons?', *Astropart. Phys.* **7**, 161.

Borrill, J., [1996], 'On the absence of open strings in a lattice-free simulation of cosmic string formation', *Phys. Rev. Lett.* **76**, 3255.

Borrill, J., Kibble, T.W.B., Vachaspati, T., & Vilenkin, A. [1995], 'Defect production in slow first order phase transitions', *Phys. Rev.* **D52**, 1934.

Brandenberger, R.H., Carter, B., Davis, A.-C., & Trodden, M. [1996] 'Cosmic vortons and particle physics constraints', *Phys. Rev.* **D54**, 6059.

Brandenberger, R., Davis, A.-C., Prokopek, T., & Trodden, M. [1996], 'Local and nonlocal defect mediated electroweak baryogenesis', *Phys. Rev.* **D53**, 4257.

Brandenberger, R., Davis, A.-C., & Trodden, M. [1994], 'Cosmic strings and electroweak baryogenesis', *Phys. Lett.* **B335**, 123.

Brandenberger, R., & Riotto, A. [1998], 'A mechanism for baryogenesis in low energy supersymmetry breaking models', *Phys. Lett.* **B445**, 323.

Bucher, M., & Spergel, D. [1999], 'Is the dark matter a solid?', *Phys. Rev.* **D60**, 043505.

Buonanno, A., & Damour, T. [1998a], 'On the gravitational, dilatonic and axionic radiative damping of cosmic strings', *Phys. Rev.* **D60**, 023517.

Buonanno, A., & Damour, T. [1998b], 'Effective action and tension renormalization for cosmic and fundamental strings', *Phys. Lett.* **432B**, 51.

Caldwell, R.R., Battye, R.A., & Shellard, E.P.S. [1996], 'Relic gravitational waves from cosmic strings: updated constraints and opportunities for detection', *Phys. Rev.* **D54**, 7146.

Carter, B. [1997], 'Brane dynamics for treatment of cosmic strings and vortons', in

Proceedings of 2nd Mexican School on Gravitation and Mathematical Physics, Garcia, A., *et al.* eds. (Science Network Publishing) (see also hep-th/9705172).

Carter, B. [2000], 'Cancellation of linearised axion-dilaton self-interaction divergence in strings', to appear in *Int. J. Theor. Phys.* (hep-th/0001136).

Carter, B., & Battye, R.A. [1998], 'Non-divergence of gravitational self-interactions for Goto–Nambu strings', *Phys. Lett.* **430B**, 49.

Carter, B., & Davis, A.-C. [1999], 'Chiral vortons and cosmological constraints on particle physics', DAMTP preprint (hep-ph/9910560).

Carter, B., & Peter, P. [1999], 'Dynamics and integrability property of the chiral string model', *Phys. Lett.* **466B**, 41.

Cho, I. [1998], 'Inflation and nonsingular spacetimes of cosmic strings', *Phys. Rev.* **D58**, 103509.

Cho, I., & Vilenkin, A. [1997], 'Spacetime structure of an inflating global monopole', *Phys. Rev.* **D56**, 7621.

Cho, I., & Vilenkin, A. [1999a], 'Vacuum defects without a vacuum', *Phys. Rev.* **D59**, 021701.

Cho, I., & Vilenkin, A. [1999b], 'Gravitational field of vacuumless defects', *Phys. Rev.* **D59**, 063510.

Cline, J., Espinosa, J., Moore, G.D., & Riotto, A. [1999], 'String-mediated electroweak baryogenesis: a critical analysis', *Phys. Rev.* **D59**, 065014.

Cohen, A.G., Kaplan, D.B., & Nelson, A.E. [1993], 'Progress in electroweak baryogenesis', *Ann. Rev. Nucl. Part. Sci.* **43**, 27.

Contaldi, C., Hindmarsh, M.B., & Magueijo, J. [1999a], 'The power spectra of CMB and density fluctuations seeded by local cosmic strings', *Phys. Rev. Lett.* **82**, 679.

Contaldi, C., Hindmarsh, M.B., & Magueijo, J. [1999b], 'CMB and density fluctuations from strings plus inflation', *Phys. Rev. Lett.* **82**, 2034.

Copeland, E.J., Kibble, T.W.B., & Steer, D.A. [1998], 'The evolution of a network of cosmic string loops', *Phys. Rev.* **D58**, 043508.

Copeland, E.J., Saffin, P.M., & Tornkvist, O. [1999], 'Phase equilibration and magnetic field generation in $U(1)$ bubble collisions', preprint (hep-ph/9907437).

Coulson, D., Ferreira, P., Graham, P., & Turok, N. [1994], 'Microwave anisotropies from cosmic defects', *Nature* **368**, 27.

Crittenden, R., & Turok, N. [1995], 'The Doppler peaks from cosmic texture', *Phys. Rev. Lett.* **75**, 2642.

Damour, T., & Vilenkin, A. [1997], 'Cosmic strings and the string dilaton', *Phys. Rev. Lett.* **78**, 2288.

Davis, A.-C., & Brandenberger, R.H., eds. [1995], *Formation and Interactions of Defects* (Plenum Press).

Davis, A.-C., & Dimopoulos, K. [1998], 'Friction domination with superconducting strings', *Phys. Rev.* **D57**, 692.

Davis, A.-C., & Lilley, M. [1999], 'Cosmological consequences of slow-moving bubbles in first-order phase transitions', preprint (hep-ph/9908398).

Davis, A.-C., & Perkins, W.B. [1997], 'Dissipating cosmic vortons and baryogenesis', *Phys. Lett.* **B393**, 46.

de Laix, A.A., Trodden, M., & Vachaspati, T. [1998], 'Topological inflation with multiple winding', *Phys. Rev.* **D57**, 7186.

de Laix, A., & Vachaspati, T. [1999], 'On random bubble lattices', *Phys. Rev.* **D59**, 045017.

D'Hoker, E., & Jackiw, R. [1982], 'Classical and quantal Liouville field theory', *Phys. Rev.* **D26** 3517.

Dimopoulos, K., & Davis, A.-C. [2000], 'Cosmological consequences of superconducting string networks', *Phys. Lett.* **B446**, 238.

Dodd, M.E. *et al.* [1998], 'Non-appearance of vortices in fast mechanical expansions of liquid ^{4}He through the lambda transition', *Phys. Rev. Lett.* **81**, 3703.

Dubovsky, S.L., & Tinyakov, P.G. [1998], 'Galactic anisotropy of ultrahigh energy cosmic rays produced by CDM-related mechanisms', preprint (hep-ph/9810401).

Durrer, R., Gangui, A., & Sakellariadou, M. [1996], 'Doppler peaks in the angular power spectrum of the cosmic microwave background: a fingerprint of topological defects', *Phys. Rev. Lett.* **76**, 579.

Durrer, R., Howard, A., & Zhou, Z.-H. [1994], 'Microwave anisotropies from texture seeded structure formation', *Phys. Rev.* **D49**, 681.

Durrer, R., Kunz, M., Melchiorri, A. [1999], 'Cosmic microwave background anisotropies from scaling seeds: global defect models', *Phys. Rev.* **D59**, 123005.

Dvali, G., Liu, H., & Vachaspati, T. [1998], 'Sweeping away the monopole problem', *Phys. Rev. Lett.* **80**, 2281.

Eardley, D., Horowitz, G., Kastor, D., & Traschen, J. [1995], 'Breaking cosmic strings without monopoles', *Phys. Rev. Lett.* **75**, 3390.

Ferreira, P. [1995], 'Cosmic strings in an open universe with baryonic and nonbaryonic dark matter', *Phys. Rev. Lett.* **74**, 3522.

Ferrera, A. [1998], 'Defect formation in first order phase transitions with damping', *Phys. Rev.* **D57**, 7130.

Ferrera, A. [1999], 'How does the geodesic rule really work for global symmetry breaking first order phase transitions?', *Phys. Rev.* **D59**, 123503.

Ferrera, A., & Melfo, A. [1996], 'Bubble collisions and defect formation in a damping environment', *Phys. Rev.* **D53**, 6852.

Gilbert, A.M., & Perivolaropoulos, L. [1995], 'Spectra and statistics of cosmic string perturbations on the microwave background: a Monte Carlo approach', *Astropart. Phys.* **3**, 283.

Gregory, R. [1996], 'Nonsingular global strings', *Phys. Rev.* **D54**, 4955.

Gregory, R., & Hindmarsh, M.B. [1995], 'Smooth metrics for snapping strings', *Phys. Rev.* **D52**, 5598.

Hagmann, C., Chang, S., & Sikivie, P., [1999], 'Axions from string decay', *Nucl. Phys. Proc. Suppl.* **72**, 81.

Hara, T., Yamamoto, H., Mahonen, P., & Miyoshi, S.J. [1999], 'Hierarchical structure of astronomical objects in the cosmic string scheme' *Prog. Theor. Phys.* **102**, 51.

Hawking, S.W., & Ross, S.F., [1995], 'Pair production of black holes on cosmic strings', *Phys. Rev. Lett.* **75**, 3382.

Hill, C.T. [1983], 'Monopolonium', *Nucl. Phys.* **B224**, 469.

Hindmarsh, M. [1993], 'Semilocal topological defects', *Nucl. Phys.* **B392**, 461.

Hindmarsh, M.B., & Kibble, T.W.B. [1995], 'Cosmic strings', *Rep. Prog. Phys.* **58**, 477.

Huguet, E., & Peter, P. [1999], 'Bound states in monopoles: sources for UHECR?', preprint (hep-ph/9901370).

Karra, G., & Rivers, R.J. [1998], 'A re-examination of quenches in ^{4}He', *Phys. Rev. Lett.* **81**, 3707.

Kaspi, V.M., Taylor, J.H., & Ryba, M.F. [1994], 'High-precision timing of millisecond pulsars. 3. Long-term monitoring of PSRS B1885+09 and B1937+21 ', *Ap. J.* **428**, 713.

Kasuya, S., & Kawasaki, M. [1997], 'Can topological defects be formed during preheating?', *Phys. Rev.* **D56**, 7597.

Kasuya, S., & Kawasaki, M. [1998], 'Topological defect formation after inflation on lattice simulations', *Phys. Rev.* **D58**, 083516.

Kibble, T.W.B., & Vilenkin, A. [1995], 'Phase equilibration in bubble collisions', *Phys. Rev.* **D49**, 679.

Kofman, L., Linde, A., & Starobinsky, A.A. [1996], 'Non-thermal phase transitions after inflation', *Phys. Rev. Lett.* **76**, 1011.

Kuzmin, V.A., Rubakov, V.A., & Shaposhnikov, M.E. [1985], 'On the anomalous electroweak baryon number nonconservation in the early universe', *Phys. Lett.* **B155**, 36.

Laguna, P., & Zurek, W.H. [1997], 'Density of kinks after a quench: When symmetry breaks, how large are the pieces?', *Phys. Rev. Lett.* **78**, 2519.

Linde, A.D. [1994], 'Monopoles as big as the universe', *Phys. Lett.* **327B**, 208

Linde, A.D., & Linde, D. [1994], 'Topological defects as seeds for eternal inflation', *Phys. Rev.* **D50**, 2456.

Lyth, D.H. [1992], 'Estimates of the cosmological axion density', *Phys. Lett.* **275B**, 279.

Magueijo, J., Albrecht, A., Coulson, D., & Ferreira, P. [1996], 'Doppler peaks from active perturbations', *Phys. Rev. Lett.* **76**, 2617.

Mahonen, P., Hara, T., Voll, T., & Miyoshi, S.J. [1997], 'Statistics of cosmic microwave background radiation with the cosmic string model', *Int. J. Mod. Phys.* **6**, 535.

Martin, X., & Vilenkin, A. [1996], 'Gravitational wave background from hybrid topological defects', *Phys. Rev. Lett.* **77**, 2879.

Martin, X., & Vilenkin, A. [1997], 'Gravitational radiation from monopoles connected by strings', *Phys. Rev.* **D55**, 6054.

Martins, C.J.A.P., & Shellard, E.P.S. [1996], 'Scale-invariant string evolution with friction', *Phys. Rev.* **D53**, R1–5. Martins, C.J.A.P., & Shellard, E.P.S. [1996], 'Quantitative string evolution', *Phys. Rev.* **D54**, 2535.

Martins, C.J.A.P., & Shellard, E.P.S. [1998a], 'Evolution of superconducting string currents', *Phys. Lett.* **432B**, 58.

Martins, C.J.A.P., & Shellard, E.P.S. [1998b], 'Vorton formation', *Phys. Rev.* **D57**, 7155. Martins, C.J.A.P., & Shellard, E.P.S. [1998b], 'Limits on cosmic chiral vortons', *Phys. Lett.* **B445** 43.

Martins, C.J.A.P., & Shellard, E.P.S. [2000a], 'The velocity-dependent one-scale string evolution model', DAMTP preprint.

Martins, C.J.A.P., & Shellard, E.P.S. [2000b], 'The fractal properties of cosmic string networks', DAMTP preprint.

Masperi, L., & Silva, G. [1998], 'Cosmic rays from decaying vortons', *Astropart. Phys.* **8**, 173.

McGraw, P. [1996], 'Dynamical simulation of nonabelian cosmic strings', preprint (hep-th/9603153).

McGraw, P. [1998], 'Evolution of a nonabelian cosmic string network', *Phys. Rev.* **D57**, 3317.

McHugh, M.P., Zalamansky, G., Vernotte, F., & Lantz, E. [1996], 'Pulsar timing and the upper limits on a gravitational wave background: a Bayesian approach', *Phys. Rev.* **54**, 5993.

Melfo, A., & Perivolaropoulos, L. [1995], 'Formation of vortices in first order phase transitions', *Phys. Rev.* **D52**, 992.

Moore, J.N., & Shellard, E.P.S. [1998], 'On the evolution of abelian-Higgs string networks', DAMTP preprint (hep-ph/9808336).

Olum, K.D., & Blanco-Pillado, J.J. [1999a], 'Field theory simulation of Abelian-Higgs cosmic string cusps', *Phys. Rev.* **D60**, 023503.

Olum, K.D., & Blanco-Pillado, J.J. [1999b], 'Radiation from cosmic string standing waves', Tufts preprint (astro-ph/9910354).

Peacock, J.A., & Dodds, S.J. [1994], 'Reconstructing the linear power spectrum of cosmological mass fluctuations', *M.N.R.A.S.* **267**, 1020.

Pen, U-.L., Seljak, U., & Turok, N. [1997], 'Power spectra in global defect theories of cosmic structure formation', *Phys. Rev. Lett.* **79**, 1611.

Pen, U-.L., & Spergel, D. [1997], 'Cosmology in a string-dominated universe', *Ap. J.* **491** L67.

Pen, U-.L., Spergel, D., & Turok, N. [1994], 'Cosmic structure formation and microwave background anisotropies from global field ordering', *Phys. Rev.* **D49**, 692.

Perivolaropoulos, L. [1992], 'COBE vs cosmic strings: an analytical model', *Phys. Lett.* **298B**, 305.

Perivolaropoulos, L. [1995], 'Spectral analysis of microwave background perturbations induced by cosmic strings', *Ap. J.* **451**, 429.

Pogosian, L., & Vachaspati, T. [1999a], 'Cosmic microwave background anisotropy from wiggly strings', *Phys. Rev.* **D60**, 083504.

Pogosian, L., & Vachaspati, T. [1999b], 'Interaction of magnetic monopoles and domain walls', preprint (hep-ph/9909543).

Riazuelo, A., Deruelle, N., & Peter, P. [1999], 'Topological defects and CMB anisotropies: are the predictions reliable?', to appear in *Phys. Rev. D* (astro-ph/9910290).

Riotto, A. & Trodden, M. [1999], 'Recent progress in baryogenesis', preprint (hep-ph/9901362).

Rivers, R.J. [1999], 'Slow ^{4}He quenches produce fuzzy, transient vortices', to appear in *Phys. Rev. Lett.* (cond-mat/9909249).

Robinson, J., & Yates, A. [1996] 'Cosmic string formation and the power spectrum of field configurations', *Phys. Rev.* **D54**, 5211.

Rubakov, V.A., & Shaposhnikov, M.E. [1996], 'Electroweak baryon number non-conservation in the early universe and in high energy collisions', *Phys. Usp.* **39**, 461.

Ruutu, V.M. *et al.* [1996], 'Big bang simulation in superfluid ^{3}He-B – vortex nucleation in neutron-irradiated superflow', *Nature* **382**, 334.

Ruutu, V.M. *et al.* [1998], 'Defect formation in quench-cooled superfluid phase transition', *Phys. Rev. Lett.* **80**, 1465.

Sakai, N. [1996], 'Dynamics of gravitating magnetic monopoles', *Phys. Rev.* **D54**, 1548.

Sakai, N., Shinkai, H., Tachizawa, T., & Maeda, K. [1996], 'Dynamics of topological defects and inflation', *Phys. Rev.* **D53**, 655.

Scherrer, R.J., & Vilenkin, A. [1997], 'Cosmic string formation from correlated fields', *Phys. Rev.* **D56**, 647.

Seljak, U., Pen, U-.L., & Turok, N. [1997], 'Polarization of the microwave background in defect models', *Phys. Rev. Lett.* **79**, 1615.

Shellard, E.P.S [2000], 'Large-scale structure and CMB anisotropy from cosmic defects', DAMTP preprint.

Sigl, G., Schramm, D., & Bhattacharjee, P. [1994], 'On the origin of highest energy cosmic rays', *Astropart. Phys.* **2**, 401.

Sornborger, A., Brandenberger, R.H., Fryxell, B., & Olson, K. [1997], 'The structure of cosmic string wakes', *Ap. J.* **482**, 22.

Stebbins, A., & Dodelson, S. [1997], 'On the computation of CMBR anisotropies from simulations of topological defects', Fermilab report (astro-ph/9705177).

Stephens, G.J., Calzetta, E.A., Hu, B.L., & Ramsey, S.A. [1999], 'Defect formation and critical dynamics in the early universe', *Phys. Rev.* **D59**, 045009.

t'Hooft, G. [1976], *Phys. Rev. Lett.* **37**, 37.

Thorsett, S.E., & Dewey, R.J. [1996], 'Pulsar timing limits on very low frequency stochastic gravitational radiation', *Phys. Rev.* **53**, 3468.

Tkachev, I. [1996], 'Phase transitions at preheating', *Phys. Lett.* **376B**, 35.

Tkachev, I., Khlebnikov, S., Kofman, L., & Linde, A. [1998], 'Cosmic strings from preheating', *Phys. Lett.* **440B**, 262.

Trodden, M., Davis, A.-C., & Brandenberger, R. [1995], 'Particle physics models, topological defects and electroweak baryogenesis', *Phys. Lett.* **B349**, 131.

Turok, N. [1992], 'Electroweak baryogenesis', in *Perspectives on Higgs physics*, ed. G.L. Kane.

Uzan, J.-P., Deruelle, N., & Riazuelo, A. [1998], 'Cosmic microwave background anisotropies seeded by coherent topological defects: a semi-analytic approach', preprint (astro-ph/9810313).

van de Bruck, C. [1998a], 'Cosmic string network evolution in arbitrary Friedmann–Lemaitre models', *Phys. Rev.* **D57**, 1306.

van de Bruck, C. [1998b], ' Cosmic strings and structure formation', in *Proceedings of the Second International Workshop on Dark Matter*, Klapdor-Kleingrothaus, H.V., & Baudis, L., eds.

Vilenkin, A. [1994], 'Topological inflation', *Phys. Rev. Lett.* **72**, 3137.

Vincent, G., Antunes, N.D., & Hindmarsh, M.B. [1998], 'Numerical simulations of string networks in the Abelian-Higgs model', *Phys. Rev. Lett.* **80** 2277.

Vincent, G., Hindmarsh, M.B., & Sakellariadou, M. [1997], 'Scaling and small-scale structure in cosmic string networks', *Phys. Rev.* **D56**, 637.

Volovik, G.E. [1999], '^{3}He and universe parallelism', preprint (cond-mat/9902171).

Volovik, G.E., & Vachaspati, T [1996], 'Aspects of ^{3}He and the standard electroweak model', *Int. J. Mod. Phys.* **B10**, 471.

Wandelt, B.J. [1998], 'Primordial nongaussianity: baryon bias and gravitational collapse of cosmic string wakes', *Ap. J.* **503**, 67.

Weiler, T.J., & Kephart, T.W. [1996], 'Are we seeing magnetic monopole cosmic rays at $E \gtrsim 10^{20}$ eV?', *Nucl. Phys. Proc. Suppl.* **51B**, 218.

Wick, S.D., Kephart, T.W., Weiler, T.J., & Biermann, P.L. [2000], 'Signatures for a cosmic flux of magnetic monopoles', preprint (astro-ph/0001233).

Yamaguchi, M. [1999], 'Scaling property of the global string in the radiation dominated universe', *Phys. Rev.* **D60**, 103511.

Yamaguchi, M., Yokoyama, J., & Kawasaki, M. [1999], 'Evolution of a global string network in a matter dominated universe', preprint (hep-ph/9910352).

Yates, A., & Zurek, W.H. [1998], 'Vortex formation in two dimensions: when symmetry breaks, how large are the pieces?', *Phys. Rev. Lett.* **80**, 5477.

Zurek, W.H. [1996], 'Cosmological experiments in condensed matter systems', *Phys. Rep.* **276**, 177.

1
Introduction

1.1 Topological defects

The hypothesis of vortices is pressed with many difficulties [Newton, 1713]. So began a scathing critique of Descartes' cosmological theory containing, perhaps, the first cosmic strings of modern science; ostensibly, the planets were swept through space by a giant vortex-line, oriented perpendicular to the plane of the solar system. It was a model which gained a surprising number of adherents on the continent, evidenced by Voltaire's remark, 'In Paris you see the universe full of vortices, whereas in London you find it empty!' Although the Newtonian view ultimately prevailed, recent developments in particle physics and cosmology have cast doubt on whether the vacuum is quite as simple or as empty as was once thought. The well-established notion of spontaneous symmetry breaking in particle physics and improvements in the understanding of cosmological phase transitions have together compelled us to confront the possibility of vortex-strings or other topological defects appearing and playing a significant role in our Universe.

A familiar starting point for our discussion is provided by condensed matter systems in which topological defects are a much-studied phenomenon: In a low temperature context, there are the magnetic flux lines in a Type II superconductor, the quantized vortex-lines in superfluid ^{4}He and the more complex textures in ^{3}He. There are also the line defects or dislocations of crystalline substances and a wide variety of defects in various phases of liquid crystals, but probably the most accessible example is the domain structure of ferromagnetic materials. Topological defects are closely associated with some form of symmetry breaking which gives rise to a non-trivial set of degenerate ground states, such as the discrete magnetic dipole orientations in a ferromagnet. They can appear in condensed matter systems both in and out of thermal equilibrium. In the example of the ferromagnet, the magnetic energy of a large sample is minimized when it splits into domains with different magnetizations. Domain walls

1

appear at the domain boundaries, and so they must be present in the equilibrium state. Of greater interest in cosmology, however, is defect formation in non-equilibrium states, especially during phase transitions. If the temperature falls rapidly below critical, there is only enough time for equilibrium to become established in relatively small volumes, with the symmetry breaking direction being set independently in each of these volumes. This can result in the formation of a stochastic distribution of defects, such as the network of vortices which appears in the liquid crystal shown in fig. 1.1. Exactly analogous effects occur when symmetries are broken in particle physics models, but their implications – cosmological and otherwise – were not recognized for some time.

1.1.1 A brief history

Skyrme [1961] presented the first three-dimensional topological defect solution arising in a nonlinear field theory, and proposed a potentially credible role for such solutions in particle physics. The original idea was to exploit the particle-like nature of defect states to provide a description of the observed spectrum of particle excitations.* While this idea is still pursued phenomenologically, the more fundamental role of defects appears to be as a by-product of spontaneous symmetry breaking, since

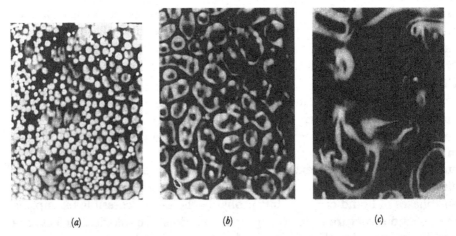

(a)   (b)   (c)

Fig. 1.1. Network of vortices formed during a phase transition in a liquid crystal. Bubbles of the new phase nucleate in (a), the bubbles grow and collide to form strings (b), and the string network coarsens (c) [Bowick *et al.*, 1994].

---

* Others, including Perring, Enz, Finklestein, Misner and Rubenstein, were also investigating this idea at around the same time.

this generally gives rise to a degenerate vacuum manifold with a non-trivial topology. Rather than explaining familiar particle excitations, the additional quantum defect states complete the true particle spectrum of a theory, providing diverse supplementary sectors.

Nambu [1966] was amongst the earliest to anticipate this role in quantum field theory, even suggesting the cosmological significance of defects in remarks which were not to be investigated for a number of years:

> If my view is correct, the universe may have a kind of domain structure. In one part of the universe you may have one preferred direction of the axis; in another part, the direction of the axis may be different.

Before these speculations could be taken seriously, however, the existence of stable topological defect solutions had to be established in realistic renormalizable theories and, in order to provide a formation mechanism, developments were required in the understanding of phase transitions. An important step in the former direction was the discovery of defect solutions in Higgs and Yang-Mills theories, notably the Nielsen–Olesen [1973] vortex-line and the 't Hooft [1974]–Polyakov [1974] monopole.

A key advance in recognising the cosmological implications of symmetry breaking was made by Kirzhnits [1972] (see also Kirzhnits & Linde [1972]) who suggested that spontaneously broken symmetries in a field theory can be restored at high temperatures, just as they are in condensed matter systems. Subsequently, Zel'dovich, Kobzarev & Okun [1974] (see also Kobzarev *et al.* [1974]) pointed out that domain walls would form at phase transitions in which a discrete symmetry was broken, emphasising that the process was causally inevitable. They studied the properties of domain walls and their severe gravitational effects. Weinberg [1974] noted independently that cosmic domains might form and this led Everett [1974] to study possible interactions of domain walls with matter. In a watershed paper, Kibble [1976] then demonstrated that the existence of domain structures – domain walls, strings, monopoles and textures – is dependent on the topology of the vacuum manifold, classifying them using homotopy theory (see also Tyupkin, Fateev & Schwarz [1975] and Coleman [1976]). He also quantitatively discussed the process of defect formation and described their cosmological evolution. Zel'dovich & Khlopov [1978] and Preskill [1979] considered the consequences of monopole production in grand unified theories, showing that monopole densities were typically unacceptably high. Cosmic strings, on the other hand, did not encounter such obvious difficulties and, in alternate scenarios, Zel'dovich [1980] and Vilenkin [1981a] suggested that strings might provide a viable spectrum for galaxy formation. The study of the cosmological implications of topological defects has since become an area of sustained interest.

1.1.2 Overview

Most of the astrophysical roles proposed for topological defects have been in the context of grand unified theories of elementary particles. The basic premise of grand unification is that the known symmetries of elementary particles resulted from a larger symmetry group G after a series of spontaneous symmetry breakings,

$$G \longrightarrow H \longrightarrow SU(3) \times SU(2) \times U(1) \longrightarrow SU(3) \times U(1)_{\text{em}}. \qquad (1.1.1)$$

In a cosmological context, this implies that the early universe has gone through a number of phase transitions, with one or several types of topological defects possibly being left behind. Unfortunately, the energy scales for grand unification are far beyond the capabilities of present accelerators and experiments have given little guidance in deciding between a large number of possible models. However, alternative models predict different types of topological defects at a variety of energy scales, thus leading to a multitude of cosmological scenarios and possible astrophysical consequences. This must be viewed in a positive light because it may provide a probe to distinguish between these models. Although individually each of these scenarios should be approached with a healthy degree of skepticism, the broader issue of the formation and evolution of defects deserves serious investigation. It is by no means unlikely that our universe is permeated by a network of defects like that shown in fig. 1.1, and that the evolution and structure of the universe have been significantly influenced by this network. Conversely, if astrophysical effects expected from a certain type of defect are not observed, particle physics models predicting such defects may be constrained or entirely ruled out.

In this monograph we attempt a systematic study of topological defects and their astrophysical implications. Our emphasis will be on the physical properties of defects and their cosmological evolution, with a view to possible observational effects that may lead to their detection. Although much of the work on topological defects has been inspired by specific cosmological applications, such as galaxy formation or baryogenesis, we shall take a broader view and discuss possible observational manifestations of a wide variety of defects with different energy scales and matter interactions, only then turning to consider special parameter values which make the defects suitable for some prominent cosmological role.

Finally, we note that the study of topological defects is greatly aided by the fact that their basic properties are rather insensitive to details of the underlying particle physics. This allows us to use simple field-theoretic models with $U(1)$ or $SU(2)$ symmetries, giving only occasional references to 'realistic' GUT models for illustration. In view of the present uncertainties in high-energy physics, we will not attempt a detailed discussion of defects in grand unified or more fundamental theories.

Before embarking on a general discussion about the structure and formation of topological defects, it is appropriate to introduce a basic framework with a brief summary of the standard Hot Big Bang cosmology, followed in the next chapter by a review of spontaneous symmetry breaking and cosmological phase transitions. After this introduction, we focus first on cosmic strings, detailing their underlying field theoretic structure, particle interactions and gravitational effects, before considering their cosmological evolution and astrophysical consequences. This pattern is then repeated for the other topological defects – domain walls, monopoles, and textures.

1.2 Units and notation

As prefigured already, discussions in this monograph will span a wide variety of length scales, ranging from the Planck length through to the size of the observed universe. In each of these different contexts, physicists choose appropriate units for historical reasons or for convenience. In order to make direct connections with the literature, we shall do likewise. However, we acknowledge that the units employed by particle physicists may appear peculiar to astronomers, and vice versa, so we shall provide a brief introduction.

The default units employed throughout this monograph are known as *fundamental* or *natural* units; the following fundamental constants are set to unity,

$$\hbar = c = k_\mathrm{B} = 1\,. \tag{1.2.1}$$

All dimensions in this system are then expressed in terms of one basic unit – energy – which is usually given in $\mathrm{GeV} = 10^9\mathrm{eV}$, that is,

$$[Energy] = [Mass] = [Temperature] = [Length]^{-1} = [Time]^{-1}\,. \tag{1.2.2}$$

For the purposes of conversion to cgs units, the Planck mass, length and time are given respectively by

$$m_\mathrm{pl} = 1.22 \times 10^{19}\mathrm{GeV} = 2.18 \times 10^{-5}\mathrm{g}\,,$$
$$\ell_\mathrm{pl} = 8.2 \times 10^{-20}\mathrm{GeV}^{-1} = 1.62 \times 10^{-33}\mathrm{cm}\,, \tag{1.2.3}$$
$$t_\mathrm{pl} = 5.39 \times 10^{-44}\mathrm{s}\,.$$

Newton's constant in these units becomes $G = m_\mathrm{pl}^{-2}$.

We shall depart from these fundamental units when comparing theoretical predictions with empirical data, notably when discussing large-scale structure formation or electromagnetic phenomena associated with superconducting strings. In the former case, we employ astronomical units which for historical reasons are based around quantities in our own solar system. For example, the distance at which the earth-sun distance (1 a.u.) subtends one second of arc is known as a *parsec* (1 pc = 3.26 light years). Typically, astronomical distances will be quoted in megaparsecs,

$$1\,\text{Mpc} = 3.1 \times 10^{24}\text{cm}. \tag{1.2.4}$$

The standard unit of mass is provided by the solar mass,

$$M_{\odot} = 1.99 \times 10^{33}\text{g}. \tag{1.2.5}$$

The transition from fundamental units to cgs or astronomical units should be explicit.

There are two systems of electromagnetic units, the Gaussian and the Heaviside–Lorentz system, which prove to be convenient for small- and large-scale applications, respectively. We employ Gaussian units (along with natural units) when discussing the microscopic electromagnetic properties of defects in chapters 5 and 14 but, in the macroscopic context of discussions in chapters 8 and 12, we use Heaviside units. The difference between the two systems amounts to a factor of 4π in the definition of the fine structure constant: $\alpha = \kappa e^2$, where e is the electron charge and $\kappa = 1/4\pi$ (Gaussian) or $\kappa = 1$ (Heaviside).

Throughout the book we adopt the following notation: Spacetime indices are denoted by Greek letters and run through the four values 0, 1, 2, 3. Spatial indices are denoted by Latin letters from the middle of the alphabet with values 1, 2, 3 (or x, y, z). Group indices or worldsheet indices are denoted by Latin letters from the beginning of the alphabet. Implicit summation over repeated indices is assumed, unless stated otherwise. Finally, the metric signature throughout the monograph is taken to be $(+, -, -, -)$.

1.3 The standard cosmology

1.3.1 The homogeneous expanding universe

The Hot Big Bang model, which describes the evolution of the universe from 1/100 second after the initial explosion through to the present day, is a theory so well-attested it is now called the *standard cosmology*.* It is based upon a simple assumption, known as the *cosmological principle*, which asserts that the universe is the same in every direction from every

---

* Standard recommended textbooks include Weinberg [1972], Zel'dovich & Novikov [1983], Peebles [1971, 1993] and Kolb & Turner [1990].

point in space, that is, it is both homogeneous and isotropic. Restated in Copernican form, our position in the universe – with respect to the largest scales – is in no sense preferred. There is considerable evidence for the homogeneity and isotropy of the universe, including the measured distributions of galaxies and faint radio sources, but by far the best evidence comes from the observed uniformity of the cosmic microwave background radiation.

The line element for any spacetime consistent with the cosmological principle can be written in the Friedmann–Robertson–Walker (FRW) form,

$$ds^2 = dt^2 - a^2(t)\, d\ell^2 \,, \tag{1.3.1}$$

where t is the physical time, $a(t)$ is the scale factor, and $d\ell^2$ represents the line element on a three-dimensional space of constant curvature. In spherical coordinates, this three-metric takes the form

$$d\ell^2 = \frac{dr^2}{1 - kr^2} + r^2\left(d\theta^2 + \sin^2\theta\, d\phi^2\right), \tag{1.3.2}$$

where the constant curvature k is determined by the spatial topology and geometry of the universe:

$$\begin{aligned} k &> 0, & S^3 & \quad \text{closed}, \\ k &= 0, & R^3 & \quad \text{flat}, \\ k &< 0, & H^3 & \quad \text{open}. \end{aligned} \tag{1.3.3}$$

Here, the three-sphere S^3 is compact but H^3, the three-space of constant negative curvature, is infinite and topologically equivalent to the Euclidean three-space R^3.

The homogeneity of the matter content of the universe implies that fluid flow lines are along constant (r, θ, ϕ) trajectories, so these are described as comoving coordinates. Note, however, that physical distances are determined by multiplying comoving distances by the scale factor $a(t)$. It is often useful to introduce an alternative time coordinate, the conformal time τ, defined by

$$d\tau = \frac{dt}{a(t)}, \tag{1.3.4}$$

so that the metric (1.3.1) takes the simple form

$$ds^2 = a^2(\tau)\left[d\tau^2 - d\ell^2\right]. \tag{1.3.5}$$

A variety of kinematic effects follow immediately from the FRW metric (1.3.1–2). Consider two comoving particles separated by a distance ℓ. It is clear that ℓ will grow in time in proportion to $a(t)$. There will be a corresponding recessional velocity given by

$$v = H\ell\,, \tag{1.3.6}$$

where the Hubble parameter H (or 'Hubble's constant') is

$$H = \frac{\dot{a}}{a}\,, \tag{1.3.7}$$

with dots denoting derivatives with respect to the cosmic time t. The relation (1.3.6) is known empirically as Hubble's law: the observation that distant galaxies are receding from our own with velocities proportional to their distance. The parameter H is today measured to be

$$H_0 \approx 50\text{--}100\,\mathrm{km\,s^{-1}\,Mpc^{-1}}\,, \tag{1.3.8}$$

with the large uncertainties arising because of the unreliability of cosmological distance measurements. We adopt the usual convention by incorporating these uncertainties in the dimensionless parameter 'little h',

$$h = H_0/100\,\mathrm{km\,s^{-1}\,Mpc^{-1}}\,, \tag{1.3.9}$$

where $0.5 \lesssim h \lesssim 1.0$.

Another implication of the metric (1.3.1–2) is that light with a frequency ν will be redshifted by the usual Doppler effect, $\dot{\nu}/\nu = -\dot{a}/a$. Consequently, the frequency will decay as $\nu(t) \propto a(t)^{-1}$, so the redshift of spectral lines from a distant galaxy can provide a measure of its relative age and distance. This redshift z is defined as

$$1 + z \equiv \frac{a(t_0)}{a(t_e)} = \frac{\nu_e}{\nu_0}\,, \tag{1.3.10}$$

where t_e is the time the radiation was emitted at a frequency ν_e and the subscript 0 denotes quantities measured at the present day.

1.3.2 The Einstein equations

The scale factor $a(t)$ in the metric (1.3.1) is constrained to obey Einstein's equations. The energy–momentum tensor $T^{\mu\nu}$ for the matter content must have the same symmetries as the homogeneous FRW-metric (1.3.1–2), so it takes the form of a perfect fluid, that is,

$$T_{\mu\nu} = (\rho + p)u_\mu u_\nu - p g_{\mu\nu}\,, \tag{1.3.11}$$

where the energy density ρ and the pressure p are functions of time t only, and u^μ is the four-velocity of the comoving matter, that is, $u^0 = 1$ and $u^i = 0$. The local conservation of energy $T^{\mu\nu}_{;\nu} = 0$ then implies

$$\dot{\rho} + 3\frac{\dot{a}}{a}(\rho + p) = 0\,. \tag{1.3.12}$$

The second term accounts for the dilution of the energy density as the universe expands, whereas the third represents the work done by the pressure of the fluid.

For a complete specification of the evolution, we must also give the equation of state for the matter, $p = p(\rho)$. If the universe is filled with 'dust', non-relativistic matter with negligible pressure $p \ll \rho$, then (1.3.12) implies $\rho \propto a^{-3}$. On the other hand, for radiation the appropriate equation of state is that of an ideal relativistic gas, $p = \rho/3$. The density then varies as $\rho \propto a^{-4}$, with the additional power of $a(t)$ due to frequency redshifting. Once matter becomes non-relativistic, it will inevitably dominate the energy density of an expanding universe because $\rho_{\rm m}/\rho_{\rm r} \propto a$.

The Einstein equations,

$$R_{\mu\nu} - \tfrac{1}{2}g_{\mu\nu}R = -8\pi G T_{\mu\nu}\,, \tag{1.3.13}$$

imply the Friedmann equation

$$\left(\frac{\dot{a}}{a}\right)^2 + \frac{k}{a^2} = \frac{8\pi G}{3}\rho\,, \tag{1.3.14}$$

which along with (1.3.12) and the equation of state determines the evolution. For radiation with $p = \rho/3$, it is clear from (1.3.14) that in a flat universe $(k = 0)$ the scale factor will depend on t as

$$a(t) \propto t^{1/2}\,. \tag{1.3.15}$$

The energy density will then be given by the simple relation

$$\rho_{\rm crit} = \frac{3}{32\pi G t^2}\,, \tag{1.3.16}$$

where the subscript indicates that this is the density of a flat universe – the critical value lying between that for open and closed universes at the same time t. The dimensionless density parameter

$$\Omega = \frac{\rho}{\rho_{\rm crit}} \tag{1.3.17}$$

summarizes this difference with $\Omega < 1$, $\Omega = 1$ and $\Omega > 1$ corresponding to open, flat and closed universes, respectively.

For a pressureless matter-dominated flat universe, the scale factor will grow as

$$a(t) \propto t^{2/3}\,. \tag{1.3.18}$$

Again, the expression for the critical density is easily found from (1.3.14)

$$\rho_{\rm crit} = \frac{1}{6\pi G t^2}\,. \tag{1.3.19}$$

It is important to note that this $\Omega = 1$ solution is a repeller from which all universes will rapidly diverge for only a marginal excess or shortfall in density. This can be seen most clearly by considering the conformal time versions of (1.3.12) and (1.3.14). In the matter era, for example, it is straightforward to derive the relation

$$\frac{d\Omega}{d\tau} = \frac{1}{a}\frac{da}{d\tau}\Omega(\Omega - 1)\,. \qquad (1.3.20)$$

Since $da/d\tau > 0$ in an expanding universe, if $\Omega > 1$ then $d\Omega/d\tau > 0$ and hence Ω will be driven towards a larger value even further from unity. Conversely, $\Omega < 1$ implies $d\Omega/d\tau < 0$ so that $\Omega \to 0$ and the universe ultimately will become curvature dominated.

Given the enormous timescale of cosmological evolution, it is remarkable how near unity Ω is placed by current observations. Using rotation curves for spiral galaxies out to a radius enclosing most visible matter, one can estimate the mass density associated with luminous matter,

$$\Omega_{\text{lum}} \lesssim 0.01\,. \qquad (1.3.21)$$

However, there is considerable evidence that galactic rotation curves extend undiminished out into a dark halo, implying that Ω is almost an order of magnitude higher. Further dynamical studies on galaxy cluster scales, such as infall to Virgo, similarly suggest

$$\Omega_{\text{clust}} \approx 0.1 - 0.2\,. \qquad (1.3.22)$$

Indeed, there is some evidence from very large-scale surveys for a smooth background with $\Omega_0 \approx 1$. This seems to indicate that there must be a significant dark matter component to the energy density of the universe. As we shall see, however, primordial nucleosynthesis arguments strongly suggest that this dark matter must be mainly non-baryonic – perhaps massive neutrinos or other more exotic forms of matter.

The above considerations allow us for the most part to assume that $\Omega = 1$. This has the benefit of considerable simplification and will be taken throughout the monograph except where explicitly stated otherwise. It is also consistent with current theoretical prejudice which suggests that an early inflationary epoch set Ω exceptionally close to unity. We should note in any case that (1.3.20) implies that $\Omega = 1$ will always be accurate in the early universe, assuming only that $\Omega_0 \sim 1$ today as in (1.3.22).

Given an expansion rate (1.3.18), the age of the universe measured from the initial singularity ($a = 0$) is

$$t_0 = \tfrac{2}{3}H_0^{-1}\,. \qquad (1.3.23)$$

Using H_0 from (1.3.8), this corresponds to an age in the range 10–20 billion years. This is corroborated by radioisotope dating of the solar system, as well as stellar and globular cluster lifetimes.

The FRW models (1.3.1) have cosmological horizons; it is only possible at a time t to have received light signals from particles lying within a radial distance

$$d_{\mathrm{H}} = a(t) \int_0^t \frac{dt'}{a(t')} . \tag{1.3.24}$$

In the radiation dominated era for a flat universe, the horizon size is $d_{\mathrm{H}} = 2t$, while in the matter dominated era $d_{\mathrm{H}} = 3t$.

1.3.3 Thermal history of the universe

The assumption that the early universe was in thermal equilibrium considerably simplifies its study, because its description then only depends on the temperature T (assuming negligible chemical potentials for all particle species). For much of the early history of the universe, the equation of state of a relativistic ideal gas applies; the energy density at a temperature T is given by

$$\rho = \frac{\pi^2}{30} \mathcal{N}(T) T^4 , \tag{1.3.25}$$

where $\mathcal{N}(T) = \mathcal{N}_{\mathrm{b}}(T) + \frac{7}{8} \mathcal{N}_{\mathrm{f}}(T)$, and $\mathcal{N}_{\mathrm{b}}(T)$ and $\mathcal{N}_{\mathrm{f}}(T)$ are respectively the effective number of distinct helicity states of bosons and fermions with masses $m \lesssim T$. For particles with $m \gg T$, the equilibrium density is exponentially suppressed.

Particle species 'freeze-out' or decouple from thermal equilibrium as the universe expands and cools. This decoupling occurs when a particle's interaction rate Γ_{int} falls below the Hubble expansion rate H, defining the decoupling temperature T_{d} when $\Gamma_{\mathrm{int}} \approx H$. Particles with $m \gg T_{\mathrm{d}}$ will 'freeze-out' at negligible densities, but massless species and those for which $m < T_{\mathrm{d}}$ will decouple with significant densities. The contribution of these out-of-equilibrium particles is not included in (1.3.25).

While the universe remains in thermal equilibrium ($\Gamma_{\mathrm{int}} \gg H$), the expansion is adiabatic, that is, entropy is conserved,

$$\frac{d}{dt} \left(sa^3 \right) = 0 , \tag{1.3.26}$$

where the entropy density s is given by

$$s = \frac{2\pi^2}{45} \mathcal{N}(T) T^3 . \tag{1.3.27}$$

For $\mathcal{N}(T) = \mathrm{const.}$, these relations imply that $aT = \mathrm{const.}$ In this case, (1.3.25) and (1.3.14) yield the radiation era expansion law (1.3.15) and the critical density (1.3.16). Combining (1.3.16) and (1.3.25) in an $\Omega = 1$ universe, we can express the cosmic time t in terms of the temperature,

$$t = \left(\frac{45}{16\pi^3 \mathcal{N} G}\right)^{1/2} T^{-2} \approx 0.30 \mathcal{N}^{-1/2} \frac{m_{\mathrm{pl}}}{T^2}. \tag{1.3.28}$$

Note, however, that the relation (1.3.28) is not valid in the vicinity of mass thresholds where the value of $\mathcal{N}$ changes.

Let us now briefly follow the thermal history of the universe backwards in time, before discussing several of the more significant events in greater detail: Until about 400,000 years after the Big Bang, all matter in the universe was fully ionized. However, when the temperature dropped below the hydrogen ionization threshold of 13.6 eV, electrons and protons began to combine to form hydrogen atoms, a process termed 'recombination'. The resulting rapid fall in the free electron density reduced the electron–photon interaction rate (Thomson scattering) and matter and radiation decoupled. This occurred at approximately

$$t_{\mathrm{dec}} \approx 5.6 \times 10^{12} (\Omega h^2)^{-1/2} \,\mathrm{s}\,,$$
$$z_{\mathrm{dec}} \approx 1100\text{--}1200\,. \tag{1.3.29}$$

The decoupled photon gas is observed today as the cosmic microwave background; it has a characteristic black-body spectrum of temperature $T = 2.7\,\mathrm{K}$ and a corresponding energy density $\rho_\gamma = 4.5 \times 10^{-34}\mathrm{g\,cm}^{-3}$.

Extrapolating further back in time, we come to a redshift z_{eq} when the matter and radiation densities were comparable. To estimate z_{eq}, we must include neutrinos (with N_ν species) in the total radiation background,

$$\rho_{\mathrm{r}} \approx (1 + 0.23 N_\nu)\rho_\gamma\,. \tag{1.3.30}$$

Taking $N_\nu = 3$ and using the relations in §1.3.2, we obtain the matter–radiation transition redshift and time,

$$t_{\mathrm{eq}} \approx 4 \times 10^{10} (\Omega h^2)^{-2} \,\mathrm{s}\,,$$
$$z_{\mathrm{eq}} \approx 2 \times 10^4 \,\Omega h^2\,. \tag{1.3.31}$$

The density of the universe then was $\rho_{\mathrm{eq}} = 3.2 \times 10^{-16} (\Omega h^2)^4 \,\mathrm{g\,cm}^{-3}$.

Prior to the matter–radiation transition, there is little of interest until the density and temperature of the primordial plasma were sufficiently high to synthesize the lightest elements: D, ^{3}He, ^{4}He and ^{7}Li. This occurred between about one second and three minutes after the Big Bang, and observed primordial abundances are consistent with theoretical predictions for this epoch. Earlier still, at $t \approx 1\,\mathrm{s}$ electron–positron pairs annihilated to leave a high temperature plasma consisting only of photons, neutrinos, electrons, protons and neutrons. Prior to $t \approx 10^{-2}\,\mathrm{s}$, a wide variety of hadronic resonances were excited, and it is here that further progress is hampered by the numerical difficulty of calculations involving strong interactions. The same applies for the preceding quark–hadron transition at about $t \approx 10^{-4}$ s, but this is not the end of the story because we do appear to have a complete description of the underlying physical theory. For $t < 10^{-4}$ s, the universe would have been filled with a plasma of quarks, leptons and gauge bosons. The common feature of such high energy particle physics models is that they exhibit asymptotic freedom, that is, as the temperature is increased, particle interactions weaken. This means that a description of the primordial plasma should simplify greatly in earlier hotter epochs. Indeed, it is generally assumed that the equation of state for a relativistic gas (1.3.25) will be valid again at times earlier than $t < 10^{-10}$ s, that is, from around electroweak symmetry breaking through to the grand unification scale at $\sim 10^{-35}$ s and even beyond to the Planck epoch at 10^{-43} s. Fig. 1.2 briefly summarizes the thermal history of the universe as it is currently understood.

At this point, it is important to emphasize the two-way interaction between particle physics and cosmology in pushing back the frontiers of the standard cosmology. Modern particle physics provides a description of the matter content of the universe which is vital to an understanding of its evolution. Indeed, cosmologists look to particle physics in the hope of unravelling a number of remaining cosmic enigmas which will be discussed in §1.3.7. From the other perspective, the Hot Big Bang is the ultimate laboratory in which to test the high energy fundamental theories which purport to describe matter up to the Planck scale. Terrestrial experiments achieving little more than a TeV are some 16 orders of magnitude adrift and, as Zel'dovich emphasized, attempting to test GUT models with a particle accelerator is like trying to verify electroweak theory with radio waves. Moreover, making a particle physics model compatible with the standard cosmology is now a highly non-trivial task, and severe cosmological constraints often result where points of contact are made. Nowhere is this confrontation more apparent than in the study of topological defects.

1.3.4 Primordial nucleosynthesis

The compatibility between nucleosynthesis calculations and the observed light element abundances is a significant triumph of the Hot Big Bang model. Nucleosynthesis began with the 'freeze out' of the neutron–proton ratio at a temperature $T \approx 0.8$ MeV at $t \approx 1$ s. Prior to this, weak interactions maintained thermal equilibrium which was governed by the

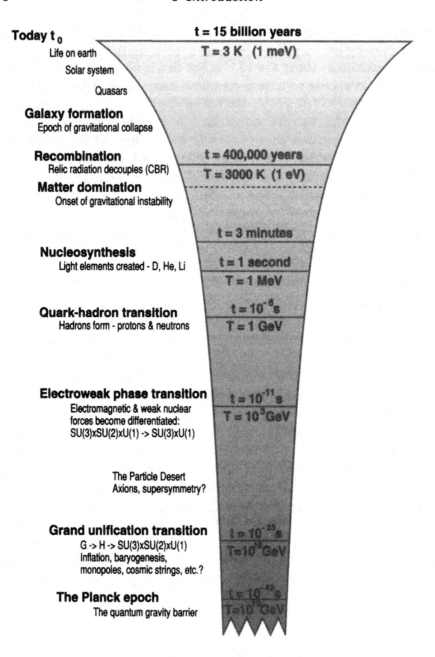

Today t$_0$ **t = 15 billion years**

Life on earth   **T = 3 K (1 meV)**

Solar system

Quasars

Galaxy formation
Epoch of gravitational collapse

Recombination   **t = 400,000 years**
Relic radiation decouples (CBR)   **T = 3000 K (1 eV)**

Matter domination
Onset of gravitational instability

t = 3 minutes

Nucleosynthesis   **t = 1 second**
Light elements created - D, He, Li   **T = 1 MeV**

Quark-hadron transition   **t = 10^{-6}s**
Hadrons form - protons & neutrons   **T = 1 GeV**

Electroweak phase transition   **t = 10^{-11}s**
Electromagnetic & weak nuclear   **T = 10^3GeV**
forces become differentiated:
SU(3)xSU(2)xU(1) -> SU(3)xU(1)

The Particle Desert
Axions, supersymmetry?

Grand unification transition   **t = 10^{-35}s**
G -> H -> SU(3)xSU(2)xU(1)   **T=10^{15}GeV**
Inflation, baryogenesis,
monopoles, cosmic strings, etc.?

The Planck epoch   **t = 10^{-43}s**
The quantum gravity barrier   **T=10^{19}GeV**

Fig. 1.2. A brief thermal history of the universe.

relation

$$\frac{n_{\rm n}}{n_{\rm p}} \approx \exp(-\Delta m/T)\,, \qquad (1.3.32)$$

where the neutron–proton mass difference is $\Delta m = m_{\rm n} - m_{\rm p} = 1.29\,{\rm MeV}$. Before the decoupled neutrons could decay, most became bound into deuterium by a temperature $T \approx 0.1\,{\rm MeV}$, that is, $\rm p + n \rightarrow D + \gamma$. Subsequently, light elements such as ^{3}He, ^{4}He and ^{7}Li were synthesized, with abundances which can be accurately estimated in a simple homogeneous model with few adjustable parameters. The observed relative abundances of the light elements are approximately: 0.25 ^{4}He, 3×10^{-5} D, 2×10^{-5} ^{3}He, and 10^{-9} ^{7}Li. Assuming a predominantly cosmological origin, these levels are compatible with theoretical predictions. However, consistency with the observational data introduces some non-trivial cosmological constraints.

The relative abundances of the light elements, particularly deuterium, depend sensitively on the baryon density $\rho_{\rm b}$. A higher density implies more collisions between the deuterium nuclei, making their demise more likely though ^{4}He formation. Given that primordial deuterium D has a relative abundance of more than 10^{-5}, this implies a fairly firm bound on the total baryonic matter of (refer to Kolb & Turner [1990])

$$\Omega_{\rm b} < 0.16\,. \qquad (1.3.33)$$

This is marginally inconsistent with dynamical astrophysical measurements of Ω and conflicts with the inflationary scenario prediction that $\Omega = 1$. It appears to imply the dominance of a non-baryonic dark matter component. Such arguments also imply a lower bound on the density of baryons, $\Omega_{\rm b} \gtrsim 0.015$.

Another constraint comes from the dependence of the freeze-out temperature $T_{\rm f}$ on the ratio of the weak interaction rate $\Gamma_{\rm W}$ and the expansion rate H. If there are additional particle species present at the time of nucleosynthesis – such as another neutrino – then by (1.3.28) the plasma will cool faster and the freeze-out temperature $T_{\rm f}$ will be higher. By (1.3.32) this will increase the neutron–proton ratio n/p, which in turn determines the final ^{4}He abundance. Given an observed helium abundance below 25%, this implies a restriction on the number of neutrino species $N_\nu < 4$. By similar reasoning, we can constrain the density of relic gravitational waves at this time to be below $\Omega_{\rm gw} < 0.054$.

1.3.5 The microwave background radiation

The discovery of an excess radio background in 1964 by Penzias and Wilson provided compelling evidence for a hot early epoch. This microwave background was subsequently shown to have a near-perfect black-body distribution with temperature $T_0 = 2.7\,\mathrm{K}$. One property of the Planck spectrum in an expanding FRW universe is that it retains its shape after radiation decoupling; the temperature merely redshifts as

$$T = T_{\mathrm{dec}} \left(\frac{1+z}{1+z_{\mathrm{dec}}} \right) . \qquad (1.3.34)$$

The effective photosphere of the universe, beyond which it is not transparent to relic photons, is called the surface of last scattering, $z = z_{\mathrm{ls}}$. If the early universe was not reionized then $z_{\mathrm{ls}} \approx z_{\mathrm{dec}}$ as given in (1.3.29). However, ionization due to the early formation of stars or quasars can shift the last scattering surface towards smaller redshifts. In general, we have

$$z_{\mathrm{dec}} > z_{\mathrm{ls}} \gtrsim 60\Omega^{1/3} \left(\Omega_b/0.06 \right)^{-2/3} \gtrsim 30 , \qquad (1.3.35)$$

where the lower bound is due to the cosmological decrease in the density of matter; at small redshifts the universe is transparent even if it is completely reionized.

Using (1.3.24), it is straightforward to show that the horizon at last scattering today subtends an angle

$$\theta_{\mathrm{ls}} \approx \Omega^{1/2} z_{\mathrm{ls}}^{-1/2} \ \mathrm{radians} . \qquad (1.3.36)$$

From (1.3.35), $2° \gtrsim \theta_{\mathrm{ls}} \gtrsim 10°$, so the backround radiation must be arriving from a multitude of causally disconnected regions on the surface of last scattering. Once a simple dipole anisotropy due to the earth's motion is subtracted, the resulting homogeneity and isotropy of the microwave background is remarkable.

The 1992 measurement of microwave background anisotropies by the COBE satellite, has probably uncovered the imprint of the primordial fluctuations which led to the formation of large-scale structure. These anisotropies were measured on scales above $10°$ with an rms amplitude [Smoot *et al.*, 1992]

$$\left. \frac{\delta T}{T} \right|_{\mathrm{rms}} \approx 1.10 \, (\pm 0.18) \times 10^{-5} . \qquad (1.3.37)$$

At the time of writing, the status of observations on scales below $10°$ is somewhat confused. Several different experiments are approaching the sensitivity of COBE and possible detections remain to be confirmed. However, all recent results on scales ranging upwards from a few arcminutes are consistent with the limit $\delta T/T \lesssim 10^{-4}$.

1.3.6 The growth of density fluctuations

It is generally believed that large-scale structures, such as galaxies and their clusters, formed by gravitational instability from an initial spectrum of primordial density perturbations. Alternative models allow the late generation of fluctuations through explosions, such as those from massive population III stars, but these usually encounter serious difficulties because of an associated spectral distortion of the microwave background.

Once density fluctuations fall within the horizon, gravitational instability will ensue unless pressure effects intervene. An object will oscillate, rather than collapse, if it can be traversed by sound waves in the collapse timescale, that is, if $\tau_G \sim (G\rho)^{-1/2} < r/v_s$ where v_s is the sound speed. The Jeans length λ_J and mass M_J define the appropriate scales above which pressure effects become subdominant and structures become unstable to collapse:

$$\lambda_J = v_s \left(\frac{\pi}{G\rho}\right)^{1/2},$$

$$M_J = \frac{4\pi}{3}\rho \lambda_J^3.$$

(1.3.38)

These quantities depend sensitively on the nature of the matter which begins to dominate the universe at t_{eq}. Non-baryonic dark matter is classified by its thermal velocities when the comoving galactic scale falls within the horizon: *Cold dark matter* (CDM) candidates are usually very heavy particles which decouple early and are non-relativistic by this time. (One exception is the axion, an extremely light particle which is created as a coherent field with negligible momentum.) *Hot dark matter* (HDM) candidates are light particles ($m \lesssim 100\,\text{eV}$) which decouple late and are still relativistic when galactic scales first come inside the horizon. The canonical example is a massive neutrino species with a mass $m_\nu \approx 100\,h^2\text{eV}$ – seemingly the most promising dark matter candidate.

The linear growth of fluctuations in a CDM universe is straightforward. Being effectively pressureless, the Jeans mass for CDM is very small and fluctuations begin to grow immediately at t_{eq} (there is only logarithmic growth during the radiation era when this component is subdominant). In the linear regime, the growing and decaying mode solutions for the density contrast are

$$\frac{\delta\rho}{\rho} \sim At^{2/3} + Bt^{-1}.$$

(1.3.39)

Consequently, by a redshift z, a fluctuation seed $(\delta\rho/\rho)_{eq}$ at t_{eq} will have grown to

$$\left(\frac{\delta\rho}{\rho}\right) \approx \left(\frac{1+z_{eq}}{1+z}\right)\left(\frac{\delta\rho}{\rho}\right)_{eq}.$$

(1.3.40)

Nonlinear structures are formed when $\delta\rho/\rho \sim 1$ on the corresponding length scale. The earliest scales to go nonlinear are those with the largest initial fluctuations. If the fluctuations become stronger towards small scales, then structures will form from the 'bottom up' through hierarchical clustering.

This picture is altered considerably if the dominant component of the dark matter consists of 'hot' neutrinos with $m_\nu \approx 100\,h^2\mathrm{eV}$. Such neutrinos become non-relativistic at $t \sim t_{\mathrm{eq}}$, but fluctuation growth is prevented on small scales by free-streaming; with high thermal velocities, neutrinos simply flow out of over-dense regions at early times, effectively erasing them. We can use the neutrino velocity $v_\nu \sim (1+z)/(1+z_{\mathrm{eq}})$ to define a free-streaming analogue of the Jeans mass,

$$M_{\mathrm{J}} \sim 10^{15} \left(\frac{1+z_{\mathrm{eq}}}{1+z} \right)^{3/2} M_{\odot}. \tag{1.3.41}$$

Consequently, galaxy cluster scales $\sim 10^{15}\,M_{\odot}$ are the first to collapse in a HDM universe, creating large sheet-like structures or 'pancakes'. Since fluctuations on smaller scales were previously erased by free-streaming, small-scale structures must form through pancake fragmentation in this 'top-down' picture.

Fluctuations in baryonic matter are prevented from collapsing before recombination because the sound speed is effectively that of the coupled radiation, $v_{\mathrm{s}} \approx 1/\sqrt{3}$, so the Jeans length remains comparable to the horizon, $\lambda_{\mathrm{J}} \sim d_{\mathrm{H}}$. However, there is a dramatic fall in the pressure at recombination with the sound speed dropping to the non-relativistic $v_{\mathrm{s}} \sim T/(m_{\mathrm{H}}T_{\mathrm{dec}})^{1/2}$ for monatomic hydrogen (mass m_{H}). The Jeans mass is then

$$M_{\mathrm{J}} \sim \frac{1}{G^{3/2}m_{\mathrm{H}}^2} \left(\frac{T}{T_{\mathrm{dec}}} \right)^{3/2} \sim 10^5 \left(\frac{1+z_{\mathrm{dec}}}{1+z} \right)^{3/2} M_{\odot}. \tag{1.3.42}$$

Consequently, at recombination $M_{\mathrm{J}} \sim 10^5\,M_{\odot}$ and so structures can begin to grow on galactic scales $\sim 10^{11}\,M_{\odot}$. In fact, however, photon viscosity prior to t_{dec} will act to erase all baryonic inhomogeneities below $M_{\mathrm{D}} \sim 10^{12}\,M_{\odot}$, so M_{D} is actually the smallest initial collapse scale – this effect is similar to HDM free-streaming. If non-baryonic matter dominates the universe, then recombination signals the beginning of the collapse of baryons into the potential wells which will have already developed in the other matter component since t_{eq}.

Depending on the nature of the fluctuations, their amplitude can also grow while outside the horizon d_{H} – or, more properly, the Hubble radius H^{-1}. This occurs for adiabatic fluctuations created by an inflationary epoch because they are true curvature perturbations. According to (1.3.14), regions with marginally different curvatures k will expand at different rates even on super-horizon scales in the radiation era. Conversely, isocurvature or pressure fluctuations, such as those created by topological defects, do not grow while outside the horizon because they are compensated by a deficit in the background radiation, that is, $\delta\rho = \delta\rho_m + \delta\rho_r = 0$. The defect or compensated seed must move away from the radiation deficit (or vice versa) before it can create a net gravitational potential. Since this is a causal process, growth on a particular scale cannot begin until the isocurvature fluctuation falls within the horizon.

A useful description of density fluctuations is provided by the power spectrum,

$$P(k) = |\delta_{\mathbf{k}}(t)|^2 \,, \tag{1.3.43}$$

where $\delta_{\mathbf{k}}$ is the Fourier transform of the density fluctuation,

$$\delta_{\mathbf{k}} = V^{-1} \int \frac{\delta\rho}{\rho}(\mathbf{x}, t)\, e^{-i\mathbf{k}\cdot\mathbf{x}} d^3x \,, \tag{1.3.44}$$

where the integration is taken over a large comoving volume V. The rms density fluctuation on a particular comoving lengthscale $l = 2\pi k^{-1} a(t)$ is given by contributions of $\delta_{\mathbf{k}}$ over a logarithmic interval $\Delta k/k \sim 1$,

$$\left(\frac{\delta\rho}{\rho}\right)^2 \approx V\frac{k^3}{2\pi^2}|\delta_{\mathbf{k}}(t)|^2 \,. \tag{1.3.45}$$

For gaussian fluctuations, the power spectrum $P(k)$ specifies all the statistical properties of the system, however, more information is required for non-gaussian fluctuations such as those produced by topological defects. It is often assumed that primordial density fluctuations are adiabatic and that the spectrum $P(k)$ on super-horizon scales exhibits a simple power law behaviour with spectral index n, that is, $P(k) \propto k^n$. An important characteristic of a density fluctuation is its magnitude at the time t_H when the corresponding scale crosses the horizon, $l(t_H) = d_H$. It can be shown that for a spectrum of index n,

$$|\delta_{\mathbf{k}}(t_H)|^2 \propto k^{n-4} \,. \tag{1.3.46}$$

The Harrison–Zel'dovich or scale-invariant spectrum has index $n = 1$. This is scale-invariant because the fluctuation amplitude is the same on any comoving scale at the time of horizon crossing,

$$\left.\frac{\delta\rho}{\rho}\right|_{\mathbf{H}} = \text{const.} \tag{1.3.47}$$

If the initial perturbations are isocurvature, they turn into adiabatic fluctuations of comparable magnitude at $t \sim t_H$, and their spectrum can also be characterized by an index n defined according to (1.3.46).

There is good reason to suppose that any realistic primordial fluctuation spectrum must closely approximate the Harrison–Zel'dovich form: If $n > 1$, there is an enhancement of power on small scales, relative to the appropriate normalization on galactic scales. This implies that black holes would have been created in abundance below a certain length scale, but there are stringent cosmological constraints precluding this possibility. Alternatively, a smaller index $n < 1$ will produce large-scale inhomogeneities which may be in conflict with the isotropy of the microwave background at large angles. Indeed, anisotropy detections by the COBE satellite are consistent with a power law spectrum with index $n = 1.15 \pm 0.5$ [Smoot *et al.*, 1992].

Models which purport to describe the formation of large-scale structure must confront a rapidly expanding and improving set of observational data. An example is given by the galaxy correlation function which characterizes galaxy clustering,

$$\xi_{\text{gg}} \approx \left(r/5\, h^{-1}\text{Mpc} \right)^{-1.8}, \quad 0.1\, h^{-1}\text{Mpc} \lesssim r \lesssim 20\, h^{-1}\text{Mpc}. \quad (1.3.48)$$

Here, ξ_{gg} is the probability in excess of random of finding two galaxies a distance r apart. Assuming that 'light traces mass', the galaxy distribution ρ_{g} will be equal to the density distribution ρ with identical correlation functions. However, there has been some difficulty making theoretical $\Omega = 1$ scenarios compatible with observation without relaxing this correspondence. This problem has led to the notion of 'biased' galaxy formation [Kaiser, 1984]; galaxies only form at the peaks of the density distribution – an assumption for which there is some physical motivation (see, for example, Rees [1985]). The effect is parametrized by a 'biasing factor' b which relates the amplitude of fluctuations in the galaxy and mass distributions,

$$\frac{\delta N}{N} = b\, \frac{\delta M}{M}. \quad (1.3.49)$$

Here, δM and δN are, respectively, the rms fluctuations in the mass and the number of bright galaxies within a randomly placed sphere of a certain radius r. The standard choice is $r = 8\, h^{-1}\text{Mpc}$, the lengthscale at which $\delta N/N = 1$.

There is similar, though less compelling, observational evidence for a cluster–cluster correlation function ξ_{cc} of the same slope as ξ_{gg}, but with a significantly higher amplitude. This indicator of the existence of very large-scale structures is corroborated on even greater scales by observations of voids, superclusters and even more exotic structures. Notable amongst these is the CfA galaxy redshift survey which revealed a 'bubbly' structure to the universe, with galaxies lying on the surfaces of voids of typical diameter $20\, h^{-1}\text{Mpc}$. Other examples include the large-scale peculiar velocities over a region of size $60\, h^{-1}\text{Mpc}$ which are ostensibly driven by the 'Great Attractor'. Given that such large scales have not yet turned fully nonlinear, they provide a direct probe into the nature of the primordial density fluctuations. As such, these observations seem to constitute a serious challenge to current orthodoxy because of the apparent lack of consistency between amplitudes on large and small scales.

1.3.7 Shortcomings of the standard cosmology

As we have seen, the standard cosmology successfully explains the Hubble expansion law, the cosmic background radiation, and the abundances of the light elements, while also providing a framework in which to understand large-scale structure formation. We have outlined reasons for believing that this model gives an adequate description of the history of the universe back to a time $t \approx 1/100\,\text{s}$ and beyond. However, if the standard cosmology is extrapolated all the way back to the singularity at $t = 0$, we are compelled to assume a very special initial state with some apparently inexplicable features. Here we shall briefly review some of these outstanding cosmic enigmas:

1) *The horizon problem*: This problem stems from the large-scale homogeneity and isotropy of the universe and, in particular, of the cosmic background radiation. Two microwave antennas pointed in opposite directions receive radiation from regions which, according to the standard model, have never been in causal contact. Nevertheless, the temperature of the radiation from the two regions is observed to be the same to within at least one part in 10^4. The standard model can only account for this fact by making the rather unnatural assumption that the early universe was highly homogeneous and isotropic on scales much greater than the causal horizon.

2) *The flatness problem*: The density of the present universe is within one order of magnitude of the critical density ρ_{crit}, that is, $\Omega \sim 1$ as in (1.3.22). It is known, however, that the critical density is a point of unstable equilibrium – (1.3.20) shows that deviations from $\Omega = 1$ grow in time. Hence, in order to have $\Omega \sim 1$ today we must assume that Ω was fine-tuned to within $|1 - \Omega| \lesssim 10^{-58}$ at the Planck time t_{pl}.

3) *The density fluctuation problem*: It is now generally believed that galaxies and clusters of galaxies evolved by gravitational instability from small density fluctuations in the early universe. The required magnitude of fluctuations on galactic scales at the Planck epoch is $\delta\rho/\rho \sim 10^{-56}$ (assuming they are adiabatic). The origin of these tiny fluctuations is mysterious in the standard cosmology and they simply have to be postulated.

4) *The thermal state problem*: The initial thermal state postulated in the standard model is also inexplicable, because for temperatures above $T \gtrsim 10^{16}\,\text{GeV}$ the expansion of the universe is too fast for thermal equilibrium to be established.

5) *The singularity problem*: The cosmological singularity at $t = 0$ corresponds to a state of infinite energy density and signals the breakdown of classical general relativity. This is an indication that quantum gravity is probably essential in understanding the initial state of the universe.

6) *The cosmological constant problem*: In §1.3.2, we neglected the freedom to add a cosmological constant Λ term to the Einstein equations (1.3.13). It is natural to expect quantum gravity effects to drive Λ up to the Planck scale. Empirical evidence today, however, shows that Λ is over 10^{120} times smaller than this expectation.

In addition to these initial state problems, there are also a number of persistent difficulties in explaining the matter content of the universe:

7) *The baryon asymmetry problem*: The creation of more matter than antimatter in the early universe requires some out-of-equilibrium process with baryon-number violating interactions. This is a twofold problem because any subsequent baryon-number violating interactions will tend to erase a pre-existing asymmetry. The alternative is to have no baryon-number violating processes and to postulate a tiny initial excess of baryons over antibaryons.

8) *The dark matter problem*: In §1.3.4, we noted nucleosynthesis arguments strongly indicating that non-baryonic matter dominates the universe if $\Omega \approx 1$. Though many candidates have been suggested, the exact nature of this dark matter remains unknown.

9) *The exotic relics problem*: Phase transitions and other processes in the early universe can be expected to create topological defects, exotic particles and small black holes. A notable example is the overproduction of monopoles which appears to be an inevitable prediction of grand unified theories.

Recent developments in particle physics and quantum gravity hold out the hope of explaining some of these conundrums. One of the most exciting possibilities is the inflationary paradigm in which the universe goes through a period of rapid expansion. As we shall discuss in §2.4, inflation may assist with problems 1–3 and 9, but it remains to be placed in a realistic particle physics context. Topological defects, on the other hand, do appear in realistic models and may provide a viable alternative mechanism for generating density fluctuations. Indeed, as we shall see, defects have been invoked to explain a variety of other cosmic enigmas, including the baryon asymmetry, dark matter and exotic relics problems.

2
Phase transitions in the early universe

2.1 Spontaneous symmetry breaking

2.1.1 Simple models

Spontaneous symmetry breaking was an idea that originated in condensed matter physics. A familiar example is the isotropic model of a ferromagnet which, although described by a rotationally-invariant Hamiltonian, can develop a magnetic moment pointing in some arbitrary direction. In modern theories of elementary particles, symmetry breaking is described in terms of scalar fields, usually called Higgs fields. The ground state of the theory is characterized by a non-zero expectation value of the Higgs field and does not exhibit all the symmetries of the Hamiltonian.

The essential features of spontaneous symmetry breaking can be illustrated in a simple model, first studied by Goldstone [1961]. This has the classical Lagrangian density

$$\mathcal{L} = (\partial_\mu \bar{\phi})(\partial^\mu \phi) - V(\phi) \,, \tag{2.1.1}$$

where ϕ is a complex scalar field and the potential $V(\phi)$ is given by

$$V(\phi) = \tfrac{1}{4}\lambda \left(\bar{\phi}\phi - \eta^2 \right)^2 \,, \tag{2.1.2}$$

with positive constants λ and η. This potential is illustrated in fig. 2.1. The model is invariant under the group $U(1)$ of global phase transformations,

$$\phi(x) \rightarrow e^{i\alpha}\phi(x) \,. \tag{2.1.3}$$

Here, 'global' indicates that α is independent of the spacetime location x. The minima of the potential (2.1.2) lie on a circle $|\phi| = \eta$, and the ground state (the vacuum) of the theory is characterized by a non-zero expectation value

$$\langle 0 \,|\phi|\, 0 \rangle = \eta \, e^{i\theta} \,, \tag{2.1.4}$$

23

with an arbitrary phase θ. The phase transformation (2.1.3) changes θ into $\theta + \alpha$. Hence, the vacuum state $|0\rangle$ is not invariant under (2.1.3), and the symmetry is spontaneously broken.

The state of unbroken symmetry with $\langle 0|\phi|0\rangle = 0$ corresponds to a local maximum of $V(\phi)$. Small perturbations about that state are described by the Lagrangian (2.1.1) with

$$V(\phi) \approx -\tfrac{1}{2}\lambda\eta^2\bar{\phi}\phi + \text{const.} \qquad (2.1.5)$$

The negative sign of the mass term in (2.1.5) indicates the instability of the symmetric state.

The broken symmetry vacua with different values of θ are all equivalent, and their properties can be found by studying any one of them. Choosing the vacuum with $\theta = 0$, we can represent ϕ as

$$\phi = \eta + \tfrac{1}{\sqrt{2}}\left(\phi_1 + i\phi_2\right), \qquad (2.1.6)$$

where ϕ_1 and ϕ_2 are real fields with zero vacuum expectation values. Substitution of (2.1.6) into the Lagrangian (2.1.1) gives

$$\mathcal{L} = \tfrac{1}{2}\left(\partial_\mu\phi_1\right)^2 + \tfrac{1}{2}\left(\partial_\mu\phi_2\right)^2 - \tfrac{1}{2}\lambda\eta^2\phi_1^2 + \mathcal{L}_{\text{int}}, \qquad (2.1.7)$$

where the interaction term $\mathcal{L}_{\text{int}}$ includes cubic and higher-order terms in ϕ_1 and ϕ_2. We see that ϕ_1 represents a particle with a positive mass $\mu = \sqrt{\lambda}\eta$, while ϕ_2 is massless. The reason for this is intuitively clear from fig. 2.1: ϕ_1 corresponds to radial oscillations about a point on the circle of minima, $|\phi| = \eta$, while ϕ_2 corresponds to motion around the circle. The appearance of massless scalar particles, called Goldstone bosons, is a general feature of spontaneously broken global symmetries.

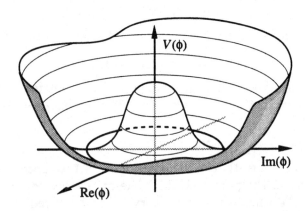

Fig. 2.1.   The Mexican-hat potential (2.1.2) for a broken $U(1)$ symmetry showing a degenerate circle of minima.

These low energy states can also be uncovered with the substitution

$$\phi(x) = \left(\eta + \tfrac{1}{\sqrt{2}}\varphi(x)\right) e^{i\vartheta(x)}, \qquad (2.1.8)$$

where we again replace the complex fields ϕ and $\bar{\phi}$ by real fields, φ and ϑ. The Lagrangian then becomes

$$\mathcal{L} = \tfrac{1}{2}(\partial_\mu\varphi)^2 + \eta^2(\partial_\mu\vartheta)^2 - \tfrac{1}{2}\lambda\eta^2\varphi^2 + \mathcal{L}_{\text{int}}, \qquad (2.1.9)$$

with φ the massive scalar particle and ϑ the massless Goldstone particle.

Although global symmetries are of considerable interest, the central role in particle physics is played by gauge theories with spontaneously broken symmetries. The simplest gauge theory example, known as the abelian-Higgs model, describes scalar electrodynamics with Lagrangian [Higgs, 1964]

$$\mathcal{L} = \bar{\mathcal{D}}_\mu\bar{\phi}\mathcal{D}^\mu\phi - V(\phi) - \tfrac{1}{4}F_{\mu\nu}F^{\mu\nu}. \qquad (2.1.10)$$

where ϕ is a complex scalar field and the covariant derivative is given by $\mathcal{D}_\mu = \partial_\mu - ieA_\mu$. The antisymmetric tensor is $F_{\mu\nu} = \partial_\mu A_\nu - \partial_\nu A_\mu$ with A_μ a gauge vector field, e the gauge coupling, and $V(\phi)$ given by (2.1.2). This model is invariant under the group $U(1)$ of local gauge transformations,

$$\phi(x) \rightarrow e^{i\alpha(x)}\phi(x), \qquad A_\mu(x) \rightarrow A_\mu(x) + e^{-1}\partial_\mu\alpha(x). \qquad (2.1.11)$$

Since the minima of $V(\phi)$ are at $|\phi| = \eta$, this symmetry is spontaneously broken, and the field ϕ acquires a non-zero vacuum expectation value. To study the properties of the broken-symmetry vacuum, it is convenient to use the gauge in which $\phi(x)$ is real. Then, representing ϕ as $\phi = \eta + \phi_1/\sqrt{2}$, we obtain

$$\mathcal{L} = \tfrac{1}{2}(\partial_\mu\phi_1)^2 - \tfrac{1}{2}\mu^2\phi_1^2 - \tfrac{1}{4}F_{\mu\nu}F^{\mu\nu} + \tfrac{1}{2}M^2 A_\mu A^\mu + \mathcal{L}_{\text{int}}, \qquad (2.1.12)$$

where

$$\mu = \sqrt{\lambda}\,\eta, \qquad M = \sqrt{2}\,e\eta, \qquad (2.1.13)$$

and $\mathcal{L}_{\text{int}}$ includes cubic and higher order terms in ϕ_1 and A_μ. We see that the breaking of a gauge symmetry is not accompanied by a massless Goldstone boson. The corresponding degree of freedom is absorbed into the vector field, which becomes massive and has three independent polarizations instead of the original two.

The models (2.1.1) and (2.1.10) have close condensed matter analogues. The non-relativistic version of (2.1.1) is used to describe superfluids with ϕ being the Bose condensate wave function. The non-relativistic version of (2.1.10) is identical to the Ginzburg–Landau model of a superconductor with ϕ, in this case, representing the Cooper pair wave function. It is possible that the Higgs fields used to describe spontaneous symmetry breaking in elementary particle theories are not fundamental and should be interpreted, instead, as phenomenological order parameters, like the magnetization vector for a ferromagnet or the Cooper pair wave function for a superconductor. Models like (2.1.1) and (2.1.10) would then be interpreted as low-energy effective theories. In our discussions, however, we shall not draw a sharp distinction between fundamental and effective theories, since this does not affect the nature of the topological defects that may arise.

2.1.2 Non-abelian gauge theories

The simple Goldstone model (2.1.1) can be generalized to possess an invariance under an arbitrary group G of global gauge transformations. These transformations are realized on an n-component scalar field ϕ_i through the action of a matrix representation of the group G,

$$\phi_i \to \phi_i' = D_{ij}(g)\phi_j \,, \tag{2.1.14}$$

where $D(g)$ is an $n{\times}n$ matrix and $g \in G$. The representation is a mapping from G to the operators D acting on the vector space spanned by the ϕ_is, which preserves group multiplication, that is, $D(g_1)D(g_2) = D(g_1 g_2)$. Elements of a Lie group G (dimension N) can be written in the form

$$g = \exp\left(-i\omega_a L^a\right) \,, \tag{2.1.15}$$

where the ω_a are arbitrary real numbers and the L^a are the N group generators. The L^a generate the Lie algebra of G and satisfy the commutation relations

$$\left[L^a, L^b\right] = -if_{abc}L^c, \tag{2.1.16}$$

where the f_{abc} are the structure constants of G. A matrix representation of arbitrary dimension n can be generated by finding a set of $n{\times}n$ matrices T^a satisfying the same commutation relations (2.1.16). We can think of G itself as being a group of matrices by taking any faithful (one-to-one) representation of G with Hermitian generators, $\{L^a\}$, the basis for the Lie algebra of G. Another example is the N-dimensional representation generated by the structure constants themselves, $(T^a)_{bc} = -if_{abc}$, which is known as the adjoint representation. For a compact Lie group G, we can normalize the generators by

$$\mathrm{Tr}\left\{T^a T^b\right\} = \delta_{ab} \,, \tag{2.1.17}$$

and we shall assume the correspondence, $T^a = D(L^a)$, between the two sets of generators.

With this apparatus in place, we now consider the generalized Lagrangian,

$$\mathcal{L} = \tfrac{1}{2}(\partial_\mu \varphi^\dagger)(\partial^\mu \varphi) - V(\varphi) , \qquad (2.1.18)$$

where we have chosen $\varphi = \{\phi_i\}$ to be n real scalar fields in a real unitary representation of G with generators T^a. Models with complex fields can be brought to this form by representing each in terms of two real fields.* The field theory described by (2.1.18) is renormalizable if the potential $V(\varphi)$ is a quartic polynomial in the ϕ_i, which we shall generally assume to be the case. The Lagrangian (2.1.18) is invariant under the transformation (2.1.14) if $V(D(g)\varphi) = V(\varphi)$, which can be re-expressed as

$$\frac{\partial V}{\partial \phi_i} T^a_{ij} \phi_j = 0 . \qquad (2.1.19)$$

If the minima of $V(\varphi)$ are at non-zero values of ϕ_i, then the symmetry will be spontaneously broken, and the ϕ_i will develop vacuum expectation values,

$$\langle 0 | \varphi | 0 \rangle = \varphi_0 . \qquad (2.1.20)$$

The elements of G corresponding to transformations which leave $\varphi_0 \ (\neq 0)$ unchanged form a group H, called the unbroken subgroup or the little group of G with respect to φ_0. Defining this in terms of the matrices $D(g)$ in the representation of G, the little group is

$$H = \{g \in G \mid D(g)\varphi_0 = \varphi_0\} . \qquad (2.1.21)$$

The generators t^α of H all annihilate φ_0,

$$t^\alpha \varphi_0 = 0 . \qquad (2.1.22)$$

We can choose the generators of the representation T^a such that the t^α become a subset of the T^a. These t^αs are referred to as the unbroken generators of G with the remainder of the T^as being the broken generators.

Representing φ as $\varphi = \varphi_0 + \varphi'$ and expanding V in powers of the ϕ_is, we find that small perturbations about φ_0 are described by the Lagrangian

$$\mathcal{L} = \tfrac{1}{2}\partial_\mu \phi'_i \partial^\mu \phi'_i - \tfrac{1}{2}\mu^2_{ij}\phi'_i\phi'_j , \qquad (2.1.23)$$

---

* Note the different normalization of φ in (2.1.1) and (2.1.18). This should be kept in mind when comparing various expressions for the two models. The same applies to the Higgs models (2.1.10) and (2.1.28).

where

$$\mu_{ij}^2 = \left[\frac{\partial^2 V}{\partial\phi_i\partial\phi_j}\right]_{\phi=\phi_0}. \tag{2.1.24}$$

Since φ_0 is a minimum of $V(\varphi)$,

$$\left[\frac{\partial V}{\partial\phi_i}\right]_{\phi=\phi_0} = 0, \tag{2.1.25}$$

and the eigenvalues of the mass matrix μ_{ij}^2 are non-negative. Differentiating (2.1.19) and using (2.1.25), we obtain

$$\mu_{ij}^2 T_{jk}^a \phi_{0k} = 0. \tag{2.1.26}$$

All vectors $T^a\varphi_0$ formed from the broken generators ($T^a\varphi_0 \neq 0$) are linearly independent, and so from (2.1.26) μ_{ij}^2 must have a zero eigenvalue for each broken generator. These zero eigenvalues correspond to massless Goldstone bosons, while the remaining masses are, in general, non-zero.

The vacuum defined by (2.1.20) is one of an infinite number of equivalent vacua. The expectation values of φ in other vacua are of the form $D(g)\varphi_0$ for some $g \in G$. (We assume that there is no accidental degeneracy, and the only minima of $V(\varphi)$ are imposed by the symmetry.) The manifold of equivalent vacua $\mathcal{M}$ can thus be identified with the coset space,

$$\mathcal{M} = G/H, \tag{2.1.27}$$

which we shall refer to as the vacuum manifold. The topology of $\mathcal{M}$ plays a crucial role in determining the nature of topological defects, as we shall see in chapter 3.

Turning now to local gauge invariance, we introduce gauge fields A_μ^a which are associated with matrices in the Lie algebra of G (2.1.16) by $\mathcal{A}_\mu = A_\mu^a L^a$. The non-abelian generalization of the Higgs model (2.1.10) is then

$$\mathcal{L} = \tfrac{1}{2}\mathcal{D}_\mu\varphi_i\mathcal{D}^\mu\varphi_i - V(\varphi) - \tfrac{1}{4}F_{\mu\nu}^a F^{a\mu\nu}. \tag{2.1.28}$$

where

$$F_{\mu\nu}^a = \partial_\mu A_\nu^a - \partial_\nu A_\mu^a + e\,f_{abc}A_\mu^b A_\nu^c \tag{2.1.29}$$

is the Yang–Mills field strength, A_μ^a is a gauge vector field, e is the gauge coupling constant, and $\mathcal{D}_\mu\varphi$ is the gauge-covariant derivative of φ,

$$\mathcal{D}_\mu\varphi = \left(\partial_\mu - ie\,A_\mu^a T^a\right)\varphi. \tag{2.1.30}$$

Note that for a real representation of G the generators T^a take the form iQ^a, where Q^a are real antisymmetric matrices, so components of $\mathcal{D}_\mu\varphi$

are real. The gauge transformations $g = g(x)$, which are now allowed to vary as functions of the spacetime coordinates, are defined by

$$\varphi \to D(g)\varphi\,,$$
$$\mathcal{A}_\mu \to g\mathcal{A}_\mu g^{-1} + i\,e^{-1}g^{-1}\partial_\mu g\,. \qquad (2.1.31)$$

The covariant derivative $\mathcal{D}_\mu\varphi$ and the field strength $\mathcal{F}_{\mu\nu} = F_{\mu\nu}^a L^a$ have simple transformation properties

$$\mathcal{D}_\mu\varphi \to D(g)\,\mathcal{D}_\mu\varphi\,,$$
$$\mathcal{F}_{\mu\nu} \to g\,\mathcal{F}_{\mu\nu}g^{-1}\,, \qquad (2.1.32)$$

and it is easily verified that the Lagrangian (2.1.28) is invariant under (2.1.31). Note also that using (2.1.17) the last term of (2.1.28) can be written as $- \mathrm{Tr}\,\mathcal{F}_{\mu\nu}\mathcal{F}^{\mu\nu}/4$.

When φ develops a non-zero vacuum expectation value φ_0, the symmetry (2.1.31) is spontaneously broken. The properties of the broken-symmetry state are most easily understood in the gauge in which the vector φ_i has vanishing components in the subspace $\mathcal{G}$ defined by the vectors $T^a\varphi_0$. Fields lying in this subspace would have corresponded to Goldstone bosons if the gauge symmetry were global. The dimension of $\mathcal{G}$ is equal to the number of broken generators (which is in turn equal to the number of gauge conditions one is allowed to impose). Using again the substitution $\varphi = \varphi_0 + \varphi'$, we obtain the Lagrangian describing the excitations above the broken-symmetry vacuum

$$\mathcal{L} = \tfrac{1}{2}\left(\partial_\mu\phi_i'\right)^2 - \tfrac{1}{2}\mu_{ij}^2\phi_i'\phi_j' - \tfrac{1}{4}\left(F_{\mu\nu}^a\right)^2 + \tfrac{1}{2}M_{ab}^2 A_\mu^a A^{b\mu} + \mathcal{L}_{\mathrm{int}}\,. \quad (2.1.33)$$

The scalar field mass matrix is given by (2.1.24) and the summation over i and j does not include components in the subspace $\mathcal{G}$. Here, the vector field mass matrix is given by

$$M_{ab}^2 = e^2 \left(T^a T^b\right)_{ij} \phi_{0i}\phi_{0j}\,. \qquad (2.1.34)$$

We see that vector fields associated with broken generators acquire non-zero masses, while the gauge fields of the unbroken subgroup H remain massless. The would-be Goldstone bosons have disappeared, and the corresponding degrees of freedom have been absorbed as additional spin states of the massive vector fields. This process is summarized in table 2.1.

Fermions can be easily incorporated into this scheme by adding appropriate kinetic and interaction terms to the Lagrangian (2.1.28),

$$\mathcal{L}_{\mathrm{F}} = i\bar{\psi}\gamma^\mu \mathcal{D}_\mu\psi - \bar{\psi}\Gamma_i\psi\phi_i\,, \qquad (2.1.35)$$

Table 2.1. The Higgs mechanism for the
symmetry breaking, $G \to H$.

Particle species	Number of fields
Massive Higgs ϕ_i	$n - K$
Massive gauge A_μ^i	K
Massless gauge A_μ^α	$N - K$

Here, N is the number of generators for the
group G, $N - K$ is the number for H and n is
the dimension of the *real* representation of G.

where ψ is a fermion multiplet transforming under a representation of
G, $D_\mu\psi$ is its gauge-covariant derivative and Γ_i is the Yukawa coupling
matrix. The fermion mass matrix in the broken-symmetry state is given
by

$$m = \Gamma_i \phi_{0i} \, . \tag{2.1.36}$$

2.1.3 Symmetry breaking in SU(2) models

We shall illustrate these points with some symmetry breaking patterns for
the simplest non-abelian group $SU(2)$, elements of which are the unitary
2×2 matrices U with $\det(U) = 1$. Most features of the topological defects
we shall be discussing become evident in this model. In more realistic
situations, defects are often directly associated with a simple embedding
of $SU(2)$ in a larger symmetry group.

The most familiar example of a global $SU(2)$ gauge invariance is isospin
in which a two-component nucleon spinor transforms as

$$\begin{pmatrix} \psi_p \\ \psi_n \end{pmatrix} \to \begin{pmatrix} \psi'_p \\ \psi'_n \end{pmatrix} = U \begin{pmatrix} \psi_p \\ \psi_n \end{pmatrix} \, . \tag{2.1.37}$$

The matrix U is an element of $SU(2)$ and can be written in the general
form

$$\begin{aligned} U &= \exp\left(i\theta\,\hat{n}_a\tau_a\right) \\ &= \cos(\theta/2) + 2i\hat{n}_a\tau_a \sin(\theta/2) \, , \end{aligned} \tag{2.1.38}$$

where the τ_a are proportional to the Pauli matrices, $\tau_a = \sigma_a/2$, and $\hat{n}_a$
is a unit three-vector. The τ_a are Hermitian and characterize degenerate
energy levels with eigenvalues of $T_3 = \tau_3$ and the total squared isospin
operator $\mathbf{T}^2 = \tau^a\tau^a$, as for angular momentum. Since this is the faithful
representation of lowest dimensionality (denoted as **2**), it is known as the
fundamental representation of the Lie algebra of $SU(2)$.

For more than one nucleon the isospin generators T_i obey the same commutation relations as the rescaled Pauli matrices τ_a, that is,

$$[T_i, T_j] = i\epsilon_{ijk}T_k \,, \qquad (2.1.39)$$

and the eigenvalues of $\mathbf{T}^2$ take the form $T(T + 1)$ with $T = 0, \frac{1}{2}, 1, \dots$. The task of finding all the matrix representations of $SU(2)$ is then reduced to solving (2.1.39). An example of a $T = 1$ isospin system is the (π^+, π^0, π^-) triplet, which corresponds to the adjoint $\mathbf{3}$ representation of $SU(2)$ because the generators are given by the structure constants

$$(T_i)_{jk} = -i\epsilon_{ijk} \,. \qquad (2.1.40)$$

These are the rotation matrices which form a faithful representation of the group of proper rotations in three dimensions, $SO(3)$. This group is closely related to $SU(2)$, obeying the same local commutation rules, but the two groups are not identical globally. Two $SU(2)$ elements, U and $-U$, are mapped to a single rotation matrix R by the relation

$$R_{ij}(U) = \mathrm{Tr}\left(U\tau_i U^{-1}\tau_j\right). \qquad (2.1.41)$$

The parameter manifold of $SU(2)$ corresponds to the three-sphere S^3, while that of $SO(3)$ is S^3 with opposite points on the sphere identified. Unlike $SU(2)$, $SO(3)$ is not simply connected because a loop passing from one point on S^3 to its antipodal point cannot be smoothly contracted to a point.

Higher dimensional representations can also be constructed by taking direct products of lower representations, though these are generally reducible and can be further decomposed. For example, the tensor representation $\mathbf{3} \times \mathbf{3}$ found by taking the direct product of two triplet $\mathbf{3}$ representations, can be split into irreducible components as

$$\mathbf{3} \times \mathbf{3} = \mathbf{1} + \mathbf{3} + \mathbf{5} \,, \qquad (2.1.42)$$

where the singlet $\mathbf{1}$ representation corresponds to the trace of the tensor field, the triplet $\mathbf{3}$ to the three independent components of the antisymmetric part of the tensor, and the remaining $\mathbf{5}$ to the five independent components of the traceless symmetric part.

We now turn to several variations of $SU(2)$ symmetry breaking schemes which we shall draw upon in some detail in future discussions.

Example 1: $SU(2) \xrightarrow{\mathbf{2}} I$

In the first example we suppose that the Higgs fields transform as a two-component spinor in the fundamental $\mathbf{2}$ representation of $SU(2)$,

$$\sigma = \begin{pmatrix} \sigma_1 \\ \sigma_2 \end{pmatrix}, \qquad (2.1.43)$$

with σ_1 and σ_2 complex. We can break this symmetry with a quartic potential $V(\sigma)$, the zeros of which are fields such that $\sigma^\dagger\sigma$ is a positive number η^2. Taking, for example, $\sigma_0^\dagger = (0, \eta)$ we can easily verify that the little group H is trivial since no generators can annihilate σ_0. The vacuum manifold is, therefore, equivalent to that of $SU(2)$. Rewriting the two-component spinor σ as four real fields it becomes clear that this is isomorphic to a three-sphere S^3 in four Euclidean dimensions,

$$G/H = SU(2) \cong S^3. \tag{2.1.44}$$

Coupling σ to the three gauge fields W_a^μ, we have the transformation properties

$$\sigma \to U\sigma, \qquad \mathcal{G}_{\mu\nu} \to U\mathcal{G}_{\mu\nu}U^{-1}, \tag{2.1.45}$$

where $U \in SU(2)$ and the field strength $\mathcal{G}^{\mu\nu}$ is given by

$$\mathcal{G}^{\mu\nu} = G_a^{\mu\nu}\tau_a, \qquad G_a^{\mu\nu} = \partial^\mu W_a^\nu - \partial^\nu W_a^\mu + e\epsilon_{abc}W_b^\mu W_c^\nu. \tag{2.1.46}$$

Counting components for this complete symmetry breaking using table 2.1, we see that the three gauge fields will become massive, having absorbed the three Goldstone bosons, and one massive Higgs boson will remain.

Example 2: $SU(2) \xrightarrow{\ 3\ } U(1) \xrightarrow{\ 3\ } Z_2$

Now consider the Higgs fields transforming as a real triplet φ in the adjoint **3** representation of $SU(2)$ (or $SO(3)$). Again we break the symmetry with a potential $V(\varphi)$ in familiar quartic form, such that its minimum occurs when φ^2 is a real positive number η_ϕ^2. We orient the reference order parameter φ_0 to point along the z-axis in group space,

$$\varphi_0 = \begin{pmatrix} 0 \\ 0 \\ \eta_\phi \end{pmatrix}. \tag{2.1.47}$$

As we would intuitively expect, φ_0 is annihilated by the generator of rotations about the z-axis,

$$T_3 = \begin{pmatrix} 0 & -i & 0 \\ i & 0 & 0 \\ 0 & 0 & 0 \end{pmatrix}. \tag{2.1.48}$$

Consequently, an $SO(2)$ subspace of rotations given by $R = \exp(-i\theta T_3)$ leaves φ_0 invariant, implying that the unbroken subgroup must be a $U(1)$ subgroup of the original $SU(2)$. The vacuum manifold is then

$$G/H = SU(2)/U(1) \cong S^2. \tag{2.1.49}$$

Making this a local gauge symmetry by coupling the gauge fields W_a^μ as in (2.1.45), the Higgs fields transform as $\varphi \to R(U)\varphi$, where the rotation matrix $R(U)$ is given by (2.1.41). The remaining massless gauge boson can be regarded as a photon because the unbroken $U(1)$ gauge theory has all the properties of electromagnetism. In general, the electric charge Q will be proportional to the generator $\varphi_{0i}T_i/\eta_\phi$, with the electromagnetic field-strength tensor in the Higgs vacuum given by

$$F^{\mu\nu} = \frac{\varphi_{0i}}{\eta_\phi}G_i^{\mu\nu}. \qquad (2.1.50)$$

We can take the above scheme a step further by introducing an additional triplet field ψ, transforming under the same local $SU(2)$ symmetry. In this case we can represent the Higgs fields and the generators respectively as

$$\Phi = \begin{pmatrix} \varphi \\ \psi \end{pmatrix}, \qquad L^a = \begin{pmatrix} T^a & 0 \\ 0 & T^a \end{pmatrix}. \qquad (2.1.51)$$

We take an overall symmetry breaking potential of the form

$$V(\varphi,\psi) = \tfrac{1}{4}\lambda_\phi(\varphi^2 - \eta_\phi^2)^2 + \tfrac{1}{4}\lambda_\psi(\psi^2 - \eta_\psi^2)^2 + \tfrac{1}{8}g(\varphi.\psi)^2. \qquad (2.1.52)$$

If φ_0 lies in the z-direction, the last cross term in (2.1.52) implies that in the ground state ψ must point in an orthogonal direction, say

$$\psi_0 = \begin{pmatrix} 0 \\ \eta_\psi \\ 0 \end{pmatrix}. \qquad (2.1.53)$$

Clearly, the remaining $U(1)$ symmetry must be broken because Φ_0 is not annihilated by any of the generators L^a in (2.1.51). The unbroken subgroup, however, is not entirely trivial. While only the identity rotation $I \in SO(3)$ leaves both φ_0 and ψ_0 invariant, two $SU(2)$ elements $(\mathbf{1}, -\mathbf{1})$ are mapped to I due to the double covering (2.1.41) in this degenerate representation. Thus we are left with a discrete little group $H = Z_2$ by the definition (2.1.21). All three gauge fields acquire masses and three massive Higgs bosons remain.

Alternatively, at the second stage we could instead have introduced another spinor field σ, as in the first example, coupled through the gauge fields and terms in the potential of the form $(\sigma^\dagger\sigma)^2$. When the spinor field acquires the non-zero expectation value σ_0, the local gauge symmetry will be completely broken as before. There is significance to which field, **2** or **3**, acquires a non-zero vacuum expectation value (VEV) first since it determines whether an intermediate unbroken $U(1)$ emerges.

Example 3: $SU(2) \xrightarrow{\mathbf{5}} Pin(2) \cong U(1) \otimes Z_2$

Consider the spin-2 tensor representation of $SO(3)$ taking Φ as a traceless real symmetric 3×3 matrix, discussed previously in (2.1.42). This **5** representation was formed from the direct product $\mathbf{3} \times \mathbf{3}$, so under a gauge transformation Φ transforms as

$$\Phi \to R(x)\, \Phi\, R^{-1}(x)\,, \qquad (2.1.54)$$

where the $R(x)$ are rotation matrices in the adjoint $\mathbf{3}$ representation of $SO(3)$. We can now arrange to break this symmetry with a potential $V(\Phi)$ – a sum of terms proportional to $\mathrm{Tr}\,\Phi^2$, $\mathrm{Tr}\,\Phi^3$ and $\mathrm{Tr}\,\Phi^4$ – such that zeros of $V(\Phi)$ take the form

$$\Phi = \eta(I - 3\hat{\varphi}^\dagger \hat{\varphi})\,, \qquad (2.1.55)$$

with $\hat{\varphi}$ a unit vector. Taking $\hat{\varphi}$ along the z-axis as in (2.1.47), Φ acquires the expectation value

$$\Phi_0 = \eta\, \mathrm{diag}\,(1, 1, -2)\,, \qquad (2.1.56)$$

where diag represents a diagonal matrix with the indicated eigenvalues. Locally, the structure of symmetry breakdown is the same as in the first stage of *Example 2* since Φ_0 is annihilated by the generator of rotations about the z-axis T_3. Thus an $SO(2)$ subgroup of $SO(3)$ remains unbroken. However, the global structure of the vacuum manifold is very different since the little group also contains another disconnected component generated by

$$R_0 = \mathrm{diag}\,(1, -1, -1)\,, \qquad (2.1.57)$$

corresponding to a 180° rotation about the x-axis. The rotation matrices leaving Φ_0 invariant, therefore, form an $O(2)$ subgroup of $SO(3)$. Using the covering groups, the symmetry breaking can be described as $SU(2) \to Pin(2) \cong U(1) \otimes Z_2$, where $\otimes$ denotes a semi-direct product since the $U(1)$-generator τ_3 does not commute with the discrete rotation [Preskill, 1985]. The group $Pin(2)$ is a double covering of $O(2)$ and can be parametrized as

$$\{\exp(i\theta\tau_3)\}\,, \quad \{i\tau_2 \exp(i\theta\tau_3)\}\,, \qquad 0 \le \theta < 4\pi\,. \qquad (2.1.58)$$

Again the vacuum manifold G/H appears to be the surface of a two-sphere, but now antipodal points (represented by $\hat{\varphi}$ and $-\hat{\varphi}$) are identified. This difference in the global topology has important implications for the type of topological defects that can be formed in this model, as we shall see in chapter 3.

We can introduce a further triplet Higgs ψ to break the discrete Z_2 symmetry. This is achieved by giving a VEV ψ_0 in the z-direction, because ψ_0 will change sign under the discrete $O(2)$ reflection, while remaining invariant under the previous $SO(2)$ rotations. The scheme then becomes

$$SO(3) \xrightarrow{\;5\;} O(2) \xrightarrow{\;3\;} SO(2)\,. \qquad (2.1.59)$$

2.1.4 Grand unification: an SO(10) example

Here we consider a grand unified model based on the simple group $SO(10)$ which has certain aesthetic and empirical advantages over the minimal $SU(5)$ model. The descents from $SO(10)$ of chief physical interest are through the subgroups,

$$SO(10) \longrightarrow \begin{cases} \text{'}SU(5)\text{'} & \text{Georgi–Glashow,} \\ \text{'}SU(4) \times SU(2)_L \times SU(2)_R\text{'} & \text{Pati–Salam,} \end{cases}$$
(2.1.60)

where quotes signify that the unbroken subgroup may be only locally isomorphic to the indicated group. Following Kibble [1990], we shall briefly describe a particular symmetry breaking scheme with a Higgs field in the **126** spinor representation, taking the former path through $SU(5)$. The model exhibits many features relevant for subsequent discussions of topological defects, but this subsection is by no means an essential prerequisite.

Strictly, we should be discussing $Spin(10)$, the simply-connected covering group for $SO(10)$, since we will be considering spinor representations. $Spin(10)$ has a direct relationship to $SO(10)$ analogous to that described above for $SU(2)$ and $SO(3)$. The generators of $Spin(10)$ can be defined as

$$\sigma_{ij} = \frac{1}{2i} [\Gamma_j, \Gamma_k], \quad j, k = 1, 2, ..., 10,$$
(2.1.61)

where the Γ_j are the Dirac matrices in ten dimensions; these are 32×32 matrices which can be constructed from direct products of the Pauli matrices σ_k. Each family of left-handed fermions belongs to a 16-dimensional spinor representation (denoted by **16**). These families transform under the symmetry $SU(5) \times U(1)_P \in Spin(10)$, having the $SU(5)$ decomposition

$$\mathbf{16} \longrightarrow \mathbf{1}_5 + \mathbf{10}_1 + \mathbf{5}_{-3},$$
(2.1.62)

where the subscripts are the eigenvalues of the operator

$$P = \sigma_{12} + \sigma_{34} + \sigma_{56} + \sigma_{78} + \sigma_{9\,10}.$$
(2.1.63)

The decomposition (2.1.62) corresponds to the usual $SU(5)$ particle assignments plus a neutral singlet which plays the role of a right-handed neutrino (with a large mass). For later reference, we also note that the charge operator Q is given by

$$Q = \tfrac{1}{3}(\sigma_{12} + \sigma_{34} + \sigma_{56}) - \sigma_{9\,10}.$$
(2.1.64)

As in (2.1.42), we consider symmetry breaking in Higgs representations formed from direct products, here from **16** and also its conjugate $\overline{\mathbf{16}}$,

$$(\mathbf{16} \times \mathbf{16})_S = \mathbf{10} + \mathbf{126}\,,$$
$$\mathbf{16} \times \overline{\mathbf{16}} = \mathbf{1} + \mathbf{45} + \mathbf{210}\,, \tag{2.1.65}$$

where the subscript S denotes the symmetric product. The pertinent decompositions for this scheme are then

$$\mathbf{126} \longrightarrow \mathbf{1}_{10} + \mathbf{10}_6 + \mathbf{50}_2 + \overline{\mathbf{5}}_2 + \overline{\mathbf{45}}_{-2} + \overline{\mathbf{15}}_{-6}\,,$$
$$\mathbf{45} \longrightarrow \overline{\mathbf{10}}_4 + \mathbf{24}_0 + \mathbf{1}_0 + \mathbf{10}_{-4}\,, \tag{2.1.66}$$
$$\mathbf{10} \longrightarrow \overline{\mathbf{5}}_2 + \mathbf{5}_{-2}\,.$$

Symmetry breaking proceeds first with a Higgs field in the **126**-representation acquiring a non-zero vacuum expectation value $\phi_{\mathbf{126}}$ lying in the direction of the $\mathbf{1}_{10}$ component. Phenomenological constraints require that this occur at a superheavy mass scale $\sim 10^{16}$–10^{17}GeV, followed by another Higgs in the adjoint **45**-representation acquiring a VEV at approximately 10^{14}GeV. Finally, at the electroweak scale, a further Higgs field in the **10**-representation acquires a non-zero VEV to give the overall scheme [Kibble, Lazarides & Shafi, 1982a],

$$Spin(10) \xrightarrow{\ \mathbf{126}\ } SU(5) \times Z_2 \tag{2.1.67a}$$
$$\xrightarrow{\ \mathbf{45}\ } SU(3)_c \times SU(2)_L \times U(1)_Y \times Z_2 \tag{2.1.67b}$$
$$\xrightarrow{\ \mathbf{10}\ } SU(3)_c \times U(1)_Q \times Z_2\,. \tag{2.1.67c}$$

The unbroken discrete Z_2 symmetry which emerges at the first stage in (2.1.67a) is generated by rotations through 2π in the P-direction, that is, by $\exp(2\pi i P/10)$. This is because $\exp(2\pi i P/5)$ is contained in $SU(5)$ whereas $\exp(2\pi i P/10)$ is not. We note that hypercharge Y belongs to the $\mathbf{24}_0$ of the **45** and so commutes with rotations by $\exp(2\pi i P/10)$, while the **10**-representation has no components in (2.1.66) with eigenvalue $P = 0$.

Although favoured by many, the scheme (2.1.67) is not the only way of descending from $SO(10)$ to the standard model: The minimal route simply employs a Higgs field in the **16**-representation at the first stage, $Spin(10) \rightarrow SU(5)$. In contrast with (2.1.67), a discrete symmetry does not appear and the unbroken subgroup is globally $SU(5)$. Alternatively, we can introduce an earlier stage of symmetry breaking with an additional Higgs field in the **45**- or **210**-representation going first through the maximal subgroup $Spin(10) \rightarrow SU(5) \times U(1)_P$ and, thence, exactly following (2.1.67a–c).

To take the second Pati–Salam symmetry breaking pathway in (2.1.60), one can break $SO(10)$ first with a Higgs field in the **54**-representation, and then follow this with a $\phi_{\mathbf{126}}$ and a $\phi_{\mathbf{10}}$ to yield [Kibble, Lazarides &

Shafi, 1982b]

$$Spin(10) \xrightarrow{54} ([Spin(6) \times Spin(4)]/Z_2) \times Z_2^C \quad (2.1.68a)$$
$$\xrightarrow{126} SU(3)_c \times SU(2)_L \times U(1)_Y \times Z_2 \quad (2.1.68b)$$
$$\xrightarrow{10} SU(3)_c \times U(1)_Q \times Z_2 , \quad (2.1.68c)$$

where we note the local isomorphism $Spin(6) \times Spin(4) \cong SU(4) \times SU(2) \times SU(2)$. A Z_2 symmetry is factored out in (2.1.68a) because globally the subgroup is not a direct product – $Spin(4)$ and $Spin(6)$ have a non-trivial intersection. The overall discrete Z_2^C symmetry in (2.1.68a) is unrelated to this, being generated by the charge conjugation operator C. Subsequently, this is broken by the **126** Higgs and another different Z_2 symmetry emerges as in (2.1.67).

2.2 The effective potential

The discussion of symmetry breaking in the previous section has been somewhat simplistic because we have used purely classical potentials like (2.1.2) to determine the expectation value of the Higgs field ϕ. In reality, ϕ is a quantum field interacting with itself and with other quantum fields, and the classical potential $V(\phi)$ is modified by radiative corrections. The corrected potential for ϕ, which is called the effective potential, can be evaluated perturbatively as an expansion in powers of coupling constants,

$$V_{\text{eff}}(\phi) = V(\phi) + V_1(\phi) + V_2(\phi) + \cdots , \quad (2.2.1)$$

where $V(\phi)$ is the classical potential and $V_n(\phi)$ is the contribution of Feynman diagrams with n closed loops.

In some models the radiative corrections are negligible, while in others they can completely alter the character of symmetry breaking. As an example of the latter, consider the Higgs model (2.1.10) with

$$V(\phi) = \mu_0^2 |\phi|^2 . \quad (2.2.2)$$

This model of massive scalar electrodynamics appears to be unremarkable, however, radiative corrections due to the gauge coupling generate a non-trivial effective potential for ϕ. The one-loop contribution to $V_{\text{eff}}(\phi)$ is given by [Coleman & E. Weinberg, 1973]

$$V_1(\phi) = \frac{3e^4}{16\pi^2} |\phi|^4 \ln \left(\frac{|\phi|^2}{\sigma^2} \right) , \quad (2.2.3)$$

where σ is the renormalization scale.[*] Introducing a dimensionless quantity

$$\nu = \frac{16\pi^2 \mu_0^2}{3e^4\sigma^2}, \tag{2.2.4}$$

it is easily shown that for $\nu > 0.45$ the effective potential has a single minimum at $\phi = 0$. For $\nu < 0.45$ there is another minimum at a non-zero value of ϕ, and for $\nu < 0.37$ this minimum plunges deeper than the one at $\phi = 0$. In the latter case the absolute minimum of $V_{\text{eff}}(\phi)$ is at a non-zero value of ϕ, and the symmetry is spontaneously broken. The shape of the effective potential for different values of ν is shown in fig. 2.2.

The one-loop contribution to V_{eff} for the general gauge model (2.1.28) and (2.1.35) is given by [Coleman & E. Weinberg, 1973; S. Weinberg, 1973]

$$V_1(\phi) = \frac{1}{64\pi^2}\left[\text{Tr}\,(\mu^4\ln\frac{\mu^2}{\sigma^2}) + 3\,\text{Tr}\,(M^4\ln\frac{M^2}{\sigma^2}) - 4\,\text{Tr}\,(m^4\ln\frac{m^2}{\sigma^2})\right], \tag{2.2.5}$$

where again σ is the renormalization scale and μ, M and m are the scalar, vector and spinor mass matrices respectively. Disregarding logarithms, we can estimate the three terms in (2.2.5) as $\lambda^2\phi^4$, $e^4\phi^4$, and $\chi^4\phi^4$, where λ and χ are the typical Higgs and Yukawa couplings, respectively. If the

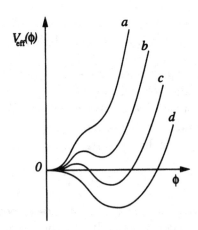

Fig. 2.2.   The Coleman–Weinberg effective potential (2.2.3) for the parameters (a) $\nu > 0.45$, (b) $0.45 > \nu > 0.37$, (c) $0.37 > \nu > 0$, and (d) $\nu < 0$.

---

[*] A change in σ results in a renormalization of the quartic Higgs coupling. In (2.2.3) it is assumed that σ is chosen so that this coupling is equal to zero.

following conditions are satisfied,

$$\lambda \ll 1, \qquad e^4 \ll \lambda, \qquad \chi^4 \ll \lambda, \qquad (2.2.6)$$

then $V_1(\phi)$ will only be a small correction to the classical potential $V(\phi)$. There are some subtleties we should not neglect to mention in this cursory discussion of the effective potential: (i) While in gauge theories V_{eff} is not generally gauge-invariant, the value of V at the minimum is a gauge-invariant quantity; (ii) V_{eff} can also be complex, but again this does not affect the minimum.

2.3 Cosmological phase transitions

2.3.1 Symmetry restoration at high temperature

In condensed matter systems, symmetries which are spontaneously broken can be restored when the system is heated to a sufficiently high temperature. For example, in a superfluid the condensate wave function ψ gradually decreases as the temperature is increased, and vanishes completely at the critical temperature when the liquid becomes normal. Kirzhnits [1972] was the first to recognize that elementary particle symmetries can be restored in a similar manner (see also Kirzhnits & Linde [1972]). He pointed out the analogy between the field theory models (2.1.1) and (2.1.10) and the theories of superfluidity and superconductivity. The expectation value of the Higgs field ϕ can be thought of as describing a Bose condensate of Higgs particles. At a non-zero temperature, in addition to this condensate, there is a thermal distribution of various particles and antiparticles. The particle masses are determined by the Higgs expectation value and, consequently, the free energy of the system,

$$F = E - TS, \qquad (2.3.1)$$

is a non-trivial function of ϕ. The equilibrium value of ϕ is found by minimizing the free energy and is, in general, temperature-dependent.* At low temperatures the second term in (2.3.1) is unimportant and F is minimized by the ordered state of minimum energy. At higher temperatures, the entropy term becomes increasingly significant. If the Higgs field becomes smaller, the particle masses typically decrease, the available phase space becomes larger, and the entropy grows. Consequently, there is a tendency for the Higgs field to decrease as a function of temperature and to vanish completely above some critical temperature T_c. If η is the

---

* We assume that the chemical potentials of all particle species are equal to zero, $\mu_n = 0$. Otherwise, thermal equilibrium corresponds to the minimum of the thermodynamic potential $\Omega = F - \Sigma \mu_i N_i$.

characteristic energy scale of symmetry breaking and the couplings of the Higgs field are not too small, then on dimensional grounds it is clear that T_c should be of the order η.

In the Hot Big Bang cosmological model, the universe starts at a very high temperature, and so the initial equilibrium value of the Higgs field will be at $\phi = 0$. As the universe expands and cools down, it undergoes a phase transition at $T = T_c$, when the symmetry is spontaneously broken. A grand unified model with a sequence of symmetry breakings, such as

$$G \to H \to \ldots \to SU(3) \times SU(2) \times U(1) \to SU(3) \times U(1)_{\text{em}}, \quad (2.3.2)$$

predicts a series of phase transitions in the early universe with critical temperatures related to the corresponding symmetry breaking scales.[*]

To describe high-temperature symmetry restoration quantitatively, one has to calculate the free energy as a function of the Higgs field ϕ and temperature T. The necessary formalism was developed by Weinberg [1974], Dolan & Jackiw [1974] and Kirzhnits & Linde [1974]. They found that the free energy density $F(\phi, T)/\mathcal{V}$ is given by the same diagrammatic expansion as the effective potential $V_{\text{eff}}(\phi)$ with all the Green's functions replaced by finite-temperature Green's functions. For this reason the free energy per unit volume is called the finite-temperature effective potential,

$$V_{\text{eff}}(\phi, T) = F(\phi, T)/\mathcal{V}, \quad (2.3.3)$$

where $\mathcal{V}$ is the volume. In this section we shall give a simple derivation of the lowest-order (one-loop) temperature-dependent contribution to V_{eff}.

To lowest order in the coupling constants, thermal particles can be considered to be non-interacting, and so we can write the effective potential as

$$V_{\text{eff}}(\phi, T) = V(\phi) + \sum_n F_n(\phi, T). \quad (2.3.4)$$

Here, $V(\phi)$ is the Higgs potential (or, more exactly, the zero-temperature effective potential). The summation is over the particle spin states, and the free energy contributions of different spin states are given by

$$F_n = \pm T \int \frac{d^3k}{(2\pi)^3} \ln\left(1 \mp \exp(-\epsilon_k/T)\right), \quad (2.3.5)$$

where $\epsilon_k = (k^2 + m_n^2)^{1/2}$ and the upper sign is for bosons and the lower for fermions.

---

[*] Typically, the low-temperature phase has less symmetry than the high-temperature phase. This rule is not without exceptions, however, because models can be constructed in which a symmetry broken at high temperatures is restored at low temperatures [Weinberg, 1974; Mohapatra & Senjanović, 1976].

For $T \ll m_n$, the free energy F_n is exponentially small and can be neglected. For bosons at high temperatures $T \gg m_n$, we have

$$F_n = -\frac{\pi^2}{90}T^4 + \frac{m_n^2 T^2}{24} + \mathcal{O}\left(m_n^4\right), \qquad (2.3.6)$$

while for fermions

$$F_n = -\frac{7\pi^2}{720}T^4 + \frac{m_n^2 T^2}{48} + \mathcal{O}\left(m_n^4\right). \qquad (2.3.7)$$

In many interesting cases, symmetry restoration occurs at a temperature which is much higher than all the relevant mass thresholds. For these, V_{eff} is well-approximated by

$$V_{\text{eff}}(\phi, T) = V(\phi) + \frac{1}{24}\mathcal{M}^2 T^2 - \frac{\pi^2}{90}\mathcal{N}T^4, \qquad (2.3.8)$$

where

$$\mathcal{N} = \mathcal{N}_{\text{B}} + \tfrac{7}{8}\mathcal{N}_{\text{F}}, \qquad (2.3.9)$$

$$\mathcal{M}^2 = \sum_{\text{B}} m_n^2 + \tfrac{1}{2}\sum_{\text{F}} m_n^2, \qquad (2.3.10)$$

with $\mathcal{N}_{\text{B}}$ and $\mathcal{N}_{\text{F}}$ representing the number of bosonic and fermionic spin states, and the sums $\sum_{\text{B}}$ and $\sum_{\text{F}}$ are taken over bosonic and fermionic states respectively. For a general gauge theory (2.1.28), the coefficient $\mathcal{M}^2$ is given by

$$\mathcal{M}^2 = \text{Tr}\,\mu^2 + 3\,\text{Tr}\,M^2 + \frac{1}{2}\,\text{Tr}\left(\gamma^0 m \gamma^0 m\right) \qquad (2.3.11)$$

where the mass matrices, μ, M and m are given as functions of the Higgs field ϕ by (2.1.24), (2.1.34) and (2.1.36), respectively.

As an illustration of this formalism, consider the abelian-Higgs model (2.1.10) with $V(\phi)$ given by (2.1.2). We shall assume that $\lambda \gg e^4$, so that the radiative corrections to the Higgs potential can be neglected. Using the representation $\phi = (\phi_1 + i\phi_2)/\sqrt{2}$, we find the Higgs mass matrix

$$\mu_{ij}^2 = \frac{\partial^2 V}{\partial \phi_i \partial \phi_j} = \frac{\lambda}{2}\left(|\phi|^2 - \eta^2\right)\delta_{ij} + \frac{\lambda}{2}\phi_i \phi_j, \qquad (2.3.12)$$

where $i, j = 1, 2$ and $|\phi|^2 = \phi_k \phi_k/2$. The gauge boson mass is $M = \sqrt{2}\,e|\phi|$, and (2.3.8) and (2.3.11) yield the following expression for the effective potential at high temperature,

$$V_{\text{eff}}(\phi, T) = V(\phi) + \frac{\lambda + 3e^2}{12}T^2|\phi|^2 - \frac{2\pi^2}{45}T^4. \qquad (2.3.13)$$

2.3.2 Second-order phase transitions

Like phase transitions in condensed matter systems, cosmological phase transitions can be of first or second order. The physics involved is rather different in the two cases, so we shall discuss them separately.

For an example of a second-order phase transition, we consider the Goldstone model (2.1.1–2). The high-temperature effective potential for this model can be written as

$$V_{\text{eff}}(\phi, T) = m^2(T)|\phi|^2 + \frac{\lambda}{4}|\phi|^4 , \qquad (2.3.14)$$

where

$$m^2(T) = \frac{\lambda}{12}\left(T^2 - 6\eta^2\right) , \qquad (2.3.15)$$

and we have omitted the ϕ-independent terms in V_{eff}. (This expression for V_{eff} can be obtained from (2.3.13) by setting $e = 0$.) The quantity $m(T)$ is the effective mass of the field ϕ in the symmetric state, $\langle\phi\rangle = 0$. This mass is equal to zero at $T = T_{\text{c}}$, where

$$T_{\text{c}} = \sqrt{6}\,\eta . \qquad (2.3.16)$$

For $T > T_{\text{c}}$, the effective mass-squared term $m^2(T)$ is positive, the minimum of V_{eff} is at $\phi = 0$, so the expectation value of ϕ vanishes and the symmetry is restored. With $T < T_{\text{c}}$, $m^2(T) < 0$, the symmetric state becomes unstable and ϕ develops a non-zero expectation value. Minimizing V_{eff} in (2.3.14) we obtain, for $T < T_{\text{c}}$,

$$|\phi| = \frac{1}{\sqrt{6}}\left(T_{\text{c}}^2 - T^2\right)^{1/2} . \qquad (2.3.17)$$

The defining feature of second-order phase transitions is that the 'order parameter' $|\phi|$ grows continuously from zero as the temperature is decreased below T_{c}. The shape of the effective potential above and below the critical temperature T_{c} is illustrated in fig. 2.3.

In a cosmological context, when the universe cools through the critical temperature, the field ϕ develops an expectation value of magnitude (2.3.17). However, the phase θ of ϕ, is not determined solely by local physics. Its choice depends on random fluctuations, and θ takes different values in different regions of space. Since the free energy is minimized by a homogeneous field ϕ, the spatial variations in θ will gradually die out. We can define the correlation length $\xi(t)$ to be the length scale above which the values of θ are uncorrelated. The rate at which $\xi(t)$ grows depends on details of the relaxation processes which are involved but, in

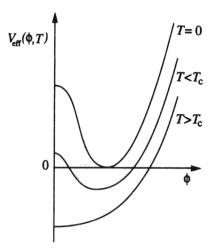

Fig. 2.3. The temperature-dependent effective potential $V_{\text{eff}}(\phi, T)$ for a second-order phase transition near the critical temperature T_{c}.

any case, ξ should satisfy the simple causality bound:

$$\xi(t) < d_{\text{H}}(t). \qquad (2.3.18)$$

Here, d_{H} is the causal horizon defined to be the distance travelled by light during the lifetime of the universe (1.3.24). The upper limit (2.3.18) is a simple consequence of the fact that correlations cannot be established faster than the speed of light. For a power-law expansion, we have $d_{\text{H}} \sim t$ and so (2.3.18) takes the form

$$\xi(t) \lesssim t. \qquad (2.3.19)$$

Features of a general second-order phase transition are similar to the example above. The critical temperature T_c is typically comparable to the symmetry breaking energy scale η, although in general there is some dependence on the coupling constants. For example, in the Higgs model (2.1.10) we have

$$T_{\text{c}} = \left(\frac{6\lambda}{\lambda + 3e^2}\right)^{1/2} \eta. \qquad (2.3.20)$$

For $T < T_{\text{c}}$, the scalar field develops an expectation value corresponding to some point in the manifold $\mathcal{M}$ of the minima of V_{eff}. Since all such points are equivalent, the choice will be determined by random fluctuations with ϕ varying in space with some correlation length $\xi \lesssim t$. In the following chapters we shall see that this spatial variation of ϕ is responsible for the formation of topological defects.

The discussion up to this point has been based on the one-loop effective potential (2.3.13). This is a good approximation while ϕ can be thought of as a homogeneous classical field with small Gaussian fluctuations. However, near the critical temperature the fluctuations in ϕ become large and higher-order corrections to V_{eff} become important. Formally, the problem manifests itself as an infrared divergence of the higher-order diagrams at $T \to T_c$. It can be shown [Kirzhnits & Linde, 1976] that the leading n^{th} order contribution to V_{eff} (for $n > 3$) is proportional to $[\lambda T/m(T)]^{n-3}$. As $T \to T_c$, the effective mass $m(T) \to 0$ and perturbation theory breaks down. The one-loop approximation is only reliable, therefore, if

$$|m(T)|/\lambda T \gg 1. \tag{2.3.21}$$

This condition (2.3.21) is known as the Ginzburg criterion [Ginzburg, 1960], and the temperature below T_c at which $|m(T)| \sim \lambda T$ is called the Ginzburg temperature T_G. Using (2.3.15) we find that

$$T_c - T_G \sim \lambda T_c. \tag{2.3.22}$$

For models with $\lambda \ll 1$, T_G is very close to T_c and the Ginzburg criterion is satisfied everywhere except in a very narrow range of temperatures near T_c.

The Ginzburg criterion (2.3.21) has a simple physical interpretation. The fluctuations of ϕ in a region of radial dimension $\sim r$ can be estimated from

$$r^3 m^2(T)|\phi\delta\phi| \sim T, \tag{2.3.23}$$

where $|\phi(T)|$ is given by (2.3.17). The characteristic length scale of the system at a temperature T is

$$r_c(T) \sim \left| m^{-1}(T) \right|. \tag{2.3.24}$$

With $r \sim r_c$, the relation (2.3.23) yields

$$\left| \frac{\delta\phi}{\phi(T)} \right| \sim \frac{\lambda T}{|m(T)|}. \tag{2.3.25}$$

The right-hand side of (2.3.25) is equal to unity at the Ginzburg temperature T_G. Hence, T_G can be defined as the temperature at which fluctuations of $|\phi|$ on the characteristic scale r_c become comparable to the equilibrium value of $|\phi|$. (In fact, this definition coincides with that originally given by Ginzburg.)

A quantitative description of the phase transition in the temperature range

$$|T - T_c| \lesssim \lambda T_c, \tag{2.3.26}$$

requires some new physics, and is the subject of the modern theory of critical phenomena. For a review, the reader is referred to Wilson & Kogut [1974] and Ma [1976].

2.3.3 First-order phase transitions

For an example of a first-order phase transition, we consider the abelian-Higgs model with a quadratic potential (2.2.2) and symmetry breaking induced by radiative corrections (2.2.3). We shall assume that the parameter ν, defined in (2.2.4), is less than the critical value 0.37, so that at zero temperature the shape of the effective potential is similar to the curve (c) or (d) in fig. 2.2. The finite-temperature effective potential for this model is given by

$$V_{\text{eff}}(\phi, T) = V_{\text{eff}}(\phi, 0)$$
$$+ \frac{3T}{2\pi^2} \int_0^\infty dk\, k^2 \ln\left(1 - e^{-\sqrt{k^2 + 2e^2|\phi|^2}/T}\right). \quad (2.3.27)$$

If $|\phi|$ is not too large, $|\phi| \ll T/e$, then the gauge boson mass is well below the temperature and we can use the high-temperature expansion (2.3.8). Omitting ϕ-independent terms, we obtain

$$V_{\text{eff}}(\phi, T) = m^2(T)|\phi|^2 + \frac{3e^4}{16\pi^2}|\phi|^4 \ln\left(\frac{|\phi|^2}{\sigma^2}\right), \quad (2.3.28)$$

where the temperature-dependent mass is

$$m^2(T) = \mu_0^2 + \tfrac{1}{4}e^2 T^2. \quad (2.3.29)$$

At very high temperatures, $T \gg e\sigma$, the effective potential is dominated by the $e^2 T^2 |\phi|^2$ term and has a single minimum at $\phi = 0$. The evolution of the shape of V_{eff} as the temperature is lowered is illustrated in fig. 2.4. When the symmetry-breaking minimum first appears, it is not favoured thermodynamically, but below some critical temperature T_c it becomes deeper than the symmetric minimum at $\phi = 0$. An order of magnitude estimate gives the critical temperature $T_c \sim e\sigma$. Note that the high-temperature expansion (2.3.8) is not valid at $\phi \sim \sigma$ with $T \sim T_c$ and so it cannot be used to calculate T_c analytically.

The distinguishing feature of first-order phase transitions is that the symmetric phase remains metastable below the critical temperature. If $\mu_0^2 < 0$, then the symmetric phase is destabilized below the temperature $T_d = (-4\mu_0^2/e^2)^{1/2}$. However, for $\mu_0^2 > 0$ it remains metastable all the way down to $T = 0$. This metastable vacuum state at $\langle \phi \rangle = 0$ is often called the 'false vacuum'. First-order phase transitions occur in a wide variety of elementary particle models, and specific examples have been

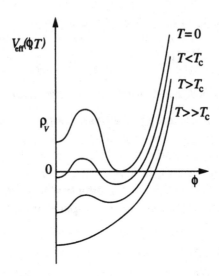

Fig. 2.4. In a first-order phase transition, the value of ϕ at the minimum of $V_{\text{eff}}(\phi, T)$ changes discontinuously near the critical temperature T_c.

discussed by Guth & Tye [1980], Daniel & Vayonakis [1981], Guth & E. Weinberg [1981] and Linde [1981].

If supercooling is not very strong, the metastable phase decays by thermal fluctuations which occasionally take the field ϕ over the potential barrier. The resulting islands of the new phase grow as the domain walls at their boundaries advance into the old phase. The thermodynamics and hydrodynamics of the phase boundaries are rather complicated and first-order transitions can give rise to astonishingly complex patterns of the two phases.

The situation is greatly simplified, on the other hand, for strong supercooling, when the decay of the metastable state requires quantum tunnelling through the potential barrier. Spherical bubbles of the new phase appear spontaneously and expand at a rate approaching the speed of light. Energy is released as the old phase is converted into the new and this energy remains concentrated in the vicinity of the accelerating bubble walls. The phase transition is completed when the bubbles coalesce, the energy of the bubble walls is thermalized, and the new phase fills the entire universe. Bubble expansion and coalescence in an expanding universe has been studied by Guth & Tye [1980], Sato [1981a,b], Einhorn & Sato [1981] and Guth & E. Weinberg [1983]. It follows from their analysis that the time of bubble coalescense can be roughly estimated from the condition

$$\Gamma H^{-4} \sim 1. \tag{2.3.30}$$

Here, $\Gamma(t)$ is the bubble nucleation rate (per unit volume per unit time) and the Hubble distance H^{-1} is defined by $H = \dot{a}/a$. For a power-law expansion, $H \sim t^{-1}$ and (2.3.30) gives $t \sim \Gamma^{-1/4}$.

Barrier penetration processes can be notoriously slow, and in many models the universe supercools to very low temperatures, approaching the false vacuum state. In this case the thermal energy density (1.3.25) becomes small, and gravitational effects of the false vacuum energy become important. The false vacuum is characterized by a constant positive energy density, ρ_v – this is positive, since it must be greater than the energy density of the present vacuum, which is very close to zero. When the energy density of the universe is dominated by ρ_v, the evolution equation (1.3.14) with $k=0$ takes the form

$$\left(\frac{\dot{a}}{a}\right)^2 = H^2,$$
(2.3.31)

where

$$H = \left(\frac{8\pi G \rho_v}{3}\right)^{1/2}.$$
(2.3.32)

The solution of (2.3.31) is

$$a(t) = e^{Ht},$$
(2.3.33)

and so a universe dominated by false vacuum energy expands exponentially. A flat FRW metric (1.3.1–2) with an exponentially growing scale factor describes a de Sitter space.

We now consider bubble coalescence in a supercooled universe. Since the Hubble distance H^{-1} is constant in de Sitter space, the quantity $\epsilon \equiv \Gamma H^{-4}$ in (2.3.30) does not change with time (at $T \to 0$ the nucleation rate Γ approaches its vacuum value which is also time-independent). If $\epsilon \gtrsim 1$, then the phase transition is completed in one Hubble time or less. However, for $\epsilon \ll 1$ the condition (2.3.30) is never satisfied. Expanding bubbles of the new phase cannot catch up with the expansion of the universe. Bubbles separated by a distance greater than H^{-1} will never collide.* Consequently, a first-order phase transition is either over very quickly or else it is never completed at all, with either outcome dependent on the bubble nucleation rate Γ.

The calculation of Γ is outside the scope of this monograph, so we shall only mention some standard references. A semiclassical method

---

* For a value of ϵ which is not too small ($\gtrsim 10^{-2}$) bubble coalescence may eventually be completed, but the resulting distribution of bubble sizes is grossly inhomogeneous. Gigantic bubbles present in this distribution can lead to unacceptable cosmological consequences [Hawking, Moss & Stewart, 1982; Guth & E. Weinberg 1983].

for calculating Γ in flat spacetime at zero temperature was developed by Coleman [1977] and Callan & Coleman [1977], following the earlier work by Voloshin, Kobzarev & Okun' [1975] (see Coleman [1979] for a review). The method was extended to curved spacetime by Coleman & de Luccia [1980]. Bubble nucleation at finite temperature has been discussed by Linde [1977], Guth & E. Weinberg [1980, 1981] and Witten [1981].

2.4 Inflationary scenarios

2.4.1 The inflationary paradigm

In chapter 1 we reviewed the successes of the standard cosmology, but we also pointed out some serious shortcomings. Amongst others, these included the horizon and flatness problems and the origin of density fluctuations. These difficulties, together with the magnetic monopole problem discussed later in §14.3, led to the development of the inflationary paradigm, in which the early evolution of the universe is radically different from that in the standard model. The crucial step was made by Guth [1981] who pointed out that the horizon and flatness problems can be resolved if the early universe went through a period of very rapid expansion, which he termed inflation. As a result of inflation, regions initially within the causal horizon are blown up to sizes much greater than the present Hubble radius. The initial curvature radius of the universe is also increased by an enormous factor, so that the universe becomes locally indistinguishable from a flat ($k = 0$) universe with $\rho = \rho_c$. As a specific implementation of inflation, Guth suggested a first-order phase transition with strong supercooling. As the universe supercools and becomes dominated by false vacuum energy, it starts expanding exponentially, according to (2.3.33). The expansion factor during inflation is

$$Z = e^{H\tau}, \qquad (2.4.1)$$

where H is given by (2.3.32) and τ is the duration of the inflationary phase. The horizon and flatness problems are solved if

$$H\tau \gtrsim 100. \qquad (2.4.2)$$

Inflation is ended by bubble nucleation and coalescence, the false vacuum energy is thermalized, and from this point onwards the evolution of the universe conforms with that of the standard Hot Big Bang model.

Guth himself pointed out a flaw in this scenario. As discussed in the previous section, bubble coalescence in a first-order phase transition can be completed only if it occurs fairly rapidly within a few Hubble times, $\tau \lesssim H^{-1}$. However, in this case the expansion factor is small, $Z \sim 1$, and the horizon and flatness problems will not be solved. Alternatively, for large τ, bubble coalescence is never completed and the universe never reaches the radiation-dominated phase. This is the so-called 'graceful exit' problem.

Several other models of inflation have been suggested which do not suffer from this difficulty; however, none of these models is particularly compelling. At present, therefore, inflation should be viewed as a paradigm, rather than a specific realistic model. The essential feature of the inflationary paradigm is that there was a period of very rapid expansion in the very early universe. In most models the expansion is exponential, but there are exceptions to this rule. The key requirement is that the scale factor $a(t)$ grows faster than the Hubble radius, $H^{-1} = a/\dot{a}$, and so it is possible, for example, to have power-law inflation, $a(t) \propto t^n$ with $n > 1$ [Abbott & Wise, 1984]. Inflation must then be followed by thermalization in which the vacuum energy (or any other form of energy driving inflation) is converted into the thermal energy of relativistic particles and radiation. After thermalization, the evolution of the universe becomes identical to that of the standard scenario.

Before reviewing specific models, it is worthwhile making a few general comments about the inflationary paradigm:

(i) Unless the parameters of the underlying particle theory are carefully fine-tuned, the expansion factor Z is typically far in excess of the minimal value required to solve the horizon and flatness problems. Consequently, we expect that the present value of the density parameter Ω will be very close to unity. This is the main observational prediction of inflation.

(ii) Primordial baryon number density will be hopelessly diluted during inflation, and so inflationary scenarios require baryon-number and CP-violating interactions which will generate an excess of baryons over antibaryons after inflation.

(iii) Inflation results in a thermalized universe, but it is not necessary to assume a thermal state before inflation. The nature of the initial state is a very speculative subject, even by cosmological standards. An attractive possibility is that the origin of the universe is similar to a quantum tunnelling event which can be described by a wave function of the universe. This wave function determines the probability distribution for the initial states of the universe [Hawking 1982a; Hartle & Hawking, 1983; Vilenkin, 1982a, 1986a; Linde, 1984]. Other possibilities include a chaotic initial state [Linde, 1983] and an eternally inflating universe [Linde, 1986]. This wide range of possibilities is mainly due to the fact that inflation erases almost all memory of the initial state.

(iv) Inflation may also help to solve the density fluctuation problem. Small density perturbations are produced due to quantum fluctuations of scalar or gravitational fields during inflation, though it has proved difficult to fine-tune these appropriately in specific models. If these density fluc tuations are too small, then another fluctuation-generating mechanism is required in the post-inflationary era. Such mechanisms could be provided by topological defects, notably cosmic strings, as will be discussed in chapter 11.

We now turn to a brief review of specific inflationary models. For details and references to the original literature the reader is referred to review articles by Blau & Guth [1987] and Olive [1990] and to the monographs by Linde [1990] and Kolb & Turner [1990].

2.4.2 Slow-rollover models

This class of models assumes the existence of a very weakly coupled scalar field ϕ, called the inflaton, which is initially displaced from the minimum of its potential $V(\phi)$, as illustrated in fig. 2.5. As ϕ slowly rolls downhill toward the minimum, its potential energy density $V(\phi)$ drives the inflationary expansion of the universe. If ϕ starts near a local maximum of the potential, the scenario is referred to as 'new inflation' [Linde 1982a; Albrecht & Steinhardt, 1982]. An alternative version of the scenario assuming a very large initial value of ϕ is called 'chaotic inflation' [Linde, 1983]. Assuming that the evolution of ϕ is slow compared to the expansion, we can write

$$a(t) = \exp\left(\int H(t)dt\right),\qquad(2.4.3)$$

where

$$H(t) = \frac{\dot{a}}{a} \approx [8\pi G V(\phi)/3]^{1/2}.\qquad(2.4.4)$$

The spacetime will be approximately de Sitter with horizon scale H^{-1}.

The evolution of the scalar field is described by the equation

$$\ddot{\phi} + 3H\dot{\phi} + V'(\phi) = 0,\qquad(2.4.5)$$

where we have omitted the gradient term, since it has a rapidly decreasing factor a^{-2}. Under the slow-rollover assumption, the first term in (2.4.5) is much smaller than the second, and so we can write

$$\dot{\phi} \approx -V'(\phi)/3H.\qquad(2.4.6)$$

Using this equation it is easy to verify that the slow-rollover assumption is valid if

$$|V''(\phi)| \ll H^2.\qquad(2.4.7)$$

As ϕ approaches the minimum of the potential, the condition (2.4.7) is violated, either because the potential becomes steeper or because $V(\phi)$

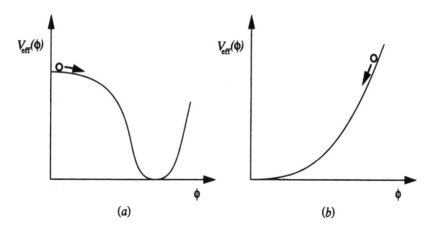

Fig. 2.5. Slow-rollover phase transitions potentially causing (a) new inflation and (b) chaotic inflation.

and, therefore, H becomes smaller. The scalar field ϕ then begins to oscillate about the minimum. These oscillations are damped by the production of particles coupled to ϕ. The corresponding decay constant is $\Gamma \sim \tau_\phi^{-1}$, where τ_ϕ is the lifetime of the ϕ-quanta, which depends on the strength of its couplings. Interactions between the newly created particles rapidly establish a thermal state, and the subsequent evolution coincides with the standard cosmology. The temperature at thermalization is of the order

$$T_{\text{th}} \sim (\Gamma m_{\text{pl}})^{1/2} . \tag{2.4.8}$$

The thermal energy density $\rho = (\pi^2/30)\mathcal{N}T_{\text{th}}^4$ cannot exceed the vacuum energy density $V(\phi)$ during inflation. This gives an upper bound on T_{th},

$$T_{\text{th}} < T_{\text{max}} \approx \tfrac{1}{4}(Hm_{\text{pl}})^{1/2} , \tag{2.4.9}$$

where we have assumed $\mathcal{N} \sim 100$ massless degrees of freedom at $T \sim T_{\text{th}}$. In slow-rollover models the couplings of ϕ are assumed to be very small, so that the decay constant Γ is typically much smaller than H and $T_{\text{th}} \ll T_{\text{max}}$.

An important role in the slow-rollover scenarios is played by quantum fluctuations of the field ϕ. On the horizon scale H^{-1}, the typical amplitude of the fluctuations is $\delta\phi \sim H/2\pi$ and their characteristic timescale is $\sim H^{-1}$. As the expansion of the universe stretches the wavelength of the fluctuations beyond the horizon, their amplitude 'freezes' and the fluctuations become part of the classical scalar field. The fluctuations of $\dot\phi$ on the horizon scale are of the order $\delta\dot\phi \sim H^2$. Comparing this with the classical expression (2.4.6), we see that on flat portions of the potential,

where

$$|V'(\phi)| \ll H^3 , \qquad (2.4.10)$$

the dynamics of ϕ is dominated by quantum fluctuations. In this regime the evolution of ϕ can be pictured as Brownian motion. In each horizon-sized region, the magnitude of ϕ fluctuates independently by an amount $\pm H/2\pi$ per Hubble time H^{-1}. When the scalar field 'diffuses' to a steeper part of the potential, where $|V'(\phi)| > H^3$, classical evolution takes over and eventually leads to thermalization and a radiation-dominated expansion. However, because of the spatial variation of ϕ introduced by quantum fluctuations, thermalization does not occur simultaneously in different parts of the universe. Such fluctuations in the thermalization time give rise to a nearly scale-invariant spectrum of density fluctuations. The amplitude of these fluctuations at the time when their comoving scale crosses the Hubble radius in the late universe is

$$\left.\frac{\delta\rho}{\rho}\right|_{\mathrm{H}} \sim \frac{H^3}{V'(\phi)} . \qquad (2.4.11)$$

Here, the right-hand side is evaluated at the time of de Sitter horizon crossing during inflation. The structure of the universe on scales much greater than the present Hubble radius is also determined by quantum fluctuations of the scalar field. It can be shown that in slow-rollover models inflation never ends completely, and at any time there are parts of the universe which remain in the de Sitter phase.

The amplitude of the density fluctuations needed to seed galaxy formation is approximately $(\delta\rho/\rho)_{\mathrm{H}} \sim 10^{-4}$. For a potential with a quartic Higgs self-coupling, $V(\phi) \sim \lambda\phi^4$, this value is obtained with $\lambda \sim 10^{-12}$. Much larger values of the coupling would contradict the observed isotropy of the microwave background, yielding the cosmological constraint,

$$\lambda \lesssim 10^{-12} . \qquad (2.4.12)$$

This has important consequences for the couplings of ϕ to other fields: ϕ should be a gauge singlet, its Higgs couplings to other scalars should not exceed $\sim 10^{-6}$, and its Yukawa couplings to fermions should not be greater than $\sim 10^{-3}$. If this were not the case, the self-coupling induced by radiative corrections would be greater than 10^{-12} and the constraint (2.4.12) would require a fine-tuning of these couplings.

In addition to density fluctuations, inflation produces a nearly flat spectrum of gravitational waves with initial amplitudes of the order

$$h \sim \frac{H}{m_{\mathrm{pl}}} , \qquad (2.4.13)$$

where the value of H is taken at the moment when the corresponding wavelength crossed the de Sitter horizon, and the amplitude h is given at the time when this wavelength comes back within the Hubble radius. Gravitational waves induce background temperature fluctuations of the order $\delta T/T \sim h$, and the observational bound $\delta T/T \leq 10^{-5}$ gives a constraint on the scale of inflation,

$$H \leq 10^{-5} m_{\mathrm{pl}}. \tag{2.4.14}$$

Combining this with (2.4.9) we find

$$T_{\mathrm{th}} \lesssim 10^{16} \mathrm{GeV}. \tag{2.4.15}$$

2.4.3 Other inflationary models

La & Steinhardt [1989] have pointed out that the flaws in Guth's original scenario based on a first-order phase transition can be circumvented in a Brans–Dicke-type theory with a variable gravitational constant. When the scalar field gets stuck in the false vacuum, the solution of the modified equation for the scale factor is a power law, rather than exponential. As a result, bubble nucleation can catch up with the expansion of the universe and the graceful exit problem can be resolved. However, the bubble distribution tends to be dominated by the very largest bubbles, and parameter fine-tuning is required to avoid conflict with the observed homogeneity and isotropy of the universe.

We should also mention the Starobinsky model [Starobinsky, 1980] in which inflation is driven by nonlinear curvature corrections to the Einstein equations, along with the Kaluza–Klein models studied by Shafi & Wetterich [1983] in which the scalar field represents the radius of the compactified dimensions. However, as we emphasized previously, none of the models suggested to date is particularly compelling and all require an unnatural adjustment of parameters. We conclude, therefore, that inflation remains an attractive paradigm in search of an acceptable model.

3

Topological defects

3.1 Introducing defects

The Goldstone model encountered in chapter 2 admits a variety of topological defects of different dimensionalities. It is illuminating to introduce the basic concepts by considering some of these, so we shall discuss special cases of a general Goldstone model,

$$\mathcal{L} = \tfrac{1}{2} \left(\partial_\mu \phi^\dagger \right) (\partial^\mu \phi) - V(\phi), \tag{3.1.1}$$

where ϕ represents a multi-component scalar field transforming under a chosen global symmetry group G. For the sake of clarity, the introduction of gauge fields will be deferred until the next section.

3.1.1 Topological conservation laws

Perhaps the most familiar topological defects are the ϕ^4-kink and the sine-Gordon soliton in one spatial and one time dimension. The ϕ^4-kink arises in a Goldstone model with a single real scalar field by choosing the potential $V(\phi)$ in (3.1.1) to have the form

$$V(\phi) = \frac{\lambda}{4} \left(\phi^2 - \eta^2 \right)^2. \tag{3.1.2}$$

This double-well potential is illustrated in fig. 3.1(a). The field equation arising from (3.1.1–2) has the analytic solution,

$$\phi(x) = \eta \tanh \left((\lambda/2)^{1/2} \eta \, x \right). \tag{3.1.3}$$

This corresponds to a localized 'kink' centred about $x = 0$, which takes ϕ from $-\eta$ at $x = -\infty$ to η at $x = +\infty$. For the sine-Gordon soliton, we take the periodic potential

$$V(\phi) = 2\lambda\eta^4 \left[1 - \cos \left(\phi/\eta \right) \right], \tag{3.1.4}$$

54

for which the analytic soliton solution is

$$\phi(x) = 4\eta \tan^{-1} \exp\left(\sqrt{\lambda}\,\eta x\right). \qquad (3.1.5)$$

These remarkable non-dissipative solutions are time-independent, of finite energy and, by the Lorentz invariance of the theory, can be boosted up to arbitrary velocities. Their classical stability can be established by a perturbative analysis and one can interpret their non-dissipative properties as a finely-tuned balance between dispersion and self-interaction. However, in a more fundamental sense their stability arises from the non-trivial topology of the vacuum manifold $\mathcal{M}$ – here, the fact that $\mathcal{M}$ is disconnected.

On the left side far from the ϕ^4-kink, the field lies in one minimum of the potential while, on the right side, it lies in the opposite minimum. The field $\phi(x)$, therefore, maps points at spatial infinity in physical space non-trivially into the vacuum manifold, $\phi = \pm\eta$ (see fig. 3.1). The energy of the configuration is due to the localized departure of ϕ from the disconnected minima as it continuously interpolates between them, rising over the potential hill of the double-well. The removal of the defect entails the infinite cost of lifting all of the field on one side over this separating barrier – hence the classical stability. Alternatively, the defect can be annihilated by an antidefect having the opposite topological orientation or 'charge'.

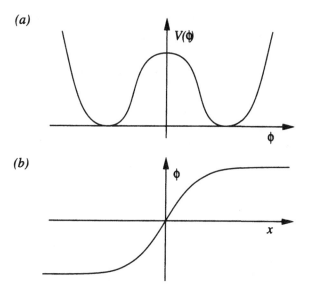

Fig. 3.1. The double-well potential (a) gives rise to the ϕ-kink solution which is illustrated in (b).

Understood in these terms, stability becomes a consequence of a topological conservation law, with each defect state corresponding to an absolutely conserved quantum number. For this one-dimensional model, the
topological current from which this law derives is [Skyrme, 1961]

$$j^\mu = \epsilon^{\mu\nu}\partial_\nu\phi. \tag{3.1.6}$$

The current j^μ is conserved trivially because it is a divergence of an antisymmetric tensor. This contrasts with a current arising from a continuous
symmetry of the Lagrangian which is conserved by Noether's theorem.
The associated conserved charge of a configuration is then simply

$$N = \int dx\, j^0 = \phi|_{x=+\infty} - \phi|_{x=-\infty}. \tag{3.1.7}$$

The presence of a ϕ^4-kink with ϕ in different vacua at $x = \pm\infty$, gives
rise to a non-zero charge N and consequently indicates the stability of
the configuration.

Such topological conservation laws, which depend only on the qualitative topological characteristics of the vacuum manifold, yield little
information about the detailed structure or energy of a particular defect solution, but they do provide strong sufficiency conditions for their
existence. In more general cases which are not tractable analytically,
particularly in higher dimensions, this economical approach can provide
powerful insight.

3.1.2 Evading Derrick's theorem

The quest in higher dimensions for further topological defects in Goldstone models must confront an elegant scaling argument suggesting their
non-existence [Derrick, 1964]. This concludes that there are no stable
time-independent, localized solutions for any of the scalar models (3.1.1)
in more than one dimension. Since we shall encounter this argument
later, a brief sketch of the proof is illuminating.

Consider the energy of a localized solution $\phi(\mathbf{x})$ in (3.1.1) in D-dimens-
ions,

$$E = \int d^D x \left[\tfrac{1}{2}(\nabla\phi)^2 + V(\phi)\right]$$
$$= I_1 + I_2, \tag{3.1.8}$$

where the integral over all space converges and I_1 and I_2 incorporate the
gradient and potential terms, respectively. Now suppose we transform
$\phi(\mathbf{x})$ under the radial rescaling $\mathbf{x} \to \alpha\mathbf{x}$, then the energy of the new
configuration becomes simply

$$E_\alpha = \alpha^{2-D}I_1 + \alpha^{-D}I_2. \tag{3.1.9}$$

Given that we assume $V(\phi)$ is non-negative for these models, clearly any non-trivial 'solutions' will be unstable to collapse for $D \geq 2$ (the special case with $I_2 \equiv 0$ and $D = 2$ will be discussed in §3.2.4).

This difficulty with scalar models is not as serious as it might seem, since there are several means by which it can be avoided:

(i) *Extensions*: One can consider more complicated models, for example, by adding higher derivative terms as in the Skyrme model [Skyrme, 1961]. Alternatively, the natural extension through the addition of gauge fields to the abelian-Higgs model or Yang–Mills–Higgs models has proved very fruitful and will be the focus of chief interest here. Note that gauge fields alone cannot produce non-dissipative solutions because of an analogue of Derrick's theorem, except for the possibility of instanton solutions in four Euclidean dimensions [Coleman, 1976].

(ii) *Time-dependence*: The alternative of allowing time-dependence has also produced a variety of non-dissipative solutions, notably the non-topological solitons of T. D. Lee [1976]. These solutions are stabilized by the confinement of a conserved charge Q, a result of a Noether current arising from a symmetry of the theory, rather than a topological conservation law.

(iii) *Non-locality*: The option of somewhat relaxing the localization assumption, makes several more scalar defects admissible. For example, we can argue that a physical vortex solution in two dimensions is present for the usual $U(1)$ Goldstone model with a complex scalar field. A mild logarithmic divergence in the energy of the isolated *global* string contravenes the theorem but, in a cosmological context, this is simply cutoff by the next nearest string [Vilenkin & Everett, 1982].

3.1.3 Higher dimensions

Topological defects can also appear in models where a continuous symmetry is broken. In the simplest example, we can break a $U(1)$ symmetry as in the Goldstone model discussed in §2.1.1. We take a complex scalar field ϕ and the quartic 'Mexican hat' potential (2.1.2) with a degenerate circle of minima $|\phi| = \eta$ (see fig. 2.1).

As we traverse a closed path L in physical space it is possible for ϕ to wrap once around the circle of minima, that is, for the phase of ϕ to develop a non-trivial winding, $\Delta\theta = 2\pi$. As indicated in fig. 3.2, the far field will then take the form

$$\phi \approx \eta e^{i\theta} . \tag{3.1.10}$$

By measuring the winding on smaller loops L' on a surface S bounded by the loop L, we can more accurately specify the actual location of the

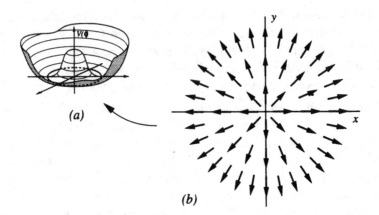

Fig. 3.2. Field configuration for a vortex string (b). A loop in physical space enclosing the vortex is mapped non-trivially into the degenerate circle of minima of the potential (2.1.2) shown in (a).

twist in the phase (see fig. 3.3). At such a point, however, the phase of ϕ varies by 2π and is no longer well-defined. This phase jump can only be resolved continuously if the field ϕ rises to the top of the potential where it takes the value $\phi = 0$. Consequently, there must be a non-zero energy density associated with the phase twist at the vortex core. We can see that such defects are constrained to be linear in three dimensions by noting that we must be able to locate the string core somewhere on any two-surface S' deformed away from S but which remains bounded by the closed path L.

If the vacuum manifold is enlarged from a circle to a two-sphere S^2, say by breaking $SO(3)$ with a three component vector φ (as in §2.1.3, *Exam-*

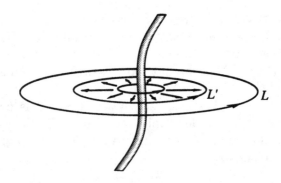

Fig. 3.3. The string in three dimensions can always be located by encircling it with a closed loop L. A non-zero winding number in the phase as the loop is traversed discloses a string within.

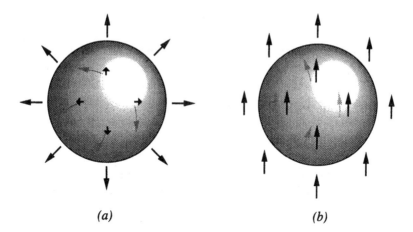

(a) (b)

Fig. 3.4. (a) A non-trivial field configuration on a two-sphere S^2 signalling the presence of a monopole within. This configuration cannot be smoothly deformed into the constant field of (b).

ple 2), then non-trivial field configurations correspond to pointlike defects or monopoles. Fig. 3.4 contrasts a constant field with that surrounding a monopole configuration.

A vacuum manifold which is a three-sphere S^3 can, in turn, give rise to further defects called textures. And, among a variety of other phenomena, successive symmetry breakings can produce hybrid defect combinations like domain walls bounded by strings or monopoles connected by strings. It is evident that the topology of the vacuum manifold is paramount in determining the nature of the defects that can appear at a symmetry breaking phase transition. For a proper understanding of this topology and the classification of defects we will review some results from homotopy theory in §3.3.

3.1.4 The Kibble mechanism

Before leaving these simple models, it is important to mention some of the implications of symmetry breakdown at a cosmological phase transition [Kibble, 1976]. Consider, first, the example of a single scalar field ϕ possessing a discrete symmetry as in (3.1.2). We anticipate from the discussions in §2.3.2 that the temperature-dependent effective potential will have a single minimum at $\phi = 0$ for temperatures above a critical temperature T_c. Below T_c, however, the potential will develop two degenerate minima (as in fig. 3.1) and ϕ will begin to fall into one or other of these ground states. The choice of minimum will depend on random fluctuations in ϕ which can be expected to differ in different regions of space. If neighbouring volumes fall into opposite minima, then a kink forms as a boundary between them, only now, given the added dimensionality, it corresponds to a two-surface of false vacuum or a domain wall. Below the Ginzburg temperature T_G (2.3.22), temperature fluctuations in the field will be insufficient to lift it from one minimum into the other, so the domain walls effectively 'freeze-out'. The typical scale of these walls at formation is set by the correlation length ξ beyond which the fluctuations in ϕ are uncorrelated. The magnitude of ξ depends on the dynamics of the phase transition (refer to §2.3). However, since correlations cannot establish on scales greater than the causal horizon d_H, it should satisfy the causality constraint, $\xi < d_H$ [Zel'dovich, Kobzarev & Okun, 1974].

In the case of continuous symmetry breaking, the Higgs field can be expected to be similarly uncorrelated, non-trivial twists in the field will develop, and defects will 'freeze-out' as the temperature falls below T_G. We shall discuss this mechanism for defect formation further in §9.1, relating this more closely to the discussion of cosmological phase transitions in chapter 2.

3.2 Basic properties

Before embarking on detailed discussions of the dynamics and interactions of different topological defects, it is appropriate to review their basic properties. While the following is in no way comprehensive, this is intended to provide an introductory appraisal of the most important characteristics of each class of defects.

3.2.1 Domain walls

The appearance of domain walls is generally associated with the breaking of a discrete symmetry. The vacuum manifold $\mathcal{M}$ then consists of several disconnected components. Domain walls occur at the boundaries between regions of space with values of the field ϕ in different components, with ϕ interpolating between these two values across the wall.

The domain wall appearing in the simplest Goldstone model with the double-well potential has already been discussed in §3.1.1. Locally, with the curvature radii of the wall much greater than its thickness, the solution corresponds to that of the ϕ^4-kink (3.1.3). The width of the wall is approximately

$$\delta \sim \left(\sqrt{\lambda}\eta\right)^{-1}. \tag{3.2.1}$$

With a vacuum energy $\rho \sim \lambda\eta^4$ at the center of the wall, the surface energy density will be

$$\sigma \sim \rho\delta \sim \sqrt{\lambda}\eta^3. \tag{3.2.2}$$

Unless the symmetry breaking scale η is very small, this surface density is extremely large and implies that cosmological domain walls would have an enormous impact on the homogeneity of the universe, as we shall see in chapter 13.

The energy–momentum tensor of a domain wall can be found by substituting the ϕ^4-kink solution (3.1.3) into

$$T_{\mu\nu} = \partial_\mu\phi\,\partial_\nu\phi - g_{\mu\nu}\mathcal{L}\,. \tag{3.2.3}$$

This yields the simple form

$$T^\mu{}_\nu = f(x)\,\mathrm{diag}\,(1,0,1,1)\,, \tag{3.2.4}$$

where $f(x)$ is a bell-shaped function of width $\sim \delta$ localized about $x = 0$,

$$f(x) = \tfrac{1}{2}\lambda\eta^4 \left[\cosh\left((\lambda/2)^{1/2}\eta x\right)\right]^{-4}. \tag{3.2.5}$$

These equations can be used to calculate the wall surface density σ exactly,

$$\sigma = \int T^0_0 dx = \frac{2\sqrt{2}}{3}\sqrt{\lambda}\,\eta^3\,. \tag{3.2.6}$$

The solution for the scalar field (3.1.3) is independent of y, z, and t and is therefore invariant under Lorentz boosts in the yz-plane. The energy–momentum tensor (3.2.4) has the same invariance. This implies that only the transverse motion of the wall is observable. We can also deduce from (3.2.4) that the wall tension in the two tangential directions is equal to the surface density σ, so static curved regions will contract and, in vacuum, will rapidly accelerate up to relativistic velocities.

3.2.2 Strings

Strings arise in models in which the vacuum manifold $\mathcal{M}$ is not simply connected; that is, $\mathcal{M}$ contains enclosed holes about which loops can be trapped, as we have seen for the degenerate circle of minima S^1. This topological property is disclosed if the fundamental group of $\mathcal{M}$ is non-trivial, $\pi_1(\mathcal{M}) \neq I$. The elements of $\pi_1(\mathcal{M})$ then classify the different types of admissible string solutions. For a connected and simply-connected symmetry group G, by the fundamental theorem of homotopy theory, strings can be classified by the disconnected components of the unbroken subgroup H, that is, by $\pi_0(H)$. Details of this general procedure will be considered in §3.3.2.

We have already encountered the global $U(1)$ string which is structurally the simplest vortex solution. However, since this has a number of subtleties related to its non-locality, we begin by discussing the breaking of a local $U(1)$-symmetry in the abelian-Higgs model (2.1.10). This also admits string solutions, the Nielsen–Olesen [1973] vortex lines, since field configurations on a closed loop in physical space can similarly correspond to non-trivial windings about the degenerate circle of minima in the Mexican-hat potential (fig. 2.1).

In the Lorentz gauge, $\partial_\mu A^\mu = 0$, the Higgs field takes the same form as the global string at large distances from the core,

$$\phi \approx \eta\, e^{in\theta}, \tag{3.2.7}$$

where n is an integer, the string winding number. The gauge field asymptotically approaches

$$A_\mu \approx \frac{1}{ie}\partial_\mu \ln \phi. \tag{3.2.8}$$

Despite this apparent 'pure gauge' form for A_μ, the solution cannot everywhere be rotated to the vacuum by a regular gauge transformation if $n \neq 0$. This can be shown with Stokes' theorem by integrating around a closed path encircling a string,

$$\Phi_B = \int \mathbf{B} \cdot \mathbf{dS} = \oint \mathbf{A} \cdot \mathbf{dl} = \frac{2\pi n}{e}, \tag{3.2.9}$$

which reveals a physical magnetic flux Φ_B flowing along the string, just as in the quantized flux lines in superconductors [Abrikosov, 1957].

With these asymptotic forms (3.2.7–8), we have $D_\mu \phi \approx 0$ and $F_{\mu\nu} \approx 0$ far from the string core. As a result, the energy density vanishes rapidly (exponentially) away from the core and the total energy per unit length is finite. The lowest energy string configuration has unit winding number, $n = \pm 1$. The width of the string is determined by the Compton wavelengths of the Higgs and gauge bosons: $\delta_\phi \sim m_\phi^{-1}$ and $\delta_A \sim m_A^{-1}$ where $m_\phi = \sqrt{\lambda}\,\eta$ and $m_A = e\eta$. For the case with $m_A > m_\phi$, the string has an inner core of false vacuum with linear mass density $\mu_\phi \sim \lambda \eta^4 \delta_\phi^2 \sim \eta^2$, and a magnetic flux tube with a smaller radius δ_A with mass density $\mu_A \sim B^2 \delta_A^2 \sim \eta^2$. The total string mass per unit length is then approximately

$$\mu \sim \eta^2. \tag{3.2.10}$$

For GUT scale strings with $\eta \sim 10^{16}$ GeV, this corresponds to the enormous mass density $\mu \sim 10^{22}\,\mathrm{g\,cm}^{-1}$, which provides a foretaste of their potential cosmological significance.

On scales much larger than the string width, the internal structure of the string becomes unimportant and physical quantities of interest, such as the energy–momentum tensor $T^\mu{}_\nu$, can be averaged over the string cross-section. For a straight string lying along the z-axis we can make the following replacement to obtain an effective energy–momentum tensor,

$$\widetilde{T}^\nu{}_\mu = \delta(x)\delta(y) \int T^\nu{}_\mu \, dx \, dy \, . \tag{3.2.11}$$

Given that the string is invariant under Lorentz boosts in the z-direction, we must have $\widetilde{T}^0_0 = \widetilde{T}^3_3$ and all other components vanishing, with the possible exception of $\widetilde{T}^i{}_i$, $i = 1,2$. To show that these remaining diagonal components are also equal to zero, we can use the conservation law, $T^{\mu\nu}{}_{,\nu} = 0$, to write

$$\int T^j{}_{i,j}\, x^k dx \, dy = 0 \, , \tag{3.2.12}$$

where all indices take the values 1 or 2. Integration by parts then yields

$$\int T^k{}_i \, dx \, dy = 0 \qquad i, k = 1, 2 \, . \tag{3.2.13}$$

Consequently, the energy–momentum tensor of the string must take the form [Vilenkin, 1981b]

$$\widetilde{T}^\mu{}_\nu = \mu \, \delta(x)\delta(y) \, \text{diag}\, (1, 0, 0, 1) \, . \tag{3.2.14}$$

The string has a large tension equal to the energy density, implying that curved static segments will contract and rapidly acquire relativistic velocities.

For the global $U(1)$ string there is no gauge field to compensate the variation of the phase at large distances and so the energy per unit length is divergent. For the $n=1$ string we have

$$\begin{aligned} \mu &\sim \eta^2 + \int_\delta^R \left[\frac{1}{r}\frac{\partial\phi}{\partial\theta}\right]^2 2\pi r \, dr \\ &\approx 2\pi\eta^2 \ln\left(\frac{R}{\delta}\right) , \end{aligned} \tag{3.2.15}$$

where δ is the core width and R is a cut-off radius that must be imposed at some large distance. Depending on the physical situation, this cut-off will be provided by the curvature radius of the string or by the distance to the nearest string in the network. The logarithmic factor in (3.2.15) reflects a coupling to the Goldstone boson field in the model, and implies long-range interactions between strings with a force law, $F \sim \eta^2/R$. Familiar condensed matter analogues for global strings are the vortex-lines in a superfluid.

Other more complicated strings, such as Z_N-strings, are encountered in phase transitions after which a discrete symmetry remains intact. A Z_2-string example appears in the GUT model discussed in (2.1.67a) [Kibble et al., 1982a],

$$SO(10) \xrightarrow{126} SU(5) \times Z_2 \, . \tag{3.2.16}$$

For $N > 2$, Z_N-strings can form networks with vertices from which several strings emanate.

Witten [1985a] pointed out the possibility and consequences of charged excitations becoming trapped on cosmic strings. These charge carriers could be either fermionic or bosonic, and would make the string super-conducting. In contrast to 'normal' cosmic strings which have subtle gravitational effects, the electromagnetic interactions of superconducting strings can vastly enrich observable string phenomena.

3.2.3 Monopoles

By analogy with strings, pointlike defects or monopoles arise if the man-ifold of degenerate vacua contains non-contractible two-surfaces (like the sphere S^2). This corresponds to the vacuum manifold $\mathcal{M}$ having a non-trivial second homotopy group $\pi_2(\mathcal{M}) \neq I$. As we shall see in the next section, it suffices to know if the unbroken symmetry group H has a non-trivial fundamental group $\pi_1(H)$ (assuming $\pi_1(G) = \pi_2(G) = I$). A symmetry group G, for example, which breaks to leave a $U(1)$ symmetry intact,

$$G \longrightarrow K \times U(1), \qquad (3.2.17)$$

must have monopole solutions ['t Hooft, 1974; Polyakov, 1974] since $\pi_1(U(1)) = Z$. Grand unified models based on simple groups which break to leave the $U(1)_{\text{em}}$ of electromagnetism must therefore produce monopoles.

The simplest monopole is the 't Hooft–Polyakov solution which ap-pears when $SU(2)$ is broken to $U(1)$ (refer to *Example 2* in §2.1.3). The three-component Higgs field ϕ far from the core of the monopole takes a 'hedgehog' configuration

$$\varphi = \eta \hat{\mathbf{r}}, \qquad (3.2.18)$$

where φ points radially outward in the direction of the unit vector $\hat{\mathbf{r}}$. As for the vortex, the gauge field aligns in such a manner as to minimize the variational energy. However, a radial magnetic field remains,

$$\mathbf{B} = \frac{\mathbf{r}}{er^3}, \qquad (3.2.19)$$

and yields an overall magnetic flux,

$$\Phi_B = \frac{4\pi}{e}. \qquad (3.2.20)$$

The monopole has two length scales δ_ϕ and δ_A which determine the radius of the core. Regardless of whether m_ϕ or m_A is the larger, the mass of these monopoles is typically

$$m \sim \frac{4\pi\eta}{e}. \qquad (3.2.21)$$

We can also envisage a global monopole consisting only of the scalar field configuration (3.2.18). Without the presence of compensating gauge fields, however, the total energy of the global monopole is linearly divergent. These are consequently strongly confined, since the force between a monopole–antimonopole pair will be independent of distance.

3.2.4 Textures

Non-trivial mappings of the three-sphere S^3 into the vacuum manifold $\mathcal{M}$ provide us with the final possibility for models exhibiting topological defects in three spatial dimensions [Kibble, 1976]. These are configurations, known as textures, which are classified by the third homotopy group $\pi_3(\mathcal{M})$. They are sometimes termed non-singular solitons because the scalar field is nowhere topologically constrained to rise from the minimum of the potential $V(\phi)$. In a condensed matter context, they appear in superfluid ^{3}He which has a highly degenerate ground state with three classes of Goldstone bosons, and also in a variety of liquid crystal phases.

In describing textures, it is simplest to begin with the related non-singular solitons in one and two dimensions. These are classified by the elements of $\pi_1(\mathcal{M})$ and $\pi_2(\mathcal{M})$, respectively. We first consider a complex scalar field $\phi(x,t)$ in one dimension with a Mexican-hat potential (2.1.2), so that the vacuum manifold is a circle and $\pi_1(\mathcal{M}) \neq I$. Suppose now that ϕ is required to approach the same value at $x \longrightarrow \pm\infty$. This boundary condition makes the line topologically equivalent to a circle and, as illustrated in fig. 3.5(a), the field can develop a non-trivial winding. Note that in contrast to the kink solution (3.1.3), here the field never departs from the vacuum manifold. A two-dimensional texture for a triplet field φ is illustrated in fig. 3.5(b). Here, the vacuum manifold is S^2 and uniform boundary conditions are imposed at infinity. Textures arise similarly in three dimensions by again imposing uniformity at spatial infinity. This boundary condition effectively compactifies physical space to S^3, and non-trivial mappings from S^3 into the vacuum manifold are possible if $\pi_3(\mathcal{M}) \neq I$.

Topology, however, only gives sufficiency conditions for the existence of stable defects, and textures provide an example in which stability should not be taken for granted. Returning to Derrick's theorem (3.1.9), we see that the energy of a texture behaves as $E_\alpha = \alpha^{2-D}I_1$ under the radial rescaling $\mathbf{x} \to \alpha\mathbf{x}$, where D is the relevant spatial dimension. The

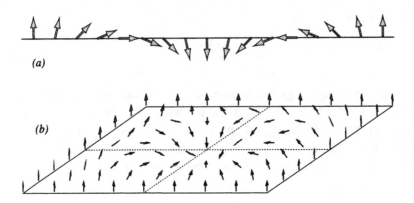

Fig. 3.5. (*a*) A one-dimensional texture corresponding to the winding of a complex scalar field ϕ about a circular vacuum manifold. (*b*) A two-dimensional texture for a triplet field φ winding on S^2.

potential energy vanishes since the Higgs field always remains very close to the degenerate vacuum. Clearly, the one-dimensional soliton will be unstable to expansion and the three-dimensional texture to collapse, while the energy of the two-dimensional soliton is apparently indifferent to a change in width. However, despite this shrinking instability for textures, speculations about their potential cosmological role will be discussed in chapter 15.

Since only the massless degrees of freedom are relevant for textures, such models can be effectively reduced to nonlinear sigma models. A vortex in the two-dimensional nonlinear σ-model corresponds to the two-dimensional texture discussed above. In three dimensions, the skyrmion [Skyrme, 1961] corresponds to a pointlike texture stabilized against collapse by the addition of higher derivative terms,

$$\mathcal{L} = \frac{1}{2}\,\mathrm{Tr}\,\left(\partial_\mu \Phi \partial_\mu \Phi^\dagger\right) + \frac{1}{g^2}\,\mathrm{Tr}\,\left[(\partial_\mu \Phi)\,\Phi^\dagger, (\partial_\nu \Phi)\,\Phi^\dagger\right]^2,$$

where Φ is an $SU(2)$ matrix transforming under a chiral $SU(2){\times}SU(2)$ symmetry. Although phenomenologically useful for modelling baryons in QCD, skyrmions will not be discussed in detail here.

3.2.5 Transient defects

Hybrid topological defects

A series of phase transitions can give rise to hybrid defect configurations when defects of different dimensionalities become joined together. Strings corresponding to a particular discrete symmetry will become attached to domain walls when this symmetry is broken. For example, we can abbreviate the first two stages of the GUT symmetry breaking in (2.1.68) as [Kibble *et al.*, 1982b]

$$G \longrightarrow H_1 \times Z_2^C \longrightarrow H_2 \,. \qquad (3.2.22)$$

The first stage will produce Z_2^C-strings associated with the charge conjugation operator C. Subsequently, the breaking of the discrete Z_2^C symmetry will result in the appearance of domain walls between regions with opposite values of C. These walls will terminate on strings of the previous H_1 phase, because one can pass continuously between the vacua by going around a string.

Another example is provided by the Goldstone model with the global $U(1)$-string (3.1.10) but in which the symmetry is only approximate [Vilenkin & Everett, 1982]. This occurs in axion models, when the axion mass 'switches on'. The degeneracy of the circular vacuum in (2.1.2) is disturbed by tilting it with a mass term dependent on the phase, $V(\theta) \sim M^4(\cos\theta - 1)$ as in (3.1.4). The phase θ will attempt to roll down to the minimum ($\theta = 0$) everywhere, but this is not possible in the vicinity of a string since there must be a net winding of 2π as the string is encircled. Most variation will collapse into a localized surface about $\theta = \pi$ – the domain wall to which the global string becomes attached. The interesting dynamics of such hybrid systems, which can lead to the rapid demise of both strings and domain walls, will be taken up again in chapter 13. We should note that such hybrid configurations are not necessarily transient. Creating a number of minima in the circular vacuum manifold will result in each string becoming attached to several domain walls. In the context of a random string network, such a complex topology may prove difficult to unravel.

Similarly, monopoles can become attached to strings. If the monopole-producing symmetry breaking (3.2.17) were to be continued by breaking the remaining $U(1)$ symmetry,

$$G \longrightarrow K \times U(1) \longrightarrow K \,, \qquad (3.2.23)$$

then strings would also arise. At the second transition the magnetic field is squeezed into flux tubes connecting monopoles with antimonopoles, although closed string loops can also form. Correlating monopoles and antimonopoles with such strings leads to rapid monopole annihilation; we shall discuss this in further detail when the cosmological implications of monopoles are addressed in chapter 14. We also note that the strings in this model are not stable, since there is a small quantum mechanical probability of the string breaking due to the nucleation of a monopole–antimonopole pair. A semiclassical calculation for this tunnelling process yields a probability P which is typically exponentially suppressed,

$$P \sim \exp\left(-\pi^2 m^2/\mu\right), \qquad (3.2.24)$$

where m is the monopole mass and μ is the string tension. In a model with strings connected to domain walls, the walls are similarly unstable to the nucleation of a string loop.

Defects and symmetry restoration

A symmetry broken at one phase transition can be restored at a subsequent phase transition. The possibility of such symmetry restoration at low temperatures was first pointed out by Weinberg [1974]. The properties of topological defects arising in this type of model are somewhat different from the usual case [Vilenkin, 1981c, 1985], and we shall briefly discuss the distinction here. Consider a finite temperature effective potential of the form (refer to §2.3.2)

$$V_{\rm eff}(\phi, T) = \tfrac{1}{2} m^2(T)|\phi|^2 + \tfrac{1}{4}\lambda|\phi|^4, \qquad (3.2.25)$$

with

$$m^2(T) = m^2(0) + AT^2. \qquad (3.2.26)$$

The parameter A is determined by the interaction of ϕ with effectively massless degrees of freedom. It can be approximately constant over a wide range of temperatures, but can also change at mass thresholds and phase transitions. It is usually assumed that $m^2(0) < 0$ and $A > 0$; hence the symmetry is broken at low temperatures. Suppose instead that $m^2(0) > 0$ and that A becomes negative below some temperature T_1 satisfying $T_1 \gg |A|^{-1/2}m(0) \equiv T_2$. Then for $T_1 > T > T_2$ the symmetry breaks and ϕ develops a vacuum expectation value,

$$|\langle\phi\rangle| = (|A|/\lambda)^{1/2}T(1 - T_2^2/T^2)^{1/2}. \qquad (3.2.27)$$

At $T = T_2$, $\langle\phi\rangle = 0$, and below T_2 the symmetry is restored.

   If ϕ is a real scalar field, then (3.2.25) exhibits a discrete symmetry $\phi \to -\phi$, and its breaking at $T = T_1$ results in the formation of domain walls. However, stable wall solutions exist only in the range $T_1 > T > T_2$. For $T \gg T_2$, the wall thickness is $\delta \sim |m(T)|^{-1} \sim |A|^{-1/2}T^{-1}$, the energy density of the symmetric state is $\rho \sim \lambda\langle\phi\rangle^4 \sim A^2\lambda^{-1}T^4$, and the wall tension is given by

$$\sigma \sim \rho\delta \sim |A|^{3/2}\lambda^{-1}T^3. \qquad (3.2.28)$$

At $T \to T_2$, we have $\delta \to \infty$, $\sigma \to 0$, and the walls 'dissolve'.

   If the Higgs field ϕ in (3.2.25) is complex, the model gives rise to transient strings which are only stable for $T_1 > T > T_2$. For $T \gg T_2$, the string tension is

$$\mu \sim \rho \delta^2 \sim (|A|/\lambda) T^2 \,. \qquad (3.2.29)$$

This crucial dependence of defect properties on temperature is characteristic of models with transient symmetry breaking, in contrast to the usual case.

3.2.6 Time-dependent solitons

Localized oscillating solutions to nonlinear classical field theories were first presented by Rosen [1968] and T. D. Lee with collaborators [1974-76]. These are often called non-topological solitons since their existence depends on a conserved Noether charge Q, arising from an unbroken continuous symmetry, rather than the topology of the vacuum. Scalar models require a specially shaped potential in which a charged field is, in some sense, on the verge of a first-order transition (see fig. 3.6). For a sufficient charge Q, a localized soliton of the form

$$\phi(\mathbf{x}, t) = \phi(\mathbf{x}) \, e^{-i\omega t} \qquad (3.2.30)$$

has a lower energy than widely separated free ϕ-particles of the same total charge.

The simplest example of this phenomenon is the Q-ball of the $U(1)$ Goldstone model [Rosen, 1968; Coleman, 1985b; see also T. D. Lee, 1976]. The potential $V(\phi)$ is assumed to be of the form shown in fig. 3.6 with the absolute minimum at $\phi = 0$ and, for small ϕ, $V(\phi) \approx \frac{1}{2} m_\phi^2 |\phi|^2$. Roughly speaking, a Q-ball can be pictured as a spherical volume $\mathcal{V}$ inside which

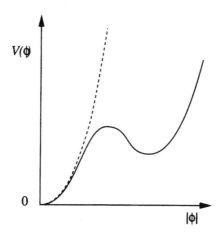

Fig. 3.6. Example of a potential $V(\phi)$ that can give rise to Q-balls. The dashed line is $V(\phi) = \frac{1}{2} m_\phi^2 |\phi|^2$.

$\phi(x)$ takes a constant non-zero value ϕ and outside of which $\phi(x) = 0$. The charge of this configuration is then

$$Q = i \int d^3x \left[\bar{\phi}\dot{\phi} - \dot{\bar{\phi}}\phi \right] \approx \omega |\phi|^2 \mathcal{V}, \qquad (3.2.31)$$

and the energy is given by

$$\begin{aligned} E &= \left[\tfrac{1}{2}\omega^2 |\phi|^2 + V(\phi) \right] \mathcal{V} \\ &\approx \tfrac{1}{2}Q^2/(|\phi|^2 \mathcal{V}) + V(\phi)\mathcal{V}. \end{aligned} \qquad (3.2.32)$$

The minimum energy of this state occurs at a volume

$$\mathcal{V} = Q/ \left[2|\phi|^2 V(\phi) \right]^{1/2}, \qquad (3.2.33)$$

which corresponds to

$$E = Q \left[2V(\phi)/|\phi|^2 \right]^{1/2}. \qquad (3.2.34)$$

Since the energy of Q free particles is $E = Q m_\phi$, the condition for Q-ball stability is that there exists some non-zero ϕ for which [Coleman, 1985b]

$$\frac{2V(\phi)}{|\phi|^2} < m_\phi^2, \qquad (3.2.35)$$

which is satisfied for the potential in fig. 3.6.

The surface energy of these Q-balls has been neglected in this discussion; it determines a $Q_{\min}$ below which Q-balls are unstable to collapse and ensures the spherical shape of a Q-ball. For large Q, however, the volume energy dominates and (3.2.33–34) give an accurate estimate for the Q-ball size and energy.

Some alternative models for these time-varying solitons depend upon an additional broken symmetry, with the charge confined by a domain wall of the extra field [Friedberg *et al.*, 1976]. Such time-dependent solutions are semi-topological in that they combine a conserved Noether current along with topological conservation laws to ensure stability. For a more comprehensive treatment, the reader is referred to review articles by T.D. Lee [1976] and Jetzer [1992].

3.2.7 Quantum corrections

Finally, since we will not be focussing upon quantization issues, these discussions based on classical field theories require some justification. Taking the example of strings, a string segment of length ℓ can be treated classically if its Compton wavelength $(\mu\ell)^{-1}$ is much smaller than the string thickness δ. If this condition is satisfied for $\ell \sim \delta$,

$$\delta^2 \mu \gg 1, \qquad (3.2.36)$$

then we expect the string to be adequately described by classical field theory. For the model (3.1.1–2), $\delta \sim (\sqrt{\lambda}\eta)^{-1}$, $\mu \sim \eta^2$, and (3.2.36) requires simply that the coupling constant λ should be small,

$$\lambda \ll 1. \tag{3.2.37}$$

For domain walls, (3.2.36) is replaced by the condition $\delta^3\sigma \gg 1$, which again reduces to the weak coupling requirement (3.2.37).

A more detailed analysis shows that for $\lambda \ll 1$ the amplitude of quantum fluctuations of strings and walls on scales $\ell \gtrsim \delta$ is in fact much larger than the corresponding Compton wavelength [Garriga & Vilenkin, 1992]. However, this amplitude is still much smaller than the thickness of the defects δ, implying that a classical treatment is justified.

The close correspondence between the classical limit $\hbar \to 0$ and the weak coupling limit $\lambda \to 0$ can be seen by employing a rescaling $\phi = \lambda^{-1/2}\tilde{\phi}$ [Coleman, 1976]. The quantum mechanically significant quantity $\mathcal{L}/\hbar$ then becomes

$$\frac{\mathcal{L}}{\hbar} = \frac{1}{\lambda\hbar}\left(\tfrac{1}{2}(\partial_\mu\tilde{\phi})^2 + \tfrac{1}{4}\left(|\tilde{\phi}|^2 - \lambda\eta^2\right)^2\right). \tag{3.2.38}$$

One can systematically construct a power series expansion in λ using (3.2.38), though quantization about soliton solutions involves a number of subtleties. Generally, the leading order term in the expansion is computed classically – it exhibits the nonlinearity of the theory – while the remaining terms yield appropriate finite quantum corrections (providing the model is renormalizable).

By simply assuming a weak coupling limit and taking this semiclassical approximation, we leave aside a wealth of interesting phenomena, such as how fermionic soliton states can emerge from a purely bosonic theory. We merely note that soliton quantization can be tackled by a variety of methods ranging from the functional integral approach to canonical quantization and, for further discussion, reviews by Coleman [1976], Jackiw [1977] and Rajaraman [1982] should be consulted.

3.3 Classification

It should be evident already that the topology of the vacuum manifold $\mathcal{M}$ determines whether defects appear at a particular symmetry breaking. Domain walls can form if $\mathcal{M}$ has disconnected components, strings can form if $\mathcal{M}$ is not simply connected (that is, if it contains unshrinkable loops), and monopoles can form if $\mathcal{M}$ contains unshrinkable surfaces. The relevant properties of the manifold $\mathcal{M}$ are most conveniently studied using homotopy theory; the nth homotopy group $\pi_n(\mathcal{M})$ classifies qualitatively

Table 3.1. Topological classification of defects
with the homotopy groups $\pi_n(\mathcal{M})$.

Topological defect	Dimension	Classification
Domain walls	2	$\pi_0(\mathcal{M})$
Strings	1	$\pi_1(\mathcal{M})$
Monopoles	0	$\pi_2(\mathcal{M})$
Textures	—	$\pi_3(\mathcal{M})$

distinct mappings from the n-dimensional sphere S^n into the manifold $\mathcal{M}$. A non-trivial first homotopy group or *fundamental group*, for example, indicates that $\mathcal{M}$ is not simply connected and, hence, signals the potential existence of strings. More generally, the topological defects of a particular dimensionality arising in a given model can be classified by the elements of the appropriate homotopy group of the vacuum manifold $\mathcal{M}$. Table 3.1 summarizes the correspondence between the defects we have already encountered and the homotopy groups [Kibble, 1976].

This correspondence between group elements and defects can be a matter of some subtlety, though it is one-to-one in most of the cases we shall be considering. The reader whose interest lies elsewhere, however, is advised to proceed directly to the next two chapters on the microphysics of strings. Alternatively, if macroscopic astrophysical effects are of primary concern, chapter 6 would provide an appropriate place to restart.

3.3.1 The fundamental group

Some definitions

In a manifold $\mathcal{M}$ consider closed paths that pass through a point x. These can be defined in terms of continuous mappings f from the interval $0 \leq t \leq 1$ into $\mathcal{M}$, requiring that $f(0) = f(1) = x$. Two such closed paths f and g are said to be *homotopic at* x if we can continuously deform one into the other while keeping contact with x. Demonstrating this entails the construction of a *homotopy*; this is an intermediate family of paths $h(s,t)$, continuous on the interval $0 \leq s, t \leq 1$, satisfying

$$\begin{aligned} h(0,t) &= f(t)\,, \\ h(1,t) &= g(t)\,, \\ h(s,0) &= h(s,1) = x\,. \end{aligned} \qquad (3.3.1)$$

Examples of paths in a space topologically equivalent to the S^1 vacuum manifold of the Mexican-hat potential (fig. 2.1) are illustrated in fig. 3.7.

It is clear that only f_1 and f_2 are homotopic, since g cannot be continuously deformed into either.

The space of such closed paths or loops can be further endowed with a product defined by

$$f \circ g(t) = \begin{cases} f(2t), & 0 \le t \le \frac{1}{2} \\ g(2t-1), & \frac{1}{2} \le t \le 1. \end{cases} \tag{3.3.2}$$

The product $f \circ g$ corresponds to traversing first the loop f and then g. The inverse $f^{-1}(t) = f(1-t)$ amounts to traversing f in the opposite direction. These definitions prefigure a group-theoretic structure but, first, the loops must be partitioned further.

All possible loops at x can be subdivided into distinct *homotopy classes*. Any loop homotopic to g is placed in the homotopy class $[g]$. Under class multiplication

$$[f][g] = [f \circ g], \tag{3.3.3}$$

generalized from the loop product (3.3.2), the homotopy classes define a group. The identity element $[I]$ consists of all loops contractible to the point x (the constant map), and the inverse is defined by $[f]^{-1} = [f^{-1}]$.

This group $\pi_1(\mathcal{M}, x)$ is called the fundamental group of the manifold $\mathcal{M}$ at x. Definition with respect to a location or base point x is usually omitted because on a connected space one can define an isomorphism* between $\pi_1(\mathcal{M}, x)$ and the fundamental group at any other base point y; to any loop at x we merely add a path from x to y and its inverse. It is clear, then, that all based groups $\pi_1(\mathcal{M}, x)$ are identical, so they can be described simply as the fundamental group $\pi_1(\mathcal{M})$.

The simple example with circular topology S^1 (fig. 2.1) has already been discussed. Elements of $\pi_1(S^1)$ are characterized by a winding num-

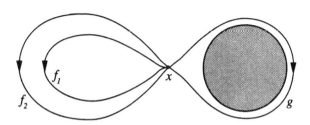

Fig. 3.7. Closed paths in a manifold $\mathcal{M}$. Only the loops f_1 and f_2 are homotopic to each other.

---

* An isomorphism is a one-to-one homomorphism – a mapping which preserves the group structure.

ber n, and the product of two elements n and m yields a winding number $n + m$. Clearly, then, the fundamental group is isomorphic to the group of integers

$$\pi_1(S^1) = Z. \tag{3.3.4}$$

Strings and elements of π_1

It is natural to consider classes of loops which are homotopically equivalent when characterizing string solutions. We can illustrate this again with the $\mathcal{M} \cong S^1$ example. The presence of a string can be detected by encircling it with a closed path $\gamma(t)$ as in fig. 3.3. The field values ϕ at points along the loop $\gamma(t)$ take values in $\mathcal{M}$, and so they map points from physical space R^3 into the vacuum manifold $\mathcal{M}$. Hence, ϕ completes a mapping from S^1 onto a path $g(t)$ in $\mathcal{M}$, $g(t) = \phi(\gamma(t))$. If the path $g(t)$ possesses a non-trivial winding n in $\mathcal{M}$, then a string must be present in physical space, but this is exactly the criterion for $g(t)$ to belong to the homotopy class n. This correspondence is one-to-one since each type of string can be identified with a unique element of $\pi_1(S^1) \cong Z$. One might suppose, therefore, that string solutions in general can be classified by the elements of $\pi_1(\mathcal{M})$. This is only true, however, if $\pi_1(\mathcal{M})$ is abelian, as the next example illustrates.

The figure of eight space with two regions excised, as in fig. 3.8, has a fundamental group which is non-abelian. Here, we can see the significance of defining π_1 using a base point x. The two loops f and g shown are not homotopic at x, since they cannot be deformed smoothly into each other while keeping a point fixed at x. Consequently, they fall into different homotopy classes in the fundamental group, even though they clearly delineate the same topological defect – a string resulting from a winding about the second void in $\mathcal{M}$.

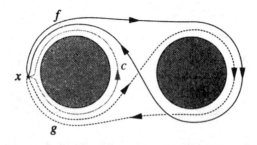

Fig. 3.8. The figure of eight space which gives rise to a non-abelian fundamental group. The two loops f and g while not homotopic at x, are freely homotopic and characterize the same defect. The loop c can be used to show that f and g are conjugate, $cfc^{-1} \sim g$.

To determine whether two loops in $\mathcal{M}$ identify topologically equivalent defects, then, is to ascertain whether *any* continuous deformation can be constructed between them, irrespective of the base point restriction in (3.3.1). Given the existence of such a deformation, the loops are termed *freely homotopic*. The classes of freely homotopic loops correspond to a subdivision of the elements of $\pi_1(\mathcal{M})$ into *conjugacy classes*; two loops are said to be conjugate if there exists a loop c such that

$$cfc^{-1} \sim g, \tag{3.3.5}$$

where $\sim$ here signifies homotopic at x. The freely homotopic loops f and g in fig. 3.8 can be seen to be conjugate because the illustrated loop c can be employed in (3.3.5) to 'release' f from x and take it around the opposite side of the topological obstruction, that is, the first void. The addition of the loop c as in (3.3.5) corresponds to an automorphism on $\pi_1(\mathcal{M})$, because it is an isomorphism of the group onto itself. Clearly, such automorphisms will only be distinct if the loops c lie in different homotopy classes, so automorphisms will, in turn, be classified by the group $\pi_1(\mathcal{M})$. Thus, the different loop addition operations (3.3.5) can be described as π_1 acting on π_1.

More precisely, then, we conclude that strings are classified by the conjugacy or *loop automorphism classes* of the fundamental group $\pi_1(\mathcal{M})$. If π_1 is abelian these classes are just the elements of π_1, however, in the non-abelian case this corresponds to a tangible subdivision.

String crossings

This classification is important when combining defects and determining the topologically allowed outcomes. If we were attempting to draw one string α across another β as illustrated in fig. 3.9(a), the final outcome is determined by the element associated with the loop γ. Through careful distortion (keeping a base point fixed) it can be shown that γ corresponds to the element $\alpha\beta\alpha^{-1}\beta^{-1}$. For an abelian $\pi_1(\mathcal{M})$, we can commute elements and this product becomes equivalent to the identity e. From this perspective, such strings have no topological constraint preventing them from passing freely through each other. Alternatively in this abelian case, it is also topologically possible for the product element $(\alpha\beta)^{-1}$ to meet at a vertex from which both α and β also emanate. If $\alpha = \beta^{-1}$, this can result in complete reconnection or the 'exchange of partners' – sometimes termed intercommutation. While the topologically acceptable outcomes can be identified by these arguments, distinguishing between them becomes a dynamical question.

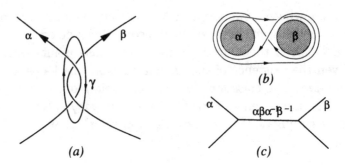

Fig. 3.9. String crossing: (a) The loop γ encloses the element $\alpha\beta\alpha^{-1}\beta^{-1}$ which determines the topologically allowed outcomes. (b) A loop γ corresponding to the element $\alpha\beta\alpha^{-1}\beta^{-1}$ in the figure of eight space, where α and β correspond to a single winding about the distinct voids. (c) In this non-abelian case, the attempt to pass α through β results in entanglement by a third string.

In the non-abelian case, the element $\alpha\beta\alpha^{-1}\beta^{-1}$ is not necessarily trivial and this has important consequences for the topologically allowed inter-actions of such strings. For example, in the figure of eight space (fig. 3.8), with α and β corresponding to windings about the different voids, one obtains the non-trivial 'cat's cradle' of fig. 3.9(b). The interaction of these two strings must leave a third string $\alpha\beta\alpha^{-1}\beta^{-1}$ stretching between them (fig. 3.9(c)).

3.3.2 Lie groups and the fundamental theorem

The actual computation of the fundamental group is greatly simplified if the vacuum manifold $\mathcal{M}$ can be expressed in terms of compact Lie groups. This is precisely the situation we have already discussed in chapter 2 where a compact Lie group G is spontaneously broken to a smaller subgroup H. With ϕ_0, a vacuum expectation value chosen in $\mathcal{M}$, we can generate the remainder of $\mathcal{M}$ by transformations of the form $\phi = D(g)\phi_0$ with $g \in G$ and $g \notin H$, since $\phi_0 = D(h)\phi_0$ for $h \in H$ (refer to §2.1.2). Consequently, we can identify any ϕ in $\mathcal{M}$ with the coset gH, that is, the vacuum manifold $\mathcal{M}$ is equivalent to the space of cosets of H in G, $\mathcal{M} = G/H$.

Covering groups

Before presenting a simple rule for finding the fundamental group of such a vacuum manifold we introduce an important restriction. We consider only connected, simply-connected groups G. It is possible to embed any compact Lie group G in a larger group $\tilde{G}$ which is simply-connected, $\pi_1(\tilde{G}) \cong I$, known as the *universal covering group* of G. The map from

$\tilde{G} \to G$ is group-preserving and onto. The most familiar example is the three-dimensional group of proper rotations $SO(3)$ which is not simply connected but which has the universal covering group $SU(2)$. We can visualize this double covering as a map of the two copies of $SO(3)$ in $SU(2)$ onto $SO(3)$ (refer to §2.1.3). Given that we have enlarged G to $\tilde{G}$, the unbroken subgroup H must also be appropriately enlarged to $\tilde{H}$ by including any additional elements in $\tilde{G}$ that leave the vacuum expectation value ϕ_0 invariant. These changes factor out of the vacuum G/H because such redundancies cancel.

Unlike G, the subgroup H can be disconnected (see fig. 3.10). The component H_0 which is connected to the identity e is a normal subgroup of H, that is, $hh_0h^{-1} \in H_0$ for any $h \in H$. The disjoint connected components of H are then just the cosets of the subgroup H_0 and, since H_0 is normal, these cosets H/H_0 form a group called the *quotient group*. This is often denoted as $\pi_0(H)$ because it corresponds to the equivalence classes of maps from a point S^0 into H, in an analogous manner to the maps from S^1 defining $\pi_1(H)$. We now have the necessary definitions to state the fundamental theorem.

Theorem

Let G be a connected and simply-connected Lie group, having a subgroup H with a component H_0 connected to the identity. The quotient group $\pi_0(H) \equiv H/H_0$, labelling the disconnected components of H, is then isomorphic to the fundamental group of the coset space $\pi_1(G/H)$, that is,

$$\pi_1(G/H) \cong \pi_0(H). \tag{3.3.6}$$

The existence of string solutions reduces, therefore, from the complicated procedure of characterizing the topology of the vacuum manifold $\mathcal{M}$, to a simple question of whether the unbroken symmetry group H is connected or not. Since applications of the theorem are aided by an understanding of its origin, we briefly sketch the steps involved in its proof. A rigorous proof can be found in a standard text such as Hilton [1953].

We suppose that the group H has several disconnected components in G as shown in fig. 3.10, with H_0 containing the identity element e. Taking the base point to be the element eH in the coset space G/H, a loop $f(t)$ in G/H can be represented as a path $g(t)$ in G by ensuring that

$$f(t) = g(t)H. \tag{3.3.7}$$

We can restrict the set of paths $g(t)$, which must begin and end in H, by always taking the identity e to be the initial point in H_0, that is, $g(0) = e$. This construction, then, provides a correspondence between closed loops

$f(t)$ in G/H and open paths $g(t)$ in G from the identity e to some other element in H.

There are two distinct alternatives for the location of the endpoint $g(1)$ of these paths: In the first case, if the path $g_0(t)$ ends in the identity component H_0 as in fig. 3.10, it can be continuously deformed within H_0 by shifting the endpoint until $g_0(1) = e$. The corresponding loop $f_0(t)$ in G/H from (3.3.7) will be unaffected by this procedure. However, since G is simply-connected, the newly closed path $g_0(t)$ is homotopic to the constant map. Hence, the loop $f_0(t)$ must also be contractible to a point. Alternatively in the second case, the path $g_1(t)$ ends in a component of H other than H_0, say H_1 (fig. 3.10). This path cannot be continuously deformed into a closed loop without moving the endpoint $g_1(1)$ out of the subgroup H, and thus breaking the correspondence (3.3.7) with the associated coset space loop $f_1(t)$. Clearly, then, loops such as $f_1(t)$ must be non-contractible. It is not difficult, therefore, to establish that the homotopy classes of loops $[f]$ in G/H – the elements of $\pi_1(G/H)$ – correspond to the disconnected components of H, labelled by the elements of $\pi_0(H)$.

Some applications

The significance of this theorem for defect classification is that attention can be focussed solely upon the remaining unbroken symmetry group H, provided we have appropriately chosen a simply-connected covering group G in the first place. The nature of the subgroup H, of course, depends on details of the symmetry breaking process – the representation of the group G and the choice of the direction of the expectation value ϕ_0. The

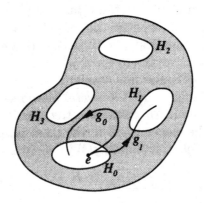

Fig. 3.10. The subgroup H of the group G may have several disconnected components H_i. We take H_0 to be that component connected to the identity.

relatively simple cases of symmetry breaking discussed in chapter 2 are enlightening in this regard:

(i) $U(1) \longrightarrow I$: In §2.1.1 we discussed the breaking of both global and local $U(1)$ gauge symmetries which we already know possess vortex solutions. The need to find a covering group $\tilde{G}$ for the multiply-connected $U(1)$ explains the apparent failure of a naive application of (3.3.6) to reveal strings. The group of one-dimensional translations (the real line R) suffices for this purpose, since we can have a one-to-one map from every interval $[2\pi(n-1), 2\pi n)$ onto $U(1)$, that is, $U(1) \cong R/Z$. The unbroken symmetry group $\tilde{H}$ will then be the set of translations by $2\pi n$, that is, the integers Z. From this we recover the expected result $\pi_1(U(1)) \cong \pi_0(Z) \cong Z$.

(ii) $SU(2) \xrightarrow{3} U(1)$: The first stage of symmetry breaking in *Example 2* of §2.1.3 has a triplet scalar field φ transforming in the adjoint representation of $SU(2)$ (or $SO(3)$), leaving an unbroken $U(1)$ subgroup. Since $SU(2)$ is simply-connected, we can immediately infer from the fact that $U(1)$ is connected $(\pi_0(U(1)) = I)$ that this symmetry breaking does not produce strings. This could have been observed directly since the vacuum is isomorphic to a two-sphere S^2 on which loops are always contractible to a point. We merely remove one point on S^2, not covered by the loop, and the surface becomes topologically equivalent to a disc on which we can contract the loop. A similar argument applies for *Example 1* in §2.1.3 which also produces no strings.

(iii) $SU(2) \xrightarrow{3+3} Z_2$: The second stage of the previous symmetry breaking was achieved with an additional triplet scalar field ψ. Since $H \cong Z_2$ is disconnected, string solutions are expected. There are only two topologically distinct ground states, either with or without a Z_2-string. While it may appear that the string and antistring are identical, they are actually related by a singular transformation that costs finite energy (refer to §4.2.2). The realistic GUT symmetry breaking $Spin(10) \xrightarrow{126} SU(5) \times Z_2$ in §2.1.4, also leaves a discrete Z_2 symmetry unbroken and can produce strings.

(iv) $SU(2) \xrightarrow{5} U(1) \otimes Z_2$: Finally, we consider the peculiar symmetry breaking in *Example 3* of §2.1.3 for a traceless symmetric 3×3 tensor field Φ. In this case we are left with an unbroken $U(1)$ associated with rotations about the z-axis (say) and an additional invariance generated by $R_0 = \mathrm{diag}(1, -1, -1)$, a 180° rotation about the x-axis. The little group H includes this discrete symmetry and so strings are again classified by

$$\pi_0 (U(1) \otimes Z_2) \cong \pi_0 (U(1)) + \pi_0(Z_2) \cong Z_2 \,. \tag{3.3.8}$$

These Z_2-strings, however, are very different in character to those in (iii) as we shall see in the next section. R_0 acts as a charge conjugation

operator, so if a charged particle travels in a circle about a string it will
return with opposite charge – hence the name, Alice string.

3.3.3 Higher homotopy groups and exact sequences

The second homotopy group

The second homotopy group $\pi_2(\mathcal{M})$ is the set of homotopically equivalent classes of maps from the two-sphere S^2 into the manifold $\mathcal{M}$. Group
structure is imposed, analogously to the fundamental group $\pi_1(\mathcal{M})$, by
considering continuous deformations of two-surfaces that keep a base
point x fixed. Two-surfaces that can be continuously shrunk to a point
are homotopic to the trivial constant map. An isomorphism can be defined between the second homotopy group $\pi_2(\mathcal{M}, x)$ (at x) and that based
anywhere else in $\mathcal{M}$, so we usually only consider $\pi_2(\mathcal{M})$. The second homotopy group is always abelian as can be easily demonstrated by the
pictorial argument in fig. 3.11.

The determination of the second homotopy group is again greatly simplified if the vacuum manifold can be described in terms of a coset space
of continuous Lie groups, G/H. Given that we take a connected and
simply-connected covering group G, then the *second fundamental theorem* states that

$$\pi_2(G/H) \cong \pi_1(H_0), \qquad (3.3.9)$$

where H_0 is the component of the unbroken subgroup H connected to the
identity e. The determination of $\pi_2(G/H)$, therefore, reduces to the com-

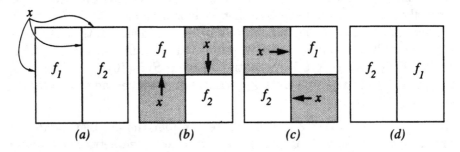

| (a) | (b) | (c) | (d) |

Fig. 3.11.   We can diagramatically represent (based) mappings, $f : S^2 \longrightarrow \mathcal{M}$,
by flattening the two-sphere into the interior of a square $0 \leq s, t \leq 1$, the edges
of which are all identified with the north pole and mapped to the base point x.
Multiplication of two-surfaces f_1, f_2 using the first coordinate s is illustrated in
(a). By appropriately shrinking in the second coordinate direction as in (b), such
that the shaded regions are also identified with x, it is possible using the steps
shown to smoothly deform the product $f_1 f_2$ into $f_2 f_1$. Thus $\pi_2(\mathcal{M})$ is always
abelian.

putation of $\pi_1(H_0)$. Since monopoles are associated with unshrinkable surfaces in G/H, (3.3.9) implies their existence if the unbroken subgroup H is multiply-connected. Note that this theorem can be stated in a more general form if we allow G to be disconnected, that is,

$$\pi_2(G/H) \cong \pi_1(H_0)/\pi_1(G_0), \qquad (3.3.10)$$

where G_0 is the component of G connected to the identity e.

By analogy with the previous arguments for strings, we must consider whether monopole classification by $\pi_2(\mathcal{M})$ is complete [Mermin, 1979]. Topological classes of monopoles can be identified with classes of freely homotopic two-surfaces in $\mathcal{M}$. These classes may not be in one-to-one correspondence with the elements of $\pi_2(\mathcal{M})$ if the manifold $\mathcal{M}$ is not simply-connected; this ambiguity is again due to the base point restriction in the definition of $\pi_n(\mathcal{M})$. An example is illustrated in fig. 3.12 for a manifold equivalent to R^3 with a sphere and infinite cylinder excised. The surfaces S and S' enclose the sphere and delineate the same monopole solution. However, it is not possible to continuously deform one into the other because of the topological obstruction, so they will correspond to different elements in π_2. As before, the addition of a loop c shown in fig. 3.12(a) can take S around the cylinder to x, making it homotopic to S'. (Note that the line in this hybrid surface can be thought of as a degenerate limit of a thin tube.)

This equivalence of two-surfaces up to the addition of a closed loop further subdivides $\pi_2(\mathcal{M})$ into loop automorphism classes, described by the action of $\pi_1(\mathcal{M})$ as a group of automorphisms on $\pi_2(\mathcal{M})$. Using the second fundamental theorem (3.3.9), this reduces to the action on $\pi_1(H_0)$ of the inner automorphisms of H, since $\pi_1(G/H) \cong \pi_0(H)$. A loop $h_0(t)$

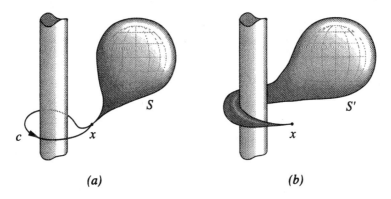

(a) (b)

Fig. 3.12. Two surfaces homotopically equivalent up to the addition of a closed loop belong to the same loop automorphism class.

in H_0 is simply taken to another loop in H_0 by $h_0'(t) = h^{-1}h_0(t)h$, $h \in H_i$. If H is abelian these automorphisms will be trivial, however, if $\pi_0(H)$ acts non-trivially on $\pi_1(H_0)$ then monopoles will be classified by enlarged automorphism classes.

Clearly, these monopole classification ambiguities will only emerge in theories in which the vacuum manifold G/H is multiply-connected, that is, in models in which there are both strings and monopoles. The significance of this result is again seen when combining defect solutions because the multiplication of elements in separate automorphism classes, in general, gives rise to a set formed from a union of automorphism classes. For example, the result of combining two defects lying in the two classes $\mathcal{A}$ and $\mathcal{B}$ will not be unique since different choices of elements from $\pi_2(\mathcal{M})$, say $a, a' \in \mathcal{A}$, may give rise to members of different automorphism classes, $ab \in \mathcal{C}$ or $a'b \in \mathcal{C}'$, and thus different point defects. This phenomenon is intimately connected with the path by which two such monopoles are brought together relative to any strings which are present, as the second example below illustrates.

Further examples

(i) $SU(2) \xrightarrow{3} U(1)$: Returning to the $SU(2)$ symmetry breaking in the triplet representation, we recall that a $U(1)$ subgroup is left unbroken. Since $SU(2)$ is simply-connected we have the isomorphism

$$\pi_2(SU(2)/U(1)) \cong \pi_1(U(1)) \cong Z, \qquad (3.3.11)$$

implying that monopole solutions will appear in this scheme and are classified by the integers Z. One example is the classical monopole solution of §3.2.3. It is now clear why grand unified theories must confront a monopole problem. If a single group G is to encompass all interactions, the multiply-connected $U(1)_{em}$ of electromagnetism must appear at some stage during symmetry breaking and, with it, the associated monopoles.

(ii) $SU(2) \xrightarrow{5} U(1) \otimes Z_2$: Again we reconsider an example from §3.3.2(iv) with the tensor field Φ. Applying the theorem (3.3.9) yields monopoles classified by the integers

$$\pi_2(SU(2)/U(1) \otimes Z_2) \cong \pi_1(U(1)) \cong Z, \qquad (3.3.12)$$

since the component connected to the identity is $H_0 \cong U(1)$. As we have seen already, however, strings also appear in this symmetry breaking. This may complicate the classification scheme because of non-trivial automorphisms arising from the action of $\pi_0(H)$ on $\pi_1(H_0)$. Here, these will be generated by the charge conjugation operator $R_0 = \text{diag}(1, -1, -1)$. Under the action of R_0, elements of H_0, $\exp(-iL_z\theta)$, simply become $\exp(iL_z\theta)$. That is, loops in H_0 under the action of this inner automorphism will be taken to loops with winding number of opposite sign.

Thus monopoles characterized by opposite windings are not topologically distinct and, since $-n$ and n belong to the same automorphism class, we can simply use positive integers to classify them. As mentioned earlier, an ambiguity remains as to the final result of combining monopoles; if we are to bring two monopoles m and n together, we can obtain either the $m + n$ or $|m - n|$ element. The two alternatives depend on the path taken relative to the Z_2-strings that arise in this model. A single winding around a string turns a monopole into an antimonopole, thus enabling the complete removal of all monopole configurations with even winding number, though at the expense of transferring their magnetic charge to the string [Preskill, 1985].

Exact sequences

The homotopy groups $\pi_1(\mathcal{M})$ and $\pi_2(\mathcal{M})$ are merely two in a series of general n^{th} homotopy groups, arising from homotopically equivalent classes of mappings from the n-sphere into the manifold $\mathcal{M}$. All homotopy groups for $n \geq 2$ are abelian, and again we can define an isomorphism between groups based at different points in $\mathcal{M}$ and restrict attention to just $\pi_n(\mathcal{M})$. Defects of the appropriate dimensionality are in one-to-one correspondence with the loop automorphism classes defined again by the action of $\pi_1(\mathcal{M})$ on $\pi_n(\mathcal{M})$.

For general n, the characterization of the manifold by Lie groups again provides an important simplification. The two fundamental theorems can be shown to be special cases of a general series of homomorphisms between homotopy groups, known as the *exact homotopy sequence*. At each stage of the scheme, exactness requires that the image of one homomorphism be the kernel of the next homomorphism:

$$\ldots \to \quad \pi_n(G) \quad \to \quad \pi_n(G/H) \quad \to \quad \pi_{n-1}(H) \quad \to \quad \pi_{n-1}(G) \quad \to \ldots$$

$$\ldots \to \quad \pi_1(G) \quad \to \quad \pi_1(G/H) \quad \to \quad \pi_0(H).$$

$$(3.3.13)$$

The two previous theorems follow from the observation that if, in the exact sequence $\mathcal{G}_1 \longrightarrow \mathcal{G}_2 \overset{\gamma}{\longrightarrow} \mathcal{G}_3 \longrightarrow \mathcal{G}_4$, the outer groups $\mathcal{G}_1$ and $\mathcal{G}_4$ are the trivial group, then the homomorphism γ is an isomorphism. One example is clear since the kernel of the homomorphism $\mathcal{G}_3 \longrightarrow \mathcal{G}_4$ must include all of $\mathcal{G}_3$, and so the mapping γ must be onto. On the other hand, the homomorphism from the identity $\mathcal{G}_1 \longrightarrow \mathcal{G}_2$ implies that the kernel of γ consists only of the identity. Hence, γ must be one-to-one, so we have the general result,

$$\pi_n(G/H) \cong \pi_{n-1}(H) \quad \text{if} \quad \pi_n(G) \cong \pi_{n-1}(G) \cong I. \quad\quad (3.3.14)$$

Finally, we note that the homotopy theory of compact Lie groups is well-understood. The following results may be helpful in determining the appropriate group topology when classifying defects. The group G, here, represents one of the compact connected Lie groups: $SO(n)$ $(n \geq 2)$, $Spin(n)$ $(n \geq 3)$, $U(n)$ $(n \geq 1)$, $SU(n)$ $(n \geq 2)$, $Sp(n)$ $(n \geq 1)$, G_2, F_4, E_6, E_7 and E_8.

The fundamental group $\pi_1(G)$:

$$\pi_1(G) \cong \begin{cases} Z & G = U(n) \ (n \geq 1), \quad SO(2) \\ Z_2 & G = SO(n) \ (n \geq 3) \\ I & \text{all other groups } G. \end{cases}$$

Higher homotopy groups $\pi_n(G)$:

$$\pi_2(G) \cong I$$

$$\pi_3(G) \cong \begin{cases} I & G = U(1), \quad SO(2) \\ Z \oplus Z & G = SO(4) \\ Z & \text{all other groups } G. \end{cases}$$

Many other useful relations can be found in standard texts such as Hilton [1953] and Maunder [1980] or, more succinctly, in a dictionary of mathematics [Iyanaga & Kawada, 1977]. A fuller exposition of this discussion of the topological classification of defects can be found in Mermin [1979].

3.3.4 Energetic considerations

Defect classification based on purely topological arguments is usually incomplete. In general, one can construct several possible defect configurations corresponding to each distinct element of the classifying homotopy group $\pi_n(\mathcal{M})$. The actual configuration to be realized in a given model is then determined by energetic considerations.

Taking the example of strings, the winding of the scalar field ϕ around a circle, parametrized by the angle θ and centred on the string, can be represented as

$$\phi(\theta) = U(\theta)\phi(0) = \exp(iT^\alpha\theta)\phi(0)\,, \tag{3.3.15}$$

where T^α is a broken generator of the appropriate representation of G, $D(G)$. Since ϕ is single-valued, a complete winding must satisfy

$$U(2\pi) = D(h)\,, \quad h \in H\,, \tag{3.3.16}$$

where, for non-trivial windings, h must lie in a disconnected component of H (see fig. 3.10). For each disconnected component, there will in general be several string solutions corresponding to rotations by different broken generators T^α.

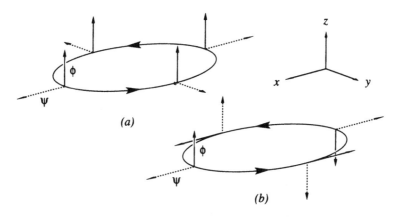

Fig. 3.13.   There are two distinct ansatze for the Z_2-string formed at an $SO(3)$ symmetry breaking with two triplet fields φ and ψ: (a) The lowest energy ansatz generated by T_3 which leaves φ invariant and (b) the ansatz generated by T_2 which rotates both fields as we encircle the string.

A simple example is provided by the Z_2-string mentioned in §3.3.2. According to the $SU(2)$ symmetry breaking scheme of §2.1.3 *Example 2*, two triplet fields φ and ψ acquire vacuum expectation values in orthogonal directions. Supposing we take the directions at $\theta = 0$ to be $\varphi = \eta_\phi \hat{z}$ and $\psi = \eta_\psi \hat{x}$, then we can consider two distinct ansatze for the string winding (3.3.15). First, we can generate transformations with T_3, leaving φ invariant but causing ψ to rotate as shown in fig. 3.13(a). Alternatively, we could rotate both vectors by employing T_2 as shown in fig. 3.13(b). In this case, the first ansatz clearly will be the lowest energy configuration (here we have $\eta_\phi > \eta_\psi$), though, in general this must be demonstrated explicitly. A more complex example of such dynamical arguments in the $SO(10)$ model (2.1.67) is discussed in detail by Aryal & Everett [1987].

4

String field theory

4.1 The abelian-Higgs model

We begin discussion of string structure and properties with the Nielsen–Olesen vortex-line of the abelian-Higgs model, a model studied in some detail in §2.1.1 and §3.2.2. With a charged complex scalar field ϕ and gauge fields A^μ, the Lagrangian density is

$$\mathcal{L} = (\partial_\mu + ieA_\mu)\bar{\phi}(\partial^\mu - ieA^\mu)\phi - \tfrac{1}{4}F_{\mu\nu}F^{\mu\nu} - \tfrac{1}{4}\lambda(|\phi|^2 - \eta^2)^2. \quad (4.1.1)$$

The energy of a static vortex configuration in two dimensions is comprised of the following three non-negative terms,

$$E = \int d^2r \left\{ |(\nabla - ieA)\phi|^2 + \tfrac{1}{2}(\mathbf{E}^2 + \mathbf{B}^2) + V(|\phi|) \right\}. \quad (4.1.2)$$

The requirement that the vortex energy be finite imposes some conditions on the asymptotic field configuration. Clearly, far from the core the Higgs field $|\phi|$ must approach its vacuum expectation value η to minimize the potential term. The covariant derivative also must vanish at infinity, implying that fields take the form of a 'pure' gauge rotation,

$$A_\theta \to \frac{1}{er}\frac{d\vartheta}{d\theta} + \dots, \quad r \to \infty, \quad (4.1.3)$$

where ϑ is the phase of ϕ. This 'Higgs screening' by the gauge fields prevents the logarithmic divergence of the global string (3.2.15). In traversing a closed path about a vortex, the field ϕ must return to its original value $\phi(2\pi) = \phi(0)$, so any winding in the phase ϑ must be an integer,

$$n = \frac{1}{2\pi} \int_0^{2\pi} \frac{d\vartheta}{d\theta} d\theta = \frac{e}{2\pi} \oint \mathbf{A} \cdot \mathbf{dl}. \quad (4.1.4)$$

A non-trivial winding $n \neq 0$ can only be resolved continuously if ϕ has a zero somewhere in the vortex core, $\phi(x) = 0$ (the location, x, of this zero is gauge-invariant). Finally, we recall that Stokes' theorem implies that a vortex configuration (4.1.4) has a net magnetic flux which is 'quantized', $\Phi_B = 2\pi n/e$.

We can estimate the properties of a vortex – its size and mass – without recourse to detailed calculations. The two characteristic scales of the vortex are determined by the regions over which the scalar and gauge fields depart from their asymptotic values. One might correctly suppose that these distance scales, r_s and r_v, are related to the Higgs and vector particle masses off the string, and a crude variational argument can be used to justify this [Preskill, 1985]. To order of magnitude, we can rewrite the three terms in the energy (4.1.2) using these length scales,

$$E \approx 2\pi\eta^2 \left[\ln\left(\frac{r_v}{r_s}\right) + \frac{1}{e^2\eta^2 r_v^2} + \lambda\eta^2 r_s^2 \right] . \qquad (4.1.5)$$

Here, we assume $r_v > r_s$ which corresponds to a Type II superconductor regime. The first term represents the gradient energy of the scalar field, the second is a consequence of magnetic pressure, since the flux does not wish to be confined, and the last term is the potential energy cost of the departure of $|\phi|$ from the minimum at $|\phi| = \eta$. The energy expression (4.1.5) is minimized by taking

$$
\begin{aligned}
r_s &\approx m_s^{-1} = (\sqrt{\lambda}\,\eta)^{-1} , \\
r_v &\approx m_v^{-1} = (\sqrt{2}\,e\eta)^{-1} .
\end{aligned}
\qquad (4.1.6)
$$

These two string core widths corrrespond to the Compton wavelengths of the scalar and vector particles respectively. With these values, the mass of the vortex configuration becomes approximately

$$\mu \approx 2\pi\eta^2 \ln\left(\frac{m_s}{m_v}\right) , \qquad (4.1.7)$$

which in three dimensions is the string mass per unit length.

4.1.1 Detailed vortex structure

A more detailed picture of vortex structure, particularly the asymptotic behaviour, is necessary if we are to consider the nature of inter-vortex interactions and in order to derive an effective string action. To this end, we consider the Euler–Lagrange equations arising from (4.1.1),

$$(\partial_\mu - ieA_\mu)(\partial^\mu - ieA^\mu)\phi + \frac{\lambda}{2}\phi(\bar{\phi}\phi - \eta^2) = 0 , \qquad (4.1.8)$$

$$\partial_\mu F^{\mu\nu} = j^\nu , \qquad (4.1.9)$$

where

$$j^\nu \equiv 2e\,\mathrm{Im}\left[\bar{\phi}(\partial^\nu - ieA^\nu)\phi\right] . \qquad (4.1.10)$$

Throughout most of this discussion we adopt the Lorentz gauge, $\partial_\mu A^\mu = 0$, along with the convenient rescaling $\phi \to \eta^{-1}\phi$, $A^\mu \to \eta^{-1}A^\mu$ and $x \to \eta x$, making these quantities dimensionless and allowing us to set $\eta = 1$ in (4.1.1). The further rescaling $x \to ex$ would reveal that the only significant parameter in the model is the ratio of the Higgs and vector masses, $\beta \equiv m_s^2/m_v^2 = \lambda/2e^2$.

To demonstrate the existence of vortex-strings we seek static cylindrically-symmetric solutions $(\phi_s, \mathbf{A}_s)$ of the form [Abrikosov, 1957; Nielsen & Olesen, 1973]

$$\phi_s(\mathbf{r}) = e^{in\theta}f(r)\,, \tag{4.1.11}$$

$$A_{s\,a}(\mathbf{r}) = -\epsilon_{ab}x_b\frac{n}{er^2}\alpha(r)\,, \quad a,b = 1,2\,, \tag{4.1.12}$$

that is, $A_{s\,\theta} = -n\alpha(r)/er$, where n is the vortex winding number. This ansatz for a straight string lying along the z-axis is dependent only on the polar coordinates (r,θ) with $r^2 = x^2 + y^2$. Appropriate asymptotic boundary conditions for a vortex solution of finite energy must take the form

$$\begin{aligned} f(r) &\to 1\,, \quad r \to \infty\,, \\ \alpha(r) &\to 1\,, \quad r \to \infty\,, \end{aligned} \tag{4.1.13}$$

with the Higgs field rapidly approaching the true vacuum. Conversely, at the vortex centre, ϕ must smoothly attain the symmetric state,

$$f(0) = \alpha(0) = 0\,, \tag{4.1.14}$$

thus ensuring regularity at the origin.

With the substitutions (4.1.11–12) the static field equations take the form

$$\frac{d^2f}{dr^2} + \frac{1}{r}\frac{df}{dr} - \frac{n^2f}{r^2}(\alpha - 1)^2 - \frac{\lambda}{2}f(f^2 - 1) = 0 \tag{4.1.15}$$

$$\frac{d^2\alpha}{dr^2} - \frac{1}{r}\frac{d\alpha}{dr} - 2e^2f^2(\alpha - 1) = 0\,. \tag{4.1.16}$$

Explicit solutions to these equations are not known, but it is straightforward to obtain approximate asymptotic solutions for large r. If we first treat $|\phi|$ as a constant, then (4.1.16) reduces to that for a modified Bessel function with solution

$$\alpha(r) \approx 1 - rK_1\left(\sqrt{2}\,er\right) \approx 1 - \mathcal{O}\left(\sqrt{r}\exp\left(-\sqrt{2}\,er\right)\right)\,. \tag{4.1.17}$$

The asymptotic form for ϕ depends on the parameter $\beta = \lambda/2e^2$ [Perivolaropoulos, 1993]. For $\beta > 4$, the large-distance fall-off of f is controlled by the gauge field contribution $\propto (1-\alpha)^2$. For $\beta \lesssim 4$, the gauge fields can

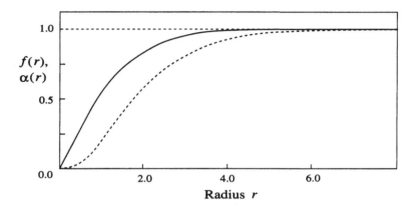

Fig. 4.1. Functions $f(r)$ (solid) and $\alpha(r)$ (dashed) for the cylindrically symmetric Nielsen–Olesen vortex solution at critical coupling, $\beta \equiv \lambda/2e^2 = 1$.

be ignored as $r \to \infty$ and the linearized version of (4.1.15) yields

$$f(r) \approx 1 - K_0\left(\sqrt{\lambda}\, r\right) \approx 1 - \mathcal{O}\left(\exp\left(-\sqrt{\lambda}\, r\right)\right). \qquad (4.1.18)$$

Close to the vortex core, the leading order dependence for the Higgs field is $f(r) \sim r^n$, that is, we have an nth-order zero of ϕ.

For intermediate and small values of r, however, the solutions must be obtained numerically; this is achieved most efficiently by solving the boundary value problem with appropriate relaxation techniques. The solutions for the functions $f(r)$ and $\alpha(r)$ for an $n = 1$ vortex are illustrated

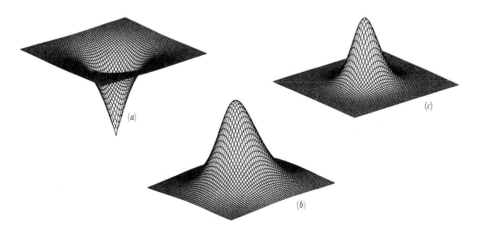

Fig. 4.2. Nielsen–Olesen vortex-line cross-section: (a) Higgs scalar $|\phi|$, (b) Magnetic field D_z, (c) Energy density.

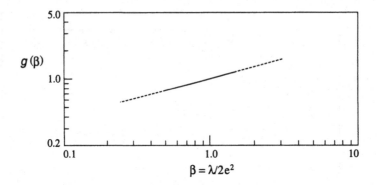

Fig. 4.3. Dependence of the string energy per unit length on the Higgs and vector particle masses, $g(\beta) = \mu/2\pi\eta^2$, where $\beta = (m_s/m_v)^2 = \lambda/2e^2$.

in fig. 4.1. Rotation of this solution about the z-axis allows us to visualize the string in cross-section as in fig. 4.2: The Higgs field, magnetic field and energy density are plotted to illustrate how well-localized the vortex solution is. Using such numerical techniques, the static vortex configuration energy can be found to high precision,

$$\mu = 2\pi\eta^2 g\left(\beta\right) , \qquad (4.1.19)$$

where we have reintroduced the symmetry breaking scale η and the function g is slowly varying and equal to unity when $\beta = \lambda/2e^2 = 1$ [Bogomol'nyi, 1976; Jacobs & Rebbi, 1979]. The weak dependence of the string energy on these parameters is illustrated in fig. 4.3. The more complex internal structure of strings with higher winding number is contrasted in fig. 4.4, showing the higher-order zero in the core with the energy density distributed away from the centre in a 'ring'.

4.1.2 Critical coupling

Significant advances in understanding the mathematical properties of vortices (and other defects in gauge models) have arisen by considering the

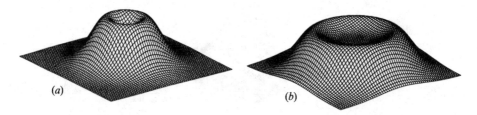

Fig. 4.4. Energy density for the cross-section of (*a*) an $n = 2$ vortex and (*b*) an $n = 4$ vortex.

critical coupling limit in which the Higgs and vector masses become identical, $\beta = 1$. In the theory of superconductivity, $\beta = 1$ corresponds to the interface between Type I and Type II behaviour, while in quantum field theory the model can be extended to become supersymmetric. The chief simplification is that the second-order field equations (4.1.8–9) reduce to a coupled first-order system which is more tractable.

By considering the vortex energy (4.1.2), Bogomol'nyi [1976] was able to show that it could be rewritten via an integration by parts as

$$E = \int d^2x \left\{ |[(\partial_x - iA_x) \pm i(\partial_y - iA_y)] \phi|^2 + \frac{1}{2} [B_z \pm (\phi\bar{\phi} - 1)]^2 \right\}$$
$$\pm \int d^2x \, B_z + (\beta - 1) \int d^2x \, (\phi\bar{\phi} - 1)^2,$$
(4.1.20)

where $B_z = F_{12} = \partial_x A_y - \partial_y A_x$ and we take the upper sign for $n \geq 0$ and the lower for $n < 0$. For the remainder of this discussion we will consider critical coupling $\beta = 1$, at which the last term vanishes.

The first integrand in (4.1.20) is positive definite and given the magnetic flux for a vortex solution (3.2.9), the second last term implies a lower bound for the energy

$$E \geq 2\pi |n|.$$
(4.1.21)

Clearly, this bound will be saturated if and only if the first integrand in (4.1.20) vanishes, implying the following first-order (Bogomol'nyi) equations:

$$[(\partial_x - iA_x) \pm i(\partial_y - iA_y)] \phi = 0,$$
(4.1.22)
$$B_z = \mp(\phi\bar{\phi} - 1).$$
(4.1.23)

At critical coupling, then, any solution to (4.1.22–23) is automatically a solution to the second-order field equations (4.1.8–9). The converse can also be established, though this is considerably more difficult (see, for example, Jaffe & Taubes [1980]).

Cylindrically-symmetric solutions to (4.1.22–23) have been studied by de Vega & Schaposnik [1976] who constructed a power series expansion for vortices with arbitrary winding number n. In this case, equations (4.1.22–23) reduce to

$$\frac{df}{dr} = \frac{nf}{r}(1 - \alpha),$$
(4.1.24)
$$\frac{1}{r}\frac{d\alpha}{dr} = \mp\frac{1}{n}\left[f^2 - 1\right]$$
(4.1.25)

For small r, the $n = 1$ vortex solution takes the form

$$\alpha(r) = -\frac{1}{n} \sum_{k=1}^{\infty} \beta_k \left(\frac{r}{\sqrt{2}} \right)^{2k} , \qquad (4.1.26)$$

with a recursion relation for the coefficients β_k given by

$$\beta_k = \frac{1}{k(k-2)} \sum_{l=2}^{k-1} l\beta_l \beta_{k-l} , \qquad \beta_1 = -1, \quad \beta_2 \approx 0.7279... \qquad (4.1.27)$$

The function f can be easily found using f^2 from (4.1.25). The asymptotic expansions for f and α at large distances from the core have identical exponential fall-off to first order

$$f(r) \approx 1 - Z_1 \left\{ \pi/2\sqrt{2}r \right\}^{1/2} e^{-\sqrt{2}r} + \mathcal{O}(e^{-2\sqrt{2}r})$$
$$\approx \alpha(r) + \mathcal{O}(e^{-2\sqrt{2}r}) \qquad (4.1.28)$$

with the coefficient $Z_1 \approx 1.7079...$, like β_2, being obtained numerically.

Several other results for static multi-vortex configurations are of interest for later discussions of vortex dynamics and interactions. A $2|n|$-parameter family of zero-energy fluctuation modes around the cylindrically symmetric n-vortex solution was found by E. Weinberg [1979]. These are solutions to a linearized version of (4.1.8–9) in the vortex background and can be written in the form

$$\delta\phi = n\, h(r,\theta)\, f(r)e^{in\theta} ,$$
$$\delta \mathbf{A} = \frac{n}{r} \left(b(r,\theta)\hat{\theta} + c(r,\theta)\hat{r} \right) , \qquad (4.1.29)$$

where we now choose the $A_0=0$ gauge following Ruback [1988]. Fourier expanding the scalar perturbations as

$$h(r,\theta) = \sum_{k=0}^{n} \omega_k h_k(r)e^{-ik\theta} , \qquad (4.1.30)$$

where ω_k is an arbitrary (small) complex constant and the real function $h_k(r)$ must satisfy

$$-\frac{1}{r}\frac{d}{dr}\left(r\frac{dh_k}{dr} \right) + \left(f^2 + \frac{k^2}{r^2} \right) h_k = 0 . \qquad (4.1.31)$$

The truncation of this series for $k \leq n$ is enforced by ensuring regularity at the origin, since only an nth-order pole in $h_k(r)$ is acceptable when compensating the nth-order zero of $f(r)$ in (4.1.29). The corresponding vector mode functions $b(r,\theta)$, $c(r,\theta)$ can be easily found in terms of the $h_k(r)$.

The two-vortex ($n = 2$) perturbations decomposed into orthogonal modes are the following: Fluctuations representing overall translation of the vortex configuration take the form

$$\delta\phi = -\epsilon\frac{df}{dr}e^{i\theta}, \quad \delta\phi' = i\epsilon\frac{df}{dr}e^{i\theta}, \tag{4.1.32}$$

where, here, $h_1(r) = f^{-1}(df/dr)$ and ϵ is a small constant. Fluctuations representing the splitting into two separated $n = 1$ vortices can be decomposed into the two modes

$$\delta\phi = 2h_2(r)f(r), \quad \delta\phi' = 2ih_2(r)f(r), \tag{4.1.33}$$

with $h_2(r)$ a solution of (4.1.31) with $k=2$.

Finally, we note Taubes [1980] result that the only smooth (C^∞) solutions to the Bogomol'nyi equations (4.1.22–23) are determined uniquely up to gauge equivalence by choosing $|n|$ points in space which are zeros of the Higgs field. If we represent these as points in the complex plane, $z = x + iy$, then locally about any zero z_i the solution takes the form

$$\phi(z) \approx g_i(z)(z - z_i)^{n_i}, \tag{4.1.34}$$

where n_i is the multiplicity of z_i in the set of zeros $\{z_i\}$ and $g_i(z)$ is a non-vanishing C^∞ function. These solutions represent stable static multi-vortex configurations, indicating that there are no forces acting between static vortices at critical coupling.

4.1.3 Global U(1)-strings

Another limit already encountered in the abelian-Higgs model is the Goldstone model with global string solutions (§3.2.2) for which the gauge fields are removed (the $e = 0$ limit). No Higgs mechanism operates at the breaking of this global symmetry and the massless Goldstone boson that remains introduces a long-range force. Without Higgs screening, the energy per unit length is logarithmically divergent because the angular gradient of the scalar field is never compensated away from the core. Moreover, the exponential approach of ϕ to its vacuum expectation value for the local string will be replaced by a power law decay mode.

The field equations for this Goldstone model with a complex scalar field ϕ are now simply

$$\partial_\mu\partial^\mu\phi + \frac{\lambda}{2}\phi(\phi\bar{\phi} - 1) = 0. \tag{4.1.35}$$

Again one can look for axisymmetric vortex solutions with the ansatz (4.1.11) and boundary conditions (4.1.13–14) chosen to minimize the energy in some bounded region. The equation for the cross-section then

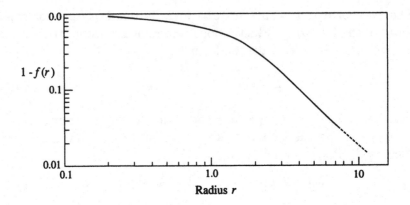

Fig. 4.5. Polynomial decay mode for the global vortex function $f(r)$.

becomes

$$\frac{d^2 f}{dr^2} + \frac{1}{r}\frac{df}{dr} - \frac{n^2}{r^2}f - \frac{\lambda}{2}f\left(f^2 - 1\right) = 0. \qquad (4.1.36)$$

As before, approximate asymptotic solutions can be obtained

$$\begin{aligned} f(r) &\approx c_n r^n + \ldots, & r &\to 0, \\ f(r) &\approx 1 - \mathcal{O}(r^{-2}), & r &\to \infty, \end{aligned} \qquad (4.1.37)$$

however, the large r expansion becomes asymptotically divergent. Numerical methods can be employed to find more accurate solutions and reveal a coefficient $c_1 = 0.58\ldots$ and the non-exponential decay shown in fig. 4.5 for the $n = 1$ global string – here we have set $\lambda = 2$. We should note that for $\beta > 1$ the interior of a local string exhibits global string behaviour because $\alpha \approx 0$ for $r \lesssim m_{\rm v}^{-1}$.

4.2 Non-abelian strings

4.2.1 Structure and zero modes

By the fundamental theorem (3.3.6), strings can arise at a symmetry breaking $G \to H$ for which the unbroken subgroup H is disconnected. For a non-abelian group G, a variety of broken generators can usually be employed to pass between the disconnected components of H. It is a dynamical question which is the actual broken generator T^S corresponding to the minimum energy string configuration. At large distances from an axisymmetric string, as we wind around by an angle θ, the Higgs field ϕ at $\theta = 0$ will rotate,

$$\phi(\theta) = U(\theta)\phi(0) = \exp(iT^S\theta)\phi(0). \qquad (4.2.1)$$

Because G is non-abelian, this rotation will generally affect the embedding of H in G, yielding a continuous family of different, but isomorphic subgroups $H(\theta)$. In the general case, then, the generators of the algebra rotate non-trivially and the local unbroken symmetry group H cannot be globally extended. The subgroup of globally well-defined symmetries, called the centralizer $\hat{H}$, can be defined as

$$\hat{H} = \{h \in H \mid [U(2\pi), D(h)] = 0\}\,, \tag{4.2.2}$$

where D is the appropriate Higgs representation of G and $H = H(0)$.

The expression (4.2.1) is only valid in a particular gauge. To find a more general form for $\phi(\theta)$, we note that outside the string core the gauge-covariant derivative of ϕ vanishes,

$$\mathcal{D}_\mu \phi = (\partial_\mu - ie A_\mu^a T^a)\phi = 0\,. \tag{4.2.3}$$

From the above, it follows that the group element $U(\theta)$ is given by the path-ordered exponential,

$$U(\theta) = P \exp\left(ie \int_0^\theta dx^\mu\, A_\mu^a T^a\right)\,. \tag{4.2.4}$$

A complete winding about a string must yield $U(2\pi) = D(h)$, where $h \in H$ lies in a component of H not connected to the identity. In the gauge (4.2.1), $A_\mu^a T^a = A_\mu^S T^S$, path-ordering is unnecessary and we can write

$$U(2\pi) = \exp\left(ie T^S \Phi^S\right)\,, \tag{4.2.5}$$

where Φ^S is the (non-abelian) magnetic flux associated with A_μ^S. Since $U(2\pi)$ is non-trivial, it is clear that $\Phi^S \neq 0$.

It is also possible to have scalar condensates in the interior of a non-abelian string. The Higgs field for the axisymmetric string solution, $\phi(r, \theta) = U(\theta)\phi(r, 0)$, does not necessarily have to vanish at $r = 0$. For regularity at the origin, we only have to require that $U(\theta)$ leaves $\phi(0,0)$ invariant, $U(\theta)\phi(0,0) = \phi(0,0)$. Hence, a more general ansatz for a global non-abelian string can be written as [Hindmarsh, 1991; see also Aryal & Everett, 1987; Everett, 1988; Alford *et al.*, 1991]

$$\phi(r, \theta) = U(\theta)\left[f(r)\phi(\infty, 0) + \sigma(r)\right]\,, \tag{4.2.6}$$

where $f(r) \to 1$ and $|\sigma(r)| \to 0$ as $r \to \infty$. Now consider transformations of the form

$$h(\theta) = g(\theta)\, h\, g(\theta)^{-1}\,, \tag{4.2.7}$$

where $U(\theta) = D(g(\theta))$ and $h \in H$. This leaves the solution $\phi(\infty, 0)$ at spatial infinity invariant, since $D(h)\phi(\infty, 0) = \phi(\infty, 0)$, but it may affect the σ-fields in the string interior. If $D(h)\sigma(r) = 0$ for all r, then the transformation acts trivially. However, it is possible for $D(h)$ to act non-trivially on σ, if the string generator remains invariant, $hL^Sh^{-1} = L^S$, that is, $h \in \hat{H}$. In this case, it is straightforward to show that the energies of these inequivalent configurations are the same, thus signalling the existence of a zero mode. These arguments apply equally for the gauged non-abelian string [Everett, 1988; Alford *et al.*, 1991], so it is clear they will in general possess zero modes – examples of which are given in §5.1.8. The transformation (4.2.7) can be promoted to become spacetime dependent $h(z,t)$ and, as we shall see in chapter 5, these correspond to massless excitations which propagate along the string at the speed of light.

4.2.2 Z_N-strings

The simplest manner in which to form non-abelian strings is through the maximal breakdown of an $SU(N)$ symmetry, $SU(N) \to Z_N$. This can be achieved by taking N Higgs fields in the adjoint representation and choosing an appropriate potential, as demonstrated for $SU(2)$ in §2.1.3.

The Z_2-string, which was also discussed by Nielsen & Olesen [1973], is perhaps the most cosmologically interesting of these, so we consider the **3 + 3** $SU(2)$ symmetry breaking of §2.1.3 in more detail. The Lagrangian for this Higgs model is

$$\mathcal{L} = \tfrac{1}{2}\mathcal{D}_\mu\varphi \cdot \mathcal{D}^\mu\varphi + \tfrac{1}{2}\mathcal{D}_\mu\psi \cdot \mathcal{D}^\mu\psi - \tfrac{1}{4}\mathrm{Tr}\,(\mathcal{G}_{\mu\nu}\mathcal{G}^{\mu\nu}) - V(\varphi,\psi)\,, \quad (4.2.8)$$

where the two triplet fields φ and ψ transform in the adjoint representation of $SU(2)$ and, with generators T^a and gauge fields W_μ^a, the covariant derivative $\mathcal{D}_\mu$ and field-strength tensor $\mathcal{G}_{\mu\nu}$ are defined in (2.1.30) and (2.1.46), respectively. Again, we take the maximal symmetry breaking potential

$$V(\varphi,\psi) = \tfrac{1}{4}\lambda_\phi(\varphi^2 - \eta_\phi^2)^2 + \tfrac{1}{4}\lambda_\psi(\psi^2 - \eta_\psi{}^2)^2 + \tfrac{1}{8}g(\varphi.\psi)^2\,. \quad (4.2.9)$$

The field equations are then

$$\mathcal{D}_\mu\mathcal{D}^\mu\varphi = \frac{\partial V}{\partial\varphi}\,, \qquad \mathcal{D}_\mu\mathcal{D}^\mu\psi = \frac{\partial V}{\partial\psi}\,,$$
$$\mathcal{D}_\mu G_a^{\mu\nu} = j_a^\nu\,, \qquad\qquad\qquad (4.2.10)$$

where

$$j_a^\nu = e\epsilon_{abc}\left((\mathcal{D}^\nu\varphi)_b\phi_c + (\mathcal{D}^\nu\psi)_b\psi_c\right)\,. \quad (4.2.11)$$

We look for cylindrically-symmetric vortex solutions using an ansatz similar to (4.1.11–12) [de Vega & Schaposnik, 1986],

$$\varphi = \eta_\phi \hat{\mathbf{z}}$$
$$\psi = \eta_\psi f(r)(\cos\theta, \sin\theta, 0)$$
$$\mathbf{W}_\theta = -\frac{1}{e}\alpha(r)\hat{\mathbf{z}}.$$

(4.2.12)

The minimum energy configuration with φ remaining in the vacuum state depends on the condition $\lambda_\phi \eta_\phi^4 > \lambda_\psi \eta_\psi^4$, otherwise φ and ψ would exchange roles.

With this ansatz the boundary conditions on $f(r)$ and $\alpha(r)$ are the same as those in (4.1.13–14). Indeed, substitution of (4.2.12) into the field equations (4.2.10) yields a set of equations essentially identical to (4.1.15–16) for the $n=1$ vortex in the abelian-Higgs model. Consequently, there is a close correspondence between the qualitative properties of these different vortices: The magnetic field decays exponentially from a maximum at the core on a characteristic length-scale $(e\eta_\psi)^{-1}$, while the Higgs field ψ rises from zero to its vacuum expectation value η_ψ on a length scale $(\sqrt{\lambda_\psi}\eta_\psi)^{-1}$.

To determine the magnetic flux of the vortex solution an 'electromagnetic' tensor $F^{\mu\nu}$ can be defined by exploiting the orthogonality of $G_a^{\mu\nu}$ and ψ in the ansatz (4.2.12),

$$F^{\mu\nu} = \frac{1}{\eta_\phi}\phi_a G_a^{\mu\nu}.$$

(4.2.13)

With φ lying along the z-axis so that $B_z = G_{12}^3$, Stokes' theorem yields the magnetic flux of the Z_2-string

$$\Phi_B = \frac{2\pi}{e}.$$

(4.2.14)

These strings have an interesting internal structure because they appear at a second-stage symmetry breaking after the monopoles of §3.2.3 have emerged. The monopole flux $\Phi_B = 4\pi/e$ is confined and so the monopole becomes attached to two Z_2-strings, each taking away half of the original magnetic flux. Strings and antistrings are not identical and segments of string with opposite flux directions are separated by these beadlike solitons. Note that multiple winding number configurations $|n| > 1$ are possible and, depending on the coupling constants, may correspond to a local minimum of the energy. However, a single bead passing along the string will change the winding by 2, hence strings with odd windings $n = \pm1, \pm3, \dots$ belong to the same topologically non-trivial sector, while strings with even windings $n = \pm2, \pm4, \dots$ are topologically equivalent to the vacuum.

4.2.3 Phenomenological strings

The $SO(10)$ symmetry breaking scheme (2.1.67) provides a useful example of a 'realistic' grand unified model in which strings are expected to appear. These strings are associated with an unbroken discrete Z_2-symmetry and, although they are analogous to the Z_2-strings discussed above in §4.2.2, there are some significant differences. Here, they are associated with the $SO(10)$ operator P in (2.1.63), which generates rotations from the identity to the non-trivial element of Z_2. The string will have a radius $\delta \sim (e\eta)^{-1}$ and energy $\mu \sim \eta^2$ where $\eta \sim 10^{16}$GeV corresponds to the first **126** Higgs symmetry breaking scale in (2.1.67). A careful examination of the true minimal energy configuration as discussed in §3.3.4, however, reveals a richer internal structure on this scale [Aryal & Everett, 1987; Everett, 1988]. Since the Higgs and gauge sectors are so large and complicated in GUT models, it is not entirely surprising to find that additional fields are excited and, in this case, these include both a scalar and a charged vector field condensate.

Subsequent symmetry breaking at the electroweak phase transition employs a Higgs field in the **10**-representation with components given in (2.1.66). None of these has eigenvalue $P = 0$, indicating that they would introduce an unacceptable phase factor on passing around a string. To avoid this, we must rotate to an alternative operator R also generating the non-trivial element of Z_2, but such that $R\phi_{10} = 0$. Thus at the electroweak scale, some additional flux will be squeezed into the string core, as discussed by Stern & Yajnik [1986] and Alford & Wilczek [1989]. The layered internal structure of these $SO(10)$ strings is illustrated in fig. 4.6. Note, however, that these properties depend on specific details of the symmetry breaking scheme, and other GUT strings can be expected to have a very different internal structure.

Finally, we should mention some further references where strings are considered in 'realistic' phenomenological models. Strings in GUT models have, for example, been discussed by Everett [1981], Kibble *et al.* [1982a,b], Olive & Turok [1982], Vilenkin & Shafi [1983], Witten [1985a] and Lazarides, Panagiotakopoulos & Shafi [1987a,b] (refer also to chapter 5). We should also note that, although strings do not appear at electroweak symmetry breaking in the standard model, it is possible to construct models in which string production is deferred until intermediate energy scales, even approaching M_W [Lazarides *et al.*, 1987b; Stern & Yajnik, 1986]. The properties of some specific string solutions in extensions of the standard model have been investigated by Peter [1992].

Realistic models in which global strings appear can also be constructed. Global strings are found in axion models [Vilenkin & Everett, 1982], since

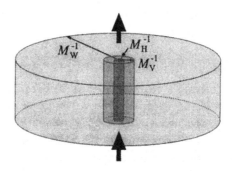

Fig. 4.6. Diagrammatic cross-section of the $SO(10)$ string arising in the scheme (2.1.67), showing concentric layers of excited scalar and gauge fields. Typically the inner GUT-scale core has a width $M_V^{-1} \sim (e\eta)^{-1} \sim 10^{-28}$cm and, in this case, the string also has an electroweak surround of width $M_W^{-1} \sim 10^{-16}$cm.

the axion is associated with the breaking of a global Peccei–Quinn $U(1)$ symmetry. Axion strings also appear to be generic in superstring models [Witten, 1985b]. Another example is the minimal $SU(5)$ model with an 'accidental' global B–L $U(1)$ symmetry which can also be broken to produce global strings [Shafi & Vilenkin, 1984b].

4.2.4 Alice strings

Alice strings [Schwarz, 1982] provide an example of the role of the rotation of the embedding of H in G as we traverse a closed loop about a non-abelian string. Recall that Alice strings arose from the symmetry breaking $SO(3) \xrightarrow{5} O(2)$ which was introduced in §2.1.3 *Example 3* and in §3.3.2(iv). This symmetry breaking leaves an unbroken $SO(2)$ component connected to the identity containing rotations about the z-axis. There is also a disconnected component which is generated from the first by a 180° rotation about an axis lying in the xy-plane (say the x-axis using T_1).

In the presence of a string, the symmetric tensor field Φ rotates as

$$\Phi(\theta) = \exp(\theta T_1/2) \, \Phi(0) \, \exp(-\theta T_1/2). \qquad (4.2.15)$$

This results in a corresponding rotation of the unbroken $SO(2)$ within $SO(3)$ since the initial generator, $Q(0) = T^3$, will become

$$Q(\theta) = \exp(\theta T_1/2) \, T^3 \, \exp(-\theta T_1/2). \qquad (4.2.16)$$

This 'charge' operator Q is double-valued, since $Q(2\pi) = -Q(0)$, which creates difficulties in interpreting the 'electrodynamics' of this model. Defined relative to $Q(\theta)$, the electromagnetic gauge potential, $A_\mu^q =$

$\cos(\theta/2)\,W_\mu^3 + \sin(\theta/2)\,W_\mu^2$, is likewise double-valued, as are the electric and magnetic fields $F^{\mu\nu}$ given by

$$F_{\mu\nu} = \cos(\theta/2)\,G_{\mu\nu}^3 + \sin(\theta/2)\,G_{\mu\nu}^2 . \qquad (4.2.17)$$

In order to study Maxwell's equations it is appropriate to introduce a branch cut from the string centre at $x = 0$ to $x = \infty$, along which we impose the boundary condition that $F_{\mu\nu}^{\text{above}} = -F_{\mu\nu}^{\text{below}}$. This apparent change in the sign of the electromagnetic field is unobservable because the charge Q of a test particle transported across the cut will also change sign to $-Q$. Consequently, there will be no discontinuous change in the force on the particle.

The presence of an Alice string affects a nearby charged particle. Consider a particle on the negative x-axis, near a string lying along the z-axis. By reflection symmetry it is clear that E_x and E_z must be equal above and below the cut, so the boundary condition, $\mathbf{E}^{\text{above}} = -\mathbf{E}^{\text{below}}$, implies that only the perpendicular component E_y can be non-zero. This is precisely the boundary condition for a conducting plate, so we can expect the particle to experience a force from its image charge and, thus, to be attracted towards the string.

It will only be possible to globally define the charge within a Gaussian surface which encloses two strings (or any even number), since a finite branch cut can pass between them as illustrated in fig. 4.7. If a charged particle were to travel between two strings as in fig. 4.7, its charge would change from Q to $-Q$ while the strings would acquire a charge $2Q$. This is manifested as a non-local 'Cheshire' charge which emanates from the branch cut between the strings, with the branch cut now acting as if

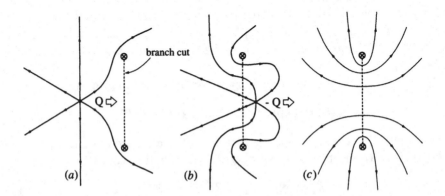

Fig. 4.7. Acquistion of 'Cheshire' charge by two Alice strings. The particle Q reappears charge conjugated $-Q$ from the branch cut and leaves a residual delocalized charge of $2Q$ associated with the two strings.

it were a charged conducting plate. Magnetic monopoles in the model cause similar effects (refer to §3.3.3). A more complete discussion of Alice strings can be found in Alford *et al.* [1990, 1991], Preskill & Krauss [1990] and Bucher, Lo & Preskill [1992]. We should note, however, that we do not expect Alice strings in our universe because charge conjugation is not an observed symmetry.

4.2.5 Semilocal strings

This string type occurs in models possessing both local and global symmetries. A simple example is given by the Lagrangian [Vachaspati & Achucarro, 1991]

$$\mathcal{L} = |(\partial_\mu - ieA_\mu)\Phi|^2 - \tfrac{1}{4}F_{\mu\nu}F^{\mu\nu} - \tfrac{1}{4}\lambda(\Phi^\dagger\Phi - \eta^2)^2 , \qquad (4.2.18)$$

where $\Phi = (\phi, \chi)$ is a complex scalar doublet. This model is invariant under global $SU(2)_g$ transformations, $\Phi \to \exp(i\omega^a\tau^a)\Phi$, with the same notation as in §2.1.3, and under local $U(1)_l$ transformations, $\Phi \to \exp[i\alpha(x)I]\Phi$. Note that without a gauge coupling the symmetry group of (4.2.18) would be $O(4)$.

When Φ acquires a vacuum expectation value, the symmetry is broken to a global $U(1)$,

$$SU(2)_g \times U(1)_l \to U(1)_g . \qquad (4.2.19)$$

In the basis where $\Phi = (\phi, 0)$, the unbroken $U(1)_g$ is the group of global phase rotations of χ. The vacuum manifold $\mathcal{M}$ of Φ is a three-sphere, $\Phi^\dagger\Phi = \eta^2$. Since $\pi_1(S^3) = I$, all loops in $\mathcal{M}$ can be contracted to a point, and the model has no topological strings. However, when global symmetries are present, one should also consider the gradient energy cost associated with the variation of Φ along a loop. Only for special loops which are generated by acting with elements of $U(1)_l$ on some $\Phi \in \mathcal{M}$ can the scalar gradient energy be compensated by the gauge field. A string configuration in which Φ traverses one of these loops on a path around the string can be stabilized by the gradient energy barrier that must be overcome if the loop is to be contracted to a point [Vachaspati & Achucarro, 1991].

The existence of string solutions in the model (4.2.18) is easily verified. If we set $\Phi = (\phi, 0)$, the Lagrangian (4.2.18) reduces to that of the abelian-Higgs model, and the solution is

$$\Phi = (\phi_s, 0) , \qquad A^\mu = A_s^\mu , \qquad (4.2.20)$$

where (ϕ_s, A_s^μ) is the Nielsen–Olesen solution (4.1.11–12). Equivalent solutions can be obtained by acting on (4.2.20) with global rotations and gauge transformations.

An important difference between semilocal strings (4.2.20) and topological strings is that the vanishing of Φ on the string axis in (4.2.20) is not enforced for topological reasons. One can construct field configurations which never leave the vacuum manifold and have the same asymptotic behaviour as (4.2.20). For example, we can have [Hindmarsh, 1992],

$$\Phi = (f(r)e^{i\theta}, g(r)), \quad A_\theta = v(r)/er, \qquad (4.2.21)$$

where $f(r) = r(a^2+r^2)^{-1/2}$, $g(r) = a(a^2+r^2)^{-1/2}$, $v(r) = r^2(a^2+r^2)^{-1}$ and a is a constant parameter. The stability of the string solution (4.2.20) depends on whether or not its energy is smaller than that of configurations like (4.2.21).

Numerical tests of semilocal string stability have been performed by Hindmarsh [1992] and by Achucarro *et al.* [1992]. They found that for $\beta \equiv \lambda/2e^2 < 1$ the strings are stable, while for $\beta > 1$ they are unstable with respect to the formation of a scalar condensate as in (4.2.21), with subsequent growth of the condensate radius a. In the case of critical coupling, $\beta = 1$, Hindmarsh [1992] has shown that the strings are only neutrally stable and that the Nielsen–Olesen-type string belongs to a two-parameter family of degenerate vortex solutions. The properties and dynamics of semilocal strings in this Bogomol'nyi limit have been studied by Gibbons, Ortiz, Ruiz Ruiz & Samols [1992] and Leese & Samols [1992].

4.2.6 'Electroweak' strings

An interesting question about semilocal models is what happens to stable string solutions when the global part of the symmetry group is gauged. By continuity one expects that, if the corresponding gauge coupling is sufficiently weak, stable string solutions should still exist and should be close to the semilocal strings of §4.2.5 [Vachaspati, 1993].

As an example we consider again the $SU(2) \times U(1)$ model (4.2.18). If the $SU(2)$ symmetry is gauged, it becomes identical to the Higgs sector of the standard electroweak model. The corresponding Lagrangian can be written as

$$\mathcal{L} = |\mathcal{D}_\mu \Phi|^2 - \tfrac{1}{4}F_B{}^2 - \tfrac{1}{4}F_W{}^2 - \tfrac{1}{4}\lambda(\Phi^\dagger \Phi - \eta^2)^2, \qquad (4.2.22)$$

where

$$\mathcal{D}_\mu \Phi = (\partial_\mu - ig\tau^a W_\mu^a - \frac{i}{2}g'B_\mu)\Phi \qquad (4.2.23)$$

and the factor $1/2$ in (4.2.23) has been introduced in order to conform with the standard notation [see, for example, Taylor, 1976]. Vachaspati [1993] has shown that this model has a vortex solution of the form

$$\Phi = (0, \phi_s), \quad Z_\mu = A_{s\mu}, \quad A_\mu = W_\mu^1 = W_\mu^2 = 0, \qquad (4.2.24)$$

where

$$Z_\mu = W_\mu^3 \cos\theta - B_\mu \sin\theta, \quad A_\mu = W_\mu^3 \sin\theta + B_\mu \cos\theta, \quad (4.2.25)$$

and $\tan\theta = g'/g$ with $(\phi_s, A_{s\mu})$ given for the Nielsen–Olesen string in (4.1.11–12). This ansatz was first proposed by Nambu [1977]. The semilocal string solution (4.2.20) is recovered in the limit $g \to 0$. Since semilocal strings are stable for $\beta \equiv 2\lambda/g'^2 < 1$, we expect the 'electroweak' string (4.2.24) to be stable when the parameters g and β are sufficiently close to $g = 0$, $0 < \beta < 1$. This expectation has been confirmed by the numerical results of James, Perivolaropoulos & Vachaspati [1993]. Somewhat disappointingly, they also found that the strings are unstable for realistic parameter values in the electroweak model. It is possible, however, that there are extensions of the standard model for which the string solutions are stable.

It should be noted that 'electroweak' strings can never be absolutely stable because, unlike semilocal strings, they are separated from the vacuum by only a finite gradient energy barrier. Preskill [1992] has argued that classically stable electroweak strings can decay quantum-mechanically by the nucleation of monopole–antimonopole pairs.* This decay probability has been estimated by Preskill & Vilenkin [1993]. For $g \ll g'$ the tunnelling action is proportional to g^{-2} and becomes infinite in the semilocal limit $g \to 0$.

4.3 String–string interactions

4.3.1 The inter-vortex potential

For simplicity, this discussion of string interaction properties returns to the abelian-Higgs model (4.1.1) with the rescalings of §4.1.1. To estimate the interaction energy of a stationary multi-vortex configuration, we suppose that the true solution $(\Phi, \mathbf{A})$ can be accurately approximated by a superposition of individual cylindrically symmetric $n=1$ vortex solutions, $(\hat{\Phi}, \hat{\mathbf{A}})$, provided that the vortex zeros are widely separated. The n-vortex solution with zeros at the points $\mathbf{r}_k$ will then be given by the ansatz [Abrikosov, 1957]

$$\hat{\Phi}(\mathbf{r}) = \prod_{k=1}^{n} \phi_k(\mathbf{r}) = \prod_{k=1}^{n} \phi_s(\mathbf{r} - \mathbf{r}_k),$$

$$\hat{\mathbf{A}}(\mathbf{r}) = \sum_{k=1}^{n} \mathbf{A}_k(\mathbf{r}) = \sum_{k=1}^{n} \mathbf{A}_s(\mathbf{r} - \mathbf{r}_k). \quad (4.3.1)$$

---

* The possibility that Z-strings in the standard model can end on magnetic monopoles was first suggested by Nambu [1977].

In order to evaluate the potential between two vortices using (4.3.1), the asymptotic accuracy of (4.3.1) must first be established for large separations, $d \gg m_{\mathrm{s}}^{-1}, m_{\mathrm{v}}^{-1}$ [Müller-Hartmann, 1966]. Rewriting the true solution in terms of the residual fields $(\varphi, \mathbf{a})$

$$\Phi = \hat{\Phi} + \varphi, \quad \mathbf{A} = \hat{\mathbf{A}} + \mathbf{a}, \tag{4.3.2}$$

we require that the additional contribution $\epsilon_{\varphi,\mathbf{a}}(\hat{\Phi}, \hat{\mathbf{A}})$ to the vortex energy be negligible,

$$\mu(\Phi, \mathbf{A}) = \mu(\hat{\Phi}, \hat{\mathbf{A}}) + \epsilon_{\varphi,\mathbf{a}}(\hat{\Phi}, \hat{\mathbf{A}}). \tag{4.3.3}$$

We can represent the left-hand side of the field equation (4.1.8) in the form $L_\phi(\phi, \mathbf{A})$. For an $n=2$ vortex configuration, the substitution of the ansatz $(\hat{\Phi}, \hat{\mathbf{A}})$ into (4.1.8) yields the result

$$L_\phi(\hat{\Phi}, \hat{\mathbf{A}}) = \phi_1 L_\phi(\phi_2, \mathbf{A}_2) + \phi_2 L_\phi(\phi_1, \mathbf{A}_1) + \mathcal{O}(\exp(-\sqrt{2}\,ed)), \tag{4.3.4}$$

because of the asymptotic fall-off (4.1.17) of the individual solutions $(\phi_i, \mathbf{A}_i)$. Here, we adopt the assumption that we are in a Type II regime with $\beta = \lambda/2e^2 > 1$. Similarly, using (4.1.9) we can show that the gauge field equations are accurate to $\mathbf{L}_\mathbf{A}(\hat{\Phi}, \hat{\mathbf{A}}) \approx \mathcal{O}(\exp(-\sqrt{2}\,ed))$. These expressions imply that the residual fields behave as $\varphi, \mathbf{a} \approx \mathcal{O}(\exp(-\sqrt{2}\,ed))$, so it can be easily verified that the residual energy consists entirely of terms of second order or higher

$$\epsilon_{\varphi,\mathbf{a}}(\hat{\Phi}, \hat{\mathbf{A}}) \approx \mathcal{O}(\exp(-2\sqrt{2}\,ed)). \tag{4.3.5}$$

Thus we can place considerable confidence in Abrikosov's ansatz (4.3.1) at large vortex separation, a conclusion which is unchanged for a vortex-antivortex pair. This is an important result for creating string configurations as initial data for dynamical numerical simulations.

To evaluate the energy of the two vortices, the expression is split into that of the free solutions plus an overlap integral,

$$E(\hat{\Phi}, \hat{\mathbf{A}}) \approx 2\mu(\phi_1, \mathbf{A}_1) + E_{\mathrm{int}}(\hat{\Phi}, \hat{\mathbf{A}}). \tag{4.3.6}$$

It is straightforward to show, using the asymptotic properties of the vortex solution (4.1.17), that each contribution to $E_{\mathrm{int}}(\hat{\Phi}, \hat{\mathbf{A}})$ is of the order $\mathcal{O}(\exp(-\sqrt{2}\,ed))$. In the case of strong Type II behaviour, $\beta \gg 1$, this overlap integral has been approximately evaluated to yield [Abrikosov, 1957; Müller-Hartmann, 1966; Bhattarcharjee, 1982]

$$E_{\mathrm{int}}(\hat{\Phi}, \hat{\mathbf{A}}) \approx \left(\frac{\sqrt{2}\pi^3}{\varrho d}\right)^{1/2} \exp(-\sqrt{2}\,ed), \tag{4.3.7}$$

indicating that at large distances vortices repel each other with an exponentially decreasing force. This repulsion confirms the expectation for Type II behaviour where vector field effects dominate.

The covariant derivative term, however, provides a negative contribution to the overlap integral, suggesting that the attraction from the scalar field might be greater for $\beta < 1$, below critical coupling at which the interaction energy vanishes. These results have been confirmed numerically using a constrained variational approach, and the interaction energy of two-vortex configurations has been accurately estimated [Jacobs & Rebbi, 1979]. The inter-vortex potential for a variety of parameters is illustrated in fig. 4.8 – note that near vortex superposition the potential becomes flat. This derivation is easily repeated for a vortex–antivortex pair but, with the magnetic fluxes cancelling, the vector contribution is now also attractive, so the vortex and antivortex always attract at large distances with an exponentially small force. As one might expect for a theory with exclusively massive fields, vortices only have short-range Yukawa-type interactions. Local $U(1)$-strings in three dimensions, therefore, are effectively non-interacting, except when the strings intersect.

The derivation of the vortex–antivortex (or vortex–vortex) interaction can be easily repeated for global $U(1)$-strings. We take the analogue of the Abrikosov ansatz,

$$\hat{\Phi} = \phi_s(\mathbf{r} - \mathbf{r}_1)\bar{\phi}_s(\mathbf{r} - \mathbf{r}_2)\,, \qquad (4.3.8)$$

which is simply the product of two isolated axisymmetric solutions. It is straightforward to show at large fixed separation that (4.3.8) will closely

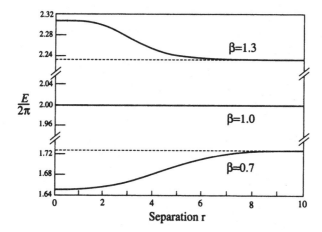

Fig. 4.8. Energy of a two-vortex configuration as a function of inter-vortex separation for the couplings $\beta = 0.7$, 1.0, 1.3 [Jacobs & Rebbi, 1979].

approximate the minimum energy configuration [Shellard, 1987]. From the field equations (4.1.35) we can estimate that $L_\phi(\hat{\Phi}) \lesssim \mathcal{O}(d^{-1})$ with the residual field $\varphi \lesssim \mathcal{O}(d^{-1})$, and hence the residual energy is

$$\epsilon_\varphi(\hat{\Phi}) \lesssim \mathcal{O}(d^{-2}) \,. \tag{4.3.9}$$

Subtracting the individual vortex energy densities of (3.2.15), yields the axisymmetric interaction energy density [Perivolaropoulos, 1992]

$$\mathcal{E}_{\text{int}}^\pm \approx \pm 2\eta^2 \frac{r^2 - d^2}{(r^2 + d^2)^2} \,, \tag{4.3.10}$$

where the upper sign is for a vortex–vortex pair and the lower for a vortex and antivortex. Assuming a cut-off distance $R \gg d$ ($\gg \delta$), this can be integrated to give,

$$E_{\text{int}}^\pm(\hat{\Phi}) \approx \pm 4\pi\eta^2 \ln\left(\frac{R}{d}\right) \,, \tag{4.3.11}$$

from which we obtain the inter-vortex force, $F = -\partial E^\pm/\partial d$,

$$F \approx \pm \frac{4\pi\eta^2}{d} \,. \tag{4.3.12}$$

This implies that two vortices feel a strong repulsive force, and we can expect an $n > 1$ global vortex configuration to be unstable. On the other hand, a vortex–antivortex pair will feel a strong attraction, being effectively 'confined'. These long-range interactions are due to a massless Goldstone boson in the model, and indicate that global and local string dynamics will differ substantially.

4.3.2 Vortex scattering in two dimensions

While the nonlinearity of the field equations (4.1.8–9) governing vortex dynamics makes analytic progress difficult, their study is readily accessible using standard numerical techniques. Numerical studies have also indicated directions leading to significant analytic advances, a cross-fertilization which has become commonplace in the study of solitons.

Numerical simulations have been performed for vortices in both the Goldstone model [Shellard, 1986a,b, 1987; Hecht & de Grand, 1990; Perivolaropoulos, 1992] and the abelian-Higgs model [Matzner, 1988; Moriarty, Myers & Rebbi, 1988a–c; Shellard & Ruback, 1988]. Global vortex interactions are the most accessible because the hyperbolic equations have no constraints and gauge-fixing problems. Specific finite-difference algorithms and other detailed information about the numerical study of topological defects can be found in Shellard [1992]. Initial data was created by Lorentz-boosting numerical solutions to the axisymmetric static equation (4.1.36). Multi-vortex configurations were obtained by matching with the Abrikosov ansatz (4.3.1) and memory limitations necessitated the imposition of reflective or periodic boundary conditions.

The time evolution of an initially static global vortex–antivortex pair is illustrated in fig. 4.9. The pair move together rapidly under the influence of the strong attractive force (4.3.12); they coalesce and then annihilate as the opposite phase twists superpose and unravel. Radiation from the annihilation chiefly emanates transversely and is predominantly in mass-less phase waves. Complete annihilation was not always the ultimate outcome of a direct vortex–antivortex encounter. Given sufficient initial kinetic energy, a threshold was reached beyond which another pair was created [Shellard, 1987]. As before, the vortices coalesce and the zeros temporarily disappear, however, the unravelling phase twist 'overshoots' and ϕ oscillates back through zero as a vortex–antivortex pair re-emerge. In contrast to a pair simply passing through each other, the new vortices emerge from this intermediate state with a phase shift of π and after a time lag, $\Delta t \sim \delta/2$, in the centre-of-mass frame. As might be expected intuitively, this threshold was achieved when the kinetic energy was of the same order as the vortex mass μ.

Vortex–vortex interactions have also been extensively studied numeri-cally. An axisymmetric $n = 2$ global vortex is unstable to small pertur-bations. After slow initial evolution, the splitting proceeds rapidly under the strong repulsive force. The relatively slow initial breakdown indi-cates that the potential must be almost flat near minimum separation,

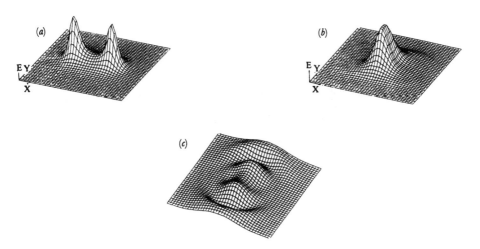

Fig. 4.9. Annihilation of an initially static global vortex–antivortex pair. The amplitude of the energy density is plotted.

as illustrated in fig. 4.8. If separated global vortices are given sufficient
initial velocity to overcome their mutual repulsion and to collide head-on,
then they will scatter at right-angles in the centre-of-mass frame [Shel-
lard, 1986b], an analytic expectation to be discussed in §4.3.3. Fig. 4.10
illustrates the generality of this result even for highly relativistic vortex
collisions. In the collision aftermath, the original two vortices have scattered at right-angles, but carry away only a small fraction of the initial
kinetic energy. Two additional vortex–antivortex pairs have been created
in the central region with the two new vortices taking away the majority
of the kinetic energy, as if to replace the original pair. This interesting
particle-like creation of vortex pairs deserves further study.

Analogous studies for vortices in the abelian-Higgs model have been
qualitatively in accord with these results. Standard finite-difference techniques in the Lorentz gauge $\partial_\mu A^\mu = 0$ were employed by Matzner [1988]
and Shellard & Ruback [1988], while Moriarty *et al.* [1988a] adopted the
$A_0 = 0$ gauge in a constrained Hamiltonian approach, employing techniques used in lattice gauge theories. Vortex–vortex scattering has received special attention and, as with global vortices, right-angled scattering in a head-on collision was observed to be generic regardless of the
β-parameter choice. Despite the vanishing inter-vortex potential at critical coupling ($\beta = 1$), the two vortices are anything but non-interacting
as shown in fig. 4.11. Here, as the two vortices draw closer together their
energy density becomes redistributed in such a manner that they appear
to exchange 'halves', the right half of the first vortex joins with the left
half of the second and vice versa. The intermediate 'ring' of energy, very
closely approximates the static $n=2$ vortex (fig. 4.4).

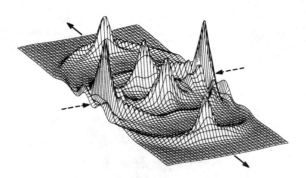

Fig. 4.10.   Aftermath of a highly energetic vortex collision ($v \approx 0.98$): The
original vortices have scattered at right-angles and, subsequently, two vortex–
antivortex pairs have been created [Shellard, 1986b].

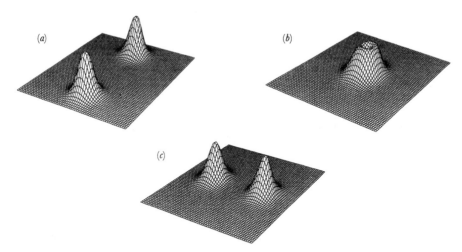

Fig. 4.11.   Right-angled scattering in a head-on vortex–vortex collision in the abelian-Higgs model at critical coupling, $\lambda = e^2$ [Shellard & Ruback, 1988]. The two vortices exchange 'halves' after forming an intermediate 'ring'.

By following the location of the zeros of the Higgs field, one can observe vortex scattering at a variety of impact parameters. Several paths are shown in fig 4.12 for critical coupling, illustrating a diminishing scattering angle with increasing impact parameter. Scattering is only significant when the central cores actually overlap, circumstances in which there is a substantial exchange of energy between the vortices. For velocities $v < 0.4$, the scattering behaviour is remarkably velocity-independent and elastic. However, at larger velocities higher-order modes become excited and the vortices radiate an increasing proportion of their initial kinetic energy. For $\beta \neq 1$, the scattering angle was altered because of the additional potential term, though this difference gradually disappeared as the impact parameter approached $b=0$. We also note that for a static initial configuration, dynamical simulations confirm that with $\beta < 1$ two vortices attract, with $\beta > 1$ they repel, and with $\beta = 1$ they remain stationary.

4.3.3 The two-vortex moduli space

With inter-vortex potential terms playing no role at critical coupling, the origin of the non-trivial scattering is somewhat puzzling. Substantial analytic progress has been made on this problem by Ruback [1988] and Samols [1990, 1992] using a collective coordinate approach. This work exploits an idea, originally proposed by Manton [1982] in the context of the Bogomol'nyi–Prasad–Sommerfeld monopole, that dynamics greatly simplify at low energy in the Bogomol'nyi limit. We have already ob-

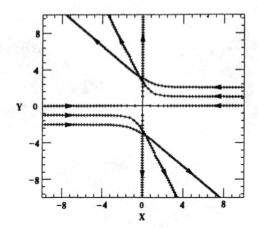

Fig. 4.12.   The locus of the zeros of the Higgs field during vortex scattering for a variety of impact parameters ($\lambda/e^2 = 1$) [Shellard & Ruback, 1988].

served in §4.1.2 that at critical coupling, the energy of n-vortex solutions is bounded below by $2\pi|n|$. The set of static solutions which achieve this bound is completely determined by choosing n points in R^2 (locations of the n zeros of the Higgs field ϕ). This static n-vortex solution space (factoring out gauge transformations) is then a $2|n|$-dimensional submanifold or moduli space M_n within all field configuration space.

The moduli space M_n corresponds to degenerate energy minima for the configuration space of all time-dependent solutions with the same winding number n. A slowly moving n-vortex configuration, therefore, will lie close in field space to M_n. Conservation of energy then implies that its trajectory must always remain confined near M_n, since this is a global minimum of the potential energy. For the same reason, in the vicinity of M_n gradients of the potential energy are small and the dynamics will be governed by the kinetic energy. These kinetic terms, $\dot{\mathbf{A}}\cdot\dot{\mathbf{A}}$, $\dot{\phi}\dot{\bar{\phi}}$, in the Lagrangian $\mathcal{L}$ induce a metric on M_n and, in the absence of potential terms, the variational principle implies that motion will be geodesic on M_n. Non-trivial scattering is possible only because M_n is, in general, a curved submanifold.

In the two-vortex case, the appropriate coordinates describing the vortex motion are the two collective coordinates describing their overall translation and a further two describing their relative motion, that is, $M_2 \cong R^2 \times M_2^0$ (a split demonstrated in (4.1.32–3) for small perturbations). Only the latter is of interest, so we consider the metric $^{(2)}g_{ij}$ on this two-dimensional subspace M_2^0. Rotation and parity invariance constrain this metric to take the form

$$ds^2 = F^2(r)dr^2 + G^2(r)d\theta^2 \,, \qquad (4.3.13)$$

where $(r\cos\theta, r\sin\theta)$ is half the separation vector of the two vortices. Since these configurations are equivalent under a 180° rotation, this suggests that θ should be identified by π rather than 2π, that is, $0 \le \theta < \pi$. At large r the vortices do not interact and the metric must be flat,

$$ds^2 \approx dr^2 + r^2 d\theta^2 \,. \qquad (4.3.14)$$

For small r when the two vortices almost coincide, the two-vortex system can be viewed as a perturbation (4.1.33) about a single vortex of winding number $n = 2$. By carefully examining such perturbations, Ruback [1988] was able to show that near the origin the metric on M_2^0 took the form

$$ds^2 \approx (c_0 r)^2 \left(dr^2 + r^2 d\theta^2 \right) \,. \qquad (4.3.15)$$

Under a coordinate transformation $\rho = c_0 r^2/2$, the metric becomes

$$ds^2 \approx d\rho^2 + 4\rho^2 d\theta^2 \,, \qquad (4.3.16)$$

and we see that, with θ identified by π, M_2^0 is a smooth manifold with no conical singularity at $r = 0$. M_2^0 can be represented as a surface of revolution in R^3, which approaches a cone of semi-angle $\pi/6$ at large r and which becomes smoothed out ('snub-nosed') at small r. A geodesic corresponding to a head-on collision passes over the smoothed peak of this cone and thus θ changes by $\pi/2$, given its identification by π. This is the prediction of right-angled scattering.

The metric (4.3.13) was studied numerically using geodesics corresponding to the trajectories of vortex zeros, such as those in fig. 4.12 [Shellard & Ruback, 1988]. The embedding of the metric in R^3 is illustrated in fig. 4.13, along with two actual geodesics wrapping around the 'snub-nosed' cone. Samols [1990] furthered the analytic study of the metric on M_2^0 by transforming (4.3.13) to a conformally flat form with $F(r) = G(r)/r$, and by finding an integral constraint for the remaining unknown function $F(r)$. This was determined more accurately numerically by considering field distortions of static separated vortices using relaxation techniques [Samols, 1992] (see also Myers, Rebbi & Strilka [1992]).

Perhaps the simplest way to observe right-angled scattering for a head-on collision is to consider the fluctuations (4.1.33), corresponding to a two-vortex splitting mode. For one of these orthogonal modes, the Higgs field takes the form

$$\phi(\mathbf{r}) = f(r)e^{2i\theta} + \kappa f(r)h(r) \,, \qquad (4.3.17)$$

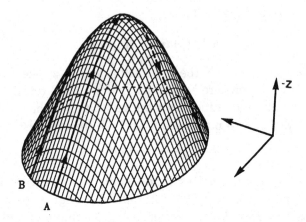

Fig. 4.13. Embedding in R^3 of the metric on the two-vortex moduli space M_2^0 showing two geodesic paths [Shellard & Ruback, 1988].

where $h \approx f^{-1}$ near $r = 0$, and κ parametrizes the vortex separation. With κ negative but increasing, it is trivial to show that the zeros lie on the x-axis and approach the origin. When κ changes sign, the zeros meet at the origin but then separate along the y-axis. It is interesting to note that the position of the zeros is constrained to move as the square root of the energy density, $r \approx \sqrt{\kappa}$, unlike at large separations where $r \approx \kappa$ to high precision. This implies that the zeros move superluminally during close vortex encounters, which is consistent with the fact that the metric function $F(r)$ in (4.3.13) vanishes at the origin. The 'ring-like' distribution of the energy density when the vortices superpose, makes the superluminal motion of the central zeros of little energetic consequence.

While this discussion has been based upon the assumption of critical coupling, right-angled scattering was observed to be generic for all head-on collisions, irrespective of coupling or velocity. To what extent, then, do these moduli space arguments reveal a universal topological constraint governing all vortex scattering? In the weakest generalization, we noted in §4.3.1 that the inter-vortex potential became flat as two vortices neared superposition, regardless of coupling. Hence, with a vanishing interaction force, at small separations the vortices will move in what approximates a degenerate moduli space. The previous arguments should remain valid in this small patch about $r=0$, most notably the identification of the angular coordinate in (4.3.13) by π which underlies right-angled scattering. At larger separations, slowly moving vortices are confined as before near the base of a steep potential 'valley' in field configuration space. For $\beta \neq 1$, the submanifold of these minima is no longer degenerate in energy and the configurations accelerate or decelerate depending on the sign of $\beta - 1$.

With $\beta \sim 1$ near critical it should be possible to reproduce this motion merely by introducing a small potential on the original moduli space M_2^0.

4.3.4 Reconnection in three dimensions

Reconnection at string intersection – the 'exchange of partners' or inter-commuting – is the key to making cosmic strings non-pathological in the early universe. It is to this question that these discussions have been chiefly directed. The homotopic classification of strings discussed in §3.3 delineated the topologically allowed outcomes at string crossing, but did not distinguish the actual physical result. For this we require detailed knowledge of the nonlinear behaviour of the vortex fields during a three-dimensional string encounter. To date this has only proved possible through numerical simulation with $U(1)$ models, but the results have been unambiguous. However, a compelling intuitive picture undergirding these results can be developed using the two-dimensional results discussed above.

Dynamical global string interactions have been studied numerically for a wide variety of collision parameters [Shellard, 1987]. These simulations generalized to three dimensions the numerical methods discussed in §4.3.2. Ignoring the inter-string potential, in §6.4.1 we show with appropriate Lorentz transformations that we require only two parameters to characterize all possible straight string intersections – the angle of incidence θ and the centre of mass velocity v. A careful study of this parameter space demonstrated that string reconnection occurred for all angles and velocities for which reliable numerical results could be obtained.

The results of a relativistic string–string intersection are illustrated in fig. 4.14. A relatively small initial separation of approximately 10

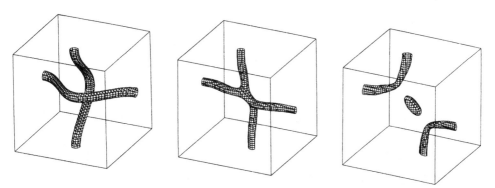

Fig. 4.14. Intercommuting of two global strings. In this energetic intersection, the creation of an 'interaction loop' can be observed in the aftermath [Shellard, 1992].

string widths was a consequence of memory limitations typical of any analysis in three dimensions. The straight global strings soon become distorted under the influence of the strong long-range force (4.3.12) acting between them and the nearest sections of string realign so as to become nearly antiparallel. When the strings meet, reconnection proceeds rapidly: The strings coalesce and locally annihilate, leaving the topological flux rerouted to the opposite string segments. The reconnected strings then decouple and move apart under the influence of their tension. In this relativistic interaction, however, there is sufficient energy remaining after reconnection to create new zeros of the Higgs field in the form of a closed loop – the higher dimensional analogue of vortex–antivortex pair creation in §4.3.2. This ephemeral interaction loop emerges from the intersection point, grows rapidly and then collapses and annihilates. Radiation in both massive and massless modes accompanied all such global string intersections.

The conclusion from these results that global strings inevitably intercommute has only minor caveats. At some level, extremely relativistic interactions (here $v > 0.98$) will always test numerical stability and, for very small collision angles, the interaction region extends beyond the numerical lattice. For parameters of cosmological interest, we can be confident that the 'intercommuting probability' is effectively unity, $p = 1$.

String reconnection has also been extensively studied in the abelian-Higgs model by Matzner [1988] and Moriarty *et al.* [1988a] using the methods discussed in §4.3.2. Again, reconnection has been observed for all collision parameters and for both strong and weak coupling (subject to the caveats mentioned previously). This has involved studies up to velocities $v \approx 0.95$, and for couplings in the range $0.25 < \beta < 4$. Qualitatively, string reconnection proceeds in much the same manner as for global strings – the Higgs field appears largely identical in the string interior. However, the absence of long-range forces in the abelian-Higgs model implies that the interacting strings do not become distorted until they are nearly overlapping. As the loci of zeros reconnect, the magnetic flux also reroutes. In the aftermath of large-angle or relativistic interactions, significant amounts of massive Higgs and gauge field radiation is produced. For sufficiently high energies, an interaction loop can also be temporarily created.

A numerical study of high winding number intercommuting has been performed by Laguna & Matzner [1989]. They observed such interacting strings to partially intercommute, as allowed topologically, and the interconnected segments to split apart into lower winding number configurations – behaviour which was anticipated heuristically [Vachaspati,

Fig. 4.15. The 'peeling' of a Type I $n = 2$ Nielsen–Olesen vortex-line (intially on the left) after intersection with an $n = 1$ vortex-line [Laguna & Matzner, 1989].

1988c]. An example of this 'peeling' interaction is shown in fig. 4.15 for an n=1 string meeting an n=2 string. This phenomenon is only relevant for Type I strings ($\lambda < e^2$) because Type II strings with $n > 1$ are unstable and will split of their own accord. These results indicate that an interacting network with high winding number strings will rapidly break down into n=1 strings.

Understanding reconnection

In fluid dynamics, an understanding of vortex reconnection has been hampered by difficulties in describing the circulation singularity in the vortex core. In contrast, for the simple models discussed here, the vortex solutions are regular and we can make considerable progress by exploiting their analytically well-understood two-dimensional interactions [Shellard, 1988, 1993]. This is most easily seen for slowly-moving strings at critical coupling which are nearly parallel. In contrast to parallel strings for which there are no forces, straight string segments misaligned by a very small angle ϵ will experience a short-range force of only $\mathcal{O}(\epsilon^2)$. Consequently, low energy dynamics will closely approximate adiabatic parallel (or two-dimensional) string interactions. In a transverse slice at the point of intersection, the strings will 'exchange halves' and scatter at right-angles as in fig. 4.11. However, above and below this point where there is a smaller overlap, there will be a correspondingly smaller exchange of energy between the two strings. At distances larger than δ/ϵ, there is effectively no interaction. The only consistent manner in which to resolve this string splitting and exchange is string reconnection. This outcome at intersection, therefore, is a consequence of the non-trivial geometry of the underlying field configuration space, specifically the moduli space M_2.

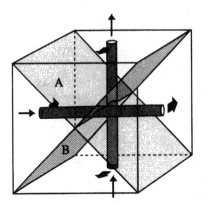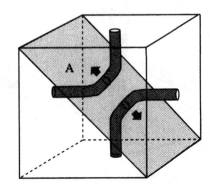

Fig. 4.16.   Configuration before and after intercommuting. Bisecting planes *A* and *B* reveal the two-dimensional basis of the interaction.

As we have seen, the qualitative form of two-dimensional vortex scattering is largely parameter-independent being unaffected by the complex aftermath at high energies. This conclusion, therefore, appears remarkably robust and can be expected in less adiabatic encounters [Shellard, 1988]. For an arbitrary string intersection, we can transform into a centre-of-mass frame and choose two planes perpendicular to the plane in which the strings intersect (as shown in fig. 4.16 [Copeland & Turok, 1986b]). Using the two-dimensional analogy, it is clear that intercommuting is topologically allowed since the 'vortex–antivortex' pair in plane *B* can annihilate, whereas the 'vortex–vortex' pair is constrained to remain in plane *A*. Add to this observation, the *inevitability* of right-angled scattering in plane *A* and annihilation in plane *B* and it appears that reconnection is the only possible outcome. Further qualitative topological arguments to the same end have been presented by Albrecht & York [1988] and Vachaspati [1988b] who consider the concept of linkages in a loop self-intersection.

4.4 Effective string actions

At low temperature we do not expect massive excitations to play a significant role in determining string dynamics, except in highly curved regions or at string crossing. In order to reduce the complexity of these field theoretic models, it is appropriate to integrate out these massive modes to obtain a smaller number of collective degrees of freedom in a low energy effective theory. Here we shall take the Nielsen–Olesen vortex line and the global $U(1)$-string to perform what is actually a remarkably model-independent reduction, which yields respectively the Nambu and the Kalb–Ramond string actions.

4.4.1 The Nambu action

Nielsen & Olesen [1973] and Tassie [1973, 1974] argued that the appropriate action for a relativistic magnetic flux line was proportional to the area of the worldsheet swept out by its motion. The heuristic justification for this is quite general. In a Lorentz-invariant theory the string is invariant with respect to longitudinal Lorentz boosts; hence, one has to consider only transverse motions of the vortex. A transverse velocity $v_\perp$ results in a Lorentz contraction of the vortex thickness and thus, by the Lorentz invariance of $\mathcal{L}$, the vortex action must take the form

$$S = \int d^4x\, \mathcal{L} \propto \int dt\, d\ell\, \left(1 - v_\perp^2\right)^{1/2},$$

where $d\ell$ is the length element along the string. It can be shown that this is equivalent to the Nambu action for a string (6.1.9). (The equivalence is easily verified using the transverse gauge (6.2.11–12).) We will now proceed to derive this more carefully for the Nielsen–Olesen vortex-line [Förster, 1974; see also Gregory, 1988a; Maeda & Turok, 1988; Turok, 1988b].

The action for the abelian-Higgs model in a general spacetime is

$$S = \int d^4y \sqrt{-g} \left\{ |\mathcal{D}_\mu \phi|^2 - \tfrac{1}{4} F_{\mu\nu} F^{\mu\nu} - V(\phi) \right\}. \tag{4.4.1}$$

To obtain an effective string action, we need to find an approximate solution for a moving and curved string and then integrate out the massive transverse degrees of freedom. A Lorentz-boosted version of the static cylindrically-symmetric solution (4.1.11–12) provides an appropriate starting point if the string curvature is small relative to the string width. The approximate solution is then constructed around each point $x^\mu(\zeta^a)$ on a curved worldsheet swept out by the zeros of the Higgs field. The worldsheet coordinate ζ^0 is chosen to be timelike, with the other, ζ^1, spacelike. This parametrization and the worldsheet metric will be discussed in considerably more detail in §6.1.

Now at each point on the worldsheet we can choose two spacelike vectors n_μ^A, $A = 1, 2$, which are normal to the two tangent vectors $x_{,a}^\mu$, that is, $n_\mu^A x_{,a}^\mu = 0$. We also require these vectors to be orthonormal, $g^{\mu\nu} n_\mu^A n_\nu^B = -\delta^{AB}$. With this choice, we can reparametrize any point y^μ near the worldsheet using two additional radial coordinates ρ^A,

$$y^\mu(\xi) = x^\mu(\zeta) + \rho^A n_A^\mu(\zeta), \tag{4.4.2}$$

where $\xi^\mu = (\zeta^a, \rho^A)$ (refer to fig. 4.17). The new coordinates ξ^μ will be single-valued if the point y^μ is closer to the string than its curvature radius $\mathcal{R}$.

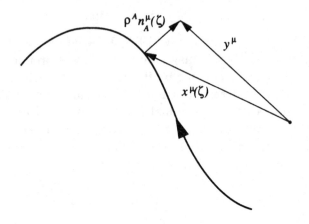

Fig. 4.17.   Local coordinate reparametrization in the vicinity of the string.

The approximate ansatz for the field configuration near the worldsheet is then

$$\phi\left(y(\xi)\right) = \phi_s(r)\,,$$
$$A^\mu\left(y(\xi)\right) = n^\mu_B(\zeta)A_s\,_B(r)\,,$$

$$(4.4.3)$$

where the solution (ϕ_s, A_s) is given in (4.1.11–12) and $r^2 = (\rho^1)^2 + (\rho^2)^2$. To change coordinates from y^μ to ξ^μ in the action (4.4.1) we must calculate the Jacobian

$$\sqrt{-g}\det\left(\frac{\partial y}{\partial \xi}\right) = (-\det M_{\alpha\beta})^{1/2}\,,$$

$$(4.4.4)$$

where

$$M_{\alpha\beta} = \left(g_{\mu\nu}\frac{\partial y^\mu}{\partial\xi^\alpha}\frac{\partial y^\nu}{\partial\xi^\beta}\right) = \text{diag}(\gamma_{ab}, -\delta_{AB}) + \mathcal{O}(r/R)\,.$$

$$(4.4.5)$$

Here, $\gamma_{ab} = g_{\mu\nu}x^\mu_{,a}x^\nu_{,b}$ is the worldsheet metric and so, defining $\gamma = \det\gamma^{ab}$, we clearly have $\det M_{\alpha\beta} \approx \gamma$. Similarly, for the other terms in (4.4.1), we must have $|\mathcal{D}\phi|^2 \approx |\mathcal{D}_s\phi_s|^2$ and $F^2 \approx F_s^2$ to an accuracy $\mathcal{O}(r/R)$. Integration over the two normal coordinates ρ^A, then yields the string energy per unit length μ as in (4.1.2). Because of the rapid exponential fall-off in the string profile (4.1.17), the apparent inaccuracies of order r/R are cut off at the string width δ and are reduced to order δ/R; these are two widely disparate scales, especially in a late cosmological context. To lowest order, then, the effective string action becomes

$$S = -\mu\int d^2\zeta\sqrt{-\gamma}\,.$$

$$(4.4.6)$$

We shall examine the dynamics predicted by this action in chapter 6. Higher-order corrections to the Nambu action have also been studied, but are the subject of some controversy [Gregory, 1988a; Maeda & Turok, 1988; Garfinkle & Gregory, 1990]. First-order corrections $\mathcal{O}(\delta/R)$ vanish by symmetry.

4.4.2 The Kalb–Ramond action

An analytic treatment of the global $U(1)$-string of §4.1.3 is hampered by difficulties in treating its 'topological' coupling to the Goldstone boson field, that is, the requirement that the phase ϑ changes by 2π as the string is encircled. The fact that the energy in this massless scalar field is considerably larger than that in the string core makes an understanding of global string dynamics intuitively hazardous. The role of this peculiar coupling, however, can be clarified by going over to an equivalent antisymmetric tensor representation in which the Kalb–Ramond action appropriately describes a global $U(1)$-string. This was first derived in the low energy limit by Witten [1985b] and Vilenkin & Vachaspati [1987] by considering only massless modes far from the string (see also Lund & Regge [1976]). However, here we follow an alternative derivation which is valid into the string core, similar to that presented by Davis & Shellard [1988c].

The well-known equivalence of a real massless scalar field and a two-index antisymmetric tensor $B^{\mu\nu}$ can be readily seen by considering the relation

$$\eta\partial_\mu\vartheta = \tfrac{1}{2}\epsilon_{\mu\nu\lambda\rho}\partial^\nu B^{\lambda\rho}\,. \tag{4.4.7}$$

The massless wave equation for ϑ, $\partial_\mu\partial^\mu\vartheta = 0$, is then trivially satisfied. On the other hand, a similar identity for ϑ, $\epsilon^{\mu\nu\lambda\rho}\partial_\lambda\partial_\rho\vartheta = 0$, implies the dual equation of motion for $B^{\mu\nu}$,

$$\partial_\mu H^{\mu\nu\lambda} = 0\,, \tag{4.4.8}$$

where $H^{\mu\nu\lambda}$ is the field strength tensor

$$H^{\mu\nu\lambda} = \partial^\mu B^{\nu\lambda} + \partial^\lambda B^{\mu\nu} + \partial^\nu B^{\lambda\mu}\,. \tag{4.4.9}$$

The action for the antisymmetric field is

$$S_H = \frac{1}{6}\int H^2 d^4x\,. \tag{4.4.10}$$

The coefficients in (4.4.7) have been chosen so that the energy–momentum tensors for the two dual fields coincide,

$$\begin{aligned}
T^\nu_\mu &= H_{\mu\alpha\beta}H^{\nu\alpha\beta} - \tfrac{1}{6}\delta^\nu_\mu H_{\alpha\beta\gamma}H^{\alpha\beta\gamma}\\
&= 2\eta^2\left(\partial_\mu\vartheta\partial^\nu\vartheta - \tfrac{1}{2}\delta^\nu_\mu\partial_\alpha\vartheta\partial^\alpha\vartheta\right),
\end{aligned} \tag{4.4.11}$$

where the action for the massless scalar field is (compare with (2.1.9))

$$S_\vartheta = \eta^2 \int d^4x \, (\partial_\mu \vartheta)^2 \,. \tag{4.4.12}$$

Note that the action (4.4.10) cannot be obtained by a direct substitution of (4.4.7) into (4.4.12) because the relation (4.4.7) only holds if the field equations for ϑ and $B_{\mu\nu}$ are satisfied – (4.4.7) does not admit arbitrary variations. However, the transition from (4.4.10) to (4.4.12) can be accomplished with a canonical transformation of the field variables [Davis & Shellard, 1988].

The periodic nature of ϑ and the presence of the massive field $|\phi|$ complicates the relationship between ϑ and $B^{\mu\nu}$ in the Goldstone model,

$$S = \int d^4x \left\{ (\partial_\mu |\phi|)^2 + |\phi|^2 (\partial_\mu \vartheta)^2 - V(\phi) \right\} \,. \tag{4.4.13}$$

The Euler–Lagrange equation for ϑ,

$$\partial_\mu (|\phi|^2 \partial^\mu \vartheta) = 0 \,, \tag{4.4.14}$$

suggests that in this model the relation (4.4.7) should be replaced by

$$|\phi|^2 \partial_\mu \vartheta = \tfrac{1}{2} \eta \epsilon_{\mu\nu\lambda\rho} \partial^\nu B^{\lambda\rho} \,. \tag{4.4.15}$$

As before, it is now easily verified that the action (4.4.13) is equivalent to

$$S = \int d^4x \left\{ (\partial_\mu |\phi|)^2 + \frac{\eta^2}{6|\phi|^2} H^2 - V(|\phi|) \right\} \,. \tag{4.4.16}$$

This equivalence, however, breaks down in the presence of vortices. The field equation for $H^{\mu\nu\sigma}$ that follows from (4.4.15) is

$$\partial_\sigma \left(\frac{\eta^2}{|\phi|^2} H^{\mu\nu\sigma} \right) = -\eta \epsilon^{\mu\nu\sigma\tau} \partial_\sigma \partial_\tau \vartheta \equiv 4\pi j^{\mu\nu} \,, \tag{4.4.17}$$

while the field equation obtained from the action (4.4.16) has zero on the right-hand side. In the presence of a vortex, ϑ is a multi-valued function of the coordinates and $j^{\mu\nu} \neq 0$ on the vortex core. Integrating over a two-surface orthogonal to the string yields the winding number of ϑ,

$$\int_S d^2x \, \epsilon^{\mu\nu} \partial_\mu \partial_\nu \vartheta = \int_{\partial S} d\vartheta = 2\pi \,. \tag{4.4.18}$$

Hence for a string lying along the z-axis,

$$j^{03} = \frac{\eta}{2} \delta(x) \delta(y) \,. \tag{4.4.19}$$

More generally, the commutator $\epsilon^{\mu\nu\sigma\tau}\partial_\sigma\partial_\tau\vartheta$ performs a projection onto the worldsheet swept out by the line of Higgs zeros in the core of the string. If the worldsheet is parametrized as $x^\mu(\zeta^a)$, then we can write

$$j^{\mu\nu} = \frac{\eta}{2}\int \delta^{(4)}\left[x - x(\zeta^a)\right]d\sigma^{\mu\nu}, \qquad (4.4.20)$$

where

$$d\sigma^{\mu\nu} = \epsilon^{ab}x^\mu_{,a}x^\nu_{,b}d^2\zeta, \qquad (4.4.21)$$

is the worldsheet area element.

The source term on the right-hand side of (4.4.17) can be accounted for in the action (4.4.16) by adding

$$4\pi\int B_{\mu\nu}j^{\mu\nu}d^4x = 2\pi\eta\int B_{\mu\nu}d\sigma^{\mu\nu}. \qquad (4.4.22)$$

The complete effective action for a global string in the antisymmetric tensor representation is then

$$S = -\mu_0\int\sqrt{-\gamma}\,d^2\zeta + \frac{1}{6}\int H^2 d^4x + 2\pi\eta\int B_{\mu\nu}d\sigma^{\mu\nu}. \qquad (4.4.23)$$

This action with an arbitrary coupling constant in the last term was originally introduced by Kalb & Ramond [1974]. The first term is the Nambu action (4.4.6) with a string tension μ_0 found by integrating over the massive $|\phi|$ modes in (4.4.16) for a static string solution, as discussed in §4.4.1. Although $|\phi|$ does not rise exponentially to its vacuum expectation value, its asymptotic properties (4.1.37) indicate that this transverse integral will converge rapidly. The apparent singularity in (4.4.16) at the vortex core when $|\phi| = 0$ is adequately compensated by the behaviour of $H^{\mu\nu\sigma}$ from (4.4.15).

The last term in the Kalb–Ramond action (4.4.23) now represents the interaction of the string with the antisymmetric field $B_{\mu\nu}$, replacing the earlier topological coupling of the Goldstone boson. This string coupling to $B_{\mu\nu}$ is local and linear and bears a striking resemblance to the pointlike interaction of the electron. Indeed, by exploiting the analogy of electromagnetism, the global string becomes immediately more accessible analytically, particularly for classical radiation problems. As such, the approach also presents similar difficulties like a divergent string self-energy which necessitates the introduction of cut-offs at both small and large scales. We shall address these issues in more detail in §8.3.

5

Superconducting strings

Superconductivity can be understood as a spontaneously broken electromagnetic gauge invariance. When the gauge invariance is broken, the photon acquires a mass and any magnetic field applied at the boundary of the superconductor decays exponentially towards its interior. The magnetic field is screened by a non-dissipative superconducting current flowing along the boundary – the well-known Meißner effect.

Cosmic strings can be turned into superconductors if electromagnetic gauge invariance is broken inside the strings. This can occur, for example, when a charged scalar field develops a non-zero expectation value in the vicinity of the string core. The electromagnetic properties of such strings are very similar to those of thin superconducting wires, but they are different from the properties of bulk superconductors. The string thickness does not typically exceed the electromagnetic penetration depth, and superconducting strings can be penetrated by electric and magnetic fields.

The idea that strings could become superconducting was first suggested in a pioneering paper by Witten [1985a]. Later it was realized that the role of the superconducting condensate could be played not only by a scalar field, but also by a vector field whose flux is trapped inside a non-abelian string [Preskill, 1985; Everett, 1988]. If the vector field is charged, the gauge invariance is again spontaneously broken inside the string. Witten also proposed another mechanism for string superconductivity, which operates in models where some fermions acquire their masses from a Yukawa coupling to the Higgs field of the string.

5.1 Bosonic string superconductivity

5.1.1 Scalar condensate models

The simplest example of scalar string superconductivity occurs in a toy model with two complex scalar fields ϕ and σ interacting with separate

$\tilde{U}(1)$ and $U(1)_Q$ gauge fields $\tilde{A}_\mu$ and A_μ, respectively [Witten, 1985a]. The first $\tilde{U}(1)$ is broken and gives rise to vortices as in §3.2.2. The second $U(1)_Q$ which we identify with electromagnetism, although unbroken in vacuum, can provide a charged scalar condensate in the string interior. The Lagrangian is merely a replicated version of the abelian-Higgs model (2.1.10),

$$\mathcal{L} = |\tilde{D}_\mu \phi|^2 + |D_\mu \sigma|^2 - V(\phi, \sigma) - \tfrac{1}{4}\tilde{F}^{\mu\nu}\tilde{F}_{\mu\nu} - \tfrac{1}{4}F^{\mu\nu}F_{\mu\nu}, \qquad (5.1.1)$$

where $\tilde{D}_\mu \phi = \partial_\mu \phi - ig\tilde{A}_\mu \phi$ and $D_\mu \sigma = \partial_\mu \sigma - ieA_\mu \sigma$. Additionally, the potential $V(\phi, \sigma)$ contains an appropriate ϕ–σ interaction term,

$$V(\phi, \sigma) = \tfrac{1}{4}\lambda_\phi \left(|\phi|^2 - \eta_\phi^2\right)^2 + \tfrac{1}{4}\lambda_\sigma \left(|\sigma|^2 - \eta_\sigma^2\right)^2 + \beta|\phi|^2|\sigma|^2. \qquad (5.1.2)$$

Particular parameter choices are required to engineer the respective roles of the scalar fields; for sufficiently large β, the last cross-coupling term prevents both $U(1)$s from being broken simultaneously.

We choose $\tilde{U}(1)$ to be broken in the ground state of the theory, which is ensured by the constraint

$$\lambda_\phi \eta_\phi^4 > \lambda_\sigma \eta_\sigma^4. \qquad (5.1.3)$$

The effective mass term for the sigma particles is then

$$m_\sigma^2 = \beta\eta_\phi^2 - \tfrac{1}{2}\lambda_\sigma \eta_\sigma^2. \qquad (5.1.4)$$

If electromagnetism $U(1)_Q$ is to remain unbroken, this m_σ^2 must be positive, ensuring that the vacuum with $|\phi| = \eta_\phi$, $|\sigma| = 0$ is stable. This requirement implies a second constraint

$$\frac{2\beta}{\lambda_\sigma} \geq \frac{\eta_\sigma^2}{\eta_\phi^2}. \qquad (5.1.5)$$

At the centre of the vortex the ϕ-field becomes small, thus eliminating the positive mass-squared term in (5.1.4) which drives σ to zero. It may become energetically favourable for the $U(1)_Q$ to be broken by the formation of a bound state solution with $\sigma \neq 0$. One can determine whether such a condensate will actually form by considering the stability of the $\tilde{U}(1)$ string solution with $\sigma = 0$ [Witten, 1985a]. For a string lying in the z-direction we look at the equation for small fluctuations about $\sigma = 0$, with solutions of the form $\sigma(x, y, z, t) = |\sigma(x, y)|e^{i\omega t}$,

$$\left(-\partial_x^2 - \partial_y^2\right)|\sigma| + V(r)|\sigma| = \omega^2|\sigma|, \qquad (5.1.6)$$

where $V(r) = \beta|\phi(r)|^2 - \lambda_\sigma \eta_\sigma^2/2$ and $r^2 = x^2 + y^2$. There are two regimes to consider depending on the magnitude of $m_\sigma(\infty)$, which determines the degree to which the condensate wavefunction is localized.

If m_σ is large, with $\beta > \eta_\phi$, then the walls of the vortex potential are steep and the problem can be treated by using standard results for a harmonic oscillator. Given that $|\phi(r)| \approx (\sqrt{\lambda_\phi}\,\eta_\phi{}^2)\,r$ near the vortex centre,* there are negative eigenvalue solutions to (5.1.6) only if [Haws, Hindmarsh & Turok, 1988; Thompson, 1988b]

$$\frac{\beta}{\lambda_\sigma} \lesssim \frac{1}{4}\frac{\lambda_\sigma \eta_\sigma^4}{\lambda_\phi \eta_\phi^4} \ . \tag{5.1.7}$$

For parameters satisfying (5.1.7), the $\sigma = 0$ state will be unstable and a lower energy configuration with a $\sigma \neq 0$ bound state will form. The condensate thickness in this case is $\delta_\sigma \lesssim \delta_\phi$, where $\delta_\phi \sim \lambda_\phi^{-1/2}\eta_\phi^{-1}$ is the width of the scalar string core, and the long-distance fall-off of the condensate is of the form $\sigma \propto \exp(-m_\sigma r)$.

One can best illustrate the restricted condensate solution space by employing the dimensionless parameters $\tilde{\lambda} = \lambda_\sigma/\lambda_\phi$, $\tilde{m}^2 = \lambda_\sigma \eta_\sigma^2/(2\lambda_\phi \eta_\phi^2)$ and $\tilde{\beta} = \beta/\lambda_\phi$, in terms of which the constraints (5.1.3), (5.1.5) and (5.1.7) become

$$\frac{\tilde{\lambda}}{4} > \tilde{m}^4 \gtrsim \tilde{\beta} > \tilde{m}^2 \ . \tag{5.1.8}$$

Note that it follows from (5.1.8) that $\tilde{m}$, $\tilde{\lambda}$ and $\tilde{\beta}$ should all be greater than unity. The parameter space for this toy $\tilde{U}(1) \times U(1)_Q$ model has been studied with numerical relaxation techniques using the cylindrically-symmetric field equations by Amsterdamski & Laguna [1988], Babul, Piran & Spergel [1988a,b], Haws *et al.* [1988] and Davis & Shellard [1988a,b] and, prior to this, with a variational approach using an approximate ansatz by Hill, Hodges & Turner [1987, 1988] and MacKenzie [1987]. All studies have been qualitatively in accord, revealing the possibility of the formation of a scalar condensate within a fairly natural parameter space.

Equations (5.1.7) and (5.1.8) do not apply when the two terms in (5.1.4) are fine-tuned to be almost equal, so that $m_\sigma^2 \ll \beta\eta_\phi^2$. The condensate thickness is then much greater than the width of the string core, and the field $\sigma(r)$ can be treated as a constant in the string interior. Using the standard result that the energy level in a weak two-dimensional potential well is exponentially small [Landau & Lifshitz, 1977], it can be shown that negative eigenmode solutions to (5.1.6) exist only when $m_\sigma^2/\beta\eta_\phi^2 \lesssim (\lambda_\phi/\beta)\exp(-C\lambda_\phi/\beta)$, where C is a numerical constant [Haws

---

* We should note that in this and the ensuing qualitative discussions we assume a Type II regime with $\lambda_\phi/2g^2 > 1$, and neglect the effect of the vortex gauge field $\tilde{A}_\mu$ on the scalar width δ_ϕ.

et al., 1988]. We see that for small values of β/λ_ϕ the parameter space for condensate solutions becomes extremely restrictive. The bulk of the allowed parameter space is therefore given by the condition (5.1.8).

An example of a scalar condensate rising up in the presence of a vortex is shown in fig. 5.1. In this case, the bounds (5.1.3) and (5.1.5) are near saturation, resulting in a significant back-reaction on the underlying vortex and a condensate which is not particularly well-localized. This solution illustrates a state between the limiting cases of small and large mass m_σ.

In summary, then, while we have shown that solutions exist in these scalar condensate models, it is evident that this form of string super-conductivity depends sensitively on the detailed quantitative structure of the Higgs sector. In contrast, other forms of superconductivity which we shall discuss later in this chapter are less dependent on such energetic considerations. Non-abelian strings may possess charged vector conden-sates for essentially topological reasons. Trapped fermionic modes, arising from particular Yukawa couplings, are also ultimately topological in ori-gin. Nevertheless, here we will continue employing the simplest scalar model (5.1.1) because it exhibits most of the essential features of string superconductivity.

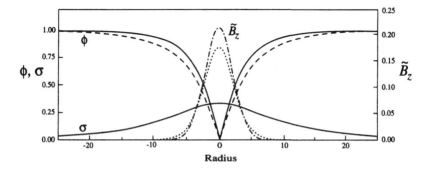

Fig. 5.1. Profile of the fields of a superconducting vortex solution. The un-derlying vortex solution with no condensate is shown with solid (ϕ) and dot-dashed ($\widetilde{B}_z$) lines. The perturbed vortex with a condensate σ shows signifi-cant back-reaction on ϕ (dashed) and $\widetilde{B}_z$ (dotted) for this choice of parameters ($\lambda_\phi - 0.05, \eta_\phi - 1, e = 1, \lambda_\sigma = 1, \eta_\sigma = 0.45, g = 1$ and $\beta = 1.1$).

5.1.2 Persistent currents

From the Lagrangian (5.1.1) we can derive the usual electromagnetic current density (4.1.10),

$$j_\mu = ie\left(\bar{\sigma}D_\mu\sigma - \sigma\bar{D}_\mu\bar{\sigma}\right). \qquad (5.1.9)$$

Since $\sigma \to 0$ away from the string core, the only significant currents on scales larger than the condensate width δ_σ, will flow in the longitudinal string direction. For a string lying along the z-axis, we take the ansatz $\sigma = |\sigma|e^{i\theta(z)}$, where $|\sigma|$ depends only on x and y. The total current J can then be found by integrating over the string cross-section. Assuming that the vector potential A_μ remains approximately constant across the string we have

$$J = 2\Sigma e\left(\partial_z\theta + eA_z\right), \qquad (5.1.10)$$

where

$$\Sigma = \int dxdy\, |\sigma|^2. \qquad (5.1.11)$$

Analytic estimates and numerical calculations show that, for small couplings, $\Sigma \gg 1$ in a large portion of the parameter space [Witten, 1985a; Hill, Hodges & Turner, 1988]. For example, if the two terms in (5.1.4) for m_σ^2 are comparable, $\beta\eta_\phi^2 \sim \lambda_\sigma\eta_\sigma^2 \sim m_\sigma^2$, then the condensate thickness is $\delta_\sigma \sim m_\sigma^{-1}$ and

$$\Sigma \approx \delta_\sigma^2\eta_\sigma^2 \sim \lambda_\sigma^{-1}. \qquad (5.1.12)$$

The assumption that θ and A_μ do not change over the condensate cross-section is justified if the electromagnetic penetration depth $\delta_A \sim (e\eta_\sigma)^{-1}$ is large compared to δ_σ, that is, when $e^2\Sigma \ll 1$. In the opposite limit, the current flows in a thin outer layer of the condensate [Bucher, 1992], and the effective value of Σ in (5.1.10) is $\Sigma_{\rm eff} \sim \eta_\sigma^2\delta_\sigma\delta_A \sim \sqrt{\Sigma}/e$.

For $\delta_A \lesssim \delta_\sigma$, the condensate and electromagnetic fields of a current-carrying string can be found using a more general ansatz [Alford *et al.*, 1991]

$$\sigma(\mathbf{x}) = \sigma_0(r)e^{-is(r)z}, \quad A_\mu(\mathbf{x}) = e^{-1}z\,\partial_\mu s(r), \qquad (5.1.13)$$

where $\sigma_0(r)$ is the unperturbed σ-field and the functions $\sigma_0(r)$ and $s(r)$ are assumed to be real. This ansatz has the form of a gauge transformation in the xy-plane, but because of the z-dependence it represents a true physical excitation of the system. We can use a gauge transformation to remove the phase of $\sigma(\mathbf{x})$, resulting in

$$\sigma = \sigma_0(r), \quad A_\mu = -e^{-1}\delta_\mu^3 s(r). \qquad (5.1.14)$$

The function $s(r)$ is determined by substituting (5.1.13) or (5.1.14) in the field equations and linearizing in $s(r)$. This gives

$$\frac{1}{r}\partial_r(r\partial_r s) = 2e^2\sigma_0^2(r)s \,, \tag{5.1.15}$$

which is essentially the London equation for a cylindrical superconductor with a penetration depth $\delta_A(r) = [e\sigma_0(r)]^{-1}$. At large distances from the string, $s(r) \propto \ln r$, which corresponds to the magnetic field produced by a dc current, $B \propto r^{-1}$. Inside the condensate, the field decays towards the string axis on a length scale $\sim \delta_A$.

We should emphasize that a detailed analysis of string electrodynamics in the regime when $\delta_A \lesssim \delta_\sigma$ has not yet been given, and we shall continue with (5.1.10), assuming that Σ_{eff} is used instead of Σ when $e^2\Sigma > 1$.

A closed loop of string can possess a net winding number in θ,

$$N = \frac{1}{2\pi} \oint \frac{d\theta}{d\zeta} d\zeta \,. \tag{5.1.16}$$

N is conserved because, being an integer, it cannot change continuously. For a fixed N, the lowest energy configuration of the field σ in a stationary loop will be time-independent with a constant tangential electromagnetic current J. Since N is conserved, this current is persistent and the string will behave like a superconducting wire.

To estimate the expected electromagnetic current J for a given winding number density $d\theta/d\zeta$, we assume the string to be locally straight relative to the condensate width. We can employ the standard vector potential solution for a static wire in the Coulomb gauge, $\nabla \cdot \mathbf{A} = 0$,

$$A_i(x) = -\frac{1}{4\pi} \oint dx^i \frac{J}{|\mathbf{x} - \mathbf{x}(\zeta)|} \,. \tag{5.1.17}$$

The solution $\mathbf{A}(x)$ diverges as $\mathbf{x} \to \mathbf{x}(\zeta)$ but only because of a zero string width idealization. An appropriate cut-off at the actual string thickness δ can be introduced to provide a realistic estimate. The physical situation we envisage is a current-carrying loop of astrophysical proportions with a circumference $2\pi R$. Since this scale is so much greater than δ, it is clear that the logarithmically growing part of the integral (5.1.17) will dominate. At the string core, therefore, we must have

$$A_z = -\frac{1}{2\pi} \ln\left(\frac{R}{\delta}\right) J \,. \tag{5.1.18}$$

For a constant current J flowing along the string, the winding number density will also be constant in this gauge,

$$n \equiv \frac{d\theta}{d\zeta} = \frac{N}{R} \,. \tag{5.1.19}$$

Using the expression for the current (5.1.10) and substituting (5.1.18) and (5.1.19), we obtain the total current flowing around the loop [Witten, 1985a]

$$J = \frac{2\Sigma e}{1 + (\Sigma e^2/\pi) \ln (R/\delta)} \frac{N}{R}.$$

(5.1.20)

For cosmological length scales, $\ln (R/\delta) \sim 100$, and with $\Sigma \gg 1$ this becomes

$$J = \frac{2\pi}{e \ln (R/\delta)} \frac{N}{R}.$$

(5.1.21)

This is considerably smaller than might be anticipated naively, since $J = 2\Sigma e\, N/R$ for a 'neutral' current-carrying string without gauge fields. However, for superheavy strings, the current (5.1.21) can build up to levels of astrophysical significance.

5.1.3 Current quenching

Given (5.1.20), one might imagine the string carrying arbitrarily large currents simply by inducing a high winding number per unit length $n = N/R$. This would occur, for example, during loop contraction and collapse. However, there is inevitably a critical value beyond which the current saturates. This can be most simply seen in the original Lagrangian (5.1.1) since the current for a locally straight string contributes an effective positive mass-squared term

$$\mathcal{L}_N \equiv |D_z\sigma|^2 = -\sigma^2(n - eA_z)^2.$$

(5.1.22)

For a sufficiently large winding number density n, this counteracts the symmetry breaking term in the potential, thus quenching the condensate down toward $\sigma = 0$ at the centre of the vortex and suppressing current flow. Such shrinkage of the condensate, partially through radial 'pinching', is illustrated in fig. 5.2. In fig. 5.3, the total current J is illustrated as n is increased, simulating the adiabatic collapse of a large circular loop with fixed winding number N. The current rises to a maximum, beyond which it falls increasingly dramatically.

We can make a crude analytic estimate of this critical current by considering the σ-dependent terms (5.1.22) in the original Lagrangian (5.1.1). Choosing the z-axis to align with the string, we can represent the relevant θ-derivative term as

$$D_z\theta \equiv \frac{d\theta}{dz} - eA_z.$$

(5.1.23)

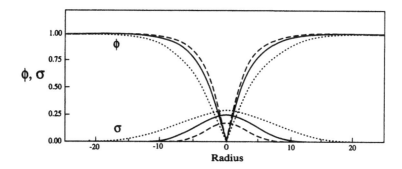

Fig. 5.2. Profiles of the scalar condensate as it is quenched by increasing the current or winding number density n (dashed lines). The broadening effect of increasing charge density ω is noted with the dotted lines (refer to §5.1.3) [Davis & Shellard, 1988b].

Since $(D_z\theta)^2$ acts as a negative mass-squared term, the σ-field will not attain its original expectation value $\sigma_0 \approx \eta_\sigma$ in the vortex core but, instead,

$$\sigma_0{}^2 \approx \eta_\sigma{}^2 - 2(D_z\theta)^2/\lambda_\sigma\,. \qquad (5.1.24)$$

The primary effect on the current will be through the integral of $|\sigma|^2$ over the cross-section (5.1.10), so in terms of the original Σ the current becomes

$$J \approx 2\Sigma e(D_z\theta)\left[1 - 2\Sigma(D_z\theta)^2/\eta_\sigma^2\right]\,. \qquad (5.1.25)$$

This has a maximum at approximately $(D_z\theta)_{\text{max}} \approx \eta_\sigma/\sqrt{\Sigma}$, thus yielding the maximum current [Thompson, 1988b],

$$J_{\text{max}} \lesssim \sqrt{\Sigma}\,e\eta_\sigma\,. \qquad (5.1.26)$$

The previously quoted numerical investigations indicate that the magnitude of the critical current is generally lower than these crude expectations (5.1.26) in a natural parameter range. The highest critical currents are found for the largest condensates with small values of the quartic coupling λ_σ which corresponds to the narrow region when the two bounds (5.1.3) and (5.1.5) are simultaneously near saturation.

As illustrated in fig. 5.3, beyond a certain point, the current begins to fall as the winding number increases. This is analogous to anti-Lenz behaviour in an applied electric field, a response which signals the onset of instability. Indeed, an instability can be easily demonstrated by considering inhomogeneous perturbations in $D_z\theta$ of the form [Thompson, 1988b]

$$D_z\theta - \langle D_z\theta\rangle + \epsilon\cos(k_z z)\,, \qquad (5.1.27)$$

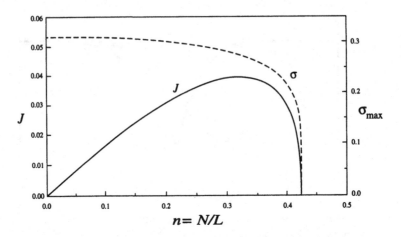

Fig. 5.3. Dependence of the total current on the winding number density n. The current begins to increase linearly with n, but as the condensate σ is quenched the current reaches a maximum and begins decreasing.

where $\langle D_z\theta\rangle$ remains constant and k_z is an arbitrary wavenumber. The string will be unstable to this perturbation if the second derivative of the current energy W becomes negative, $\partial^2 W/\partial\epsilon^2 < 0$. Now the equation for the current density, $\mathbf{j} = e\delta W/\delta\mathbf{A}$, implies that $\delta^2 W/\delta(D_z\theta)^2 \propto \delta J/\delta(D_z\theta)$. By (5.1.27) this is in turn proportional to $\delta^2 W/\delta\epsilon^2$ and, thus, will become negative when J becomes a decreasing function of $D_z\theta$ after attaining its maximum J_{max}. Consequently, we can expect the decreasing branch of the current diagram shown in fig. 5.3 to be unstable. Rather than a uniform quench for these super-critical currents, inhomogeneities in the winding number density will grow and the condensate will become preferentially pinched in localized regions. This classical 'neck' or 'sausage' instability can be demonstrated with numerical simulations of current-carrying strings as shown in fig. 5.4. A super-critical current flowing along a string breaks up and becomes temporarily punctuated by 'normal' regions ($\sigma = 0$) where the winding number is able to unravel. Once winding number is lost the condensate can spring back to a superconducting state with sub-critical current, creating a variety of massive excitations which may leave the string.

Even before reaching these super-critical levels, a superconducting current in a string loop can decay by more quiescent magnetic flux-line tunnelling [Witten, 1985a]. This is directly analogous to the process by which the flux through a superconducting wire loop decays by one quantum. A short flux tube in the σ-condensate is created on one side of a superconducting string by quantum or thermal effects and then migrates across it (see fig. 5.5). Such a process changes the total winding number N

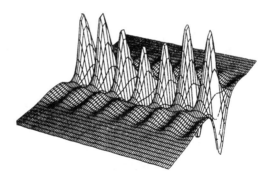

Fig. 5.4. Instability onset for a super-critical current in a 'neutral' current-carrying string [Shellard, 1989]. The winding of the condensate field σ is shown by plotting the real part $\mathrm{Re}(\sigma)$. The amplitude of σ indicates that it is becoming 'pinched' in the central region.

around the loop (5.1.16) by one, thus providing a current decay mechanism. There is a significant energy cost to this process, however, since a vortex of length δ_σ and mass per unit length $\sigma_0^2 \sim \eta_\sigma^2$ must be created in the condensate field. For small currents the action for this process will be approximately

$$S_{\mathrm{E}} \approx \sigma_0^2 \delta_\sigma^2 \sim \Sigma. \tag{5.1.28}$$

Estimating the tunnelling rate from the expression $\Gamma \simeq m_\sigma^2 \, e^{-S_{\mathrm{E}}}$ implies a further constraint on physically interesting bosonic models if superconducting currents are to persist over astronomical timescales [Zhang, 1987; Haws *et al.*, 1988]. Consequently, the model parameters must be adjusted such that $\Sigma \gg 1$ in order to maintain stability – parameter values which yield the largest currents in any case.

Finally, we note that it is in principle possible to choose parameters for $J_{\max}$ in (5.1.26) such that the energy in the superconducting current exceeds that in the underlying vortex. This longitudinal current contributes a positive pressure counteracting the vortex-string tension, so we can envisage loops with large currents being stabilized against collapse and effectively becoming 'springs' [Ostriker, Thompson & Witten, 1986; Copeland *et al.*, 1988]. For a large loop $R \gg \delta$, the dominant energy contribution associated with the current will be given by logarithmic terms from the magnetic fields $|B|^2$. Spring solutions will exist if

$$J_{\max}^2 \ln(R/\delta) \gtrsim \mu, \tag{5.1.29}$$

where $\mu \sim \eta_\phi^2$ is the vortex energy. Clearly, springs will be associated with the same parameter values which provide large stable currents, that is, for $\lambda_\sigma \ll 1$. Models for which we expect (5.1.29) to hold must confront the potential problem that stable spring loops may come to dominate the energy density of the universe at an early epoch – an analogue of the 'monopole problem'. The degree to which springs are a generic phenomenon has been discussed by several authors [Hill, Hodges & Turner, 1988; Babul *et al.*, 1988a; Haws *et al.*, 1988]. However, the stability of these objects to non-radial perturbations and the processes by which they would form dynamically remain to be studied in detail.

5.1.4 Worldsheet charge

As well as persistent currents, superconducting strings can accumulate net electric charge – the phase θ acquires a time-dependence as in the Q-balls discussed in §3.2.6. Indeed, if cosmic strings are expected to carry high current densities, then charge densities of comparable magnitude must also be considered. Not only will charge become trapped on the strings through the Kibble mechanism at formation, the continual reconnection of strings with unequal currents, will ensure the presence of significant charge densities at all times.

The electric charge Q on a closed loop of string is given by

$$Q = \Sigma \oint d\zeta \, (\partial_t \theta - eA_0) \,. \tag{5.1.30}$$

The additional time-dependent terms in the Lagrangian (5.1.1) will act as a negative mass-squared term,

$$\mathcal{L}_Q = \sigma^2 (\omega - eA_0)^2 \,, \tag{5.1.31}$$

where $\omega \equiv \partial\theta/\partial t$. Unlike the current terms $\mathcal{L}_N$ in (5.1.22), $\mathcal{L}_Q$ will tend to increase the expectation value of $|\sigma|$ in the string core and to enlarge the condensate. Fig. 5.2 contrasts condensate quenching due to an increase in $n = N/2\pi R$ with the growth due to an increase in $\omega = Q/2\pi\Sigma R$.

Like superconducting currents, such string charge densities cannot be expected to grow indefinitely. For a sufficiently large ω, the effective mass m_σ off the string is driven toward zero and there is only a small energy barrier preventing the σ-particles from leaving. The large electric field near the string will further enhance tunnelling processes.

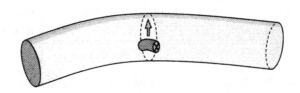

Fig. 5.5. Creation and migration of a flux tube through the condensate in a superconducting string loop allows the current J to relax by one unit.

It is interesting to note that the presence of string charge will tend to counteract current quenching; since Q must be conserved, the $\sigma = 0$ state is classically unattainable. Idealized simulations of contracting string loops maintaining fixed quantum numbers N and Q, show that a state with equal charge and current density is approached, $\omega/n \to 1$, as the radius is reduced [Davis & Shellard, 1988b]. For the attractor 'chiral' current state with $\omega = n$, quenching is eliminated completely because the effects of $\mathcal{L}_N$ and $\mathcal{L}_Q$ cancel. This indicates that for a sufficient initial charge to current ratio, currents may grow to considerably higher levels than previously anticipated. Moreover, despite large electric fields near the string, for a chiral current, pair production is suppressed because the electromagnetic field is null, $|E| = |B|$ (refer to §5.3.1). Another consequence is that strings which initially trap charges, say at a phase transition, may become superconducting even if the presence of a non-zero condensate is not energetically favoured on a neutral string.

A string possessing both charge and current densities will have a longitudinal momentum per unit length,

$$P_z = \Sigma(\omega - eA_0)(n - eA_z).\qquad(5.1.32)$$

In the context of a circular superconducting loop of radius R, this implies that it will possess a total angular momentum

$$M = 2\pi R^2 \Sigma(\omega - eA_0)(n - eA_z).\qquad(5.1.33)$$

It is possible that stationary vortex ring solutions or 'vortons' may exist which are stabilized against further collapse by the conservation of this angular momentum. Stability against quantum tunnelling is anticipated for a sufficiently large radius [Davis, 1988]. Vortons, like springs, may create a cosmological dark matter problem and so potentially constrain realistic models for superconducting strings [Davis & Shellard, 1989a].

5.1.5 Microphysical string interactions

The dynamics of bosonic superconducting strings have been studied numerically in three-dimensional simulations [Laguna & Matzner, 1990]. These employed the same finite-difference techniques in the Lorentz gauge which were described in §4.3.2, but faced additional difficulties in matching initial data for the electromagnetic gauge fields because these fall off slowly, $A(r) \sim \mathcal{O}(r^{-1})$. Perpendicular string collisions were studied for a variety of initial currents and impact velocities. Irrespective of the direction and magnitude of the superconducting currents at string crossing, reconnection always occurred. This was governed wholly by the topology of the underlying vortex fields, as described for 'normal' strings in §4.3.4. However, the supercurrents and magnetic fields had a significant effect on string dynamics as can be seen from fig. 5.6. One can define the current **J** to be 'parallel' or 'antiparallel' on the basis of whether its direction coincides with the topological flux direction $\widetilde{\mathbf{B}}$ of the underlying string. The most striking back-reaction effects are seen when strings with parallel and antiparallel currents meet because, after reconnection, each string will possess opposite currents feeding into the interaction region. This leads to current cancellation and a net build-up of charge between the reconnection 'kinks' (refer to §6.4.1). The presence of the charge causes the string to 'fatten' (fig. 5.6) as described above in §5.1.4. There is no reason to expect the directions of supercurrent and the string topology to be correlated on different strings, so an interacting network of superconducting strings can be expected to exhibit large charge fluctuations.

We have already alluded to simulations with 'neutral' current-carrying global strings with no $U(1)_Q$ gauge field [Shellard & Davis, 1988]. Similar qualitative results were obtained without boundary and gauge matching problems. It is of interest to note that these global $\widetilde{U}(1) \times U(1)_Q$ models possess stable Q-ball solutions [Shellard & Davis, 1988; Frieman & Lynn, 1990]. The end product of the collapse of a current-carrying charged loop may be a relic Q-ball possessing a net angular momentum.

5.1.6 An effective action

In deriving an effective action for the superconducting string, we assume that the current is much smaller than the critical current. The energy of a straight string lying along the z-axis is minimized with a scalar condensate described by $\sigma_0(x, y)$. Low-energy excitations will take the form,

$$\sigma(x, y, z, t) = e^{i\theta(z,t)}\sigma_0(x, y), \qquad (5.1.34)$$

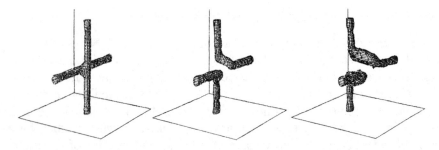

Fig. 5.6. Reconnection of two superconducting strings with antiparallel currents [Laguna & Matzner, 1990]. Plots of the scalar condensate width are shown. The current missmatch causes a net charge build-up between the propagating reconnection kinks – note the condensate 'fattening' due to charge back-reaction.

where $\theta(z,t)$ is a slowly-varying function on scales comparable to the condensate thickness. All the θ-dependent terms in the Lagrangian (5.1.1) come entirely from $|D_\mu\sigma|^2$, and we employ the ansatz (5.1.34) in evaluating the contribution of these to the action. We make the additional assumption that the vector potential A_μ is also slowly varying on the scale of the condensate thickness. Following the same procedure detailed in §4.4.1, we substitute (5.1.34) in the action and integrate over the transverse directions to obtain a low-energy effective action [Witten, 1985a],

$$S = \Sigma \int dz\, dt\, (\partial_a\theta + eA_a)^2 \,. \tag{5.1.35}$$

Here, θ and A_a are functions of t and z only, the index a runs over the two string worldsheet coordinates (t,z), and Σ is the cross-sectional integral over $|\sigma|^2$ defined in (5.1.11).

As we discussed in §5.1.2, the ansatz (5.1.34) with A_μ slowly varying across the condensate is justified only when $e^2\Sigma \ll 1$. In the opposite limit of large $e^2\Sigma$ the current is confined to a thin surface layer and the effective value of Σ is $\Sigma_{\text{eff}} \sim \sqrt{\Sigma}/e$. We shall assume that the effective action (5.1.35) can still be used in this case if A_a is understood to be the electromagnetic potential near the condensate surface and Σ is replaced by Σ_{eff}.

From the action (5.1.35) we can compute the electromagnetic current density,

$$j^\mu = -\frac{\delta S}{\delta A_\mu} = J^\mu(z,t)\delta(x)\delta(y)\,, \tag{5.1.36}$$

where $J_1 = J_2 = 0$ and

$$J_a = -2\Sigma e\,(\partial_a\theta + eA_a)\,. \tag{5.1.37}$$

The two-component vector J_a is a conserved current on the string worldsheet,

$$\partial_a J^a = 0\,, \tag{5.1.38}$$

since this conservation law is just the Euler–Lagrange equation for θ. The total electric current through the string cross-section is $J = J^z$.

An alternative and more useful form of the effective action (5.1.35) can be obtained if we note that in two dimensions a conserved current such as J^a can always be written as a derivative of a scalar field φ (also confined to the worldsheet). That is, we take

$$J^a = q\epsilon^{ab}\partial_b\varphi\,, \tag{5.1.39}$$

where the rescaled charge coupling is given by

$$q = e\sqrt{2\Sigma}\,. \tag{5.1.40}$$

Note that for $e^2\Sigma \gtrsim 1$ this gives a large value for the charge, $q \gtrsim 1$. However, as we shall see in §5.1.7 and §8.4.3, electromagnetic self-interactions renormalize the charge such that $q_{\text{ren}} \ll 1$. A comparison of the currents (5.1.39) and (5.1.37) yields

$$\partial^a \varphi = -q \epsilon^{ab} \left(\partial_b \theta + e A_b \right) . \tag{5.1.41}$$

Differentiating the last equation with respect to x^a, we obtain

$$\partial_a \partial^a \varphi = -q \epsilon^{ab} \partial_a A_b . \tag{5.1.42}$$

Both this equation and the current (5.1.39) can be derived from a variation of the action [Witten, 1985a]

$$S = \int dz \, dt \left[\tfrac{1}{2} (\partial_a \varphi)^2 - q A_a \epsilon^{ab} \partial_b \varphi \right] . \tag{5.1.43}$$

This form of the action is equivalent to (5.1.35) but is more convenient for most applications. The generalization of this action for an arbitrary worldsheet $x^\mu(\zeta^a)$ is then

$$S = \int d^2\zeta \left\{ -\mu\sqrt{-\gamma} + \tfrac{1}{2}\sqrt{-\gamma}\gamma^{ab}\varphi_{,a}\varphi_{,b} - q A_\mu x^\mu_{,a} \epsilon^{ab} \varphi_{,b} \right\} , \tag{5.1.44}$$

where γ_{ab} is the usual worldsheet metric defined in (4.4.6). The first term in (5.1.44) is the familiar Nambu action from §4.4.1, the second is the worldsheet supercurrent correction, the third is the current coupling to the electromagnetic gauge potential, and we have omitted the usual $\tfrac{1}{4}F^2$ contribution of the external electromagnetic fields.

5.1.7 Current growth in an electric field

As a final exercise in scalar vortex superconductivity, we shall use the action (5.1.43) to derive the fundamental equation of string superconductivity and discuss some qualitative aspects. Consider the current induced in a z-directed string by a homogeneous (z-independent) electric field **E**. The symmetry of the problem suggests that the current J^a is also z-independent, and the equation for φ (5.1.42) takes the form

$$\ddot{\varphi} = -qE . \tag{5.1.45}$$

where $E = E_z$ is the component of **E** directed along the string. The electric current through the string cross-section is then

$$J = J^z = -q\dot{\varphi} . \tag{5.1.46}$$

The combination of these last two equations yields

$$\frac{dJ}{dt} = q^2 E , \tag{5.1.47}$$

from which we can immediately observe that the string is superconducting. The current in the string builds up if an electric field is applied and remains constant if the field is removed. In contrast, for a wire of finite conductivity σ, the current is $J = \sigma E$ when the field is applied, and it decays when the field is switched off.

It is important to note the significance of the self-inductance implicit in (5.1.47). The electric field E in (5.1.47) includes both the applied field E_0 and the induced field due to the current itself. With this effect taken into account (5.1.47) takes the form

$$\frac{dJ}{dt} = q^2 \left(E_0 - L\frac{dJ}{dt} \right) , \qquad (5.1.48)$$

where L is the string inductance per unit length given by

$$L \approx 2\ln\left(\frac{R}{\delta}\right) . \qquad (5.1.49)$$

The cut-off radius R depends on the physical situation. It can be set by the typical scale of variations in the electric field (for example, the wavelength of an electromagnetic wave) or else by the size of a closed loop of string. Substituting L from (5.1.49) into (5.1.48) we obtain

$$\frac{dJ}{dt} = q^2_{\text{ren}} E_0 , \qquad (5.1.50)$$

where the renormalized charge q_{ren} is given by

$$q^2_{\text{ren}} = \frac{q^2}{1 + 2q^2 \ln(R/\delta)} . \qquad (5.1.51)$$

In a cosmological setting, the logarithm in (5.1.50) is large (~ 100). Hence, for natural parameter values, when q is not very small, we will typically have $q^2_{\text{ren}} \sim 10^{-2}$.

5.1.8 Gauge boson superconductivity

In non-abelian string models, the elements of the unbroken group H do not generally commute with the string generator T^S (refer to §4.2). If T^S does not commute with the electromagnetic charge generator Q, then the gauge field of the string is charged and can play the role of a superconducting condensate. This type of string superconductivity arises as a consequence of the non-abelian group structure and, therefore, appears to be more natural than the scalar model of §5.1.1 which depends on the details of the Higgs potential. Everett [1988] presented the first working example of this phenomenon, the $SO(10)$-strings of §2.1.4 and §4.2.3, although the possibility was previously suggested by Preskill [1985].

For $[T^S, Q] \neq 0$, the electromagnetic generator varies with the angle θ around the string,

$$Q(\theta) = U(\theta)Q(0)U^{-1}(\theta), \qquad (5.1.52)$$

where $U(\theta) = \exp(iT^S\theta)$. We shall assume that $[U(2\pi), Q(0)] = 0$, so that $Q(\theta)$ is single-valued. Otherwise the string would be an Alice string and its electromagnetic properties would be rather unusual (see §4.2.4).

By analogy with (5.1.13), the Higgs and gauge fields of a current-carrying string can be sought using the ansatz [Alford *et al.*, 1991]

$$\phi = \Omega\phi^{(0)}, \quad \mathcal{A}_\mu = \Omega\mathcal{A}_\mu^{(0)}\Omega^{-1} + e^{-1}f(z,t)\,\partial_\mu s(r,\theta), \qquad (5.1.53)$$

where $\mathcal{A}_\mu = A_\mu^a T^a$, T^a are the group generators in the appropriate representation to act on the field ϕ, $\phi^{(0)}$ and $\mathcal{A}_\mu^{(0)}$ are the fields of the unperturbed string, and

$$\Omega = \exp[-if(z,t)\,s(r,\theta)]. \qquad (5.1.54)$$

The function $s(r,\theta)$ takes values in the Lie algebra and is proportional to $Q(\theta)$ at large r. A gauge transformation by Ω^{-1} brings (5.1.53) to an equivalent form, which is similar to (5.1.14),

$$\phi = \phi^{(0)}, \quad \mathcal{A}_\mu = \mathcal{A}_\mu{}^{(0)} - e^{-1}s(r,\theta)\partial_\mu f(z,t). \qquad (5.1.55)$$

Substituting (5.1.55) into the field equation for ϕ one finds to linear order that

$$\partial_\mu\partial^\mu f(z,t) = 0, \qquad (5.1.56)$$

indicating that the perturbation describes massless excitations propagating along the string [Alford *et al.*, 1991]. A constant current corresponds to the solution $f = z$. Substitution of (5.1.55) into the gauge field equation yields an equation for $s(r,\theta)$,

$$D^i D_i s = -e^2(s\phi^{(0)})^\dagger (T^a\phi^{(0)})T^a, \qquad (5.1.57)$$

where $i = 1,2$ and $D_i = \partial_i - ie\mathcal{A}_i^{(0)}$. At large distances from the string, $D_i Q(\theta) = 0$ and the appropriate solution of (5.1.57) is

$$s(r,\theta) = (\alpha \ln r + \beta)Q(\theta). \qquad (5.1.58)$$

On the string axis $Q(\theta)$ is undefined, and $s(r = 0)$ can either vanish or be proportional to some other generator, for example, T^S.

The symmetry breaking scheme (2.1.65), $SO(10) \rightarrow SU(5) \times Z_2$ provides a realistic non-abelian model in which superconductivity appears [Everett, 1988]. Here the broken generator $T^S = T_R^1$ is the first component of right-handed weak isospin $SU(2)_R$ (refer to §4.2.3). The electromagnetic charge is a linear combination of another generator T_R^3 of $SU(2)_R$ and of right-handed weak hypercharge Y_R, $Q = Y_R + T_R^3$. It does

not commute with T^S, and thus the string must exhibit vector superconductivity.

5.2 Fermionic string superconductivity

5.2.1 Vortex zero modes

Vortex superconductivity is also possible if charged fermionic zero modes
are present on a string [Witten, 1985a]. These can arise through the
Yukawa coupling of fermions to the scalar fields which are excited in the
string interior. Such zero modes have been discussed by Kiskis [1977],
Ansourian [1977], and Nielson & Schroer [1977], though more pertinent
here is the model investigated by Jackiw & Rossi [1981]. We should
also point out that, in a condensed matter context, it has been known for
some time that low-energy fermionic excitations can exist on an Abrikosov
vortex-line [Caroli, de Gennes & Matricon, 1964].

Consider a four-component fermion field $\psi = \psi_L + \psi_R$, split into
eigenstates $\psi_{L,R}$ of $\gamma_5 = i\gamma_0\gamma_1\gamma_2\gamma_3$ with eigenvalues -1 and $+1$ respectively. Under the transformation $\psi \to \exp(i\alpha\gamma_5)\psi$, the eigenstates become $\psi_L \to \exp(-i\alpha)\psi_L$ and $\psi_R \to \exp(i\alpha)\psi_R$. We couple this to a
complex scalar field ϕ through the Lagrangian

$$\mathcal{L}_F = i\bar{\psi}_L\gamma^\mu\partial_\mu\psi_L + i\bar{\psi}_R\gamma^\mu\partial_\mu\psi_R - g\bar{\psi}_L\psi_R\phi - g\bar{\psi}_R\psi_L\bar{\phi}. \tag{5.2.1}$$

This is invariant under $\psi \to \exp(i\alpha\gamma_5)\psi$, $\phi \to \exp(-2i\alpha)\phi$, and the Dirac
equation becomes

$$\begin{aligned} i\gamma^\mu\partial_\mu\psi_L &= g\phi\psi_R \\ i\gamma^\mu\partial_\mu\psi_R &= g\bar{\phi}\psi_L \,. \end{aligned} \tag{5.2.2}$$

The fermion mass $m = g\eta$ is given by the expectation value $|\langle\phi\rangle| = \eta$ which breaks a global $U(1)$ symmetry (following Preskill [1985], for
simplicity we temporarily eliminate the gauge fields). We have already
encountered global string solutions in §4.1.3 which, for a string lying along
the z-axis, take the form

$$\phi(r, \theta, z) = f(r)e^{i\theta}, \tag{5.2.3}$$

with $f(0) = 0$ and $f(\infty) = \eta$. In this vortex-string background we look
for solutions of the Dirac equation (5.2.2) which are zero modes ψ^0, that
is, time-independent solutions of zero energy. By an index theorem due
to E. Weinberg [1981] for the abelian-Higgs model, we expect exactly one
such normalizable solution in this case, the number of solutions being
equal to the string winding number. We can find this solution explicitly
by assuming that ψ^0 has only a radial dependence. Using (5.2.3), the

Dirac equation (5.2.2) becomes

$$i\gamma_1(\cos\theta - \gamma_1\gamma_2\sin\theta)\frac{d\psi_L^0}{dr} = g\,f(r)e^{i\theta}\psi_R^0\,,$$

$$i\gamma_1(\cos\theta - \gamma_1\gamma_2\sin\theta)\frac{d\psi_R^0}{dr} = g\,f(r)e^{-i\theta}\psi_L^0\,.$$

(5.2.4)

This simplifies if $\psi_{L,R}^0$ are eigenstates of the operator $i\gamma_1\gamma_2$,

$$i\gamma_1\gamma_2\psi_L^0 = \psi_L^0$$

$$i\gamma_1\gamma_2\psi_R^0 = -\psi_R^0\,,$$

(5.2.5)

implying a form of two-dimensional chirality. Equation (5.2.4) then reduces to

$$i\gamma_1\frac{d\psi_L^0}{dr} = g\,f(r)\psi_R^0\,,$$

$$i\gamma_1\frac{d\psi_R^0}{dr} = g\,f(r)\psi_L^0\,,$$

(5.2.6)

which has solutions of the form [Callan & Harvey, 1985]

$$\psi_L^0(r) = \tau\,\exp\left(-g\int_0^r f(r')dr'\right)\,,$$

$$\psi_R^0 = -i\gamma_1\psi_L^0(r)\,.$$

(5.2.7)

Here, because of the chirality in two and four dimensions, the spinor τ must satisfy

$$\tau = i\gamma_1\gamma_2\tau = -\gamma_5\tau\,.$$

(5.2.8)

In two spatial dimensions, this solution corresponds to a bound fermion-vortex state; the fermion localized on the vortex has zero energy while, far away from the core, it has a mass m.

In four dimensions we look for more general solutions in which the fermion is free to propagate along the string by considering the separation of variables

$$\psi_L = \alpha(z,t)\,\psi_L^0(x,y)$$

$$\psi_R = -i\gamma_1\psi_L\,.$$

(5.2.9)

Now $\psi_{L,R}^0$ obeys the 'transverse' Dirac equation (5.2.6), so (5.2.2) reduces to

$$\left(\gamma_0\partial^0 + \gamma_3\partial^3\right)\alpha(z,t)\tau = 0\,.$$

(5.2.10)

By (5.2.8), we have $\gamma_0\gamma_3\tau = -\tau$, so (5.2.10) becomes the simple 'right-moving' wave equation

$$(\partial_t + \partial_z)\,\alpha = 0\,,$$

(5.2.11)

which has the solution

$$\alpha(z, t) = \alpha(z - t). \tag{5.2.12}$$

This corresponds to a massless chiral fermion, trapped in a transverse zero mode, moving at the speed of light in the positive z-direction ('right-movers' by convention). For an antivortex the fermions propagate in the opposite direction. Alternatively, if $\bar{\psi}_L \psi_R$ were Yukawa coupled to $\bar{\phi}$ instead of ϕ, then the fermions would be 'left-movers', propagating along the vortex-string in the negative z-direction. We shall generally use $\chi_{L,R}$ to denote fermion fields that couple as left-movers.

5.2.2 Fermionic currents

Let us suppose that the fermions carry charges under both the original gauged $U(1)$ which is spontaneously broken (labelled as $U(1)_Y$), as well as the electromagnetic gauge group $U(1)_Q$ which remains unbroken. For a particle of charge q, applying an electric field $\mathbf{E}$ in the positive z-direction increases the momentum at the rate

$$\frac{dp}{dt} = qE. \tag{5.2.13}$$

The momenta of right-movers will grow, while the momenta of left-movers will decrease. Thus, the electric field $\mathbf{E}$ causes upward and downward shifts in the Dirac sea as illustrated in fig. 5.7. This disturbance of the Dirac vacuum results in the creation of right-moving fermions and left-moving 'holes'. Given the one-dimensional density of states $1/2\pi$, the number of particles per unit length n produced by this process over a time t is

$$n = \frac{qEt}{2\pi}. \tag{5.2.14}$$

Accounting for both left- and right-moving modes, the charge density on the string changes at the rate

$$\frac{dQ}{dLdt} = \frac{E}{2\pi} \left[\sum_{\text{right}} q_r^2 - \sum_{\text{left}} q_l^2 \right]. \tag{5.2.15}$$

Electric charge, therefore, will only be conserved if

$$\sum_{\text{right}} q_r^2 = \sum_{\text{left}} q_l^2. \tag{5.2.16}$$

It should be noted here that particles and their corresponding antiparticles travel in the same direction. This is because the dependence of ψ on $z - t$ (as opposed to $z + t$) is unaffected by complex conjugation.

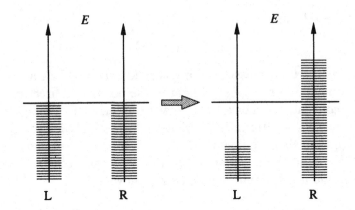

Fig. 5.7. Application of an electric field **E** along the string causes a shift in the Dirac sea for the left- and right-movers.

The contributions to the left- and right-hand sides of (5.2.16), therefore, must come from different particle species. In a realistic model, a string would be expected to have several right- and left-moving modes, each of which will carry separate quantum numbers such as lepton or baryon number. It is not immediately apparent that these quantum numbers will be conserved by such current-inducing processes but, as we shall discuss in §5.2.3, it is in fact a natural consequence of the consistency of the full four-dimensional theory.

The induced left- and right-movers make equal and opposite contributions to the charge density, but together they combine to create an overall electric current

$$J = \frac{q^2 E t}{2\pi},\qquad (5.2.17)$$

where

$$q^2 = \sum_{\text{right}} q_r^2 + \sum_{\text{left}} q_l^2. \qquad (5.2.18)$$

Current growth which is proportional to the applied electric field **E** as in (5.2.17) is characteristic of a superconducting wire. Any string moving through magnetic field lines, such as those in our galaxy, will experience an effective electric field causing such current growth.

5.2.3 Anomaly cancellation

We have already noted that unless the two terms in (5.2.15) cancel, the charge on the strings will not be conserved and the effective two-dimensional gauge theory will be inconsistent. Fortunately, such a cancellation can be expected if the effective theory derives from an anomaly-free four-dimensional theory [Witten, 1985a].

This can be illustrated by using the earlier example with a $U(1)_Y \times U(1)_Q$ symmetry. Since $\bar{\psi}_L \psi_R \phi$ is invariant under $U(1)_Y \times U(1)_Q$, we have the following charges for the right-moving modes $\psi_{L,R}$,

	$U(1)_Y$	$U(1)_Q$
ϕ	1	0
ψ_L	y_r	q_r
ψ_R	$y_r - 1$	q_r

$$(5.2.19)$$

Consequently, the contribution to the QQY triangle anomaly will have a coefficient from $\psi_{L,R}$ proportional to $q_r^2 y_r - q_r^2(y_r - 1) = q_r^2$. On the other hand, the left-moving modes $\chi_{L,R}$ which couple through $\bar{\chi}_L \chi_R \bar{\phi}$ will have charges

	$U(1)_Y$	$U(1)_Q$
χ_L	y_l	q_l
χ_R	$y_l + 1$	q_l

$$(5.2.20)$$

The left-moving contribution to the QQY anomaly will be proportional to $q_l^2 y_l - q_l^2(y_l + 1) = -q_l^2$. Hence, cancellation of the QQY anomaly in four dimensions requires that (5.2.16) be satisfied. Charge conservation in two dimensions then follows automatically.

We should emphasize that we were only compelled to add the left-moving fields $\chi_{L,R}$ to the original fermionic Lagrangian (5.2.1) in order to cancel the axial vector anomaly. For a global $U(1)_Y$ symmetry, there is no gauge field $\tilde{A}^\mu$ and so we require no extra fermions for anomaly cancellation. Consequently, all zero modes on global axion strings can have the same chirality and (5.2.16) need not be satisfied. The resulting two-dimensional anomaly will imply that the charge carried by the string will not be conserved in the presence of an electric field. The string must, therefore, exchange quantum numbers with the external world. This cannot occur by the usual emission or absorption of fermions since these have a large mass off the string. However, it can be shown that an electric field applied in the presence of a varying Goldstone boson field will induce an electric current. For strings, this current flows radially inward and exactly compensates the charge imbalance due to the anomaly [Callan & Harvey, 1985; Lazarides & Shafi, 1985]. In axion models, strings become attached to domain walls, so that variations in the axion field are confined to the wall interiors (refer to §3.2.5). These walls also possess superconducting zero modes, so that any charge appearing on the string is compensated by currents flowing radially in on the domain wall.

5.2.4 Some Fermi phenomenology

It can be illuminating to briefly consider some more realistic examples of grand unified models in which fermionic superconducting strings can arise. The $SO(10)$-strings discussed in §2.1.4, §4.2.3 and §5.1.8 exhibited interesting phenomena in the bosonic sector, but did not possess fermionic zero modes capable of carrying electromagnetic currents. However, it is possible to create these by introducing an additional scalar multiplet transforming under the **210** of $SO(10)$ [Witten, 1985a]. This causes the electroweak $SU(2) \times U(1)$ doublet Φ to change phase by 2π as a closed path encircling a string is traversed. Any fermions getting their mass from a Yukawa coupling to Φ have corresponding vortex zero-modes by the same mechanism already outlined in §5.2.1 for the simple $U(1)_Y \times U(1)_Q$-model. In this $SO(10)$ model, the up quarks u, ū obtain their masses from a coupling to Φ and the down quarks d, d̄ and electrons e⁻, e⁺ from Φ^*. Consequently, we will have zero mode charge carriers u_R, $\bar{u}_R$ travelling in one direction along the string, while going in the opposite direction we will have d_L, $\bar{d}_L$, e_L^- and e_L^+ (see fig. 5.8). Under an applied electric field the electrons, down quarks and up quarks will be created in the ratio 1:1:2. Anomaly cancellation is easily verified in (5.2.16), after summing over the three quark colours. The other generations of heavier leptons and quarks which also obtain masses from Yukawa coupling to the electroweak doublet Φ (that is, μ, τ, c, s, t and b) will also be produced in the same proportions.

A superstring-inspired E_6-model provides a phenomenological example with superheavy charge carriers [Witten, 1985a]. Here, E_6 is broken in the **78** and **27** Higgs representations,

$$E_6 \xrightarrow{\textbf{78}} SO(10) \times U(1) \xrightarrow{\textbf{27}} SO(10) \rightarrow \dots. \qquad (5.2.21)$$

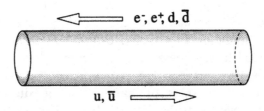

Fig. 5.8.   Charge carriers in the superconducting version of the $SO(10)$-string with up quarks travelling in one direction and down quarks and electrons travelling in the other. Note that a particle and its antiparticle move in the same direction.

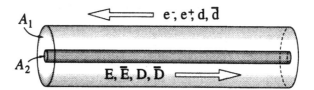

Fig. 5.9. Superheavy and light fermionic superconducting modes in the E_6-string. There is a geometric suppression of the scattering cross-section because the area $A_2 \ll A_1$.

$U(1)$-strings are produced at the second stage symmetry breaking when an $SO(10)$ Higgs singlet 1^1 transforming in the **27**-representation acquires a large vacuum expectation value. It is important to note that the **27**-representation of E_6 decomposes as

$$\mathbf{27} = 1^1 + 10^{-1/2} + 16^{1/4} . \tag{5.2.22}$$

The standard light fermions are contained in the $16^{1/4}$, obtaining their mass from coupling to a $10^{-1/2}$ Higgs field component. They provide the left-moving modes on the string which must be either u_L or d_L and e^-. Fermions in the 1^1 and $10^{-1/2}$ couple to the Higgs singlet 1^1 and so they are superheavy. These are the right-moving modes on the string, and they can be either charged E_L, D_L or neutral N_L. Thus, these superconducting E_6-strings will possess an inner core of superheavy charge carriers propagating in the opposite direction to an outer envelope of light fermions, as illustrated in fig. 5.9. This model can be easily modified so that it possesses superheavy fermions travelling in both directions.

We note finally that metastable string solutions of the standard electroweak model (§4.2.6) are also superconducting. The charge carriers in this case are electrons and quarks which get their masses from their coupling to the Higgs doublet.

5.2.5 Charge carrier scattering

Under an applied electric field, the fermionic supercurrent growth eventually saturates, as for bosonic superconductivity. Naively, if the fermion mass is m off the string, then once the Fermi momentum rises above $p_F > m$, it would appear energetically favourable for the particles to leave the string. This threshold suggests a maximum current for fermionic superconductivity,

$$J_{\text{max}} \sim \frac{qm}{2\pi} . \tag{5.2.23}$$

With electrons as charge carriers this would yield a small current of only 20A, while for GUT-scale superheavy charge carriers, enormous currents up to 10^{20}A might be anticipated. However, this estimate is flawed because the conservation of longitudinal momentum inhibits the escape of individual fermions; each fermion must pass from a zero-mode state with $E^2 = p_F^2$ to a massive state off the string with $E^2 = p_F^2 + m^2$. This barrier can only be overcome at regions of high string curvature, such as at a cusp. A crude dimensional analysis suggests that the critical momentum is instead $p_{max} \sim m^2 R$, where R is the string radius of curvature. This estimate has been confirmed by explicit calculations of the amplitude for a fermion to tunnel from the string to a massive state at larger radius, while conserving energy and angular momentum [Barr & Matheson, 1987b]. This would appear to allow much larger currents than originally anticipated, although an ultimate threshold is set by the 'spring' current, $J \leq q\eta$, where η is the symmetry breaking scale of the string.

The actual fermionic maximum current is a model-dependent quantity which is determined by inelastic scattering between left- and right-moving charge carriers. In general, gauge interactions mediated by gauge bosons will couple string zero-modes to light charged fermions which can escape from the string. In the modified $SO(10)$ model of §5.2.4 we can expect weak interactions such as $e_L^- + u_L \rightarrow d_L + \nu_L$, where the electron neutrino ν_L leaves the string. For typical scattering cross-sections, the threshold for such interactions will occur at a few GeV and the current will saturate at this level [Barr & Matheson, 1987a]. In the E_6-string of §5.2.4, some current components will be similarly restricted by weak scattering events. For example, an interaction like $E_L^- + U_L \rightarrow N_L + d_{L\,free}$ will allow light fermions to leave the string as a d-triplet, while the superheavy fermion E_L^- is scattered into the neutral N_L current components. However, there are also charged D_L current components with no weak couplings. Under an applied field, these will continue to grow beyond this threshold, as will a matching light fermion current. Interactions mediated by GUT-scale gauge bosons will ultimately intervene to restrict the remaining superheavy currents.

It is interesting to note that the highest critical currents are actually achieved with superheavy fermions travelling in one direction and light charge carriers, such as quarks and leptons, travelling in the other. As illustrated in fig. 5.9, the zero modes of the light fermions have a transverse width m_l^{-1}, while the superheavy fermions travel in a much narrower tube of width m_h^{-1}. The wavefunction overlap is small and, consequently, there is a geometric suppression for the scattering cross-section.

Lazarides *et al.* [1988] have studied E_6-string scatterings in some detail, confirming the qualitative picture presented above. They showed the existence of a number of Fermi momentum thresholds, beginning at

100 GeV, beyond which various components of the current saturate. The last remaining superheavy charge carriers D_L, $\bar{D}_L$ scatter and create particles which leave the string when $p_F \sim 10^{10}$GeV, implying a substantially lower critical current than naively anticipated with (5.2.23). They argue that this comparatively low threshold is to be expected for most realistic models, unless they are specifically tailored to carry high superconducting currents by, for example, the imposition of additional discrete symmetries. Barr & Matheson [1987b] studied a modified $SO(10)$ model which possessed superheavy charge carriers. With superheavy currents flowing in only one direction the critical current is $\sim 10^9$GeV, whereas with superheavy components in both directions the result is only $\sim 10^6$GeV. They note that the exchange of a GUT-scale boson between left- and right-moving modes can act to lower the Fermi levels of all the charge carriers proportionally, even if the end-products remain on the string.

5.2.6 Bosonization and the effective action

The equivalence of fermionic and bosonic superconductivity is apparent from the equations for current growth in an electric field, (5.1.47) and (5.2.17), if we make the substitution $\sum_i q_i^2 \leftrightarrow 4\pi e^2 \Sigma$ using (5.1.40). This can be demonstrated more formally by recasting the effective fermionic action in a form identical to (5.1.43) using the method of bosonization [Witten, 1985a]. This technique exploits the well-established relationship between fermionic and bosonic quantum field theories in two spacetime dimensions (see, for example, Coleman [1976]).

We assume that the four-dimensional fermion field ψ only possesses very low-energy longitudinal excitations, so that it remains very close to the transverse zero mode solution ψ^0 in (5.2.7). Since ψ^0 vanishes rapidly away from the vortex core, the field theory reduces to a two-dimensional fermion model restricted to the string worldsheet. This can be described with a two-dimensional worldsheet fermion Ψ which satisfies a gauged version of the longitudinal Dirac equation (5.2.10), through which it couples to the four-dimensional electromagnetic potential A^μ. For a string lying along the z-axis, the effective action will then take the form

$$S = \int dz\, dt\, \bar{\Psi}(z,t) i\gamma^a D_a \Psi(z,t)\,, \qquad (5.2.24)$$

where $D_a = (\partial_a - ieA_a)$ is the covariant derivative on the string worldsheet and $a = 0, 3$.

The action (5.2.24) is most simply studied by bosonizing the bilinear fermion term using a scalar field $\varphi(z,t)$ which is also confined to the string worldsheet. To achieve this we make the standard identification,

$$\bar{\Psi}\gamma^a \Psi = (1/\sqrt{\pi})\, \epsilon^{ab} \partial_b \varphi\,, \qquad (5.2.25)$$

and the action immediately becomes

$$S = \int dz\, dt \left[\tfrac{1}{2}(\partial_a \varphi)^2 - (q/\sqrt{\pi})A_a \epsilon^{ab}\partial_b \varphi \right].$$

(5.2.26)

This can be generalized to a string possessing several charged zero modes of the same chirality by simply taking $q^2 \to \sum_i q_i^2$. The effective action (5.2.26) describing fermionic superconductivity is clearly equivalent to the bosonic form (5.1.43). The macroscopic dynamics of superconducting strings, therefore, can be expected to be independent of the nature of the charge carriers. The fermionic current, $J^z = -q\bar{\Psi}\gamma^3\Psi = -(q/\sqrt{\pi})\dot{\varphi}$, will likewise be superconducting as in (5.1.47).

5.2.7 Massive fermionic bound states

There is no essential reason why fermion bound states on the string are required to be massless. Given two Dirac fields, the left-moving ψ and the right-moving χ, we can modify the original Lagrangian (5.2.1) by adding cross-couplings of the form [Hill & Widrow, 1987; Davis, 1987b; Hindmarsh, 1988]

$$\mathcal{L}_{\mathrm{M}} = m_\epsilon(\bar{\psi}\chi + \bar{\chi}\psi).$$

(5.2.27)

The resulting equation of motion for the right-moving excitations (5.2.11) becomes modified to

$$(\partial_t^2 - \partial_z^2 + m_\epsilon^2)\alpha = 0.$$

(5.2.28)

From this it is clear that the fermions trapped on the string have acquired a mass m_ϵ through the pairing (5.2.27). When the effective two-dimensional theory is bosonized as in §5.2.6, the mass term (5.2.27) is replaced by a cosine potential term,

$$\mathcal{L}_{\mathrm{M}}^{(2)} = m'_\epsilon \cos(\sqrt{4\pi}\,\varphi),$$

(5.2.29)

where $m_\epsilon \sim m'_\epsilon$ are related by a renormalization constant (refer to Coleman [1976]). The bosonized field equations (5.1.42) are then

$$\partial_a \partial^a \varphi + (q/\sqrt{\pi})\epsilon^{ab}\partial_a A_b + \sqrt{4\pi}m'_\epsilon{}^2 \sin(\sqrt{4\pi}\varphi) = 0.$$

(5.2.30)

This is a modified version of the sine-Gordon equation encountered in §3.1.1 which possesses the soliton solutions (3.1.5) of mass $\sim m'_\epsilon$. These solitons correspond to the massive fermions in the model.

By assuming only time-dependence, (5.2.30) can be integrated to yield

$$\dot{\varphi}^2 + (2q/\sqrt{\pi})E\varphi + 4m'_\epsilon{}^2 \sin^2(\sqrt{4\pi}\,\varphi/2) = 0,$$

(5.2.31)

where we have taken $\varphi(0) = \dot{\varphi}(0) = 0$. For an applied electric field satisfying $qE \ll 2\pi m'_\epsilon{}^2$, there is an oscillatory solution for φ with characteristic frequency $\omega \sim \sqrt{4\pi}\,m'_\epsilon$. This is a frequency at which the string can be expected to radiate electromagnetically. The string is not super-conducting in this regime and the amplitude of the current will attain a maximum value [Hill & Widrow, 1987],

$$J_{\text{max}} \approx \frac{q^2 E}{2\pi^{3/2} m'_\epsilon}, \qquad (5.2.32)$$

which is proportional to the applied field E. Conversely, for $qE > 2\pi m'^2_\epsilon$ we are above the threshold for soliton pair creation and the current will approach superconducting behaviour, $J \sim q^2 Et/\pi$. However, the current retains a small oscillatory modulation which will radiate and cause current decay if the electric field is switched off. In an astrophysical context, the triggering of supercurrent growth requires magnetic fields above the threshold,

$$B\left(\frac{v}{c}\right) \geq 10^3 \left(\frac{m'_\epsilon}{1\,\text{eV}}\right)^2 \text{Gauss}, \qquad (5.2.33)$$

where v is the transverse string velocity. Unless the mass term m'_ϵ in (5.2.27) is extremely small, fermionic superconductivity is unlikely to be of much consequence. This applies equally to bosonic superconductivity where small mass terms can also be introduced.

Worldsheet fermions can also obtain masses by coupling to scalar fields which acquire an expectation value in the string interior, as in the condensate model §5.1.1. In this case, the string may exhibit a combination of bosonic and fermionic superconductivity [Davis, 1987b; Hindmarsh, 1988].

5.3 Vacuum effects

5.3.1 Pair production

The effective action for superconducting strings (5.1.44) allows the study of their electromagnetic properties using the equations of classical electrodynamics. This is justified in regions where the electromagnetic field is not very strong. However, near a string, the electric and magnetic fields are given by

$$E = \frac{2\lambda}{r}, \qquad B = \frac{2J}{r}, \qquad (5.3.1)$$

where λ and J are the local values of charge per unit length and the current, respectively. Even for modest values of λ and J, the fields become extremely strong in the immediate vicinity of the string, and so quantum effects, such as vacuum polarization and pair production, must be taken into account.

Before we discuss vacuum effects in the strongly inhomogeneous fields near a cosmic string, it is instructive to review well-known results on pair production in slowly-varying electric and magnetic fields. In a pure electric field, the energy of a pair of particles of mass m and charge e is

$$\epsilon = 2m - e\mathbf{E} \cdot \Delta\mathbf{x}, \tag{5.3.2}$$

where $\Delta\mathbf{x}$ is the separation vector between the particle and antiparticle, which we assume are at rest. It is clear from (5.3.2) that pair creation out of the vacuum is energetically allowed if the distance between the particles is $|\Delta\mathbf{x}| \gtrsim 2m/eE$. Pairs are copiously produced if $|\Delta\mathbf{x}| \lesssim m^{-1}$, or $E \gtrsim m^2/e$, while in the opposite limit, $E \ll m^2/e$, the rate of pair production is exponentially suppressed. For spin-1/2 particles this rate is [Schwinger, 1951]

$$\frac{dN}{dV dt} \approx \frac{\alpha E^2}{\pi^2} \exp\left(-\frac{\pi m^2}{|eE|}\right). \tag{5.3.3}$$

Note that (5.3.3) only applies when the typical scale of variation in the field is much greater than $2m/eE$.

For a pure magnetic field, it can be shown that there is no spin-1/2 particle production. The energy levels for a particle in a constant magnetic field $\mathbf{B}$ are

$$\epsilon^2 = m^2 + p_z^2 + 2eB\left(n + \frac{1}{2} + s\right), \qquad n = 0, 1, 2, \ldots \tag{5.3.4}$$

where p_z and s are, respectively, the components of momentum and spin along the magnetic field. For spin-1/2 particles $s = \pm 1/2$, $\epsilon^2 \geq m^2$ and pair production is energetically forbidden. The magnetic spin dipole energy $2esB$ can be negative, but it is always compensated by the kinetic energy of the Landau levels $2eB(n + 1/2)$. This argument for vacuum stability fails, however, for particles of spin-1. For $s = -1$, the lowest energy level with $n = 0$ and $p_z = 0$ becomes negative when B exceeds the critical value [Nielsen & Olesen, 1978]

$$B_c = \frac{m^2}{e}. \tag{5.3.5}$$

This signals the beginning of particle production. An example of this effect for electroweak gauge fields is discussed in §5.3.2.

We now turn to pair production in the electromagnetic field of a string. First, consider a portion of the string where $J = 0$ and $\lambda \neq 0$, so only an electric field is present. The onset of pair production is signalled by the appearance of bound states with energies $\epsilon \leq -m$. If the particle's wave function is localized within a distance r from the string, then its energy can be roughly estimated to be

$$\epsilon \sim \left(r^{-2} + m^2\right)^{1/2} + e\lambda \ln\left(\frac{r}{R}\right). \tag{5.3.6}$$

The cut-off radius R is set by the curvature radius of the string or by the characteristic scale of variation in λ and J along the string. Minimizing ϵ with respect to r, we obtain estimates for the ground state radius and energy

$$r_0 \sim \begin{cases} (em\lambda)^{-1/2}, & \lambda \lesssim m/e, \\ (e\lambda)^{-1}, & \lambda \gtrsim m/e, \end{cases} \tag{5.3.7}$$

$$\epsilon_0 \sim m + e\lambda \ln\left(\frac{r_0}{R}\right). \tag{5.3.8}$$

Hence, the condition for particle production is

$$\lambda \gtrsim \frac{m}{e \ln (Rm)} \sim 0.1\, m. \tag{5.3.9}$$

Electron–positron pair production begins at $\lambda \sim 0.1\, m_e \sim 10^2\,\mathrm{esu/cm}$. If the string is positively charged, then electrons accumulate in bound states near the string, while positrons are repelled from the string. The pair production ceases when the number of pairs is sufficient to screen the electric charges on the string, so that the effective charge density at $r \gtrsim m_e^{-1}$ is $\lambda \sim 0.1 m_e$. In situations of astrophysical interest the charge density of the string λ changes on timescales which are much greater than the characteristic timescale of pair production, $\tau \sim m_e^{-1}$. As a result, λ_{eff} never gets substantially greater than $0.1 m_e$, and particles heavier than electrons are never produced. Pair production in the electric field of a string has been discussed by Aryal, Vachaspati & Vilenkin [1987] and by Amsterdamski [1988]. We note, however, that these earlier discussions contain some errors.

Now consider a portion of string where $\lambda = 0$, $J \neq 0$, and the field is purely magnetic. The following argument [Soffel, Muller & Greiner, 1982] shows that, as for a homogeneous field, there is no spin-1/2 particle production in the field of the string. Energy eigenstates for a charged particle in an arbitrary time-independent magnetic field $\mathbf{B}(\mathbf{x}) = \nabla \times \mathbf{A}(\mathbf{x})$ can be found from the Dirac equation

$$(\boldsymbol{\alpha} \cdot \mathbf{P} + \beta m)\,\psi = \epsilon\psi, \tag{5.3.10}$$

where $\boldsymbol{\alpha} = \gamma^0 \boldsymbol{\gamma}$ and $\beta = \gamma^0$ are Dirac matrices and

$$\mathbf{P} = -i\nabla - e\mathbf{A}(\mathbf{x}). \tag{5.3.11}$$

Acting on (5.3.10) with the operator $\boldsymbol{\alpha} \cdot \mathbf{P} + \beta m$, multiplying by $\bar{\psi}$ and integrating over $\mathbf{x}$, we obtain

$$\epsilon^2 - m^2 = \frac{\int d^3x\, \bar{\psi}(\boldsymbol{\alpha} \cdot \mathbf{P})^2 \psi}{\int d^3x\, \bar{\psi}\psi} \,. \tag{5.3.12}$$

Since $\boldsymbol{\alpha} \cdot \mathbf{P}$ is a Hermitian operator, the right-hand side of this equation is non-negative. This shows that the gap $-m < \epsilon < m$ between the positive- and negative-energy states is never crossed, and thus there is no particle production.

In the general case, when both J and λ are non-zero, we can choose a reference frame where $J = 0$ or $\lambda = 0$, depending on the sign of the invariant $J_a J^a$ (such a frame cannot be chosen only in the 'chiral' case with $|J| = |\lambda|$). A Lorentz-invariant condition for spin-1/2 particle production generalizing (5.3.9) can be written as

$$J_a J^a \gtrsim 10^{-2} m^2 \,. \tag{5.3.13}$$

Since pairs screen the fields produced by certain portions of the string, the classical equations of string electrodynamics need to be modified, for example, for calculations of electromagnetic radiation. Vacuum polarization effects can also modify the electric and magnetic fields near the string and need to be taken into account.[*]

5.3.2 W-condensation

Superconducting strings formed at the GUT-scale can carry enormous currents $J \lesssim e\eta$ of up to 10^{20} A. The corresponding magnetic fields (5.3.1) are so large they can destabilize the electroweak vacuum because of the non-abelian coupling between the vector bosons W_μ and the electromagnetic potential A_μ [Ambjørn & Olesen, 1988]. In the electroweak Lagrangian this appears as an anomalous magnetic moment term $-eF^{\mu\nu}W_\mu^\dagger W_\nu$, where $F^{\mu\nu}$ is the electromagnetic field due to the string.

In order to simplify the discussion, we first consider a homogeneous magnetic field pointing in the z-direction, $F_{12} = B_z$. Motivated by considering linearized perturbations about the $W_\mu = 0$ vacuum, Ambjørn & Olesen [1988] proposed the following ansatz for W_μ,

$$W \equiv W_1 = -iW_2\,, \qquad W_3 = W_0 = 0\,. \tag{5.3.14}$$

In this first approximation, they neglected the contribution of Z_μ and assumed that $m_H > m_W$, where the Higgs and vector boson masses are

---

[*] Vacuum polarization near a string with a constant current has been studied by Amsterdamski & O'Connor [1988].

$m_{\rm H}^2 = \lambda v^2$ and $m_{\rm W}^2 = g^2 v^2/2$ respectively. The potential for W_μ and the Higgs field ϕ then reduces to

$$V(\phi, W) = -2eB_z|W|^2 + g^2\phi^2|W|^2 + 2g^2|W|^4 + \tfrac{1}{4}\lambda(\phi^2 - v^2)^2 \,. \quad (5.3.15)$$

As expected from (5.3.5), the usual electroweak vacuum with $\phi = v$ and $W = 0$ will only pertain if $eB_z < m_{\rm W}^2$. The magnetic field effectively introduces a negative mass-squared term and, for $eB_z > m_{\rm W}^2$, there will be a new minimum with a W-condensate,

$$2g^2|W|^2 = eB_z - \tfrac{1}{2}g^2\phi^2 \,. \quad (5.3.16)$$

The presence of non-zero W-fields causes a back-reaction on ϕ reducing it to

$$\phi^2 = v^2 \frac{m_{\rm H}^2 - eB_z}{m_{\rm H}^2 - m_{\rm W}^2}\,, \quad (5.3.17)$$

so for $eB_z > m_{\rm H}^2$ we might expect the entire electroweak symmetry $SU(2) \times U(1)_Y$ to be restored. Here, we will have $\phi = 0$ and a W-Z condensate in which the $SU(2)$ field strength $G_{\mu\nu}^a$ vanishes. The original magnetic field B_z will be rotated by the condensate into a pure hyper-charge magnetic field associated with $U(1)_Y$.

Although the parameter space for these effects remains to be fully explored, these arguments do suggest that a W-condensate will exist in the range,

$$m_{\rm W}^2 < eB_z < m_{\rm H}^2 \,. \quad (5.3.18)$$

As in ordinary superconductors, different choices of the couplings are expected to give rise to Type I and II regimes. The lowest energy configuration in the Type II case appears to be a condensate of magnetic fluxes which form a periodic structure, analogous to the Abrikosov lattice of vortex-lines in a superconductor. Ambjørn & Olesen [1990a,b] were able to confirm these qualitative expectations in the Bogomol'nyi limit when $m_{\rm H} = m_{\rm Z}$. In particular, they found that the ansatz (5.3.14) with $Z_\mu = 0$ for the gauge fields and (5.3.17) for the Higgs field are quantitatively correct in this case. The general case is more complicated; for a discussion see MacDowell & Törnkvist [1991, 1992] and Perkins [1991].

In the presence of a GUT-scale superconducting string, the inhomogeneous magnetic field (5.1.18) can be expected to destabilize the electroweak vacuum. Consequently, a current-carrying string will acquire a W-condensate, a surround which can have macroscopic dimensions of up to $r \lesssim 10^{-4}$–10^{-3}cm for large currents $\sim 10^{20}$ A [Ambjørn & Olesen, 1990c].

6

String dynamics

6.1 The Nambu action

When the radius of curvature of a string is much greater than the string thickness δ, we can regard the string as a one-dimensional object. Its world history can then be represented by a two-dimensional surface in spacetime,

$$x^\mu = x^\mu\left(\zeta^a\right), \qquad a = 0,\, 1, \tag{6.1.1}$$

which is called the string worldsheet. The worldsheet coordinates ζ^0, ζ^1 are arbitrary parameters: the first ζ^0 is chosen to be timelike, while the second ζ^1 is spacelike and in some sense labels points along the string. The spacetime interval between two nearby points on the worldsheet is

$$ds^2 = g_{\mu\nu} x^\mu_{,a} x^\nu_{,b}\, d\zeta^a d\zeta^b, \tag{6.1.2}$$

where $g_{\mu\nu}$ is the four-dimensional metric. Hence the two-dimensional worldsheet metric is given by

$$\gamma_{ab} = g_{\mu\nu} x^\mu_{,a} x^\nu_{,b}. \tag{6.1.3}$$

The contravariant metric tensor γ^{ab} can be defined as usual by $\gamma^{ab}\gamma_{bc} = \delta^a_c$.

In §4.4.1 we presented a simple field theoretic derivation of an effective string action – the Nambu action – for vortex-lines in the abelian-Higgs model. Irrespective of the specific model, however, more general considerations can be employed to justify its use. For gauge strings, there is no long-range interaction between different string segments, so it should be possible to derive the equations of motion from a local action of the form

$$S = \int \mathcal{L}\sqrt{-\gamma}\, d^2\zeta, \tag{6.1.4}$$

where γ is the determinant of γ_{ab},

$$\gamma = \det(\gamma_{ab}) = \tfrac{1}{2}\epsilon^{ac}\epsilon^{bd}\gamma_{ab}\gamma_{cd}. \tag{6.1.5}$$

154

The Lagrangian $\mathcal{L}$ should be invariant under general spacetime coordinate transformations,

$$x^\mu \to \tilde{x}^\mu(x^\nu) \qquad (6.1.6)$$

and under arbitrary reparametrizations of the worldsheet,

$$\zeta^a \to \tilde{\zeta}^a\left(\zeta^b\right). \qquad (6.1.7)$$

The 'building blocks' from which we can construct the Lagrangian are then limited to the string tension and geometric quantities, such as the intrinsic and extrinsic worldsheet curvatures and their covariant derivatives. Note the absence of the string four-velocity u^μ from this list, since the local rest frame of the string, and therefore its four-velocity, is only defined up to longitudinal Lorentz boosts.

The Lagrangian $\mathcal{L}$ has the dimensions of mass-squared, so schematically we can write

$$\mathcal{L} = -\mu + \alpha\kappa + \beta\frac{\kappa^2}{\mu} + \dots \qquad (6.1.8)$$

where κ denotes the curvature (with indices suppressed) with numerical coefficients α, β. For a string with a typical radius $R \gg \delta$, we have $\kappa \sim R^{-2} \ll \mu$, recalling that the string thickness $\delta \gtrsim \mu^{-1/2}$. Hence, given our original assumptions, the curvature terms in (6.1.8) can be neglected and we obtain

$$S = -\mu \int \sqrt{-\gamma}\, d^2\zeta. \qquad (6.1.9)$$

This is the well-known Nambu action for a string [Nambu, 1969; Goto, 1971]. Up to an overall factor, it corresponds to the surface area swept out by the string in spacetime. Note the analogy of the action (6.1.9) to that for a relativistic particle,

$$S = -m \int d\tau \sqrt{\dot{x}^2}, \qquad (6.1.10)$$

which is proportional to the proper length of the particle's worldline.

The string equations of motion can be obtained by varying the action (6.1.9) with respect to $x^\mu(\zeta^a)$. Using the relation

$$d\gamma = \gamma\gamma^{ab}d\gamma_{ab}, \qquad (6.1.11)$$

and the expression (6.1.3) for γ_{ab}, we find that

$$x^\mu_{,a}{}^{;a} + \Gamma^\mu_{\nu\sigma}\gamma^{ab}x^\nu_{,a}x^\sigma_{,b} = 0, \qquad (6.1.12)$$

where $\Gamma^\mu_{\nu\sigma}$ is the four-dimensional Christoffel symbol,

$$\Gamma^\mu_{\nu\sigma} = \tfrac{1}{2}g^{\mu\tau}\left(g_{\tau\nu,\sigma} + g_{\tau\sigma,\nu} - g_{\nu\sigma,\tau}\right). \qquad (6.1.13)$$

From the two-dimensional point of view, $x^\mu(\zeta^a)$ is a set of four scalar fields and the covariant Laplacian in (6.1.12) is given by

$$x^{\mu\;;a}_{,a} = \frac{1}{\sqrt{-\gamma}}\,\partial_a\left(\sqrt{-\gamma}\gamma^{ab}x^\mu_{,b}\right).\tag{6.1.14}$$

The string energy–momentum tensor can be found by varying the action (6.1.9) with respect to the metric $g_{\mu\nu}$,

$$T^{\mu\nu}\sqrt{-g} = -2\frac{\delta S}{\delta g_{\mu\nu}} = \mu\int d^2\zeta\,\sqrt{-\gamma}\,\gamma^{ab}x^\mu_{,a}x^\nu_{,b}\,\delta^{(4)}\left(x^\sigma - x^\sigma\left(\zeta^a\right)\right).\tag{6.1.15}$$

For a straight string in a flat spacetime lying along the z-axis, taking $\zeta^0 = t$, $\zeta^1 = z$, (6.1.15) reduces to (3.2.14), so justifying the choice of the overall factor in (6.1.8).

6.2 Strings in flat spacetime

6.2.1 Equations of motion and gauge fixing

In flat spacetime with $g_{\mu\nu} = \eta_{\mu\nu}$ and $\Gamma^\mu_{\nu\sigma} = 0$, the string equations of motion (6.1.12) take the form

$$\partial_a\left(\sqrt{-\gamma}\gamma^{ab}x^\mu_{,b}\right) = 0.\tag{6.2.1}$$

Since the Nambu action is invariant under the transformations (6.1.7), we are free to choose any parametrization of the worldsheet. This can be done explicitly, for example, as in the 'light-cone' gauge

$$\zeta^0 = t,\qquad \zeta^1 = z - t,\tag{6.2.2}$$

or else by imposing two gauge conditions. A very convenient choice of gauge for a string in flat spacetime is

$$\gamma_{01} = 0,\qquad \gamma_{00} + \gamma_{11} = 0.\tag{6.2.3}$$

Using (6.1.3) for γ_{ab}, this can be rewritten as

$$\dot{x}\cdot x' = 0,\tag{6.2.4}$$

$$\dot{x}^2 + x'^2 = 0,\tag{6.2.5}$$

where dots and primes stand for derivatives with respect to ζ^0 and ζ^1, respectively. This is known as the 'conformal' gauge because the worldsheet metric takes a conformally flat form,

$$\gamma_{ab} = \sqrt{-\gamma}\,\eta_{ab},\qquad \gamma^{ab} = \frac{1}{\sqrt{-\gamma}}\,\eta^{ab}.\tag{6.2.6}$$

The string equation of motion (6.2.1) then simply becomes the two-dimensional wave equation

$$\ddot{x}^\mu - x^{\mu\prime\prime} = 0.$$ (6.2.7)

The gauge conditions (6.2.4–5) do not fix the gauge completely. There is still the freedom to perform transformations (6.1.7) such that $\dot{\zeta}^1 = \tilde{\zeta}^{0\prime}$ and $\tilde{\zeta}^{1\prime} = \dot{\zeta}^0$, implying that

$$\ddot{\zeta}^a - \tilde{\zeta}^{a\prime\prime} = 0.$$ (6.2.8)

We can remove this residual freedom by setting

$$t \equiv x^0 = \zeta^0,$$ (6.2.9)

which is consistent with (6.2.7). Employing this choice allows us to write the string trajectory as the three-vector

$$\mathbf{x}(\zeta, t),$$ (6.2.10)

where $\zeta \equiv \zeta^1$ is the spacelike parameter on the string, and (6.2.4–5) and (6.2.7) take the form

$$\dot{\mathbf{x}} \cdot \mathbf{x}' = 0,$$ (6.2.11)
$$\dot{\mathbf{x}}^2 + \mathbf{x}'^2 = 1,$$ (6.2.12)
$$\ddot{\mathbf{x}} - \mathbf{x}'' = 0.$$ (6.2.13)

The physical meaning of these equations is readily apparent. The first (6.2.11) implies that the vector $\dot{\mathbf{x}}$ is perpendicular to the string and thus represents the physically observable velocity. For this reason this gauge is also often called the 'transverse' gauge. The second constraint equation (6.2.12) can be written as $d\zeta = (1 - \dot{\mathbf{x}}^2)^{-1/2}|d\mathbf{x}| = d\mathcal{E}/\mu$, where

$$\mathcal{E} = \mu \int (1 - \dot{\mathbf{x}}^2)^{-1/2} dl = \mu \int d\zeta$$ (6.2.14)

is the energy of the string and $dl = |d\mathbf{x}|$. Thus we have chosen ζ to be proportional to the string energy measured from some arbitrary point on the string. Finally, (6.2.13) implies that the acceleration of a string element in its local rest frame ($\dot{\mathbf{x}} = 0$) is inversely proportional to the local curvature radius, $R^{-1} = |d^2\mathbf{x}/dl^2|$. The direction of $\ddot{\mathbf{x}}$ is such that a curved string tends to straighten. In doing so, of course, it develops a velocity and begins to oscillate.

The general solution of (6.2.13) is

$$\mathbf{x}(\zeta, t) = \tfrac{1}{2}\left[\mathbf{a}(\zeta - t) + \mathbf{b}(\zeta + t)\right],$$ (6.2.15)

and (6.2.11) and (6.2.12) give the following constraints for the otherwise arbitrary functions,

$$\mathbf{a}'^2 = \mathbf{b}'^2 = 1.$$ (6.2.16)

The geometric interpretation of these constraints is that ζ is the length parameter along the three-dimensional curves $\mathbf{a}(\zeta)$ and $\mathbf{b}(\zeta)$.

The string energy–momentum tensor in this gauge is then

$$T^{\mu\nu}(\mathbf{x}, t) = \mu \int d\zeta \left(\dot{x}^\mu \dot{x}^\nu - x'^\mu x'^\nu \right) \delta^{(3)} \left(\mathbf{x} - \mathbf{x}(\zeta, t) \right). \qquad (6.2.17)$$

The total energy of the string is given by

$$\mathcal{E} = \int T_0^0 d^3x = \mu \int d\zeta, \qquad (6.2.18)$$

in agreement with (6.2.14). The string momentum and angular momentum are, respectively,

$$\mathbf{P} = \mu \int d\zeta \, \dot{\mathbf{x}}(\zeta, t), \qquad (6.2.19)$$

$$\mathbf{J} = \mu \int d\zeta \, \mathbf{x}(\zeta, t) \times \dot{\mathbf{x}}(\zeta, t). \qquad (6.2.20)$$

6.2.2 Oscillating loops

The motion of a closed loop of string is described by (6.2.15–16) with ζ varying in the range

$$0 \le \zeta < L, \qquad (6.2.21)$$

where $L = \mathcal{E}/\mu$ is called the invariant length of the loop and $\mathcal{E}$ is its energy (6.2.18). For the loop to be closed we require the periodicity $\mathbf{x}(\zeta + L, t) = \mathbf{x}(\zeta, t)$ or, equivalently,

$$\mathbf{b}\left(\zeta_+ + L\right) - \mathbf{b}\left(\zeta_+\right) = -\mathbf{a}\left(\zeta_- + L\right) + \mathbf{a}\left(\zeta_-\right) \equiv \Delta \qquad (6.2.22)$$

with $\mathbf{a}$ and $\mathbf{b}$ defined in (6.2.15) and $\zeta_\pm = \zeta \pm t$. The vector Δ must be a constant vector independent of $\zeta_\pm$, and it is directly proportional to the momentum of the loop

$$\mathbf{P} = \frac{\mu}{2} \int_0^L d\zeta \, (\mathbf{b}' - \mathbf{a}') = \mu \Delta. \qquad (6.2.23)$$

In the centre-of-mass frame of the loop, $\Delta = 0$ and $\mathbf{a}$ and $\mathbf{b}$ are periodic functions

$$\mathbf{a}(\zeta + L) = \mathbf{a}(\zeta), \qquad \mathbf{b}(\zeta + L) = \mathbf{b}(\zeta). \qquad (6.2.24)$$

It is clear from (6.2.15) and (6.2.24) that the motion of the loop must also be periodic in time. The period is $T = L/2$, rather than $T = L$, since

$$\mathbf{x}\left(\zeta + L/2, t + L/2\right) = \mathbf{x}(\zeta, t). \qquad (6.2.25)$$

For some special loop trajectories this periodicity can be even shorter. The fact that the timescale of the oscillation is comparable to the loop length L indicates that the motion of the loop must be relativistic.

We can define the mean square string velocity of segments on the loop as

$$v_{\rm av} = \left\langle v^2 \right\rangle^{1/2}, \tag{6.2.26}$$

where

$$\left\langle v^2 \right\rangle = \int_0^T \frac{dt}{T} \int_0^L \frac{d\zeta}{L} \dot{\mathbf{x}}^2. \tag{6.2.27}$$

Integrating by parts and using (6.2.11) and (6.2.12) we can write

$$\int_\Sigma d^2\zeta\, \dot{\mathbf{x}}^2 = -\int d^2\zeta\, \mathbf{x}\cdot\ddot{\mathbf{x}} = -\int d^2\zeta\, \mathbf{x}\cdot\mathbf{x}''$$

$$= \int d^2\zeta\, \mathbf{x}'^2 = \int d^2\zeta\left(1 - \dot{\mathbf{x}}^2\right), \tag{6.2.28}$$

where the limits of integration on Σ are over one period as in (6.2.27). We deduce, therefore, that

$$\left\langle v^2 \right\rangle = 0.5. \tag{6.2.29}$$

6.2.3 Cusps and kinks

An interesting property of loop solutions is that particular points along the string can reach the velocity of light during each period [Turok, 1984]. From (6.2.15) we have

$$\dot{\mathbf{x}}^2(\zeta, t) = \tfrac{1}{4}\left[\mathbf{a}'(\zeta - t) - \mathbf{b}'(\zeta + t)\right]^2. \tag{6.2.30}$$

Now, from (6.2.16) and (6.2.24) it follows that the vector functions $\mathbf{a}'(\zeta)$ and $-\mathbf{b}'(\zeta)$ describe closed curves on a unit sphere as ζ runs from 0 to L. These functions should satisfy

$$\int_0^L \mathbf{a}'\, d\zeta = \int_0^L \mathbf{b}'\, d\zeta = 0, \tag{6.2.31}$$

but are otherwise arbitrary. If the two curves intersect, $\mathbf{a}'(\zeta_a) = -\mathbf{b}'(\zeta_b)$, then (6.2.30) gives $\dot{\mathbf{x}}^2(\zeta, t) = 1$ for $2\zeta = \zeta_a + \zeta_b + nL$, $2t = \zeta_b - \zeta_a + nL$, where n is an integer. Smooth loops will generically have such luminal points. By the condition (6.2.31), neither of the two curves $\mathbf{a}'$, $-\mathbf{b}'$ can completely lie in one hemisphere of the unit sphere, so a somewhat contrived choice is required to prevent any intersections. The situation is different for loops which have kinks, or sharp corners, and are described by discontinuous functions $\mathbf{a}'$ and $\mathbf{b}'$. In this case there are gaps in the two curves on the unit sphere, and it is much easier to avoid intersections.

The formation and properties of kinks will be discussed in greater detail
in §6.4.

To study the behaviour of the string near the point of luminal motion,
it is convenient to choose the origin of the spacetime coordinates and the
parametrization of the string so that the luminal point is at $\zeta = t = 0$
and $\mathbf{x} = 0$. Expanding the functions $\mathbf{a}$ and $\mathbf{b}$ near $\zeta = 0$,

$$
\begin{aligned}
\mathbf{a}(\zeta) &= \mathbf{a}_0'\zeta + \tfrac{1}{2}\mathbf{a}_0''\zeta^2 + \tfrac{1}{6}\mathbf{a}_0'''\zeta^3 + \ldots, \\
\mathbf{b}(\zeta) &= \mathbf{b}_0'\zeta + \tfrac{1}{2}\mathbf{b}_0''\zeta^2 + \tfrac{1}{6}\mathbf{b}_0'''\zeta^3 + \ldots,
\end{aligned}
\tag{6.2.32}
$$

and using (6.2.16) we obtain

$$
\begin{aligned}
\mathbf{a}_0' &= -\mathbf{b}_0' \\
|\mathbf{a}_0'| &= |\mathbf{b}_0'| = 1, \\
\mathbf{a}_0' \cdot \mathbf{a}_0'' &= \mathbf{b}_0' \cdot \mathbf{b}_0'' = 0 \\
\mathbf{a}_0''^2 + \mathbf{a}_0' \cdot \mathbf{a}_0''' &= \mathbf{b}_0''^2 + \mathbf{b}_0' \cdot \mathbf{b}_0''' = 0, \ldots.
\end{aligned}
\tag{6.2.33}
$$

The first of these relations just expresses the fact that the curves $\mathbf{a}'(\zeta)$
and $-\mathbf{b}'(\zeta)$ intersect at $\zeta = t = 0$. From (6.2.30) and (6.2.32–33) we
find

$$
\dot{\mathbf{x}}^2(\zeta, t) = 1 - \tfrac{1}{4}\left[(\zeta - t)\mathbf{a}_0'' + (\zeta + t)\mathbf{b}_0''\right]^2 + \ldots.
\tag{6.2.34}
$$

The shape of the string at $t = 0$ is given by

$$
\mathbf{x}(\zeta, 0) = \tfrac{1}{4}\left(\mathbf{a}_0'' + \mathbf{b}_0''\right)\zeta^2 + \tfrac{1}{12}\left(\mathbf{a}_0''' + \mathbf{b}_0'''\right)\zeta^3 + \ldots.
\tag{6.2.35}
$$

It is easily seen that for $(\mathbf{a}_0'' + \mathbf{b}_0'') \neq 0$ the string momentarily develops a
cusp, as shown in fig. 6.1, and it follows from (6.2.33) that the direction
of the cusp $(\mathbf{a}_0'' + \mathbf{b}_0'')$ is orthogonal to that of the luminal velocity $(-\mathbf{a}_0')$.
With an appropriate choice of axes, the shape of the string near a cusp
is given by $y \propto x^{2/3}$.

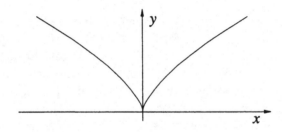

Fig. 6.1.   Generic shape of a string segment when a cusp forms.

6.2.4 Some loop examples

The general solution of (6.2.13) can be represented as an expansion in Fourier modes

$$\mathbf{x}(\zeta, t) = \mathbf{x}_0 + t\Delta + \sum_{n=1}^{\infty} \left[\mathbf{A}_n^+ \sin\left(2\pi n\zeta_+/L\right) + \mathbf{B}_n^+ \cos\left(2\pi n\zeta_+/L\right) \right]$$

$$+ \sum_{n=1}^{\infty} \left[\mathbf{A}_n^- \sin\left(2\pi n\zeta_-/L\right) + \mathbf{B}_n^- \cos\left(2\pi n\zeta_-/L\right) \right].$$

(6.2.36)

The constraints (6.2.11–12) then give nonlinear algebraic equations for the coefficients $\mathbf{A}_n^\pm$, $\mathbf{B}_n^\pm$. While difficult to solve in general, when truncated to the lowest harmonics, smooth exact solutions can be easily found.

A two-parameter family of such loop solutions involving the first and third harmonics is [Kibble & Turok, 1982; Turok, 1984].

$$\mathbf{x}(\zeta, t) = \frac{L}{4\pi} \left\{ \hat{\mathbf{e}}_1 \left[(1-\kappa)\sin\sigma_- + \tfrac{1}{3}\kappa\sin 3\sigma_- + \sin\sigma_+ \right] \right.$$

$$- \hat{\mathbf{e}}_2 \left[(1-\kappa)\cos\sigma_- + \tfrac{1}{3}\kappa\cos 3\sigma_- + \cos\varphi\cos\sigma_+ \right] \qquad (6.2.37)$$

$$\left. - \hat{\mathbf{e}}_3 \left[2\kappa^{1/2}(1-\kappa)^{1/2}\cos\sigma_- + \sin\varphi\cos\sigma_+ \right] \right\}$$

where $\sigma_\pm = (2\pi/L)\zeta_\pm$, $\hat{\mathbf{e}}_i$ are unit vectors in the directions of the Cartesian axes and the constant parameters κ and φ take values in the range $0 < \kappa \leq 1$ and $-\pi \leq \varphi \leq \pi$. Several loop configurations are illustrated in fig. 6.2 for some parameter values.

The single frequency solution with $\kappa = 0$ corresponds to an elliptical loop that collapses to a double line, a moment at which the ends move at the speed of light. The loop then returns to the original elliptical shape and repeats the cycle (refer to fig. 6.2). The two degenerate cases with $\varphi = 0$ and $\varphi = \pi$ correspond respectively to an oscillating circular loop, which collapses to a point, and a rotating double line. With a non-zero contribution from the third harmonic ($\kappa \neq 0$) the solutions never collapse to a point or a double line, and for a significant region of parameter space they never self-intersect. Chen, DiCarlo & Hotes [1988] have generalized (6.2.37) to a three-parameter family that includes all solutions constructed from the first and third harmonics in this gauge, a problem also studied by DeLaney, Engle & Scheick [1989]. Brown & DeLaney [1989], adopting a new approach to the harmonic analysis of strings, found a practical procedure for solving the constraints for any loop with a given number of harmonics.

It is worthwhile presenting another family of loop solutions in which the functions $\mathbf{a}$ and $\mathbf{b}$ describe circular motions in two planes subtending an angle φ [Burden, 1985]

$$\mathbf{a}(\zeta) = \alpha^{-1} \left(\hat{\mathbf{e}}_1 \sin \alpha\zeta + \hat{\mathbf{e}}_3 \cos \alpha\zeta \right)$$
$$\mathbf{b}(\zeta) = \beta^{-1} \left[(\hat{\mathbf{e}}_1 \cos \varphi + \hat{\mathbf{e}}_2 \sin \varphi) \sin \beta\zeta + \hat{\mathbf{e}}_3 \cos \beta\zeta \right] . \tag{6.2.38}$$

The two parameters $\alpha = 2\pi m/L$, and $\beta = 2\pi n/L$ are defined in terms of the relatively prime integers m, n (if m, n have a common factor p then the loop becomes degenerate, being traversed p times by ζ between 0 and L). The periodicity of the loop motion (6.2.38) is apparent from the two periods of $\mathbf{a}$ and $\mathbf{b}$, $T_1 = L/2m$ and $T_2 = L/2n$ respectively, which gives an actual oscillation period $T = L/2mn$. In general, with $m \neq 1$, $n \neq 1$ these solutions self-intersect, and the degenerate solutions with $\varphi = 0$ and $\varphi = \pi$ rigidly rotate, possessing persistent cusps [Burden & Tassie, 1984].

Relaxing the condition of smoothness implicit in the truncated Fourier expansion, we note that it is even simpler to construct loops which have discontinuous derivatives or kinks. One method is merely to match the ends of N straight string segments of varied lengths and velocities, requiring only that they form a closed loop. These straight segments correspond to separated point values of $\mathbf{a}'$ and $\mathbf{b}'$ on the unit sphere. A particularly simple example has four kinks,

$$\mathbf{a}(\zeta) = \begin{cases} \left(\zeta - \dfrac{L}{4} \right) \hat{\mathbf{A}}, & 0 \leq \zeta < L/2, \\[2mm] \left(\dfrac{3L}{4} - \zeta \right) \hat{\mathbf{A}}, & L/2 \leq \zeta < L, \end{cases}$$
$$\mathbf{b}(\zeta) = \begin{cases} \left(\zeta - \dfrac{L}{4} \right) \hat{\mathbf{B}}, & 0 \leq \zeta < L/2, \\[2mm] \left(\dfrac{3L}{4} - \zeta \right) \hat{\mathbf{B}}, & L/2 \leq \zeta < L, \end{cases} \tag{6.2.39}$$

where $\hat{\mathbf{A}}$ and $\hat{\mathbf{B}}$ are arbitrary unit vectors [Garfinkle & Vachaspati, 1987]. The two pairs of kinks propagate in opposite directions around the loop. In the special case $\hat{\mathbf{A}} \cdot \hat{\mathbf{B}} = 0$, the loop is planar and oscillates between a square and a double line.

More realistically, kinky loop solutions can be constructed from smooth loops by considering loop self-intersections. At the moment of intersection we have

$$\mathbf{x}(\zeta_1, t) = \mathbf{x}(\zeta_2, t), \qquad \zeta_2 > \zeta_1, \tag{6.2.40}$$

where ζ_1, ζ_2 are the two values of ζ along the loop. Clearly, on the separate domains $[\zeta_1, \zeta_2)$ and $[0, \zeta_1) \cup [\zeta_2, L)$ we have two sets of functions

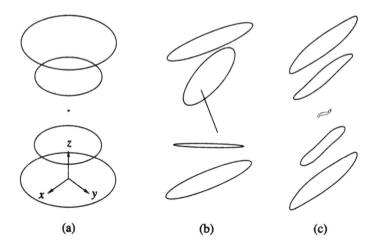

Fig. 6.2. Several of the oscillating loop solutions (6.2.37) shown at quarter period intervals for the parameters: (a) $\phi = 0$, $\kappa = 0$; (b) $\phi = \pi/3$, $\kappa = 0$; (c) $\phi = \pi/3$, $\kappa = 0.5$.

a and **b** which satisfy the closed loop boundary conditions though now with different periodicities $(\zeta_2 - \zeta_1)/2$ and $(L - (\zeta_2 - \zeta_1))/2$. We can evolve these two distinct loop solutions separately, but now each possesses two kinks, discontinuities in **a$'$** and **b$'$** resulting from the self-intersection. In general, these loop fragments will move apart with substantial relative velocities, a break-up process we shall discuss in §6.4.2.

6.3 Strings in curved spacetime

6.3.1 Gauge fixing

We have already derived the string equations of motion (6.1.12) in a general curved spacetime. As in flat spacetime, these equations can be simplified by imposing two gauge conditions, but the best choice of gauge is not necessarily the same as in flat space and various alternatives may be appropriate for different problems.

We first consider the conformal gauge,

$$\dot{x} \cdot x' = 0 \,, \tag{6.3.1}$$
$$\dot{x}^2 + x'^2 = 0 \,, \tag{6.3.2}$$

which was our preferred choice in Minkowski space. In this gauge, we have $\gamma^{ab}\sqrt{-\gamma} = \eta^{ab}$, and (6.1.12) takes the form

$$\ddot{x}^\mu - x''^\mu + \Gamma^\mu_{\nu\sigma}(\dot{x}^\nu \dot{x}^\sigma - x'^\nu x'^\sigma) = 0 \,. \tag{6.3.3}$$

In flat spacetime, the residual gauge freedom remaining after imposing the conditions (6.3.1–2) allowed us to identify ζ^0 with the time variable $x^0 (\equiv t)$. However, in a curved background with $\Gamma^0_{\nu\sigma} \neq 0$ it can be easily seen that $x^0 = \zeta^0$ is not a solution of (6.3.3).

In many applications it is convenient to abandon the second gauge condition (6.3.2) and instead to replace it by

$$\zeta^0 = t. \qquad (6.3.4)$$

The string equation of motion is then

$$\frac{\partial}{\partial t}\left[\frac{\dot{x}^\mu x'^2}{\sqrt{-\gamma}}\right] + \frac{\partial}{\partial \zeta}\left[\frac{x'^\mu \dot{x}^2}{\sqrt{-\gamma}}\right] + \frac{1}{\sqrt{-\gamma}}\Gamma^\mu_{\nu\sigma}(x'^2 \dot{x}^\nu \dot{x}^\sigma + \dot{x}^2 x'^\nu x'^\sigma) = 0, \quad (6.3.5)$$

where $\zeta \equiv \zeta^1$. This gauge will be used in the next section to study string dynamics in an expanding universe. Note that (6.3.1) and (6.3.4) do not fix the gauge completely; we still have the freedom to perform transformations

$$\zeta \to \tilde{\zeta}(\zeta). \qquad (6.3.6)$$

Further simplification is possible for a static spacetime described by the metric

$$ds^2 = f(\mathbf{x})dt^2 + g_{ij}(\mathbf{x})dx^i dx^j. \qquad (6.3.7)$$

In this metric, the $\mu = 0$ component of (6.3.5) is

$$\frac{\partial}{\partial t}\left[f\left(-\frac{x'^2}{\dot{x}^2}\right)^{1/2}\right] = 0. \qquad (6.3.8)$$

Using the residual gauge freedom (6.3.6) we can require [Basu & Vilenkin, 1989]

$$f(-x'^2/\dot{x}^2)^{1/2} = 1. \qquad (6.3.9)$$

Then the $\mu = i$ components of (6.3.5) take the form

$$\partial_t(f^{-1}\dot{x}_i) - \partial_\zeta(f x'_i) + \tfrac{1}{2}g_{\nu\sigma,i}(f^{-1}\dot{x}^\nu \dot{x}^\sigma - f x'^\nu x'^\sigma) = 0, \qquad (6.3.10)$$

where $\dot{x}_i \equiv g_{ij}\dot{x}^j$ and $x'_i \equiv g_{ij}x'^j$.

There is a simple interpretation for (6.3.9). The conserved energy of the string in a static spacetime is

$$\mathcal{E} = \int d^3x \sqrt{-g}\, T^0_0 \qquad (6.3.11)$$

where T^ν_μ is defined in (6.1.15). With our gauge conditions (6.3.1) and (6.3.4), this reduces to

$$\mathcal{E} = \mu \int d\zeta\, f\,(-x'^2/\dot{x}^2)^{1/2} = \mu \int d\zeta, \qquad (6.3.12)$$

where in the last step we have used (6.3.9). From (6.3.12), we see that our choice of gauge corresponds to choosing the spacelike parameter ζ to be equal to the invariant length along the string, $\Delta\zeta = \Delta\mathcal{E}/\mu$.

6.3.2 Strings in an expanding universe

Equations of motion

We now turn to the dynamics of strings in an expanding universe described by an FRW metric (1.2.4). It will be convenient to use conformal time τ, $d\tau = dt/a(t)$, so that the metric takes the form

$$ds^2 = a^2(\tau)\left(d\tau^2 - d\mathbf{x}^2\right), \tag{6.3.13}$$

and the gauge conditions

$$\zeta^0 = \tau, \qquad \dot{\mathbf{x}} \cdot \mathbf{x}' = 0. \tag{6.3.14}$$

The equations of motion in this gauge can be obtained from (6.3.5) or directly by varying the action (6.1.9) with respect to $\mathbf{x}(\zeta, \tau)$ [Turok & Bhattacharjee, 1984]:

$$\ddot{\mathbf{x}} + 2\frac{\dot{a}}{a}\left(1 - \dot{\mathbf{x}}^2\right)\dot{\mathbf{x}} = \epsilon^{-1}(\epsilon^{-1}\mathbf{x}')', \tag{6.3.15}$$

$$\dot{\epsilon} = -2\frac{\dot{a}}{a}\epsilon\dot{\mathbf{x}}^2. \tag{6.3.16}$$

Here, dots and primes are derivatives with respect to conformal time τ and the spacelike parameter ζ, respectively, and

$$\epsilon = \left(\frac{\mathbf{x}'^2}{1 - \dot{\mathbf{x}}^2}\right)^{1/2}. \tag{6.3.17}$$

It can be verified that (6.3.16) is not an independent equation; it is satisfied as a consequence of (6.3.14) and (6.3.15).

The string energy and momentum in an expanding universe can be defined as

$$\mathcal{E} = a(\tau)\mu\int d\zeta\, \epsilon, \tag{6.3.18}$$

$$\mathbf{P} = a(\tau)\mu\int d\zeta\, \epsilon\dot{\mathbf{x}}. \tag{6.3.19}$$

The rate of change of $\mathcal{E}$ and $\mathbf{P}$ can then be found using (6.3.16),

$$\dot{\mathcal{E}} = \frac{\dot{a}}{a}\left(1 - 2\langle v^2\rangle\right)\mathcal{E}, \tag{6.3.20}$$

$$\dot{\mathbf{P}} = -\frac{\dot{a}}{a}\mathbf{P}, \tag{6.3.21}$$

where $\langle v^2 \rangle$ is the average string velocity squared,

$$\langle v^2 \rangle = \frac{\int d\zeta \, \epsilon \, \dot{\mathbf{x}}^2}{\int d\zeta \, \epsilon} \, . \tag{6.3.22}$$

The derivation of (6.3.21) involved the integration of a total derivative, while dropping the boundary terms on the assumption that the string is a closed loop.

Equation (6.3.20) describes the manner in which the string will either gain energy through stretching or else lose it through velocity redshifting. In flat space we have already observed in §6.2.2 that $\langle v^2 \rangle = 1/2$ for any loop at rest, provided that this is time-averaged over one period. In the expanding universe, Hubble damping on large scales would be expected to reduce velocities somewhat, $\langle v^2 \rangle < 1/2$, so stretching effects should predominate in (6.3.20) for large loops ($L \gtrsim d_{\mathrm{H}}$). The damping term in (6.3.15) is of the order $(L/d_{\mathrm{H}})^2$ and becomes increasingly less significant as the loop falls inside the horizon. For $L \ll d_{\mathrm{H}}$, the evolution will be almost indistinguishable from that in flat space, and the energy of a loop will remain nearly constant. In this regime by (6.3.21), the loop momentum will be redshifted like the momentum of a point particle in an expanding universe, $\mathbf{P} \propto a^{-1}$.

String dynamics

A straight static string, $\mathbf{x}(\zeta) = \mathbf{c}\zeta$ with $\mathbf{c}$ a constant, is clearly a solution of (6.3.15). A straight string remains straight and is simply stretched by the expansion. For a moving straight string we can write

$$\mathbf{x} = \mathbf{b}(\tau) + \mathbf{c}\zeta, \tag{6.3.23}$$

with $\dot{\mathbf{b}} \cdot \mathbf{c} = 0$, and (6.3.15) gives

$$\dot{\mathbf{v}} + 2 \left(\frac{\dot{a}}{a} \right) (1 - v^2) \mathbf{v} = 0, \tag{6.3.24}$$

where $\mathbf{v} = \dot{\mathbf{b}}$ and $v = |\mathbf{v}|$. The solution of this equation is

$$v \left(1 - v^2 \right)^{-1/2} \propto a^{-2}. \tag{6.3.25}$$

Note that the string momentum per comoving volume decreases as a^{-1}, like the momentum of a loop. The difference between the two cases is that the mass of string per comoving volume grows as $a(\tau)$.

Let us now consider small perturbations on a static straight string [Vilenkin, 1981c]. Representing $\mathbf{x}(\zeta, \tau)$ as

$$\mathbf{x} = \mathbf{c}\zeta + \delta\mathbf{x}, \tag{6.3.26}$$

we can linearize (6.3.14–15) in $\delta\mathbf{x}$. For a power-law expansion,

$$a(\tau) = \tau^\alpha, \tag{6.3.27}$$

we have

$$\delta\ddot{\mathbf{x}} + \frac{2\alpha}{\tau}\delta\dot{\mathbf{x}} - \delta\mathbf{x}'' = 0, \tag{6.3.28}$$

$$\mathbf{c}\cdot\delta\dot{\mathbf{x}} = 0, \tag{6.3.29}$$

where the parameter ζ has been chosen so that $|\mathbf{c}| = 1$. The solution of (6.3.28) which is well-behaved as $\tau \to 0$ is a superposition of waves of the form

$$\delta\mathbf{x} = \mathbf{A}\tau^{-\nu}J_\nu(k\tau)e^{ik\zeta}, \tag{6.3.30}$$

where $\nu = \alpha - 1/2$ and the constant vector $\mathbf{A}$ is orthogonal to $\mathbf{c}$. Recall that x^i are comoving coordinates, so to obtain the physical wave parameters, (6.3.30) has to be rescaled by $a(\tau)$. The physical wavelength is then $\lambda = 2\pi a(\tau)/k$, which grows in proportion to the scale factor. The quantity $k\tau \sim t/\lambda$ gives the ratio of the horizon size to the wavelength (to order of magnitude).

As $k\tau \to 0$, the time-dependent factor in (6.3.30) approaches a constant, which means that waves with $\lambda \gg t$ do not move in comoving coordinates. These waves are conformally stretched: both the amplitude and the wavelength grow like $a(\tau)$, while the shape of the string remains unchanged. In the opposite limiting case $k\tau \gg 1$, $\lambda \ll t$, the wavelength still grows as $a(\tau)$, but the physical amplitude of the wave, $a(\tau)\delta\mathbf{x}$, remains constant. As the ratio of the amplitude to the wavelength decreases, the string progressively straightens.

Numerical simulations of string evolution in the early universe indicate that the results of this perturbative analysis also apply for strongly curved strings, as we shall discuss in chapter 9. Waves on scales larger than the horizon are conformally stretched, while irregularities on scales below the horizon are smoothed out by the expansion. To give some examples of this behaviour, first consider a circular loop of initial radius R_0 much greater than the horizon at time t_0, $R_0 \gg t_0$, in the radiation-dominated era, $a(t) \propto t^{1/2}$. Initially, the loop is stretched in proportion to the scale factor

$$R(t) = \left(\frac{t}{t_0}\right)^{1/2} R_0 \tag{6.3.31}$$

until it comes within the horizon. This happens at $t = t_h$ such that $R(t_h) \sim t_h$,

$$t_h \sim \frac{R_0^2}{t_0} \tag{6.3.32}$$

At this time the radius of the loop begins to decrease. When the loop is much smaller than the horizon, the effects of the expansion become unimportant, and it collapses relativistically to a point, just as it would in flat spacetime.

For another illustration, consider a large irregular loop having overall size $R_0 \gg t_0$, but locally having the shape of a three-dimensional random walk of step-length $\xi_0 \sim t_0$. The initial total length of the loop will be $L_0 \sim R_0^2/\xi$. (As we shall see in §9.1, strings formed at a phase transition in the early universe can be expected to be Brownian.) In the course of expansion the size of the loop is stretched as in (6.3.31) until the loop becomes smaller than the horizon at $t \sim t_h$. Since irregularities on scales smaller than the horizon are gradually smoothed out, at any time $t_0 \ll t \ll t_h$, the loop has the shape of a random walk of step-length $\xi \sim t$ with overall length, in a radiation dominated universe,

$$L(t) \sim \frac{R^2(t)}{t} \sim \text{const.} , \tag{6.3.33}$$

and, in a matter-dominated universe,

$$L(t) \propto t^{1/3} . \tag{6.3.34}$$

By the time the loop comes within the horizon it has lost its Brownian appearance, having become relatively smooth (this stretching will not completely remove all small-scale structure). When the size of the loop is much smaller than the horizon, the effects of the expansion can be neglected, and it will begin to oscillate.

This discussion does not take into account the effect of reconnection, which will be discussed in §6.4. For the remainder of this chapter we shall consider strings in flat spacetime (except for the next sub-section), returning to an expanding universe in chapter 9 to discuss the cosmological evolution of strings.

6.3.3 Strings in a stationary field

We first consider a string in a weak gravitational field produced by a non-relativistic distribution of masses. The corresponding metric can be written as [Landau & Lifshitz, 1975]

$$ds^2 = (1 + 2U)dt^2 - (1 - 2U)d\mathbf{x}^2 , \tag{6.3.35}$$

where the gravitational potential $U(\mathbf{x}, t)$ is a slowly-varying function of t and we have neglected quadratic and higher-order terms in U. In the same approximation, the string equations of motion (6.3.10) and the gauge condition (6.3.9) become

$$\frac{\partial}{\partial t}[(1 - 4U)\dot{\mathbf{x}}] - \mathbf{x}'' = 2\dot{\mathbf{x}}^2 \cdot \nabla U , \tag{6.3.36}$$

$$(1 - 4U)\dot{\mathbf{x}}^2 + \mathbf{x}'^2 = 1.$$ (6.3.37)

A static straight string is clearly a solution to (6.3.36–37),

$$\dot{\mathbf{x}} = 0, \qquad \mathbf{x}' = \text{const}.$$ (6.3.38)

This means, somewhat surprisingly, that a straight string does not react to a Newtonian gravitational field [Vilenkin, 1981b]. Insight into the origin of this result is provided in chapter 7, where it is shown that a static straight string produces no gravitational force on surrounding matter. The behaviour of an oscillating closed loop is quite different. The average value of $\dot{\mathbf{x}}^2$ for a loop is 0.5, and it is straightforward to show from (6.3.36) that the average acceleration of a small loop is the same as that of a test particle.

The straight string solution (6.3.38) was derived using linearized gravity. Turning now to the full Einstein theory, we consider a general stationary metric

$$ds^2 = f(dt + A_i dx^i)^2 - g_{ij} dx^i dx^j,$$ (6.3.39)

where f, A_i and g_{ij} are functions only of $\mathbf{x}$. To find an analogue of the solution (6.3.38), we look for static string configurations $\mathbf{x}(\zeta)$ in the metric (6.3.39). It is easily shown that the Nambu action (6.1.9) for such a string is independent of A_i and is given by

$$S = -\mu\Delta t \int d\zeta \left(f g_{ij} \frac{dx^i}{d\zeta} \frac{dx^j}{d\zeta} \right)^{1/2}.$$ (6.3.40)

Extremizing this action is clearly equivalent to extremizing the length in the metric,

$$\tilde{g}_{ij} = f g_{ij}.$$ (6.3.41)

Hence, static string configurations are geodesics in a three-dimensional space with metric (6.3.41) conformally related to the spatial metric g_{ij} [Frolov, Skarzhinsky, Zelnikov & Heinrich, 1989]. Explicit solutions of the corresponding geodesic equations were obtained for strings in a Kerr–Newman spacetime describing a charged, rotating black hole [Frolov *et al.*, 1989; see also Carter & Frolov, 1989].

The motion of strings in the gravitational field of a black hole has been studied by Lonsdale & Moss [1988]. They considered an initially straight string moving from infinity towards a Schwarzschild black hole. If the impact parameter r of the string is sufficiently small, the string is captured by the hole. Numerical integration of the string equations of motion in this case reveals some surprising features (see fig. 6.3). The string distortion caused by the gravity of the black hole often evolves into

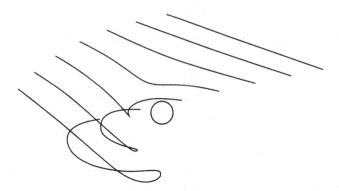

Fig. 6.3. The distortion of a string in the vicinity of a black hole. The position of the black hole is indicated by a circle with its Schwarzschild radius r_s. The string is shown at successive times for an initial velocity $v = 0.7$ and impact parameter $b = 4.55r_s$ [Lonsdale & Moss, 1988].

a cusp. This cusp continues to evolve into a loop which then forms another cusp, and the process may repeat itself several times before resulting in string capture. Note that the string in fig. 6.3 passes the black hole without being captured and that we could expect the string to lose such convoluted loop configurations through self-intersection.

6.4 String interactions

6.4.1 Reconnection

The Nambu action (6.1.9) gives an accurate description of string motion provided that the string does not intersect itself or another string. The zero-width approximation breaks down near the crossing point and so to determine the ultimate outcome of a collision the underlying nonlinear field theory must be studied. Such investigations have been discussed in §4.3. For the abelian-Higgs model (4.1.1), numerical simulations reveal that strings invariably reconnect or 'exchange partners' as illustrated in fig. 6.4(*a*). As demonstrated for a wide range of collision parameters, strings are unable to simply pass through one another as shown in fig. 6.4(*b*). These conclusions are not an inevitable outcome for strings in more complicated models because it is possible to have a topological obstruction to reconnection as discussed in §3.3.1. In this case, the strings become entangled as in fig. 6.4(*c*), perhaps with a third string stretching between them. In the context of an interacting network, such strings evolve very differently from those which reconnect and, as we shall see in §9.4, are more difficult to reconcile with the standard cosmology. Unless otherwise stated, here we shall consider strings that reconnect.

Every string intersection between locally straight segments can be completely characterized by two collision parameters, the angle θ subtended by the strings and a relative perpendicular velocity v [Shellard, 1987]. This is easy to demonstrate for two string solutions $\mathbf{x}$, $\mathbf{y}$, in the transverse gauge (6.2.11), with tangent vectors $\mathbf{x}'$, $\mathbf{y}'$, and velocities $\dot{\mathbf{x}}$, $\dot{\mathbf{y}}$. We choose a frame Σ in which the first string is stationary and lies along the x-axis, $\dot{\mathbf{x}} = 0$, $\mathbf{x}' = \hat{\mathbf{e}}_1$ (near the crossing point). Now, $\mathbf{x}(\zeta, t)$ will remain invariant under longitudinal Lorentz boosts in the x-direction and also under arbitrary rotations about the x-axis. So a boost can be employed to set $\dot{y}_1 = 0$, and then a rotation to set $\dot{y}_2 = 0$ and $y_3' = 0$ (since $\dot{\mathbf{y}} \cdot \mathbf{y}' = 0$). Thus for the second string we obtain

$$\dot{\mathbf{y}} = (0, 0, v),$$
$$\mathbf{y}' = (1 - v^2)^{1/2} (\cos\theta, \sin\theta, 0). \tag{6.4.1}$$

This corresponds to string intersection in the xy-plane with a z-directed relative velocity v.

At every intercommuting, pieces of string that were previously moving independently become connected. On opposite sides of the point of intersection the newly-connected string will point in different directions and will be moving with different velocities. That is, at the moment of intersection t_0, $\mathbf{x}'(\zeta, t_0)$ and $\dot{\mathbf{x}}(\zeta, t_0)$ change very rapidly as functions of ζ on a length scale comparable to the string thickness. This region of rapid

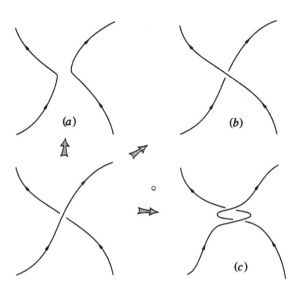

Fig. 6.4. Possible outcomes at string intersection: (a) reconnection, (b) no interaction, and (c) entanglement.

variation is called a 'kink' and, in the zero-width approximation, $\mathbf{x}'$ and $\dot{\mathbf{x}}$ are discontinuous at the kink. This implies that the functions $\mathbf{a}(\zeta - t)$ and $\mathbf{b}(\zeta + t)$ describing the motion of the string after intercommuting must have discontinuous derivatives at certain values of their arguments. The kink, therefore, does not disappear after intercommuting. Instead, it splits into two kinks running along the string in opposite directions. A single kink is present in the special case when only one of the functions $\mathbf{a}'$ and $\mathbf{b}'$ is discontinuous. The kink due to a discontinuity in $\mathbf{a}'$ has a fixed value of $\zeta - t$, with a velocity given by

$$\frac{d\mathbf{x}}{dt} = \frac{1}{2}\mathbf{b}' \cdot \left(1 + \frac{d\zeta}{dt}\right) = \mathbf{b}'. \qquad (6.4.2)$$

Since $|\mathbf{b}'| = 1$, the kink must move at the speed of light. Similarly, the kink due to a discontinuity in $\mathbf{b}'$ remains at a fixed value of $\zeta + t$ and has a velocity $\dot{\mathbf{x}} = -\mathbf{a}'$ with $|\dot{\mathbf{x}}| = 1$. The angle subtended by a single kink – its opening angle – in any frame must be greater than $\pi/2$. The sharpest kinks are created when nearly antiparallel strings reconnect, where the resulting double line separates into two right-angled kinks.

6.4.2 Loop fragmentation

Closed loops of string can self-intersect and break into smaller loops. Close examination of the loop solutions presented in §6.2.4 reveals that some self-intersect, while others do not. For example, solutions (6.2.37) with $\varphi = 0$, $\kappa \neq 0$ and solutions (6.2.38) with $m = 1$, $n \neq 1$ describe loops which never self-intersect. One could try to describe loop fragmentation as a stochastic process in which the parent loop fragments into two 'daughters' with some probability p. Each of the daughters has the same probability of further fragmentation and so on, with the fragmentation probability remaining unchanged in subsequent generations [Smith & Vilenkin, 1987a]. For $p < 1/2$ this model predicts that the end result of loop fragmentation is a finite set of non-intersecting loops, while for $p > 1/2$ the average number of fragments is infinite. However, numerical simulations of loop fragmentation indicate the inadequacy of such a stochastic model.

Scherrer & Press [1989] studied the fragmentation of loops described by the functions

$$a(\zeta) = \sum_{n=1}^{N} a_n \cos\left(n\zeta + \varphi_n\right),$$

$$b(\zeta) = \sum_{n=1}^{N} b_n \cos\left(n\zeta + \varphi_n\right), \qquad (6.4.3)$$

where $N = 10$ and the phases φ_n are chosen at random between 0 and 2π. The invariant length of a loop L is then normalized to 2π. They sampled a large number of loops belonging to two different types, for which the coefficients a_n and b_n were randomly chosen (i) between 0 and 1 and (ii) between 0 and $1/n^2$. Loops of type (i) were very convoluted, while loops of type (ii) were relatively smooth, with higher harmonics suppressed. Almost all the initial loops were self-intersecting (100% for type (i) and 94% for type (ii)), so according to the naive stochastic model one would expect an infinite number of loops in the final state. However, the fraction of self-intersecting loops became smaller in each subsequent generation, and after several generations the fragmentation process stopped completely. The number of fragments in the final state was 28 ± 6 for loops of type (i) and 8 ± 4 for loops of type (ii). In both cases, the characteristic size of the largest fragment was comparable to the typical curvature radius (or the persistence length) of the parent loop. The average velocity of the fragments was $v \sim 0.6$, with smaller fragments typically having larger velocities. Similar results were obtained by York [1989] using loops constructed of random straight segments.

The failure of the stochastic model is not very surprising if we note that loop fragmentation is a completely deterministic process. The condition that a loop described by the functions $\mathbf{a}(\zeta - t)$ and $\mathbf{b}(\zeta + t)$ self-intersects is that

$$\mathbf{a}(\zeta_1 - t_0) + \mathbf{b}(\zeta_1 + t_0) = \mathbf{a}(\zeta_2 - t_0) + \mathbf{b}(\zeta_2 + t_0) \qquad (6.4.4)$$

for some values of $\zeta_1, \zeta_2 > \zeta_1$ and t_0. At the moment of intersection, the daughter loop, corresponding to the points $\zeta_1 \leq \zeta \leq \zeta_2$, is described by the same functions $\mathbf{a}$ and $\mathbf{b}$ as the parent loop. The invariant length of the daughter loop is $\ell = \zeta_2 - \zeta_1$ and we can use the closure condition (6.2.22) to continue these functions beyond the range $\zeta_1 < \zeta < \zeta_2$:

$$\begin{aligned} \mathbf{a}(\zeta_- + \ell) &= \mathbf{a}(\zeta_-) - \Delta, \\ \mathbf{b}(\zeta_+ + \ell) &= \mathbf{b}(\zeta_+) + \Delta. \end{aligned} \qquad (6.4.5)$$

Here, $\zeta_\pm = \zeta \pm t$ and Δ is a constant vector determined by the intersection points on the original loop,

$$\Delta = \mathbf{b}(\zeta_2 + t_0) - \mathbf{b}(\zeta_1 + t_0) = \mathbf{a}(\zeta_1 - t_0) - \mathbf{a}(\zeta_2 - t_0). \qquad (6.4.6)$$

The new functions $\mathbf{a}(\zeta_-)$ and $\mathbf{b}(\zeta_+)$ have discontinuous derivatives at $\zeta_- = \zeta_1 - t_0$ and $\zeta_+ = \zeta_1 + t_0$, respectively, indicating that the daughter loop acquires a pair of kinks propagating in the opposite directions. The second daughter loop, which has the invariant length $L - \ell$, can be treated in the same manner.

In general, (6.4.4) can have several solutions for potential intersection points of the parent loop. Since the daughter loops are described by the same functions **a** and **b**, they can inherit some of these intersections. However, some of the intersections can be lost if the corresponding points on the string belong to different daughters. New intersection points can also be created, since (6.4.4) can acquire new solutions due to the changed periodicity properties of the functions **a** and **b**. Viewed in this way, the fragmentation process is a competition between the creation of new intersection points and the destruction of old ones. The stochastic model can apply only if the number of intersection points per loop remains roughly constant throughout the fragmentation process. Since the number of loops is sharply increased in each generation, this would require that the creation of intersection points be much more efficient than their destruction. As we have mentioned, numerical simulations indicate that this is not the case.

This observation becomes intuitively clearer if we consider loops consisting only of straight segments separated by kinks [York, 1989]. Since every kink subtends an angle greater than $\pi/2$, each closed loop must possess at least five kinks (four is possible (6.2.39), but highly degenerate). By following a single fragment from a parent loop with n kinks, we see that at each splitting a daughter loop must lose at least five kinks to its sibling, while gaining only two from the intersection – a net loss of ≥ 3 kinks. This loss of structure can only continue for at most $n/3$ generations, and will end when all the daughter loops have achieved relatively simple configurations with only a few kinks. It should be noted that this argument assumes that merging between loops is unlikely.

Finally, we comment on an inheritance law for cusps. Cusps on a parent loop correspond to intersections of the curves $\mathbf{a}'(\zeta_-)$ and $-\mathbf{b}'(\zeta_+)$ on the unit sphere. If the loop has no kinks and is not very convoluted, so that the typical curvature radius of the curves is ~ 1, then in general the curves will have $2m$ intersections, where m is a small integer. Daughter loops inherit parts of the $\mathbf{a}'$ and $\mathbf{b}'$ curves of the parent loop and it is easily seen that no new cusps can be created during the fragmentation process. On the other hand, old cusps can be destroyed if the intersecting parts of the $\mathbf{a}'$ and $-\mathbf{b}'$ curves belong to different daughters. Hence, the total number of cusps in the daughter loops cannot exceed the number of cusps in the parent loop [Thompson, 1988a]. Of course, conversely, in the unlikely event that loops form by mergers from loop collisions, new cusps can also be created.

6.5 Strings with small-scale structure

6.5.1 Equation of state

As we shall see in chapter 9, strings in an interacting network develop substantial substructure in the form of kinks and wiggles on scales much smaller than the characteristic length-scale of the network. To an observer who is unable to resolve this small-scale structure, the string will appear to be smooth, but the effective mass per unit length $\tilde{\mu}$ and the tension $\tilde{T}$ will be different from those of an unperturbed string. Carter [1990a,c] has argued that the effective equation of state for a wiggly string should be such that the speeds of propagation of transverse and longitudinal waves along the string should be the same. These speeds are given by

$$v_{\mathrm{T}} = \left(\tilde{T}/\tilde{\mu}\right)^{1/2}, \qquad v_{\mathrm{L}} = \left(-d\tilde{T}/d\tilde{\mu}\right)^{1/2} \tag{6.5.1}$$

and the requirement $v_{\mathrm{T}} = v_{\mathrm{L}}$ leads to a unique equation of state

$$\tilde{\mu}\tilde{T} = \mu^2. \tag{6.5.2}$$

The quantities $\tilde{\mu}$ and $\tilde{T}$ must be interpreted as the mass per unit length and the tension in the local rest frame of the string and are, in general, functions of position and time.

The wiggly string equation of state (6.5.2) can be derived from first principles, taking the Nambu equation of motion as the starting point [Vilenkin, 1990]. We first recall that a static straight string is represented by the functions $\mathbf{a}(\zeta) = \mathbf{b}(\zeta) = \hat{\mathbf{n}}\zeta$, where $\hat{\mathbf{n}}$ is a unit vector along the string. We shall consider a perturbed string

$$\begin{aligned} \mathbf{a}(\zeta) &= k_1 \zeta \hat{\mathbf{n}} + \boldsymbol{\xi}_1(\zeta), \\ \mathbf{b}(\zeta) &= k_2 \zeta \hat{\mathbf{n}} + \boldsymbol{\xi}_2(\zeta) \end{aligned} \tag{6.5.3}$$

with $k_1, k_2 = $ const. and

$$\hat{\mathbf{n}} \cdot \boldsymbol{\xi}_1(\zeta) = \hat{\mathbf{n}} \cdot \boldsymbol{\xi}_2(\zeta) = 0. \tag{6.5.4}$$

The perturbations $\boldsymbol{\xi}_1, \boldsymbol{\xi}_2$ represent the small-scale structure or 'wiggles' and can be pictured as a superposition of waves propagating along the string in opposite directions. The constraint equations (6.2.11–12) imply

$$\boldsymbol{\xi}'^2_{1,2} = 1 - k^2_{1,2} \tag{6.5.5}$$

and, therefore,

$$-1 \leq k_1, k_2 \leq 1. \tag{6.5.6}$$

We shall assume that the direction of $\hat{\mathbf{n}}$ is chosen so that $k_1 + k_2 > 0$. The assumption that k_1 and k_2 are constant means that all physical properties are uniform along the string when averaged over the wiggles.

To an observer who cannot resolve the substructure, the string appears to be a straight line. If we use Cartesian coordinates with the x-axis along the string, then $\hat{\mathbf{n}} = (1,0,0)$, $\xi^1_1 = \xi^1_2 = 0$, and the effective energy–momentum tensor of the string has the form

$$T^{\mu\nu}_{\mathrm{eff}}(\mathbf{x}, t) = \Theta^{\mu\nu}\delta(y)\delta(z)$$

The quantities $\Theta^{\mu\nu}$ can be found by averaging the microscopic energy–momentum tensor (6.2.17) over a distance d and time interval τ much greater than the typical wavelength and oscillation period of the wiggles, respectively,

$$\Theta^{\mu\nu} = (\tau d)^{-1} \int T^{\mu\nu}(\mathbf{x}, t) d^3 x \, dt \,. \tag{6.5.7}$$

Here, the spatial integration is over a region between two parallel planes perpendicular to the x-axis and separated by a distance d (see fig. 6.5), and the time integration is over a time interval $\Delta t = \tau$.

For the energy density component T^{00}, the spatial integration gives

$$\int T^{00}(\mathbf{x}, t) d^3 x = \mu \int d\zeta = \mu \Delta \zeta \tag{6.5.8}$$

and the time averaging replaces $\Delta \zeta$ by its average value $\langle \Delta \zeta \rangle$. With the aid of (6.5.3) for $\mathbf{a}$ and $\mathbf{b}$, the distance d can also be expressed in terms of $\langle \Delta \zeta \rangle$:

$$d = \left\langle \int x^{1\prime} d\zeta \right\rangle = \frac{1}{2}(k_1 + k_2) \langle \Delta \zeta \rangle \,. \tag{6.5.9}$$

Hence,

$$\Theta^{00} = 2\mu (k_1 + k_2)^{-1} \,, \tag{6.5.10}$$

and, similarly, we obtain

$$\begin{aligned} \Theta^{11} &= -2\mu k_1 k_2 (k_1 + k_2)^{-1} \,, \\ \Theta^{01} &= \mu(k_2 - k_1)(k_1 + k_2)^{-1} \,, \end{aligned} \tag{6.5.11}$$

with all other components of $\Theta^{\mu\nu}$ equal to zero.

From (6.5.10–11) we see that the string energy density Θ^{00} can take any value in the range $\mu \le \Theta^{00} < \infty$ and that $-\Theta^{00} \le \Theta^{11} \le \Theta^{00}$. The stress component Θ^{11} represents tension when k_1 and k_2 are both positive and pressure when they have opposite signs. The components Θ^{00} and

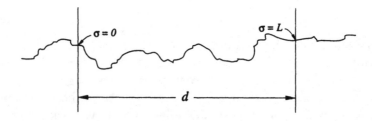

Fig. 6.5. Integration over the small-scale structure on a string yields the effective energy–momentum tensor $\Theta^{\mu\nu}$. The average is taken on a scale d much greater than the typical wavelength of the wiggles.

$-\Theta^{11}$ can be respectively identified with the proper mass density $\tilde{\mu}$ and the proper tension $\tilde{T}$ only when $k_1 = k_2 \; (\equiv k)$, so that $\Theta^{01} = 0$ and the string has zero momentum. In this case $\tilde{\mu} = k^{-1}\mu$ and $\tilde{T} = k\mu$, in agreement with the equation of state (6.5.2). More generally, to obtain $\tilde{\mu}$ and $\tilde{T}$, one has to perform a Lorentz transformation in the x-direction. Alternatively, $\tilde{\mu}$ and $-\tilde{T}$ can be invariantly defined as the eigenvalues of $\Theta^{\mu\nu}$. Their product is given by the Lorentz-invariant determinant

$$-\tilde{\mu}\tilde{T} = \Theta^{00}\Theta^{11} - \left(\Theta^{01}\right)^2. \qquad (6.5.12)$$

Substituting $\Theta^{\mu\nu}$ from (6.5.10–11) into (6.5.12), we recover the equation of state (6.5.2). It is easily verified that $\Theta^{\mu\nu}\ell_\mu\ell_\nu \geq 0$, with a null vector $\ell^\mu = (\ell, \ell, 0, 0)$, thus implying that $\tilde{\mu} \geq \tilde{T}$.

Until now we assumed that k_1 and k_2 in (6.5.3) are constant. It is easily seen that this analysis also applies to the more general situation when they are slowly varying functions of ζ. (By slow variation we mean that the function changes very little on the characteristic length and time scale of the wiggles.) In this case $\tilde{\mu}$ and $\tilde{T}$ are functions of ζ and t, and the equation of state (6.5.2) applies locally. A more formal and general treatment of string equations of state is given in Carter [1992].

6.5.2 Wiggly string dynamics

At the microscopic level, all possible string trajectories are described by (6.2.11–13). For a wiggly string, the functions $\mathbf{a}(\zeta)$ and $\mathbf{b}(\zeta)$ have small-scale components with wavelengths much smaller than the characteristic length-scale of the string. The motion of a smoothed string,

$$\langle \mathbf{x}(\zeta, t) \rangle = \tfrac{1}{2}\left[\mathbf{A}(\zeta - t) + \mathbf{B}(\zeta + t)\right], \qquad (6.5.13)$$

is then described by the averaged functions

$$\mathbf{A}(\zeta) = \langle \mathbf{a}(\zeta) \rangle, \qquad \mathbf{B}(\zeta) = \langle \mathbf{b}(\zeta) \rangle. \qquad (6.5.14)$$

As before, the averaging here is over a scale large compared to the typical wavelength of the wiggles. The derivatives $|\mathbf{A}'|$ and $|\mathbf{B}'|$ are similar to the quantities $|k_1|$ and $|k_2|$ defined in (6.5.3) and, from (6.5.6), it follows that

$$|\mathbf{A}'(\zeta)| \leq 1, \qquad |\mathbf{B}'(\zeta)| \leq 1. \qquad (6.5.15)$$

The values of $|\mathbf{A}'|$ and $|\mathbf{B}'|$ are determined by the local density and size of the wiggles.

In the most general case, (6.5.13) represents the motion of *some* wiggly string for arbitrary functions $\mathbf{A}(\zeta)$ and $\mathbf{B}(\zeta)$ satisfying (6.5.15) [Carter, 1990a,c; Vilenkin, 1990]. The variety of possible motions for wiggly strings is much richer than that for Nambu strings, since $\mathbf{A}'$ and $\mathbf{B}'$

are no longer constrained to lie on a unit sphere. Generically, the curves $\mathbf{A}'(\zeta)$ and $-\mathbf{B}'(\zeta)$ will not intersect, and the strings will not develop cusps. (Of course, the cusps do occur at the microscopic level, but their scale is determined by the wavelength of the wiggles.)

Let us now consider some simple applications. If the wiggles are statistically identical for both the **a** and **b** curves and are uniformly distributed along each, then we have

$$|\mathbf{A}'(\zeta)| = |\mathbf{B}'(\zeta)| \equiv k = \text{const.} \qquad (6.5.16)$$

These equations are similar to the constraint equations (6.2.12) for a Nambu string. Given a solution of (6.2.13), we can then obtain a solution of (6.5.16) simply as

$$\mathbf{A}(\zeta) = \mathbf{a}(k\zeta), \qquad \mathbf{B}(\zeta) = \mathbf{b}(k\zeta). \qquad (6.5.17)$$

In this case, the motion of wiggly strings is the same as that of Nambu strings, except it is slowed down by a factor $k < 1$.

Another interesting example is when one of the curves, say $\mathbf{B}(\zeta)$, shrinks to a point. We can choose coordinates such that this point is at $\mathbf{x} = 0$, so that $\langle \mathbf{x}(\zeta, t) \rangle = \frac{1}{2}\mathbf{A}(\zeta - t)$. In this situation, it is easily seen that the shape of the loop does not change in time. When $\mathbf{B}(\zeta)$ shrinks to a point, $\mathbf{B}'(\zeta)$ also vanishes, which corresponds to $k_2 = 0$ for the 'wiggly' straight string discussed in §6.5.1. From (6.5.11) we see that for $k_2 = 0$ the effective tension vanishes, explaining why the string remains static.

6.6 String thermodynamics

Consider a box of volume V containing strings with total energy U. The strings intersect and fragment, the fragments collide and merge, and after some time an equilibrium state may be established. In this section we shall discuss the properties of such a state. In later chapters we shall refer to these properties for insight into some aspects of cosmic string formation and evolution.

In statistical mechanics probabilities are found by counting the states of a system satisfying certain requirements. A classical string of any size has an infinite number of states and so, to make the state counting well-defined, the system of strings has to be 'discretized'. One way to achieve this is to actually quantize the system. The quantum statistical mechanics of strings has been studied by Huang & Weinberg [1970], Frautschi [1971], Carlitz [1972], and more recently by Mitchell & Turok [1987a,b] (see also Aharonov, Englert & Orloff [1987]). Here we shall review an alternative approach which yields very similar results [Smith & Vilenkin, 1987b; Sakellariadou & Vilenkin, 1988]. Strings can be represented by discrete points on a three-dimensional lattice: points on the strings are equally spaced in energy, so that the total energy of a closed loop is $E = N\epsilon$, where N is the number of points and ϵ is the energy between neighbouring points. A minimum loop size is imposed, being greater than or equal to the smallest possible value, $N = 2$. The strings were then numerically evolved using a discretized version of the Nambu equations, an algorithm which exactly preserved string energy and momentum. During the course of the evolution, any two strings which passed through the same lattice site were reconnected.

Let us suppose that the system of strings is in equilibrium at some temperature T. Apart from momentary reconnections, the loops do not interact and can be treated like particles in an ideal gas. Since the loops are allowed to split and join, their number is not conserved, and they will be described by a Bose distribution with zero chemical potential. The number of loops with energies between E and $E + dE$ per unit volume is

$$dn = \left(e^{\beta E} - 1\right)^{-1} \nu(E)dE, \qquad (6.6.1)$$

where $\beta = T^{-1}$ and $\nu(E)dE$ is the number of closed loop states with energies between E and $E + dE$.

The form of the density of states $\nu(E)$ for $E \gg \epsilon$ can be easily determined (the following argument is due to Turok [1989a]). A loop of energy E consists of $N = E/\epsilon$ links. If each link can be in s states, then a chain of N links has s^N states. For a closed loop this number has to be multiplied by the probability for a random walk to return to its origin after N steps. Within an unimportant numerical factor, this probability is $N^{-3/2}$. Since any point on a closed loop can be used as an origin, we have to normalize by an additional factor N^{-1}. The final expression for the density of states is then

$$\nu(E) = C\,E^{-5/2}e^{bE}, \qquad (6.6.2)$$

where $C \sim \delta^{-3}\epsilon^{3/2}$ is a constant coefficient, $\delta \sim \epsilon/\mu$ is the lattice spacing, and

$$b = \epsilon^{-1}\ln s. \qquad (6.6.3)$$

Equation (6.6.2) should not be confused with the centre-of-mass density of states $\nu_0(M)$. In the centre-of-mass frame the momenta of all the links should sum to zero, a condition which introduces an additional factor of $N^{-3/2}$ such that $\nu_0(M) \propto M^{-4}\exp(bM)$.

For large loops with $E \gg b^{-1}$, T, (6.6.1) and (6.6.2) give

$$dn = C\,E^{-5/2}e^{(b-\beta)E}dE. \qquad (6.6.4)$$

At temperatures below b^{-1}, the strings tend to fragment into the smallest allowed loops, while large loops become exponentially suppressed. However, at $T > b^{-1}$, (6.6.4) gives an exponentially increasing dependence of dn/dE on E and so an infinite total energy density. This indicates that the system of strings cannot be in equilibrium with a heat bath at a temperature above $T_{max} = b^{-1}$. The existence of a maximum temperature in systems with an exponentially growing density of states was first pointed out by Hagedorn [1965].

At the Hagedorn temperature $T = T_{max}$, (6.6.4) reduces to

$$dn = C E^{-5/2} dE.   \qquad (6.6.5)$$

Large loops have the shape of random walks with a step-length comparable to the lattice spacing δ. The overall size of a loop of energy E is of the order

$$r = \delta \left(\frac{E}{\epsilon} \right)^{1/2},   \qquad (6.6.6)$$

and we can rewrite (6.6.5) as

$$dn \sim \frac{dr}{r^4}.   \qquad (6.6.7)$$

This size distribution of loops is scale-invariant, in the sense that it depends on neither the lattice spacing δ nor the energy cut-off ϵ. The energy density at $T = T_{max}$ is finite,

$$\rho_{max} = \rho(T_{max}) \sim \epsilon \delta^{-3},   \qquad (6.6.8)$$

which corresponds to a density of about one string segment per elementary lattice cell.

Below T_{max}, the average energy density of the system of strings can be expressed as a function of temperature

$$\bar{\rho} = \int_{\epsilon}^{\infty} \left(e^{\beta E} - 1 \right)^{-1} \nu(E) E dE.   \qquad (6.6.9)$$

Fluctuations of ρ about this value vanish in the limit of infinite volume and, for this reason, the canonical ensemble can be used to study the behaviour of a closed system with a fixed value of ρ. However, as $T \rightarrow T_{max}$, fluctuations of ρ become infinite and the correspondence between the canonical and microcanonical ensembles breaks down.

The behaviour of the system at $\rho > \rho_{max}$ can be studied using the microcanonical ensemble [Mitchell & Turok, 1987a,b]. Alternatively, one can take an 'experimental' approach and observe the system evolving in numerical simulations [Sakellariadou & Vilenkin, 1988]. Regardless of the initial configuration, the system reaches the equilibrium state which depends only on the the energy density ρ. By running a series of such

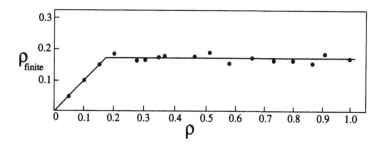

Fig. 6.6. The energy density of finite loops, $\rho_{\text{finite}} = \rho - \rho_{\text{inf}}$, is plotted as a function of the total density ρ.

simulations at various densities, one finds that for $\rho < \rho_{\text{max}}$ the size distribution of loops is in good agreement with (6.6.4). At $\rho = \rho_{\text{max}}$, the system undergoes a phase transition which marks the appearance of infinite Brownian strings. As the density is increased above ρ_{max}, the loop distribution (6.6.5) remains unchanged, and the total energy density in closed loops remains equal to ρ_{max}. All the additional energy is transformed into the infinite strings.

In fig. 6.6, the density in finite loops, $\rho_{\text{finite}} = \rho - \rho_{\text{inf}}$, is plotted as a function of the total density ρ. We see a marked change of behaviour at $\rho = \rho_{\text{max}} \approx 0.17$ (the units are those in which $\epsilon = \delta = 1$). For $\rho < \rho_{\text{max}}$, there are no infinite strings and $\rho_{\text{finite}} = \rho$, while for $\rho > \rho_{\text{max}}$, $\rho_{\text{finite}} = \rho_{\text{max}} = \text{const}$.

It should be emphasized that the critical density (6.6.8) depends crucially on the energy cut-off ϵ. If we let $\epsilon \to 0$ keeping the string tension μ constant, then $\rho_{\text{max}} \sim \mu^3/\epsilon^2 \to \infty$. Hence, in the absence of a cut-off, the system of strings is always below critical density, and tends to fragment into smaller and smaller loops. An important role in the fragmentation process is played by loop collisions. Without collisions individual loops would just fragment into several non-intersecting daughters, and the process would stop. When the daughters collide and merge, the resulting loops are likely to be self-intersecting and fragment further. Suppose the system has fragmented into loops of typical energy $\bar{E}$. The number density of the loops is $n \sim \rho/\bar{E}$ and the collision cross-section is $\sigma \sim (\bar{E}/\mu)^2$. Small loops move at relativistic speeds, and their mean collision time is given by $\tau \sim (n\sigma)^{-1} \sim \mu^2/\rho\bar{E}$ in flat space. This time τ determines the timescale for further fragmentation, so as the loops become smaller the fragmentation process slows.

7
String gravity

7.1 The straight string metric

The gravitational properties of cosmic strings are strikingly different from those of non-relativistic linear distributions of matter. To explain the origin of the difference, we note that for a static matter distribution with energy–momentum tensor,

$$T^{\mu}_{\nu} = \text{diag}\,(\rho, -p_1, -p_2, -p_3)\,,$$

the Newtonian limit of the Einstein equations become

$$\nabla^2 \Phi = 4\pi G(\rho + p_1 + p_2 + p_3)\,,$$

where Φ is the gravitational potential. For non-relativistic matter, $p_i \ll \rho$ and $\nabla^2 \Phi = 4\pi G\rho$. Strings, on the other hand, have a large longitudinal tension. For a straight string parallel to the z-axis, $p_3 = -\rho$, while p_1 and p_2 vanish when averaged over the string cross-section. Hence, the right-hand side of the last equation vanishes, suggesting that straight strings produce no gravitational force on surrounding matter. As we shall see, this conclusion is confirmed by a full general-relativistic analysis. Another feature distinguishing cosmic strings from more familiar sources is their relativistic motion. As a result, oscillating loops of string can be strong emitters of gravitational radiation.

A gravitating string is described by the combined system of Einstein, Higgs and gauge field equations. The problem of solving these coupled equations is formidable and no exact solutions have been found to date. Fortunately, for most cosmological applications the problem can be made tractable by adopting two major simplifications. First, assuming that the string thickness is much smaller than all other relevant dimensions, the string can be approximated as a line of zero width with a distributional δ-function energy–momentum tensor (6.2.17). Secondly, the gravitational field of the string is assumed to be sufficiently weak that the linearized Einstein equations can be employed. These approximations are not valid for supermassive strings with a symmetry breaking scale $\eta \gtrsim m_{\rm pl}$. However, for strings with $\eta \ll m_{\rm pl}$, linearized gravity is applicable almost everywhere, except in small regions of space affected by cusps and kinks. Most of the analysis in this chapter will be based on these thin-string and weak-gravity approximations.

7.1.1 Linearized gravity

In a weak gravitational field, the spacetime metric is almost Minkowskian,

$$g_{\mu\nu} = \eta_{\mu\nu} + h_{\mu\nu}, \qquad |h_{\mu\nu}| \ll 1, \tag{7.1.1}$$

and the Einstein equations can be linearized in $h_{\mu\nu}$. The linearized equations take a particularly simple form in the harmonic gauge,

$$\Box h_{\mu\nu} = -16\pi G S_{\mu\nu}, \tag{7.1.2}$$

where

$$S_{\mu\nu} = T_{\mu\nu} - \tfrac{1}{2}\eta_{\mu\nu} T_\sigma^\sigma, \tag{7.1.3}$$

with the energy–momentum tensor $T_{\mu\nu}$ and with indices raised and lowered using the Minkowski metric $\eta_{\mu\nu}$. The harmonic gauge is specified by the conditions

$$\partial_\nu \left(h_\mu^\nu - \tfrac{1}{2}\delta_\mu^\nu h_\sigma^\sigma \right) = 0. \tag{7.1.4}$$

In the limit of zero string thickness, the string energy–momentum tensor is given by (6.2.17),

$$T^{\mu\nu}(\mathbf{r}, t) = \mu \int d\zeta \, (\dot{x}^\mu \dot{x}^\nu - x'^\mu x'^\nu) \, \delta^{(3)}(\mathbf{r} - \mathbf{x}(\zeta, t)). \tag{7.1.5}$$

We shall first apply this formalism to find the metric of a static straight string lying along the z-axis [Vilenkin, 1981b]. In this case,

$$T_\mu^\nu = \mu \, \delta(x)\delta(y) \, \text{diag} \, (1, 0, 0, 1), \tag{7.1.6}$$

and the solution of (7.1.2) is easily found,

$$\begin{aligned} h_{00} &= h_{33} = 0, \\ h_{11} &= h_{22} \doteq 8G\mu \ln(r/r_0), \end{aligned} \tag{7.1.7}$$

where $r = (x^2 + y^2)^{1/2}$ and r_0 is a constant of integration. Note that at large distances from the string, $h_{\mu\nu}$ becomes large and the weak-field approximation breaks down. We shall see, however, that this is merely a consequence of a poor choice of coordinates.

For a better insight into the geometry of the metric (7.1.7), we rewrite it in cylindrical coordinates,

$$ds^2 = dt^2 - dz^2 - (1 - h)(dr^2 + r^2 d\theta^2), \tag{7.1.8}$$

where

$$h = 8G\mu \ln(r/r_0) \,. \tag{7.1.9}$$

Introducing a new radial coordinate r' as

$$(1 - h)r^2 = (1 - 8G\mu)r'^2 \,, \tag{7.1.10}$$

we obtain, to linear order in $G\mu$,

$$ds^2 = dt^2 - dz^2 - dr'^2 - (1 - 8G\mu)r'^2 d\theta^2 \,. \tag{7.1.11}$$

Finally, with a new angular coordinate,

$$\theta' = (1 - 4G\mu)\theta \,, \tag{7.1.12}$$

the metric takes a Minkowskian form,

$$ds^2 = dt^2 - dz^2 - dr'^2 - r'^2 d\theta'^2 \,. \tag{7.1.13}$$

We have thus arrived at a surprising conclusion; the geometry around a straight cosmic string is locally identical to that of flat spacetime. This geometry, however, is not globally Euclidean, since the angle θ' varies in the range

$$0 \leq \theta' < 2\pi(1 - 4G\mu) \,. \tag{7.1.14}$$

Hence, the effect of the string is to introduce an azimuthal 'deficit angle',

$$\Delta = 8\pi G\mu \,, \tag{7.1.15}$$

implying that a surface of constant t and z has the geometry of a cone rather than that of a plane.

The dimensionless parameter $G\mu$ plays an important role in the physics of cosmic strings. Its magnitude can be estimated using (3.2.10),

$$G\mu \sim \left(\frac{\eta}{m_{\text{pl}}}\right)^2 \,, \tag{7.1.16}$$

where η is the string symmetry breaking scale and m_{pl} is the Planck mass. The weak-field approximation (7.1.1) is justified if $G\mu \ll 1$ or, equivalently, $\eta \ll m_{\text{pl}}$. The string scenario for galaxy formation requires $G\mu \sim 10^{-6}$, corresponding to a grand-unified scale $\eta \sim 10^{16}\text{GeV}$, while observations constrain $G\mu$ to be less than $\sim 10^{-5}$ (refer to chapters 10 and 11).

7.1.2 Full Einstein gravity

Since the conical metric (7.1.11) is locally flat, it is an exact solution of the vacuum Einstein equations. The validity of this metric can therefore be extended beyond linear perturbation theory. In this section we shall show that the string spacetime is indeed asymptotically conical and we shall discuss the dependence of the conical deficit angle on the internal structure of the string.

A string lying along the z-axis in Minkowski space is invariant under time translations, spatial translations in the z-direction, rotations around the z-axis and Lorentz boosts in the z-direction. Assuming that the spacetime of a gravitating string has the same symmetries, we can write the metric as

$$ds^2 = e^{A(r)}(dt^2 - dz^2) - dr^2 - e^{B(r)}d\theta^2 . \qquad (7.1.17)$$

With this ansatz, the Einstein equations take the form

$$A'' + A'^2 + \tfrac{1}{2}A'B' = 16\pi G(T_t^t - \tfrac{1}{2}T), \qquad (7.1.18)$$

$$B'' + \tfrac{1}{2}B'^2 + A'B' = 16\pi G(T_\theta^\theta - \tfrac{1}{2}T), \qquad (7.1.19)$$

$$2A'' + B'' + A'^2 + \tfrac{1}{2}B'^2 = 16\pi G(T_r^r - \tfrac{1}{2}T), \qquad (7.1.20)$$

where $T = T_\nu^\nu$ and primes stand for derivatives with respect to r.

Asymptotic behaviour

In flat spacetime the energy–momentum tensor of a local gauge string vanishes exponentially at $r \gg \delta$, where δ is the string thickness (this follows immediately from (4.1.17) and (4.1.18)). For $r \gg \delta$, therefore, it is natural to assume that the metric of a gravitating string is accurately described by the vacuum Einstein equations. Setting $T_\mu^\nu = 0$ in (7.1.18–20), it is not difficult to show that the asymptotic form of the metric must either be a conical space,

$$ds^2 = a_1(dt^2 - dz^2) - dr^2 - (a_2 r + a_3)^2 d\theta^2 , \qquad (7.1.21)$$

or, else,

$$ds^2 = (b_1 r + b_2)^{4/3}(dt^2 - dz^2) - dr^2 - (b_1 r + b_2)^{-2/3}d\theta^2 , \qquad (7.1.22)$$

where a_i and b_i are arbitrary constants [Vilenkin, 1981b].

The second form of the metric (7.1.22) is a special case of a Kasner [1921] metric. Depending on the sign of b_1/b_2, the length of a circle $r = $ const., decreases as $r^{-1/3}$ when $r \to \infty$ or alternatively blows up at a finite value of r. In both cases this does not match the behaviour of the weak-field solution (7.1.11). Hence, for the time being we dismiss the metric (7.1.22) as being 'unphysical'. (We shall see later that it may be relevant for global strings or for supermassive strings with $G\mu \gtrsim 1$.)

Deficit angle

In the conical metric (7.1.21), the deficit angle Δ is related to the parameter a_2. It can be found by solving Einstein's equations in all space, including the string interior. It is convenient here to introduce the notation

$$C(r) = e^{B(r)/2} . \tag{7.1.23}$$

The deficit angle is then determined by the asymptotic behaviour of $C(r)$,

$$\Delta = 2\pi[1 - C'(r \to \infty)] . \tag{7.1.24}$$

The metric (7.1.17) is non-singular only if

$$A'(0) = 0, \qquad C(0) = 0, \qquad C'(0) = 1. \tag{7.1.25}$$

The geometry of a surface $(t, z) = $ const. in this metric is illustrated in fig. 7.1. The surface is asymptotically conical, but the point of the cone is smoothed on a scale comparable to the characteristic width δ of the distribution $T^\nu_\mu(r)$.

   We shall first represent the interior of the string by assuming a simple form for the energy–momentum tensor,

$$T^\nu_\mu = \sigma(r) \operatorname{diag}(1, 0, 0, 1), \tag{7.1.26}$$

with the general case discussed later. This form (7.1.26) is clearly consistent with all the symmetries of the string and reduces to (7.1.6) in the limit of negligible string thickness. The special case when $\sigma = $ const. inside the string core and $\sigma = 0$ outside has been analyzed by Gott [1985] and Hiscock [1985]. The case of arbitrary $\sigma(r)$ has been discussed by Linet [1985].

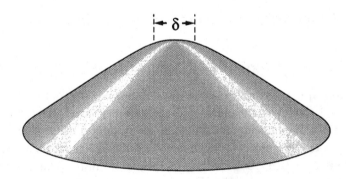

Fig. 7.1.   The snub-nosed cone, the cross-sectional geometry about a realistic string of width δ.

Substituting (7.1.17) in the Einstein equations, it is easily seen that $A(r)$ can be consistently set equal to a constant. By a suitable rescaling of t and z we can set $A(r) = 0$. The metric can then be rewritten as

$$ds^2 = dt^2 - dz^2 - dr^2 - C^2(r)d\theta^2 , \tag{7.1.27}$$

and the Einstein equations reduce to a single equation for the function $C(r)$,

$$C''/C = -8\pi G\sigma . \tag{7.1.28}$$

For the special form of the energy–momentum tensor (7.1.26), a simple relation can be derived between the deficit angle Δ and the string mass per unit length μ,

$$\mu = \int_0^\infty dr \int_0^{2\pi} d\theta \, T_0^0 \, [^{(2)}g]^{1/2} . \tag{7.1.29}$$

Here, $^{(2)}g_{ij}$ is the metric on the surface $(t, z) = \text{const.}$ and $^{(2)}g = C^2(r)$ is its determinant. With the aid of (7.1.28), equation (7.1.29) yields

$$\mu = \frac{1}{4G} \left[1 - C'(\infty) \right] , \tag{7.1.30}$$

and thus we obtain

$$\Delta = 8\pi G\mu . \tag{7.1.31}$$

In the general case, when the components T_r^r and T_θ^θ do not vanish inside the string, (7.1.31) does not apply. The deficit angle can be expressed in terms of the scalar curvature of the surface $(t, z) = \text{const.}$,

$$^{(2)}R = -2C''/C . \tag{7.1.32}$$

Integrating $^{(2)}R$ over the surface and using (7.1.24) we obtain [Ford & Vilenkin, 1981]

$$\Delta = \tfrac{1}{2} \int d^2x \, ^{(2)}R \, [^{(2)}g]^{1/2} . \tag{7.1.33}$$

This is a version of the Gauss–Bonnet theorem. Another useful relation is [Garfinkle, 1985]

$$\Delta = 8\pi G\mu + \frac{\pi}{2} \int_0^\infty dr \, A'^2 C . \tag{7.1.34}$$

This can be derived by integrating the expression

$$-8\pi GCT_t^t = C'' + \tfrac{1}{2}(CA')' + \tfrac{1}{4}CA'^2, \qquad (7.1.35)$$

which follows directly from the field equations (7.1.18–20). Note also that (7.1.34) implies that $\Delta \geq 8\pi G\mu$.

Although (7.1.31) is not valid in general, it holds for the cosmologically interesting case with $G\mu \ll 1$. Essentially, the reason is that for $G\mu \ll 1$ the metric inside the string is approximately flat and the transverse tensions T_r^r and T_φ^φ average out to zero in flat spacetime. Alternatively, it is easily seen from the Einstein equations (7.1.18–20) that the integrand in (7.1.34) is second order in $G\mu$ and, thus, to linear order Δ is given by (7.1.31).

$U(1)$-strings

The fact that (7.1.31) may be sufficient for cosmological applications has not discouraged relativists from studying string metrics in considerable detail, including those with $\Delta \gtrsim 1$. The simplest realistic model of a gravitating string is a covariant generalization of the abelian-Higgs model. The abelian-Higgs Lagrangian (4.1.1) depends on three parameters: the gauge coupling e, the Higgs self-coupling λ, and the symmetry breaking scale η. As we discussed in chapter 4, flat-space string solutions depend only on the ratio of couplings, $\beta = \lambda/2e^2$, with η determining the scale of the solution. In curved space, the equations contain an additional dimensionless parameter, $G\eta^2$, which characterizes the gravitational coupling of the string [Garfinkle, 1985]. Numerical solutions of these equations for a range of values of β and $G\eta^2$ were obtained by Laguna & Matzner [1987b] (see also Garfinkle & Laguna [1989]). Significant corrections to (7.1.31) only appear if $\Delta \gtrsim 1$.

For the special value of $\beta = 1$ (critical coupling), the Einstein–Higgs-gauge field equations can be greatly simplified since they reduce to solving a curved-space analogue of the Bogomol'nyi equations (4.1.22–23) [Linet, 1987a, Comtet & Gibbons, 1988]. In this case it can be shown, first, that the transverse tensions vanish everywhere, $T_r^r = T_\theta^\theta = 0$, so that T_μ^ν takes the form (7.1.26), that the appropriate metric is (7.1.27), and that the deficit angle is given exactly by (7.1.31). Secondly, the string mass per unit length saturates the flat-space expression (4.1.21),

$$\mu = 2\pi n\eta^2, \qquad (7.1.36)$$

where n is the winding number of the string.

Supermassive strings

Up to this point we have implicitly assumed that the deficit angle Δ is smaller than 2π. However, as the symmetry breaking scale grows, Δ increases and exceeds 2π above some critical value η_{crit}. For such values of η the conical picture of a string spacetime must be abandoned. For $\beta = 1$, it follows from (7.1.31) and (7.1.36) that $\eta_{\text{crit}} = m_{\text{pl}}/\sqrt{8\pi}$.

For a critical string, the $(t, z) = \text{const.}$ sections are asymptotically cylindrical (fig. 7.2(a)), and the asymptotic form of the metric is given by (7.1.21) with $a_2 = 0$ [Gott, 1985]. A closed form of the metric for a critical $U(1)$-string with $\beta = 1$ has been found by Linet [1990].

For an over-critical string with $\Delta > 2\pi$, one expects the exterior region to be like an inverted cone (fig. 7.2(b)) and to have a singular point at a finite distance from the core [Gott, 1985]. Numerical analysis by Ortiz [1991] demonstrated that this is indeed the case when the parameter β is sufficiently close to 1. However, Laguna & Garfinkle [1989] found another type of solution in which the exterior metric has the Kasner form (7.1.22) with $b_1/b_2 < 0$, similarly developing a singularity at a finite distance from the string axis.

We shall see in §7.3 that the global string metric also exhibits a Kasner-type singularity. This makes the behaviour of the Laguna–Garfinkle solutions less surprising, at least for small β. At present it is not clear whether both types of solutions co-exist for the same parameter values or whether there is a transition between them along some line in the β–$G\eta^2$ plane. In any case, due to the presence of such singularities, supermassive strings with $\Delta > 2\pi$ appear to be of little physical interest.

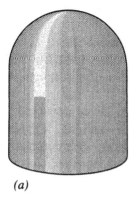
(a)

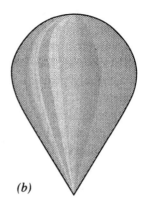
(b)

Fig. 7.2. String geometries for (a) a critical mass-scale string and (b) a super-critical string.

After this detour into the gravity of exotic supermassive strings we now return to the cosmologically interesting case with $G\mu \ll 1$. We shall assume $G\mu$ to be small for the remainder of this chapter.

Other metrics with strings

Spacetimes with conical singularities had been studied long before their relevance to cosmic strings was recognized – see, for example, Marder [1959], Israel [1977] and Sokolov & Starobinsky [1977]. Even earlier, Bach & Weyl [1922] found a solution of Einstein's equations describing a pair of black holes held apart by an infinitely thin 'strut'. This solution can be easily reinterpreted in terms of cosmic strings. With an appropriate choice of integration constants, it describes a pair of black holes held apart by two strings extending to infinity in opposite directions.

Solutions of Einstein's equations with conical singularities describing straight strings can easily be constructed [Aryal, Ford & Vilenkin, 1986]. One needs only a spacetime with a symmetry axis. If one then cuts out a wedge then a space with a string lying along the axis is obtained. This is achieved by requiring the azimuthal angle to run over the range (7.1.14) or by multiplying the corresponding metric coefficient by $(1 - 8G\mu)$.

For example, the metrics

$$ds^2 = \left(1 - \frac{2Gm}{r}\right) dt^2 - \left(1 - \frac{2Gm}{r}\right)^{-1} dr^2 - r^2 d\Omega^2 , \quad (7.1.37)$$

$$ds^2 = dt^2 - a^2(t) \left[(1 - kr^2)^{-1} dr^2 + r^2 d\Omega^2\right] , \quad (7.1.38)$$

with

$$d\Omega^2 = d\theta^2 + (1 - 8G\mu)\sin^2\theta d\varphi^2 , \quad (7.1.39)$$

describe, respectively, a black hole with a string passing through it and an FRW universe with a string.

A non-axisymmetric solution of the combined Einstein and Maxwell equations with a string has been found by Linet [1987b]. This solution describes a maximally charged black hole ($e^2 = Gm^2$) in equilibrium with an infinite string.

Spacetimes containing several strings can be constructed by cutting several wedges out of Minkowski space. For example, a space with two parallel strings is illustrated in fig. 7.3. Solutions with non-parallel strings and with strings moving with respect to one another can be obtained by applying different rotations and Lorentz boosts to regions above and below the plane P parallel to the strings. If the rotation axes are perpendicular to P and the boost velocities are parallel to P, then the plane P is mapped into itself, and the two regions still match after the transformations are applied [Gott, 1991]. This solution has been generalized for multiple non-parallel straight strings [Letelier & Gal'tsov, 1993].

Metrics describing the causal 'birth' of a cosmic string have been studied by Mageuijo [1992]. In a specific toy model for straight string formation, the metric takes account of a compensating radiation underdensity and also an outgoing gravitational shock wave where the deficit angle 'switches off'.

7.1.3 Gravity in (2+1) dimensions

Closely related to straight string metrics is the study of (2+1)-dimensional gravity [Staruszkiewicz, 1963; Deser, Jackiw & 't Hooft, 1984; Gott & Alpert, 1984]. If the dz^2 term in (7.1.11) is removed, we obtain a solution of the Einstein equations for a point mass in (2+1) dimensions. The flatness of spacetime outside a straight string can be easily explained from this (2+1)-dimensional perspective. In (2+1) dimensions, the Weyl tensor vanishes and the Riemann tensor is completely determined by the Ricci tensor,

$$R^{\mu\nu}{}_{\sigma\tau} = \epsilon^{\mu\nu\alpha}\epsilon_{\sigma\tau\beta}\left(R^{\beta}{}_{\alpha} - \tfrac{1}{2}\delta^{\beta}{}_{\alpha}R\right). \qquad (7.1.40)$$

Einstein's equations require that outside of any sources $R_{\mu\nu} = 0$, and thus the spacetime is flat. This implies that there are no gravitational waves and there is no Newtonian gravity in $(2+1)$ dimensions.

The metric for a spinning point source in $(2+1)$ dimensions has been found by Deser, Jackiw & 't Hooft [1984]. In spatially conformal coordinates, the line element is

$$ds^2 = (dt + 4GJd\theta)^2 - R^{-8G\mu}(dR^2 + R^2d\theta^2), \qquad (7.1.41)$$

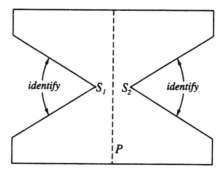

Fig. 7.3.   Identifications for a spacetime with two strings.

where μ, $J = $ const., and the energy–momentum tensor is given by

$$\sqrt{g}\, T^{00} = \mu\delta^{(2)}(\mathbf{x})\,, \qquad T^{ij} = 0,$$
$$\sqrt{g}\, T^{0i} = \tfrac{1}{2}J\epsilon^{ij}\partial_j\delta^{(2)}(\mathbf{x})\,. \qquad (7.1.42)$$

The angular momentum of the source is

$$J = \epsilon_{ij}\int d^2\mathbf{x}\sqrt{g}\, x^i\, T^{0j}\,. \qquad (7.1.43)$$

With a new radial coordinate, $r = (1 - 4G\mu)^{-1}R^{(1-4G\mu)}$, the metric (7.1.41) becomes

$$ds^2 = (dt - 4GJd\theta)^2 - dr^2 - (1 - 4G\mu)^2 d\theta^2\,. \qquad (7.1.44)$$

It is now easy to see that outside the source this metric is flat and can be brought to the Minkowski form by a coordinate transformation $t' = t + 4GJ\theta$, $\theta' = (1 - 4G\mu)\theta$.

The spacetime outside an arbitrary distribution of particles in the centre-of-mass frame can also be described by a metric of the form (7.1.44). If particles are concentrated in a region $r < r_0$ around the origin, then (7.1.44) holds for $r > r_0$. In this case, μ is the total energy of the system and J is the total angular momentum which may be due to both internal and orbital momenta of the particles. In the limit of weak gravity when all deficit angles are small, μ is just the sum of particle energies (including kinetic energy) but, in the general case, the calculation of μ is more complicated.

A curious property of the metric (7.1.41) is that it admits closed time-like curves. Consider, for example, a curve of constant $t = t_0$ and $r = r_0$ with θ changing from 0 to 2π. This curve is obviously closed and is also timelike provided that $r_0 < 4GJ(1 - 4G\mu)^{-1}$. However, it can be shown that closed timelike curves are not present in a spacetime describing any system of spinless particles with a total deficit angle smaller than 2π [Deser, Jackiw & 't Hooft, 1984, 1991; Carroll, Farhi & Guth, 1992]. In the $(3 + 1)$-dimensional case, Tipler [1976] and Hawking [1992] have shown that closed timelike curves cannot evolve in a finite region of space-time if the energy–momentum tensor satisfies the weak energy condition, $T_{\mu\nu}\ell^\mu\ell^\nu \geq 0$, for an arbitrary null vector ℓ^μ. This condition is satisfied for cosmic strings, and thus strings cannot be used to build time machines.

7.2 Propagation of particles and light

The string metric (7.1.11) describes a conical space, which is merely a flat space with a wedge of angular size $\Delta = 8\pi G\mu$ removed and the two faces of the wedge identified. The geodesics in this space are straight lines, and it is clear that a test particle initially at rest relative to the string will remain at rest and will not experience any gravitational force. However, the non-Minkowskian global structure of the string metric gives rise to a number of interesting effects.

The situation here is analogous to the Aharonov–Bohm effect in electrodynamics [Aharonov & Bohm, 1959]. The interference pattern of two coherent electron beams propagating on different sides of a thin solenoid is affected by the magnetic flux within the solenoid, despite the fact that the magnetic field vanishes everywhere along the beams. The phase shift is proportional to the line integral of the vector potential around the electron's path, which by Stokes' theorem is equal to the enclosed magnetic flux. Outside the solenoid the vector potential **A** is pure gauge and can be set equal to zero in any region not enclosing the solenoid. However, no gauge choice makes **A** vanish everywhere along both beams. For a string, the spacetime curvature is confined to the string core, but its effect is 'felt' by particles propagating in the flat spacetime region around it. As for the Aharonov–Bohm effect, a Minkowskian coordinate system can be chosen in any region on one side of the string, but such a system does not exist in a region surrounding the string.

7.2.1 Double images

Perhaps the most straightforward consequence of the conical geometry is the formation of double images of light sources located behind the string [Vilenkin, 1981b, 1984a; Gott, 1985]. This is illustrated in fig. 7.4. Light rays from the quasar Q intersect behind the string and the observer sees two images of the quasar. The observer is represented by two points, O and O', on the opposite sides of the missing wedge. The points Q, O and O' define a plane P which contains both light rays and intersects the string at a point S. If the string is perpendicular to this plane, then the missing angle $\angle OSO'$ is equal to the deficit angle Δ. More generally, it is easily understood that $\angle OSO' = \Delta \sin \theta$, where θ is the angle the string makes with the plane P. If $l = QS$ and $d = OS$ are the distances from the string to the quasar and to the observer, respectively, then it is a simple exercise in Euclidean geometry to show that the angular separation between the two images is (for $G\mu \ll 1$)

$$\delta\varphi = \frac{l\Delta \sin \theta}{l + d} . \qquad (7.2.1)$$

(The angular separation is given by the sum of the angles α and β.) Equation (7.2.1) is valid as long as the angles α and β in fig. 7.4(a) are smaller than $\delta\varphi$. If the angular distance between the string and the quasar is greater than $\delta\varphi$, then only a single image is formed.

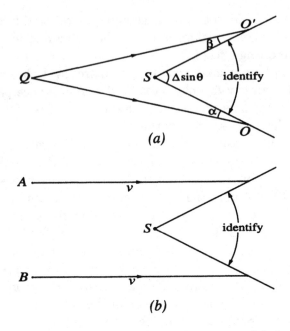

Fig. 7.4.   Consequences of the conical string geometry: (a) double images of distant objects and (b) discontinuous velocity change for objects moving past a string.

7.2.2 String Doppler shifts

This effect is illustrated in fig. 7.4(b). Here, two particles, A and B, are moving towards the string at the same speed v. We shall first assume for simplicity that the string is perpendicular to the plane of the diagram, and so the missing angle is equal to Δ. The general case will be discussed in §10.2. Initially, as the particles move, the distance between them remains constant. But once the line connecting the particles crosses the string, this distance is measured by the sum of the distances from each particle to the missing wedge. Hence, the distance decreases and the particles collide when they hit opposite points of the wedge. The transverse velocity component of the particles towards one another is $u_0 = v\Delta$. Viewed in the frame of the particles, the string approaches two static particles, and at the moment it crosses the line connecting them, the particles start moving towards one another at a velocity

$$u = \gamma v\Delta, \tag{7.2.2}$$

with $\gamma = (1 - v^2)^{-1/2}$. Note that this is found by transforming the three-velocity $v_y = dy/dt$ and is not valid for velocities very close to the speed of light. If the x-axis is chosen to lie along the cut in fig. 7.4, then the velocity components of the particles passing the string are $v_x = v\cos(\Delta/2)$ and

$v_y = \pm v \sin(\Delta/2)$. The centre-of-mass velocity of the particles is obtained by transforming to the frame moving with velocity v_x,

$$u = v \sin(\Delta/2) \left[1 - v^2 \cos^2(\Delta/2)\right]^{-1/2}.$$

For $\Delta \ll 1$ and $\gamma \ll \Delta^{-1}$, the relative motion of the particles is non-relativistic and the relative velocity reduces to (7.2.2).

If particle A carried a light source and particle B an observer, then the observer will detect a discontinuous change in the frequency of light due to the Doppler shift as the particles start moving towards each other [Kaiser & Stebbins, 1984; Gott, 1985],

$$\delta\omega/\omega = \gamma v \Delta. \tag{7.2.3}$$

More precisely, the observer will detect the appearance of a second image of the source with a shifted frequency, and shortly thereafter the first image will disappear.

7.2.3 Field theory in a conical space

This topic lies outside the scope of this monograph and so we shall confine discussion to a brief guide to the literature. Linet [1986a] and Smith [1990] have shown that, although a straight string exerts no gravitational force on a test particle, the particle's own gravitational field is distorted by the string, resulting in an attractive self-force proportional to the particle mass squared,

$$F = \tfrac{1}{4}\pi G\mu Gm^2/r^2. \tag{7.2.4}$$

A similar effect exists for the electrostatic field of a charged particle, but in that case the force is repulsive with magnitude,

$$F = \tfrac{1}{4}\pi G\mu e^2/r^2. \tag{7.2.5}$$

Note that the two forces cancel when $Gm^2 = e^2$. This explains the origin of the metric, found by Linet [1987b], describing a maximally charged black hole in equilibrium with a string. Linet's solution shows that the cancellation of self-forces in the case of $Gm^2 = e^2$ is valid beyond linear perturbation theory in $G\mu$ and Gm^2.

Solutions of the Schrödinger, Klein–Gordon and Dirac equations describing particle scattering in the gravitational field of a string have been found by t'Hooft [1988], Deser & Jackiw [1988] and de Sousa Gerbert & Jackiw [1988]. The effect of the particle self-interaction on the scattering has been analyzed by Gibbons, Ruiz Ruiz & Vachaspati [1990].

Quantum field theory in a conical space has been investigated by Dowker [1977, 1987a,b], Helliwell & Konkowski [1986], Frolov & Serebrianii [1987], Linet [1987b] and Smith [1990] amongst others. In particular, they calculated the vacuum expectation values of the stress tensor for massless conformal fields of spin 0, 1/2 and 1. The general form of $\langle T^\nu_\mu \rangle$ can be deduced from dimensionality, the symmetries of the problem, the conservation of $\langle T^\nu_\mu \rangle$, and the conformal invariance of the fields:

$$\langle T^\nu_\mu \rangle = \frac{A}{r^4} \mathrm{diag}(1, 1, -3, 1), \qquad (7.2.6)$$

where A is a dimensionless constant. A field-theoretic calculation is required only to determine the value of A. For a conformally coupled, massless scalar field one finds, to linear order in $G\mu$,

$$A = G\mu/90\pi^2 . \qquad (7.2.7)$$

Similar expressions are obtained for photons and neutrinos. The contribution of (7.2.6) to the mass per unit length of string, $\delta\mu/\mu \sim 10^{-3} G/\delta^2$, where δ is the string thickness, is small for realistic parameter values.

7.3 Gravitational field of a global string

So far in this chapter we have discussed the gravitational properties of gauge strings which have a vanishing energy–momentum tensor outside a thin core. Global strings, on the other hand, have a Goldstone boson field extending beyond the core, giving rise to a non-zero energy–momentum tensor throughout space. In this section we shall discuss the gravitational field of a straight global string.

We shall consider a global $U(1)$-string described by the scalar field ansatz

$$\phi = f(r)e^{i\theta} , \qquad (7.3.1)$$

with $f(0) = 0$ and $f(r) \approx \eta$ for $r \gg \delta$, where δ is the thickness of the string core. The global string spacetime has the same symmetries as that of a gauge string, and the metric can still be written in the form (7.1.17). Outside the core, the energy–momentum tensor is given by

$$T^0_0 = T^z_z = T^r_r = -T^\theta_\theta = \eta^2 e^{-B(r)} . \qquad (7.3.2)$$

Harari & Sikivie [1988] studied the linearized Einstein equations with T^ν_μ given by (7.3.2) and found the solution,

$$ds^2 = [1 - 4G\mu(r)](dt^2 - dz^2) - dr^2 - [1 - 8G\mu(r)]r^2 d\theta^2, \qquad (7.3.3)$$

where $\mu(r)$ is the string mass per unit length out to a distance scale r from the core,

$$\mu(r) \simeq \int_\delta^r T^0_0 2\pi r dr \simeq 2\pi\eta^2 \ln\left(\frac{r}{\delta}\right) . \qquad (7.3.4)$$

We note two qualitative differences between the metrics of global and gauge strings: (i) the effective deficit angle for a global string,

$$\Delta(r) = 8\pi G\mu(r), \tag{7.3.5}$$

increases logarithmically with distance from the core, while the gauge string deficit angle remains constant; and (ii) a global string has a repulsive gravitational potential. A non-relativistic test particle at a distance r from the string experiences a gravitational acceleration away from the string,

$$\ddot{r} = 4\pi G\eta/r. \tag{7.3.6}$$

The linear approximation to the Einstein equations breaks down at $G\mu(r) \gtrsim 1$. Gregory [1988b] and Gibbons, Ortiz & Ruiz Ruiz [1989] analyzed the field equations and argued that the metric must develop a singularity at a finite distance from the string axis. A closed-form solution of the Einstein equations (7.1.18–20) with T^ν_μ from (7.3.2) was found by Cohen & Kaplan [1988]. This solution can be written as

$$ds^2 = \left(\frac{u}{u_0}\right)(dt^2 - dz^2) - r_0^2 \left(\frac{u_0}{u}\right)^{1/2} e^{\left(u_0 - u^2/u_0\right)}(du^2 + d\theta^2), \tag{7.3.7}$$

where $u_0 = 1/8\pi G\eta^2$ and r_0 is a constant of integration. The only approximation made in deriving (7.3.7) was to set

$$f(r) \approx \eta, \tag{7.3.8}$$

so that T^ν_μ has the form of (7.3.2).

For $\eta \ll m_{pl}$, u_0 is a large parameter. To leading order in $1/u_0$, the relation between the coordinate u and the radial coordinate r in the linearized metric (7.3.3) is

$$u \approx u_0 - \ln(r/r_0). \tag{7.3.9}$$

The two metrics agree if we set $r_0 \sim \delta$. Hence, $u \sim u_0$ corresponds to the vicinity of the string core. The values of $u \gtrsim u_0$ are in the core interior $r \lesssim \delta$, where the approximation (7.3.8) is violated, and the metric (7.3.7) is no longer valid.

Large values of r correspond to small values of u. When r/r_0 becomes comparable to $\exp(u_0)$, the linearized metric (7.3.3) can no longer be used. The exact solution (7.3.7) is singular at $u = 0$, and it can be verified that this is a true geometric singularity. In the vicinity of $u = 0$, the metric can be transformed to

$$ds^2 = \xi^{4/3}(dt^2 - dz^2) - d\xi^2 - \xi^{-2/3}d\theta^2, \tag{7.3.10}$$

with $\xi \propto u^{3/4}$. Note that the surfaces of constant (t, z) do not close as in fig. 7.2(b), as might be implied by the linearized metric (7.3.3). In the nonlinear regime, the deficit angle ceases to grow with r. Instead, it starts decreasing and turns into an *excess* angle which blows up at the singularity. A comparison with (7.1.22) shows that the asymptotic form (7.3.10) is in fact a Kasner metric with $b_1/b_2 < 0$. It is worth pointing out that the approach to the singularity is driven by the nonlinearity of the *vacuum* Einstein equations and not by the string energy density, which actually vanishes at the singularity [Harari & Polychronakos, 1990].

It is clear from (7.3.10) that the singularity occurs at a finite proper distance from the string core; this distance is of order

$$r_{\max} \sim \delta \, \exp(u_0) \qquad (7.3.11)$$

which can be very large for $\eta \ll m_{\mathrm{pl}}$. In particular, curvature singularities are of little concern for global strings with $\eta < 2 \times 10^{17} \mathrm{GeV}$ in a cosmological context, since $r_{\max}$ would be greater than the present horizon.

We conclude with some open questions. (i) Although the approximation (7.3.8) is reasonable for a wide range of radii, one can question its validity near the singularity. Any deviation from (7.3.8) could affect the character of the metric and in particular change the asymptotic form (7.3.10). (ii) If global strings are formed in the early universe, the distance $r_{\max}$ will eventually become smaller than the horizon. If this leads to singularities of the form (7.3.10), we would have a contradiction with the cosmic censorship hypothesis [Penrose, 1969] which forbids the formation of naked singularities from regular initial conditions. The metric in the vicinity of $r_{\max}$ can, of course, be modified due to the presence of matter. What, then, is the character of the resulting metric and does it satisfy cosmic censorship?

7.4 Gravitational field of an oscillating loop

The most striking feature of the gravitational field of a straight string is that there is no Newtonian force. Its absence can be understood from linear perturbation theory (7.1.2–4). The source of the Newtonian potential $\Phi = \frac{1}{2}h_{00}$, is

$$S_{00} = S_0^0 = \tfrac{1}{2}(T_0^0 - T_i^i). \qquad (7.4.1)$$

For the energy–momentum tensor of a string (7.1.5), we have

$$S_0^0(\mathbf{r}, t) = \mu \int d\zeta \, \dot{\mathbf{x}}^2 \, \delta^{(3)}(\mathbf{r} - \mathbf{x}(\zeta, t)). \qquad (7.4.2)$$

In the rest frame of a straight string, $S_0^0 = 0$ and the gravitational potential vanishes. A string can be static only if it is straight, and so curved strings produce non-zero gravitational potentials. In this section we shall analyze the gravitational field produced by oscillating closed loops.

The retarded solution of the wave equation (7.1.2) can be written as

$$h^{\mu\nu}(\mathbf{r}, t) = -4G \int \frac{d^3 y}{|\mathbf{r} - \mathbf{y}|} S^{\mu\nu}(\mathbf{r}, t - |\mathbf{r} - \mathbf{y}|).$$

(7.4.3)

Using (7.1.3) and (7.1.5) and integrating over $\mathbf{y}$ we obtain

$$h^{\mu\nu}(\mathbf{r}, t) = -4G\mu \int \frac{F^{\mu\nu}(\zeta, \tau) \, d\zeta}{|\mathbf{r} - \mathbf{x}(\zeta, \tau)| \, (1 - \hat{\mathbf{n}} \cdot \dot{\mathbf{x}}(\zeta, \tau))},$$

(7.4.4)

where

$$\hat{\mathbf{n}} = \frac{\mathbf{r} - \mathbf{x}(\zeta, \tau)}{|\mathbf{r} - \mathbf{x}(\zeta, \tau)|},$$

(7.4.5)

$$F^{\mu\nu} = \dot{x}^\mu \dot{x}^\nu - x'^\mu x'^\nu + \eta^{\mu\nu} x'^\sigma x'_\sigma,$$

(7.4.6)

and the retarted time τ is defined by

$$\tau = t - |\mathbf{r} - \mathbf{x}(\zeta, \tau)|.$$

(7.4.7)

7.4.1 The average field

The motion of non-relativistic test particles is determined mainly by the time-averaged field of the loop,

$$\langle h^{\mu\nu}(\mathbf{r}) \rangle = \frac{2}{L} \int_0^{L/2} dt \, h^{\mu\nu}(\mathbf{r}, t).$$

(7.4.8)

The period of the oscillating component of the field, $T = L/2$, is much shorter than the time it takes for the particle to traverse a distance $\sim L$. The effect of the oscillating component on the particle trajectory at a distance $\lesssim L$ from the loop is to introduce a small perturbation of the position $\Delta x \sim aT^2 \sim G\mu L$. Here, $a \sim GM/L^2$ is the typical acceleration, $M = \mu L$ is the mass of the loop, and for the present we disregard the effects of cusps and kinks which will be discussed in §7.4.2.

Instead of the integration variable t in (7.4.8) it is more convenient to use the retarded time τ. Differentiation of (7.4.7) gives

$$dt = d\tau[1 - \hat{\mathbf{n}} \cdot \dot{\mathbf{x}}(\zeta, \tau)],$$

(7.4.9)

and using (7.4.4) we can rewrite the average field (7.4.8) in the form [Turok, 1983b]

$$\langle h^{\mu\nu}(\mathbf{r}) \rangle = -\frac{8G\mu}{L} \int_0^{L/2} d\tau \int_0^L d\zeta \frac{F^{\mu\nu}(\zeta, \tau)}{|\mathbf{r} - \mathbf{x}(\zeta, \tau)|}$$

$$= -4G \int \frac{dS}{|\mathbf{r} - \mathbf{x}(\zeta, \tau)|} \left(\Sigma^{\mu\nu} - \tfrac{1}{2}\eta^{\mu\nu}\Sigma \right)$$

(7.4.10)

Here, $dS = |\dot{\mathbf{x}}||\mathbf{x}'|d\zeta d\tau$ is the area element on the surface traced out by the loop and

$$\Sigma^{\mu\nu} = \frac{2\mu}{L|\dot{\mathbf{x}}||\mathbf{x}'|}(\dot{x}^\mu \dot{x}^\nu - x'^\mu x'^\nu). \qquad (7.4.11)$$

We see that the time-averaged field of the loop is equal to that of a stationary source distribution over the shell traced by the loop's motion with the surface energy–momentum tensor (7.4.11). The mass per unit area is $\Sigma^{00} = 2\mu/L|\dot{\mathbf{x}}||\mathbf{x}'|$, and the total mass of the shell is equal to the mass of the loop, M:

$$\int \Sigma^{00} dS = \frac{2\mu}{L} \int_0^{L/2} d\tau \int_0^L d\zeta = \mu L = M. \qquad (7.4.12)$$

At a large distance from the loop $r \gg L$, (7.4.10) gives

$$\langle h^{\mu\nu}(\mathbf{r}) \rangle = -\frac{A^{\mu\nu}}{r}, \qquad (7.4.13)$$

where

$$A^{\mu\nu} = 8G\mu L^{-1} \int_0^{L/2} d\tau \int_0^L d\zeta \, F^{\mu\nu}(\zeta, \tau) \qquad (7.4.14)$$

The integration in (7.4.14) can be easily performed. For example,

$$A^{00} = 4G\mu L\langle \dot{\mathbf{x}}^2 \rangle = 2GM, \qquad (7.4.15)$$

where $\langle \dot{\mathbf{x}}^2 \rangle$ is defined by (6.2.26) and we have used the result (6.2.29). Similarly, we find

$$A^{0i} = 0, \qquad A^{ij} = 2Gm\delta^{ij}. \qquad (7.4.16)$$

Combining (7.4.13), (7.4.15) and (7.4.16), we see that the long-distance behaviour of the average field coincides with that of a Schwarzschild field of the same mass [Turok, 1983a]. It should be emphasized that the Schwarzschild limit is approached only in the average sense. Since the motion of the loop is relativistic, the oscillating component of $h^{\mu\nu}$ describing gravitational waves has the same order of magnitude as the average field.

7.4.2 Effects of cusps and kinks

As we discussed in chapter 6, oscillating loops can develop cusps where the string velocity momentarily reaches the speed of light. When $|\dot{\mathbf{x}}| \approx 1$, the factor $(1 - \hat{\mathbf{n}} \cdot \dot{\mathbf{x}})^{-1}$ in (7.4.4) becomes large for the directions of $\hat{\mathbf{n}}$ close to that of $\dot{\mathbf{x}}$, indicating that the cusp creates a strong pulse in the gravitational field. Since the intergrand in (7.4.4) is taken at retarded time, it is clear that the pulse propagates at the speed of light along the beam extending from the cusp in the direction of $\dot{\mathbf{x}}$ (fig. 7.5).

Sufficiently close to the pulse, the components of $h_{\mu\nu}$ become large and linear perturbation theory breaks down. However, for $G\mu \ll 1$ we expect this to happen only in a very small region near the pulse centre.

The main contribution to $h^{\mu\nu}$ in a propagating pulse comes from integration over a small region near the cusp in (7.4.4). Taking the slowly varying factors out of the integral, we can rewrite it as

$$h^{\mu\nu}(\mathbf{r}, t) \approx -4G\mu F_0^{\mu\nu} r^{-1} \int d\zeta \, [1 - \hat{\mathbf{n}} \cdot \dot{\mathbf{x}}(\zeta, \tau)]^{-1} \,, \qquad (7.4.17)$$

where $F_0^{\mu\nu} = F^{\mu\nu}(0,0)$ and τ is defined in terms of ζ, t and $\mathbf{r}$ through (7.4.7). An estimate of $h^{\mu\nu}$ can be obtained from (7.4.17) after tedious analysis which involves the expansion of the integrand around the cusp and a fortuitous change of the integration variable to $u = \zeta/\tau$. Omitting the details of the calculation, we shall only state the final results [Vachaspati, 1987a].

It is convenient to choose coordinates in which the the cusp is at $\mathbf{x} = 0$, $t = 0$ and the beam lies in the positive z-direction. Then, for points on the beam ($x = y = 0$), one finds

$$h^{\mu\nu} \sim \frac{G\mu \, \ell^{4/3}}{z|t - z|^{1/3}} \,. \qquad (7.4.18)$$

The z-component of the acceleration of a non-relativistic particle located at $\mathbf{r} = (0, 0, z)$ is found by differentiating (7.4.18):

$$a_z \sim \frac{G\mu \, \ell^{4/3} \text{sign}(z - t)}{z(z - t)^{4/3}} \,. \qquad (7.4.19)$$

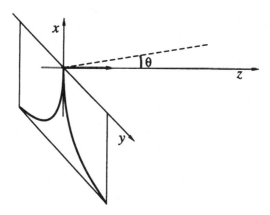

Fig. 7.5. Velocity and shape of a string cusp and the resulting direction of beamed radiation.

Here, the characteristic scale ℓ is defined by

$$|\mathbf{a}_0''|, \ |\mathbf{b}_0''| \sim \ell^{-1}, \tag{7.4.20}$$

where $\mathbf{a}_0''$ and $\mathbf{b}_0''$ are coefficients in the expansion (6.2.32) of the string trajectory near the cusp. The generic value of ℓ is $\sim L$ for a smooth loop, but it can be much smaller for a loop with small-scale structure. The force acting on the particle is initially repulsive, it grows in magnitude and suddenly switches to become a large attractive force, which then decreases. To find the x- and y-components of the acceleration, one has to calculate $h^{\mu\nu}$ slightly off the z-axis. The result is

$$a_x, \ a_y \sim (\pm)\frac{G\mu\ell}{z|z-t|}. \tag{7.4.21}$$

The signs of a_x, a_y are determined by the coefficients $\mathbf{a}_0'', \mathbf{b}_0''$ in the expansion (6.2.32); they do not experience any rapid change as functions of $(z-t)$.

To probe the shape of the pulse in the directions perpendicular to the beam, one can calculate the acceleration of a particle located in the plane $z = t$ at a small angle θ off the z-axis (as seen from the cusp). This gives

$$a_x, \ a_y \sim \frac{G\mu}{z\theta^3}, \qquad a_z \sim \frac{G\mu}{z\theta^4}. \tag{7.4.22}$$

More generally, the acceleration is given by (7.4.19), by (7.4.21) for $\theta \lesssim (|z-t|/\ell)^{1/3}$, and by (7.4.22) for $\theta \gtrsim (|z-t|/\ell)^{1/3}$. These linear perturbation results will be valid for

$$\min\left\{\theta, \ (|z-t|/\ell)^{1/3}\right\} \gg G\mu\ell/z. \tag{7.4.23}$$

Vachaspati [1987a] has studied the effect of gravitational pulses from cusps on particle trajectories. The effect is large only for a rather limited class of trajectories which pass through the pulse close to its centre. In this case the overall effect of the pulse tends to be repulsive.

Kinks on oscillating loops can also give rise to strong gravitational fields. This has been studied by Garfinkle & Vachaspati [1988] who found that kinks produce pulses of gravitational field propagating at the speed of light in the direction of the kink velocity. As the kink runs around the loop, it creates a fan-like pattern of beams (unlike a cusp which gives a single beam). The acceleration of a test particle located on one of the beams is

$$a_z \sim \frac{G\mu\ell^{2/3}}{z(z-t)^{2/3}}, \tag{7.4.24}$$

where the z-axis is chosen along the beam with $z = t = 0$ at the kink. The sign of this acceleration can be either positive or negative and can differ for different directions swept out by the fan. The x- and y-components of the acceleration are non-singular on the beam. Although the acceleration grows large near the pulse centre, the resulting velocity is not large, and the effect of kinks on the motion of particles is not dramatic.*

7.5 Gravitational radiation from a loop

The lifetime of a non-intersecting loop of string is determined by the rate at which it radiates away its energy. For macroscopic loops, the emission of massive particles is strongly suppressed (see chapter 8), and gravitational radiation can be expected to be the main energy-loss mechanism. Of course, for superconducting and global strings there are competing mechanisms due to electromagnetic and Goldstone boson radiation, respectively.

The gravitational radiation power for a loop of length L can be roughly estimated using the quadrupole formula,

$$\dot{E} \sim G \left(\frac{d^3 D}{dt^3} \right)^2 \sim G M^2 L^4 \omega^6 . \tag{7.5.1}$$

Here, $D \sim ML^2$ is the quadrupole moment, $M \sim \mu L$ is the loop's mass, $\omega \sim L^{-1}$ is the characteristic frequency, and we assume that the loop does not have substantial perturbations of size much smaller than L. Substituting all this into (7.5.1) we obtain

$$\dot{E} = \Gamma G \mu^2 , \tag{7.5.2}$$

where the numerical coefficient Γ is independent of the loop size, but does depend on its shape and trajectory. The lifetime of the loop is then

$$\tau \sim \frac{M}{\dot{E}} \sim \frac{L}{\Gamma G \mu} . \tag{7.5.3}$$

The quadrupole radiation formula applies only to slowly moving sources, and thus its validity for radiation from loops is dubious. In an oscillating loop the string moves at velocities close to the speed of light and can even become ultra-relativistic in the vicinity of cusps. Gravitational radiation from loops should therefore be studied using a full relativistic formalism.

---

* Note added in paperback edition: A closer examination has shown that the longitudinal acceleration (7.4.19) and (7.4.24) is in fact a gauge artifact and can be removed by a coordinate transformation. This is not surprising, since the metric (7.4.18) on the beam is like that of a plane gravitational wave, which is known to produce only transverse forces. As for the transverse acceleration (7.4.21), more work is needed to separate true physical effects from coordinate artifacts.

7.5.1 Radiation power

The power in gravitational radiation from a weak, isolated, periodic source to lowest order in G, can be found from the following equations, without further assumptions about the source [Weinberg, 1972],

$$P = \dot{E} = \sum_n P_n = \sum_n \int d\Omega \frac{dP_n}{d\Omega}, \qquad (7.5.4)$$

$$\frac{dP_n}{d\Omega} = \frac{G\omega_n^2}{\pi} \left\{ T_{\mu\nu}^* (\omega_n, \mathbf{k}) T^{\mu\nu} (\omega_n, \mathbf{k}) - \tfrac{1}{2} |T_\nu^\nu (\omega_n, \mathbf{k})|^2 \right\}. \qquad (7.5.5)$$

Here, $dP_n/d\Omega$ is the radiation power at frequency $\omega_n = 2\pi n/T$ per unit solid angle in the direction of $\mathbf{k}$, $|\mathbf{k}| = \omega_n$, T is the period of oscillation and

$$T^{\mu\nu} (\omega_n, \mathbf{k}) = \frac{1}{T} \int_0^T dt \, \exp(i\omega_n t) \int d^3x \, \exp(-i\mathbf{k} \cdot \mathbf{x}) \, T^{\mu\nu}(\mathbf{x}, t) \quad (7.5.6)$$

is the Fourier transform of the energy–momentum tensor.

For a closed loop, a convenient split for (7.5.4) into left- and right-moving contributions was derived by Burden [1985] (see also Garfinkle & Vachaspati [1987]):

$$\frac{dP_n}{d\Omega} = 8\pi G\mu^2 n^2 \left\{ |I_n(\hat{\mathbf{n}}_1)J_n(\hat{\mathbf{n}}_1) - I_n(\hat{\mathbf{n}}_2)J_n(\hat{\mathbf{n}}_2)|^2 \right.$$
$$\left. + |I_n(\hat{\mathbf{n}}_1)J_n(\hat{\mathbf{n}}_2) + I_n(\hat{\mathbf{n}}_2)J_n(\hat{\mathbf{n}}_1)|^2 \right\}. \qquad (7.5.7)$$

Here, $\hat{\mathbf{n}}_1$ and $\hat{\mathbf{n}}_2$ are unit vectors orthogonal to $\mathbf{k}$ and to one another, $I_n(\hat{\mathbf{n}})$ and $J_n(\hat{\mathbf{n}})$ are defined as

$$I_n(\hat{\mathbf{n}}) = \frac{1}{L} \int_0^L d\zeta \, \mathbf{a}'(\zeta) \cdot \hat{\mathbf{n}} \, \exp\left[-\frac{i}{2}(\omega_n\zeta + \mathbf{k} \cdot \mathbf{a})\right],$$
$$J_n(\hat{\mathbf{n}}) = \frac{1}{L} \int_0^L d\zeta \, \mathbf{b}'(\zeta) \cdot \hat{\mathbf{n}} \, \exp\left[\frac{i}{2}(\omega_n\zeta - \mathbf{k} \cdot \mathbf{b})\right], \qquad (7.5.8)$$

and $\mathbf{a}(\zeta), \mathbf{b}(\zeta)$ are the functions characterizing the loop trajectory (refer to (6.2.15)). Since the motion of the source is assumed to be periodic, note that (7.5.5–8) apply only in the centre-of-mass frame of the loop. A detailed derivation of (7.5.7) and its generalization can be found in Allen & Shellard [1992].

The first application of the formalism (7.5.4–6) to cosmic strings was attempted by Turok [1984] who calculated the radiation power from a rotating string segment held fixed at both ends. The radiation from simple families of oscillating loops (6.2.37–38) was studied by Vachaspati & Vilenkin [1985] and by Burden [1985]. For loops described by (6.2.38), the angular distribution of $dP_n/d\Omega$ can be expressed analytically in terms of

Bessel functions. The spectral power P_n is then calculated by a straightforward numerical integration. The resulting values of the total power, $P = \Gamma G \mu^2$, are shown in fig. 7.6 for several values of the parameters m, n and φ. Somewhat surprisingly, Γ turned out to be fairly large, typically ~ 100. Similar values of Γ are obtained for loops described by (6.2.37).

All loops in the families (6.2.37–38) have cusps and no kinks. The radiation from cuspless loops with kinks and from loops having neither cusps nor kinks was studied by Garfinkle & Vachaspati [1987]. The values of Γ they found are also typically ~ 100. In the particular case of a degenerate kinky loop (6.2.39), having the shape of a parallelogram, the radiation power can be expressed analytically,

$$\Gamma = \frac{32}{\sin^2 \alpha} \left[(1 + \cos \alpha) \ln \left(\frac{2}{1 + \cos \alpha} \right) \right.$$
$$\left. + (1 - \cos \alpha) \ln \left(\frac{2}{1 - \cos \alpha} \right) \right] . \tag{7.5.9}$$

Here, $\cos \alpha = \hat{\mathbf{A}} \cdot \hat{\mathbf{B}}$ with $\hat{\mathbf{A}}$ and $\hat{\mathbf{B}}$ defined in (6.2.39). The function (7.5.9) is plotted in fig. 7.7. The smallest value of Γ obtained at $\alpha = \pi/2$ is $\Gamma_{\min} \approx 44$. For $\alpha = 0$ and $\alpha = \pi$, the loops turn into oscillating double lines and the power diverges.

Two groups have numerically investigated gravitational radiation from loops which are the end products of a stochastic fragmentation process. Scherrer, Quashnock, Spergel & Press [1991] considered loops created in

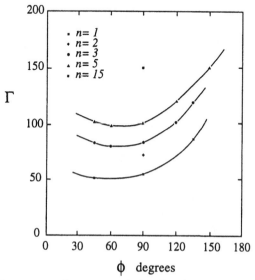

Fig. 7.6. Radiation power for some of the Burden loop trajectories (6.2.38) [Burden, 1985].

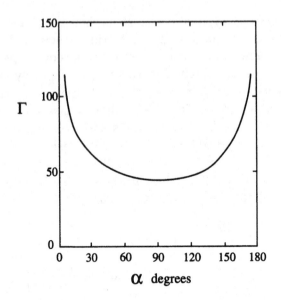

Fig. 7.7. Radiation power from the loop solutions (6.2.39) possessing four kinks [Garfinkle & Vachaspati, 1987].

flat space through the disintegration of a convoluted parent loop as discussed in §6.4.2. In a somewhat more realistic context, Allen & Shellard [1992] studied the smallest loops created by a string network during the radiation-dominated era in an expanding universe. Averaged measurements by both groups were remarkably consistent yielding

$$\langle \Gamma \rangle \approx 65 . \tag{7.5.10}$$

7.5.2 Angular distribution and spectrum

For a 'normal', non-relativistic source, the angular distribution of gravitational radiation has a characteristic quadrupole pattern and the spectral power P_n decreases exponentially at large n. Radiation from oscillating loops deviates from this standard behaviour, especially if cusps or kinks are present. Since cusps and kinks involve arbitrarily short wavelengths, they contribute to the integrals (7.5.8) with arbitrarily large values of n. As a result, the spectral power falls off only as a power law.

The behaviour of $dP/d\Omega$ and of P_n for a loop with cusps was first studied by Vachaspati & Vilenkin [1985]. For the solution described by (6.2.38) with loop parameters $m = n = 1$ (distinct from the spectral n),

$$P_n \sim G\mu^2 n^{-4/3}, \qquad dP/d\Omega \sim G\mu^2 \theta^{-1} , \tag{7.5.11}$$

where θ is the angle between the wave vector $\mathbf{k}$ and the direction of the luminal velocity at the cusp (see fig. 7.5). Although (7.5.11) was initially derived for a very special class of loops, the following argument demonstrates wider validity.

In the large n limit, the main contribution to I_n and J_n in (7.5.8) comes from the vicinity of the cusp, and we can use the expansions (6.2.32) for the functions $\mathbf{a}(\zeta)$ and $\mathbf{b}(\zeta)$. We shall assume for simplicity that the loop has a single characteristic length scale L, so that $|\mathbf{a}_0''| \sim |\mathbf{b}_0''| \sim L^{-1}$ etc. If we take $\mathbf{k}$ exactly in the direction of the luminal velocity $(-\mathbf{a}_0')$ and choose the string parameter ζ so that the cusp is at $\zeta = 0$, then $(\omega_n \zeta + \mathbf{k} \cdot \mathbf{a})$ and $(\omega_n \zeta - \mathbf{k} \cdot \mathbf{b})$ are both of order $nL^{-3}\zeta^3$. The main contribution to $I_n(\hat{\mathbf{n}})$ comes from the integration region where $|\omega_n \zeta + \mathbf{k} \cdot \mathbf{a}| \lesssim 1$, that is,

$$|\zeta| \lesssim n^{-1/3} L \equiv \zeta_n. \tag{7.5.12}$$

Since $\hat{\mathbf{n}} \perp \mathbf{a}_0'$, $\mathbf{a}' \cdot \hat{\mathbf{n}} \sim \zeta/L$, and we obtain an estimate $I_n \sim \zeta_n^2/L^2 \sim n^{-2/3}$. Similarly, we find $J_n \sim n^{-2/3}$ and

$$\frac{dP_n}{d\Omega} \sim G\mu^2 n^{-2/3}. \tag{7.5.13}$$

Now suppose there is a small angle θ between the directions of $\mathbf{k}$ and $-\mathbf{a}_0'$. The additional terms in $(\omega_n \zeta + \mathbf{k} \cdot \mathbf{a})$ and $(\omega_n \zeta - \mathbf{k} \cdot \mathbf{b})$ are then of the order $\omega_n \theta^2 \zeta$ and $\omega_n \theta \zeta^2/L$, while the additional terms in $\mathbf{a}' \cdot \hat{\mathbf{n}}$ and $\mathbf{b}' \cdot \hat{\mathbf{n}}$ are $\sim \theta$. This shows that the estimate (7.5.13) remains unchanged for

$$\theta \lesssim n^{-1/3} \equiv \theta_n. \tag{7.5.14}$$

Note that for $\theta > \theta_n$, it is not difficult to show that (7.5.12–13) are replaced by $|\zeta| \lesssim Ln^{-1}\theta^{-2}$ and $dP_n/d\Omega \sim G\mu^2\theta^{-4}n^{-2}$. From (7.5.13–14), we obtain an estimate for P_n,

$$P_n \sim \theta_n^2 \left(\frac{dP_n}{d\Omega}\right) \sim G\mu^2 n^{-4/3}. \tag{7.5.15}$$

To find the angular distribution $dP/d\Omega$, we sum (7.5.13) over n and cut the summation at $n \sim \theta^{-3}$. This gives

$$\frac{dP}{d\Omega} \sim G\mu^2\theta^{-1}, \tag{7.5.16}$$

in agreement with (7.5.11). The radiation intensity (7.5.16) diverges at $\theta = 0$, but the divergence is integrable and the total power is finite. The immediate vicinity of the beam, where linear perturbation theory breaks down, contributes very little to the total power and does not significantly affect the values of Γ quoted in the previous section.

Gravitational radiation from loops with kinks was studied by Garfinkle & Vachaspati [1987]. For the particular degenerate kinky loops (6.2.39) they were able to obtain a closed analytic expression for the radiation intensity,

$$\frac{dP}{d\Omega} = \frac{8G\mu^2}{\pi} \cdot \frac{2 - |e_1 + e_2| - |e_1 - e_2|}{(1 - e_1^2)(1 - e_2^2)} \ . \qquad (7.5.17)$$

Here, $e_1 = \hat{\mathbf{k}} \cdot \hat{\mathbf{A}}$, $e_2 = \hat{\mathbf{k}} \cdot \hat{\mathbf{B}}$ and $\hat{\mathbf{k}}$ is a unit vector in the direction of $\mathbf{k}$. The denominator of (7.5.17) vanishes when $e_1 = 1$ or $e_2 = 1$, but in these cases the numerator also vanishes, and $dP/d\Omega$ remains finite. In fact, the radiation intensity (7.5.17) is not strongly peaked in any particular direction. The asymptotic behaviour of the spectral power for this loop is

$$P_n \sim G\mu^2 n^{-2} \ . \qquad (7.5.18)$$

In the degenerate loop (6.2.39) the kinks always move along one of the vectors $\pm\hat{\mathbf{A}}, \pm\hat{\mathbf{B}}$. In a more general case, the direction of motion of the kink gradually changes, tracing a fan-like pattern. The asymptotic form of $dP_n/d\Omega$ for a general kinky loop can be studied with the same techniques employed for cusps [Garfinkle & Vachaspati, 1987]. One finds that $dP_n/d\Omega \propto n^{-2}$ in directions off the fans traced out by the kinks and $dP_n/d\Omega \propto n^{-4/3}$ for directions along the fans. If the wave vector $\mathbf{k}$ makes a small angle θ with the fan, then the $n^{-4/3}$ asymptotics apply for $n \ll \theta^{-3}$ and the n^{-2} asymptotics for $n \gg \theta^{-3}$. The radiation intensity is finite in all directions, and the contribution of large-n harmonics is small. The angular distribution of the radiation is not expected to have any strong features.

7.5.3 Radiation of momentum

Gravitational radiation from an oscillating loop carries away not only energy, but also momentum. As a result, the loop can accelerate like a rocket [Hogan & Rees, 1984]. The rate of momentum radiation is given by

$$\frac{d\mathbf{P}}{dt} = \sum_n \int \frac{dP_n}{d\Omega} \mathbf{k} \, d\Omega \ . \qquad (7.5.19)$$

It vanishes for loops described by (6.2.37–39) because of their high symmetry. In general, however, one expects $|\dot{\mathbf{P}}|$ to be of the same order as $\dot{E}$ in (7.5.2). A numerical calculation for several asymmetric loops gives [Vachaspati & Vilenkin, 1985]

$$|\dot{\mathbf{P}}| = \Gamma_P G\mu^2 \ , \qquad (7.5.20)$$

with $\Gamma_P \sim 10$.

A loop radiating momentum at the rate (7.5.20) will accelerate as

$$\dot{v} = \Gamma_P \frac{G\mu}{L} \,. \tag{7.5.21}$$

If it starts from rest, then in a time period comparable to its lifetime τ it will reach the velocity

$$v \sim \dot{v}\tau \sim \frac{\Gamma_P}{\Gamma} \sim 0.1 \,, \tag{7.5.22}$$

where τ is from (7.5.3). The rate of angular momentum radiation from loops has been found by Durrer [1989] to be

$$|\dot{\mathbf{J}}| = \Gamma_J G\mu^2 L \tag{7.5.23}$$

with $\Gamma_J \sim 10$.

7.6 Wiggly string gravity

Evolving string networks in an expanding universe are expected to have substantial small-scale structure in the form of kinks and wiggles on scales well below the characteristic length scale of the network (see chapter 9). The gravitational effects of such 'wiggly' strings will be studied in this section showing, in particular, that the gravitational field about a long wiggly string is qualitatively different from that produced by a straight string.

7.6.1 The average field

To a distant observer who cannot resolve the kinks and wiggles on a long string, the string will appear smooth, but the effective mass per unit length $\tilde{\mu}$ and tension $\tilde{T}$ will differ from their unperturbed values. We found in §6.5.1 that the effective equation of state for a wiggly string is

$$\tilde{\mu}\tilde{T} = \mu^2 \,. \tag{7.6.1}$$

Consider now a long wiggly string oriented to lie along the z-axis. At distances large compared to the scale of the wiggles, the average field of the string can be found by solving the linearized Einstein equations (7.1.2–4) with the energy–momentum tensor

$$T^\nu_\mu = \delta(x)\delta(y)\,\mathrm{diag}\,(\tilde{\mu}, 0, 0, \tilde{T})\,. \tag{7.6.2}$$

This gives

$$\begin{aligned} h_{00} &= h_{33} = 4G(\tilde{\mu} - \tilde{T})\ln(r/r_0)\,, \\ h_{11} &= h_{22} = 4G(\tilde{\mu} + \tilde{T})\ln(r/r_0)\,, \end{aligned} \tag{7.6.3}$$

where r_0 is a constant of integration.

The coordinate transformation

$$(1 - h_{11})r^2 = [1 - 4G(\tilde{\mu} + \tilde{T})]r'^2 \qquad (7.6.4)$$

brings the metric (7.6.3) to the form

$$ds^2 = (1 + h_{00})dt^2 - (1 - h_{00})dz^2 - dr^2 - (1 - b)r^2 d\theta^2 . \qquad (7.6.5)$$

Here, $b = 4G(\tilde{\mu} + \tilde{T})$, h_{00} is from (7.6.3), and we have dropped the primes from r'. We see that the surfaces of constant t and z are cones, as for a straight string, but the conical deficit angle is now $\pi b = 4\pi G(\tilde{\mu} + \tilde{T})$.

The main distinction between (7.6.5) and the straight string metric (7.1.11) is that a wiggly string produces a non-vanishing Newtonian potential,

$$\Phi = \tfrac{1}{2}h_{00} = 2G(\tilde{\mu} - \tilde{T})\ln(r/r_0) . \qquad (7.6.6)$$

This has the same form as the potential arising from a massive rod. Note also that the linearized solution (7.6.5) can only be trusted when $h_{00} \ll 1$.

7.6.2 Propagation of particles and light

We shall consider the effect of wiggles on particle and light propagation, assuming for simplicity that the direction of propagation is perpendicular to the string. It will be convenient to introduce a new radial coordinate, r'', which is related to r in (7.6.3) by

$$[1 - 8G\tilde{\mu}\ln(r/r_0)]r^2 = (1 - 8G\tilde{\mu})r''^2 . \qquad (7.6.7)$$

Then, setting $dz = 0$ and dropping the primes from r'' we obtain the metric

$$ds^2 = (1 + h_{00})[dt^2 - dr^2 - (1 - 8G\tilde{\mu})r^2 d\theta^2] , \qquad (7.6.8)$$

where h_{00} is given in (7.6.3). The conformal factor $(1 + h_{00})$ does not affect light propagation and can be dropped. The problem is thus reduced to that of light propagation in the space of a straight string with a deficit angle

$$\tilde{\Delta} = 8\pi G\tilde{\mu} . \qquad (7.6.9)$$

The formation of double images and the discontinuous Doppler shift due to a wiggly string can be described using the results of §7.2. In particular, the angular separation of the images is given by (7.2.1) with Δ replaced by $\tilde{\Delta}$ [Hindmarsh & Wray, 1990; Vachaspati & Vilenkin, 1991]. An analysis of null geodesics in the metric (7.6.5), demonstrates that the same equation is applicable when the direction of light propagation is not assumed to be perpendicular to the string [Hindmarsh & Wray, 1990].

We next consider the deflection of a particle moving past the string. We shall assume that the particle's velocity is sufficiently large and its initial distance from the string is sufficiently small, so that $v^2 \gg h_{00}$. The velocity then changes little during the course of its motion, the deflection angle is small and can be calculated perturbatively. It will be convenient to use the conformally Minkowskian coordinates

$$ds^2 = (1 + h_{00})(dt^2 - dx^2 - dy^2) \qquad (7.6.10)$$

with a missing wedge of angular width $\tilde{\Delta} = 8\pi G\tilde{\mu}$. The linearized geodesic equations in the metric (7.6.10) have the form

$$2\ddot{x} = -(1 - \dot{x}^2 - \dot{y}^2)\partial_x h_{00} ,$$
$$2\ddot{y} = -(1 - \dot{x}^2 - \dot{y}^2)\partial_y h_{00} , \qquad (7.6.11)$$

where dots stand for derivatives with respect to t. If we choose the x-axis along the initial velocity of the particle, the unperturbed trajectory can be written as

$$x = vt, \qquad y = y_0 . \qquad (7.6.12)$$

After the particle passes the string, its velocity gains a small y-component which can be found, to linear order in $G\mu$, by integrating over the trajectory (7.6.12),

$$v_y = \frac{1}{v} \int_{-\infty}^{\infty} \ddot{y}\, dx = -\frac{2\pi G(\tilde{\mu} - \tilde{T})}{v\gamma^2} , \qquad (7.6.13)$$

where $\gamma = (1 - v^2)^{-1/2}$. Note that the impact parameter y_0 has dropped out of the expression for v_y.

If two particles moving with the same initial velocity v pass on opposite sides of the string, they develop a velocity difference

$$|\Delta v_y| = v\tilde{\Delta} + \frac{4\pi G(\tilde{\mu} - \tilde{T})}{v\gamma^2} , \qquad (7.6.14)$$

where the first term is due to the conical deficit angle $\tilde{\Delta}$. In astrophysical applications we shall be interested in the relative velocity developed by the particles after the string has passed between them. This can be found from (7.6.14) by a Lorentz transformation [Vachaspati & Vilenkin, 1991; Vollick, 1992a],

$$u = \gamma|\Delta v_y| = 8\pi G\tilde{\mu}v\gamma + \frac{4\pi G(\tilde{\mu} - \tilde{T})}{v\gamma} . \qquad (7.6.15)$$

Here, the first term has the same form as (7.2.2) for a straight string, while the second term is due to the attractive gravitational force produced by the wiggles. The dependence of this term on $v, \tilde{\mu}$ and $\tilde{T}$ can be easily understood. From (7.6.6), the gravitational force on a non-relativistic particle of mass m is $F = 2mG(\tilde{\mu} - \tilde{T})/r$. A particle with an impact parameter r is exposed to this force for a time $\Delta t \sim r/v$, and the resulting velocity is $u \sim (F/m)\Delta t \sim G(\tilde{\mu} - \tilde{T})/v$.

7.6.3 Gravitational radiation

To study gravitational radiation from wiggly strings, we shall use an idealized model of an infinite string oriented along the z-axis with perturbations running along the string in opposite directions. The standard gravitational radiation formalism (7.5.4–6) was derived for localized sources and does not directly apply to infinite strings. The necessary modifications have been made by Sakellariadou [1990] who found the following expression for the radiation power per unit angle per unit length of string

$$\frac{dP}{dz d\theta} = 2G \sum_{\omega} \sum_{k_z} \omega \left\{ T_{\mu\nu}^*(\omega, \mathbf{k}) T^{\mu\nu}(\omega, \mathbf{k}) - \tfrac{1}{2}|T_\nu^\nu(\omega, \mathbf{k})|^2 \right\}, \quad (7.6.16)$$

where $\omega = |\mathbf{k}|$, $\mathbf{k} = (\mathbf{k}_\perp, k_z)$, and the energy flux is calculated in the direction of $\mathbf{k}_\perp$. It is also assumed for simplicity that the perturbations are periodic along the string with some large repeat length L, so that the spectra of ω and k_z are discrete. Sakellariadou used this formalism to calculate the gravitational radiation from a helicoidal standing wave on a string.

A more general case can be analyzed when the string is nearly straight and perturbations can be treated as a superposition of small-amplitude waves,

$$\begin{aligned} \mathbf{a}(\zeta) &= \hat{\mathbf{n}} a_z(\zeta) + \mathbf{a}_\perp(\zeta) \\ \mathbf{b}(\zeta) &= \hat{\mathbf{n}} b_z(\zeta) + \mathbf{b}_\perp(\zeta) \end{aligned} \quad (7.6.17)$$

with $\hat{\mathbf{n}}$ along the z-axis, $\hat{\mathbf{n}} \cdot \mathbf{a}_\perp = \hat{\mathbf{n}} \cdot \mathbf{b}_\perp = 0$, and $|\mathbf{a}'_\perp|, |\mathbf{b}'_\perp| \ll 1$. The conformal gauge conditions (6.2.16) then tell us that

$$a'_z \approx 1 - \tfrac{1}{2}{\mathbf{a}'_\perp}^2, \qquad b'_z \approx 1 - \tfrac{1}{2}{\mathbf{b}'_\perp}^2. \quad (7.6.18)$$

Using a formalism similar to (7.5.7–8), linearizing the integrals in $\mathbf{a}_\perp$ and $\mathbf{b}_\perp$, and finally integrating over θ, one can express the radiation power (7.6.16) in the form [Hindmarsh, 1990a; Battye & Shellard, 1993]

$$\frac{dP}{dz} = \pi G \mu^2 \sum_{n,m} (k_n + k_m) \left\{ |\mathbf{I}_n|^2 |\mathbf{J}_m|^2 - |\mathbf{I}_n^* \cdot \mathbf{J}_m|^2 + |\mathbf{I}_n \cdot \mathbf{J}_m|^2 \right\},$$

$$(7.6.19)$$

where

$$\mathbf{I}_n = \frac{1}{L} \int_0^L d\zeta \, \mathbf{a}'_\perp(\zeta) \exp(-ik_n\zeta) ,$$

$$\mathbf{J}_m = \frac{1}{L} \int_0^L d\zeta \, \mathbf{b}'_\perp(\zeta) \exp(ik_m\zeta) , \qquad (7.6.20)$$

and the summation is taken over $k_n = 2\pi n/L, k_m = 2\pi m/L$, with m and n positive integers. The wavenumbers k_n and k_m are related to ω and k_z by

$$k_n + k_m = 2\omega , \qquad k_n - k_m = 2k_z . \qquad (7.6.21)$$

It is clear from (7.6.19) that both right- and left-moving modes are required to produce gravitational radiation. If all the waves on the string travel in the same direction, then either $\mathbf{I}_n = 0$ or $\mathbf{I}_m = 0$, and the radiation power vanishes (see also §7.6.4). Equation (7.6.21) suggests that at the quantum level the radiation process can be viewed as two oppositely moving string states with momentum k_n and k_m annihilating into two gravitons.

The average perturbation energy per unit length of string can be expressed in terms of $\mathbf{I}_n$ and $\mathbf{J}_m$ using (6.5.10) and (7.6.18) with $k_1 = \langle a'_z \rangle$ and $k_2 = \langle b'_z \rangle$, where angular brackets indicate averaging over the period L. This gives

$$\tilde{\mu} - \mu = \tfrac{1}{4}\mu \left(\langle \mathbf{a}'_\perp{}^2 \rangle + \langle \mathbf{b}'_\perp{}^2 \rangle \right)$$

$$= \tfrac{1}{2}\mu \left(\sum_{n>0} |\mathbf{I}_n|^2 + \sum_{m>0} |\mathbf{J}_m|^2 \right) , \qquad (7.6.22)$$

where the extra factor of 2 in the second step is due to the fact that the summation is only taken over positive values of k_n and k_m. From (7.6.19) and (7.6.22) we can estimate the radiation power as

$$\frac{dP}{dz} \sim 2\pi G (\tilde{\mu} - \mu)^2 \bar{\omega} , \qquad (7.6.23)$$

where the characteristic frequency $\bar{\omega}$ is the average value of $(k_n + k_m)/2$ weighted by the product $|\mathbf{I}_n|^2 |\mathbf{J}_m|^2$.

7.6.4 *Travelling waves on strings*

In the previous section we saw that strings having only right- or left-moving perturbations do not emit gravitational waves. This conclusion holds even if the amplitude of the waves is not assumed to be small. Moreover, a simple analytic expression can be given for the gravitational field produced by such travelling waves [Vachaspati, 1986].

We consider a right-moving wave propagating along a string in the positive z-direction. In the gauge specified by

$$\zeta^0 = t, \quad \zeta^1 = z, \tag{7.6.24}$$

the wave is described by

$$x = f(z - t), \qquad y = g(z - t), \tag{7.6.25}$$

where f and g are arbitrary functions. The source of the linearized Einstein equations in (7.1.2) is found by a straightforward calculation using (6.1.15):

$$S^{\mu\nu} = \mu F^{\mu\nu}(z - t)\, \delta(x - f)\, \delta(y - g), \tag{7.6.26}$$

where

$$F = \begin{pmatrix} f'^2 + g'^2 & -f' & -g' & f'^2 + g'^2 \\ -f' & 1 & 0 & -f' \\ -g' & 0 & 1 & -g' \\ f'^2 + g'^2 & -f' & -g' & f'^2 + g'^2 \end{pmatrix}. \tag{7.6.27}$$

Note that since the gauge (7.6.24) is not conformal, the simplified form (6.2.17) of $T^{\mu\nu}$ cannot be used.

It is not difficult to verify that

$$h^{\mu\nu} = h F^{\mu\nu}(z - t), \tag{7.6.28}$$

with

$$h = 4G\mu \ln\left[r_0^{-2}\left\{ (x - f)^2 + (y - g)^2 \right\} \right], \tag{7.6.29}$$

and an arbitrary integration constant r_0, is a solution of (7.1.2). Clearly, $(\partial_t^2 + \partial_z^2)h^{\mu\nu} = 0$ because $h^{\mu\nu}$ depends on t and z only in the combination $(t - z)$. On the other hand, $(\partial_x^2 + \partial_y^2)$ acts only on the logarithm and gives the δ-function in $S^{\mu\nu}$.

For a static straight string parallel to the z-axis, all components of $F^{\mu\nu}$ vanish except $F^{11} = F^{22} = 1$, and we recover the metric (7.1.7). We know that this straight string metric is locally flat, but for a non-trivial choice of f and g the metric (7.6.28) has a non-vanishing Riemann tensor.

Some interesting properties of the gravitational field (7.6.28) can be seen by considering a pulse of length d propagating along a straight string. We can choose the origin of z so that at any time t the pulse is localized in the range

$$t \leq z \leq t + d. \tag{7.6.30}$$

An unusual feature of the gravitational field of the pulse is that for values of z outside this range the metric is the same as for a straight string. Hence, the pulse exerts a gravitational force on a test particle only for a short period $\Delta t \sim d$ when the particle is within the slab (7.6.30). Suppose the particle is initially at rest at a large distance r from the string (much greater than the amplitude of the pulse). Then it can be shown that, after the pulse has passed by, the particle gains a velocity directed towards the string given by

$$v = \frac{4G\mu}{r} \int \left(f'^2 + g'^2\right) dt \,. \tag{7.6.31}$$

Note that an analogous situation arises in electromagnetism when one considers the electric field of an ultra-relativistic charge. In the limit $v \to 1$, the field is confined to the plane perpendicular to the direction of motion. The impulse received as this plane passes through a test charge is also inversely proportional to the distance from the source.

The most remarkable feature of the travelling wave solutions (7.6.28) is their non-dissipative nature. The energy flux is equal to zero in directions perpendicular to the z-axis, and thus the wave does not lose any energy as it propagates along the string. This property can be shown to hold beyond linearized gravity. Garfinkle & Vachaspati [1990] presented a method for constructing exact solutions of Einstein–Higgs-gauge field equations for travelling waves on strings from a given solution describing a static straight string. These travelling waves are non-dissipative and propagate without changing shape.

7.7 Self-interaction and related issues

7.7.1 Gravitational back-reaction

The gravitational field produced by a moving loop of string acts back on the string and affects its motion. This back-reaction is responsible for the decrease of the loop's energy which is lost as gravitational radiation. The shape of the loop is also expected to gradually change.

The gravitational back-reaction problem has been studied quantitatively by Quashnock & Spergel [1990]. The string equations of motion in curved spacetime (6.3.3) and the conformal gauge conditions (6.3.1–2) can be expressed as

$$\partial_u \partial_v x^\mu = -\Gamma^\mu_{\alpha\beta} \partial_u x^\alpha \partial_v x^\beta \,, \tag{7.7.1}$$

$$\partial_u x^\alpha \partial_u x_\alpha = 0 \,, \qquad \partial_v x^\alpha \partial_v x_\alpha = 0 \,, \tag{7.7.2}$$

where we have introduced the null worldsheet coordinates $u = \zeta^0 + \zeta^1$, $v = \zeta^0 - \zeta^1$. To linear order in the metric perturbation $h_{\mu\nu}$, (7.7.1) takes the form

$$\partial_u \partial_v x^\mu = -\tfrac{1}{2}\eta^{\mu\gamma} \left(\partial_\alpha h_{\beta\gamma} + \partial_\beta h_{\alpha\gamma} - \partial_\gamma h_{\alpha\beta}\right) \partial_u x^\alpha \partial_v x^\beta \,. \tag{7.7.3}$$

In the back-reaction problem the field $h_{\mu\nu}$ is produced by the string itself and can be found by solving the linearized Einstein equations

(7.1.2–5). In terms of the variables u and v, the retarded solution for $h_{\mu\nu}$ is

$$h_{\mu\nu}(x) = 8G\mu \int du\, dv\, F_{\mu\nu}\left(z(u,v)\right) \theta(x^0 - z^0)\, \delta\left((x-z)^2\right), \quad (7.7.4)$$

where $F_{\mu\nu} = \partial_u z_\mu \partial_v z_\nu + \partial_v z_\mu \partial_u z_\nu - \eta_{\mu\nu}\partial_u z^\alpha \partial_v z_\alpha$ and $z^\mu(u,v)$ is the string worldsheet. The string equation of motion to linear order in $G\mu$ can now be written as

$$\mu \partial_u \partial_v x^\alpha = f^\alpha, \quad (7.7.5)$$

where f^α is the right-hand side of (7.7.3) with $h_{\mu\nu}$ from (7.7.4), all multiplied by the string mass per unit length, μ. The non-local character of the self-force f^α is due to the fact that the string is an extended object.

Back-reaction problems are notorious for their divergences, and string back-reaction is no exception. The infinite part of f^α has been calculated by Copeland, Haws & Hindmarsh [1990], with the result

$$f^\alpha_{\text{inf}} = -\delta\mu\, \partial_u \partial_v x^\alpha, \quad (7.7.6)$$

where $\delta\mu$ is a logarithmically divergent integral. With a large-distance cut-off at the string radius of curvature R and a short-distance cut-off provided by the string thickness δ, we have

$$\delta\mu = -8G\mu^2 \ln(R/\delta). \quad (7.7.7)$$

Since f^α_{inf} has the same form as the left-hand side of (7.7.5), it can be absorbed to renormalize the string tension,

$$\mu_{\text{ren}} \partial_u \partial_v x^\alpha = f^\alpha_{\text{ren}}, \quad (7.7.8)$$

where $\mu_{\text{ren}} = \mu + \delta\mu$ and f^α_{ren} is free of divergences.

The effect of back-reaction on loop dynamics has been studied numerically by Quashnock & Spergel [1990]. They calculated the change in

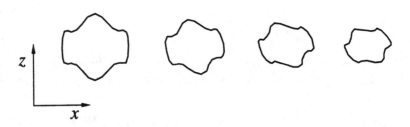

Fig. 7.8. Evolution of a Burden loop trajectory with gravitational back-reaction [Quashnock & Spergel, 1990].

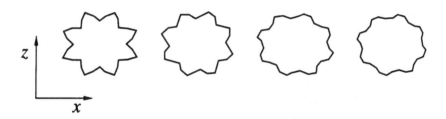

Fig. 7.9. Evolution of a kinky loop trajectory including gravitational back-reaction [Quashnock & Spergel, 1990].

$\partial_u x^\alpha$ and $\partial_v x^\alpha$ during one period of oscillation using the unperturbed string trajectory to calculate f^α in (7.7.5). No infinities are encountered with this technique, since f_{inf} vanishes for an unperturbed trajectory. Fig. 7.8 shows the evolution of a Burden trajectory (6.2.38) with $m = 1, n = 5, \psi = \pi/2$, and $G\mu = 10^{-3}$ (only $x - z$ projections of the loop are shown). The steps are separated by one loop period ($\Delta t = L/2$). For a smaller value of $G\mu$, we expect that configurations similar to those in fig. 7.8 will be separated by $10^{-3}/G\mu$ periods. Numerical results indicate that back-reaction does not prevent cusps from forming, neither does it lead to string self-intersection near cusps.

The evolution of a kinky loop is shown in fig. 7.9, with steps again separated by $10^{-3}/G\mu$ periods. The loop has a 'sawtooth' shape with eight pairs of kinks. Clearly, the sharp angles of the kinks are rounded by the back-reaction on a timescale much shorter than the loop's lifetime. Comparing the evolution of loops with different numbers of kinks, Quashnock & Spergel [1990] estimate the characteristic decay time of a kink as

$$\tau \sim L/n\Gamma G\mu, \qquad (7.7.9)$$

where L is the loop's length, n is the number of kinks, and $\Gamma \sim 50$. The estimate (7.7.9) was obtained for kinks of initial opening angle $\sim 145°$. They suggest that τ depends on this angle, but a quantitative description of the dependence has not been given.

7.7.2 Strings as distributional sources

For most of this chapter we have approximated the string energy–momentum tensor by a δ-function source (7.1.5). Although this procedure is consistent in linearized gravity, it has some technical mathematical short-comings in the full Einstein theory; the difficulty arises because general relativity is nonlinear, and a product of distributions is not generally well-defined [Israel, 1977; Geroch & Traschen, 1987].

One can try to define a δ-function using some regular string-like solution of the Einstein equations with an adjustable width, taking the limit as the width vanishes. Consider for example a straight string with T_μ^ν and the metric described by (7.1.26) and (7.1.27), respectively, with $\sigma(r)$ a bell-shaped function of variable height and width. If we let the width of $\sigma(r)$ go to zero, while keeping the string mass per unit length μ fixed, then T_μ^ν approaches the δ-function form (7.1.6) and the metric becomes conical everywhere outside the string axis, with a deficit angle given by (7.1.31). The problem with this approach is that the result depends on the limiting procedure. Geroch & Traschen [1987] have shown that by choosing different initial forms of T_μ^ν, one can obtain a variety of deficit angles for the same limiting value of μ. This is not particularly surprising if we recall that the deficit angle depends not only on T_0^0, but also on the transverse tensions T_r^r and T_θ^θ.

Alternatively, one can adopt a more cavalier approach and simply look for solutions of Einstein's equations with conical singularities, leaving the mathematical purist to resolve the question of whether or not they can be interpreted as arising from distributional sources. We have already exhibited several such solutions at the end of §7.1.2. Frolov, Israel & Unruh [1989] found a class of three-dimensional metrics describing a circular loop of string at the moment of time symmetry and analyzed in detail how the conical geometry near the string goes over into a Schwarzschild metric at large distances. The general behaviour of these metrics near the conical singularity was studied by Unruh, Hayward, Israel & McManus [1989] (see also Clarke, Ellis & Vickers [1990] and Garfinkle [1989]). As the string is approached, they found that the spacetime curvature grows as

$$|R_{\sigma\tau}^{\mu\nu}| \sim R^{-2}(R/r)^{1-\alpha}, \tag{7.7.10}$$

where R is the characteristic curvature radius of the string, r is the distance from the string, $\alpha = \Delta(2\pi - \Delta)^{-1}$, and Δ is the conical deficit angle. The divergence (7.7.10) is rather weak: the metric $g_{\mu\nu}$ and the connection $\Gamma_{\nu\sigma}^\mu$ need not be singular, and covariant differentiation is defined all the way to the string. Unruh *et al.* [1989] were able to show that in this background geometry, the string worldsheet is totally geodesic (that is, any two points on the worldsheet can be joined by a spacetime geodesic wholly within the worldsheet). The surprising conclusion is that the string deforms the background geometry in its vicinity in such a manner that the string, as well as the orbit of each of its points, become geodesics of this geometry. The curvature κ of a parallel curve at a small, constant distance r from the string tends to zero with r according to

$$\kappa \sim R^{-1}(r/R)^\alpha. \tag{7.7.11}$$

Before the reader becomes alarmed about whether it is ever possible to bend a cosmic string, we point out that these results are not relevant for finite-thickness strings with a small deficit angle. We shall see in chapter 10 that realistic strings must have $G\mu \lesssim 10^{-5}$ and thus $\alpha \lesssim 4 \times 10^{-5}$. For such a small α, the curvature κ in (7.7.11) decreases so slowly that it remains practically equal to its flat-spacetime value, $\kappa \sim R^{-1}$, at a distance r comparable to the string thickness. Finally, we note that, to leading order in $G\mu$, (7.7.10) gives $|R^{\mu\nu}_{\sigma\tau}| \sim (Rr)^{-1}$, which implies that the connections $\Gamma^{\mu}_{\nu\sigma}$ are logarithmically divergent. This explains the origin of the logarithmic divergence in the self-force derived in the previous section using linearized gravity.

8
String interactions

8.1 Particle scattering by strings

8.1.1 Scattering cross-section

The gauge and Higgs fields which have non-zero vacuum expectation values inside the string can couple to various other fields. This results in an effective interaction between the string and the corresponding particles. Naively, one would expect the scattering cross-section for a particle per unit length of string to be comparable to the string thickness δ. However, this expectation turns out to be incorrect, and the actual cross-section is determined not by the string thickness, but by the wavelength of the incident particle [Everett, 1981].

As an illuminating example we consider the coupling

$$\mathcal{L}_{\text{int}} = -\lambda_\chi \phi^\dagger \phi \chi^\dagger \chi \qquad (8.1.1)$$

between the Higgs field ϕ of the string and a scalar field χ. The field ϕ vanishes on the string axis and $|\phi| \simeq \eta$ outside the string. Using the notation $\tilde\phi = \eta - |\phi|$, the interaction Lagrangian can be written as $\mathcal{L}_{\text{int}} = \lambda_\chi \left(\tilde\phi^2 - 2\eta\tilde\phi \right) \chi^\dagger \chi$, with the remainder $\lambda_\chi \eta^2 \chi^\dagger \chi$ absorbed into the χ-particle mass term. For a straight string parallel to the z-axis,

$$\mathcal{L}_{\text{int}} = -\lambda_\chi \eta^2 \, f(r) \, \chi^\dagger \chi \,, \qquad (8.1.2)$$

where the function $f(r)$ falls off rapidly for $r \gg \delta$ and $f(0) \sim 1$. Assuming that the string thickness is much smaller than all other length-scales in the problem, we can approximate $f(r)$ by a δ-function:

$$\mathcal{L}_{\text{int}} = -\kappa \, \delta^{(2)}(\mathbf{x}) \, \chi^\dagger \chi \,, \qquad (8.1.3)$$

where $\delta^{(2)}(\mathbf{x}) = \delta(x)\delta(y)$ and

$$\kappa = \lambda_\chi \eta^2 \int f d^2 x \sim \lambda_\chi \eta^2 \delta^2 \,. \tag{8.1.4}$$

The string thickness is typically $\delta \sim \eta^{-1}$ (unless the Higgs self-coupling is very small), and so $\kappa \sim \lambda_\chi$.

To find the scattering cross-section, we have to solve the equation

$$\left[\square + m_\chi^2 + \kappa \delta^{(2)}(\mathbf{x}) \right] \chi = 0 \,, \tag{8.1.5}$$

where m_χ is the χ-particle mass. Representing χ as

$$\chi = e^{-i\omega t + ikz} A(\mathbf{x}) \tag{8.1.6}$$

we obtain an equation for $A(\mathbf{x})$:

$$\left(\nabla^2 + q^2 \right) A(\mathbf{x}) = \kappa A(0) \delta^{(2)}(\mathbf{x}) \,. \tag{8.1.7}$$

Here, $\mathbf{x}=(x,y)$, ∇^2 is the two-dimensional Laplace operator, and $q^2 = \omega^2 - k^2 - m_\chi^2$. The scattering solution for (8.1.7) is

$$A(\mathbf{x}) = e^{i\mathbf{q}\cdot\mathbf{x}} - \kappa A(0) G(\mathbf{x}) \,, \tag{8.1.8}$$

where $\mathbf{q}$ is the wave vector of the incident wave and the two-dimensional Green's function is given by

$$G(\mathbf{x}) = \int \frac{d^2 p}{(2\pi)^2} \frac{e^{i\mathbf{p}\cdot\mathbf{x}}}{\mathbf{p}^2 - q^2 - i\epsilon} \,. \tag{8.1.9}$$

The asymptotic form of $G(\mathbf{x})$ for $|\mathbf{x}| \equiv r \gg q^{-1}$ is

$$G(\mathbf{x}) \longrightarrow \left(\frac{i}{8\pi q r} \right)^{1/2} e^{iqr} \,. \tag{8.1.10}$$

Equations (8.1.8) and (8.1.10) describe isotropic scattering with cross-section (per unit length of string)

$$\frac{d\sigma}{d\theta} = \frac{\kappa^2 |A(0)|^2}{8\pi q} \,. \tag{8.1.11}$$

To determine $A(0)$, we set $\mathbf{x} = 0$ in (8.1.8), which yields

$$A(0) = [1 + \kappa G(0)]^{-1} \,. \tag{8.1.12}$$

Here, one has to proceed with care, since $G(0)$ is logarithmically divergent: $G(0) = (2\pi)^{-1} \ln(\Lambda/q)$, where Λ is an appropriate cut-off. We expect the cut-off to be due to the finite string thickness, so that $\Lambda \sim \delta^{-1}$ and

$$A(0) = \left[1 - \frac{\kappa}{2\pi} \ln(q\delta) \right]^{-1} \,. \tag{8.1.13}$$

In most cases of cosmological interest, the particle wavelength, $\lambda = 2\pi/q$, is much greater than δ and the logarithm in (8.1.13) is large. If the logarithmic term in $A(0)$ dominates, the cross-section (8.1.11) takes the form

$$\frac{d\sigma}{d\theta} = \frac{\pi}{2q\left[\ln(q\delta)\right]^2} \, . \qquad (8.1.14)$$

Note that in this limit the cross-section is independent of the coupling κ.

Everett [1981] studied the scattering of particles belonging to a scalar multiplet ψ_a with mass matrix $M^2_{ab}(\theta)$ which changes around the string. Assuming the heavy fields of the multiplet to have masses much greater than q, he found that light members of the multiplet have a scattering cross-section given by (8.1.11) and (8.1.13) with $\kappa \sim 1$. (The large value for κ is not surprising since the model has no small parameters.) In this case, the asymptotic formula (8.1.14) is valid for all wavelengths $\lambda \gg \delta$. The scattering of charged fermions coupled to the gauge magnetic flux trapped in the string core has a similar cross-section [Perkins et al., 1991]. Here, again, the effective coupling is $\mathcal{O}(1)$, since the fermion charge e is multiplied by the gauge field $A_\mu \propto e^{-1}$.

A different and potentially more important effect is the Aharonov–Bohm-type interaction of charged particles with the pure gauge field outside the string [Rohm, 1985; de Sousa Gerbert & Jackiw, 1988; Alford & Wilczek, 1989]. The phase change experienced by a particle as it is transported around the string is

$$2\pi\nu = e\Phi \, , \qquad (8.1.15)$$

where e is the particle charge relative to the gauge field of the string and Φ is the string magnetic flux. The Aharonov–Bohm scattering is present when ν has a non-integer value. The effect is essentially the same as the scattering of electrons off a thin solenoid. The corresponding cross-section is

$$\frac{d\sigma}{d\theta} = \frac{\sin^2(\pi\nu)}{2\pi q \sin^2(\theta/2)} \, , \qquad (8.1.16)$$

where, as before, q is the transverse momentum of the particle. Alford & Wilczek [1989] gave an example of a realistic $SO(10)$ grand unified theory in which light fermions have non-integer values of ν (for example, $\nu = 1/4$ for electrons and $\nu = 1/2$ for electron neutrinos).

The total cross-section obtained by integrating (8.1.16) over θ is divergent, but this does not cause any serious difficulties because the divergence comes from very small scattering angles corresponding to a negligible momentum transfer. We will see in the next section that the interaction force between particles and the string is determined by the quantity

$$\sigma_t = \int_0^{2\pi} d\theta \, \frac{d\sigma}{d\theta}(1 - \cos\theta) = 2q^{-1}\sin^2(\pi\nu), \qquad (8.1.17)$$

which is called the transport cross-section. Note that this cross-section is greater than the corresponding quantity for (8.1.14) by the extra factor $2\pi^{-2}\sin^2(\pi\nu)\ln^2(q\delta)$, which can be quite large.

Baryon-number-violating scattering

For grand unified strings, the heavy gauge field trapped in the string core can have baryon-number-violating couplings. Light fermions scattering off the string S, therefore, can change their baryon number. For example,

$$Q + S \rightarrow \ell + S, \qquad (8.1.18)$$

where Q is a quark and ℓ is a lepton. Baryon-number-violating processes can also be mediated by a scalar field φ which has a Yukawa coupling to quarks and leptons, for example,

$$\varphi\bar{Q}\ell + \varphi^\dagger\bar{\ell}Q. \qquad (8.1.19)$$

If such a field develops a condensate in the string core, it can cause baryon-number-violating scattering of the form (8.1.18). (Note that the field φ in (8.1.19) must be charged, and thus a string with a φ-condensate will be superconducting.) The cross-sections (8.1.14) and (8.1.17), which were calculated for elastic scattering, are not directly applicable to an inelastic process like (8.1.18). An initial analysis by Brandenberger, Davis & Matheson [1988b] indicated that the baryon-number-violating cross-section is much smaller than (8.1.17). However, Alford, March-Russell & Wilczek [1989] have shown that in certain cases the cross-section can be strongly amplified. In particular, for the model (8.1.19) they found

$$\sigma \sim q^{-1}(q\delta)^\beta, \qquad (8.1.20)$$

where the power $\beta \geq 0$ depends on the parameter ν for the incident particle. For half-integer values of ν, $\beta = 0$ and σ is comparable to (8.1.17). Perkins *et al.* [1991] have shown that a similar amplification can occur in models where baryon non-conservation is mediated by the gauge field of the string, but in this instance the maximum cross-section is obtained for integer values of ν. Fewster & Kay [1993] have examined the model-dependence of baryon decay by cosmic strings using a scattering length formalism which they have developed. They find enhancement processes which depend very sensitively on the internal structure of the string, interpreting this effect as a resonance phenomenon.

8.1.2 Frictional force

Particle scattering results in a frictional force acting on a string moving through the plasma of the early universe. In a plasma of temperature T, the typical particle momentum is $q \sim T$ and the cross-section (8.1.17) is $\sigma_t \sim T^{-1}$. The force per unit length can be estimated as

$$\mathbf{F} \sim n\sigma_t v_T \Delta \mathbf{p} \,, \tag{8.1.21}$$

where $n \sim T^3$ is the particle density (refer to §1.2), $v_T \sim 1$ is the thermal velocity of the particles and $\Delta \mathbf{p}$ is the average momentum transfer per collision. Here, we have assumed for simplicity that the particle masses are negligible (that is, $T \gg m$) and that the string motion is non-relativistic ($v \ll 1$, where $\mathbf{v}$ is the string velocity). In this case, $\Delta \mathbf{p}$ is proportional to $-\mathbf{v}$, and on dimensional grounds we can write $\Delta \mathbf{p} \sim -T\mathbf{v}$, so that (8.1.21) gives [Everett, 1981]

$$\mathbf{F} \sim -T^3 \mathbf{v} \,. \tag{8.1.22}$$

To go beyond this order-of-magnitude estimate, we consider a straight string moving at speed $\mathbf{v}$ through a gas of massless particles. It is convenient to do the calculation in the rest frame of the string, where the phase space distribution function of the particles is given by

$$n_k = n\left[\gamma \left(k + \mathbf{k} \cdot \mathbf{v}\right)/T\right] \,. \tag{8.1.23}$$

Here, $\mathbf{k}$ is the particle momentum, $k = |\mathbf{k}|$, $\gamma = \left(1 - v^2\right)^{-1/2}$, T is the temperature (in the rest frame of the gas) and $n(\epsilon/T)$ is a Fermi or Bose distribution function. The force per unit length on the string can be written as

$$\mathbf{F} = \int \frac{d^3k}{(2\pi)^3} n_k \cdot \frac{q}{k} \cdot \mathbf{k} \int_0^{2\pi} d\theta \, \frac{d\sigma}{d\theta} (1 - \cos\theta) \,. \tag{8.1.24}$$

Here, q is the particle momentum in the plane perpendicular to the string, $(q/k)n_k(d^3k/(2\pi)^3)$ is the flux of particles with momenta in the interval d^3k, and $\mathbf{k}(1 - \cos\theta)$ is the momentum transferred to the string by a particle scattered by an angle θ. The integral over θ in (8.1.24) is the transport cross-section calculated in the previous section. Using (8.1.17) we can write

$$F = 2\sin^2(\pi\nu) \int_0^\infty \frac{k^2 dk}{(2\pi)^3} \int_0^\pi d\xi \sin\xi \cos\xi \, n\left(\frac{\gamma k(1 + v\cos\xi)}{T}\right) \,, \tag{8.1.25}$$

where ξ is the angle between $\mathbf{k}$ and $\mathbf{v}$. The integration over k is easily performed using the relation

$$n_0 = \int \frac{d^3k}{(2\pi)^3} n\left(\frac{k}{T}\right) = b\pi^{-2}\zeta(3)T^3 \,, \tag{8.1.26}$$

where n_0 is the number density of particles in the rest frame of the gas, $b = 3/4$ for fermions and $b = 1$ for bosons. The remaining angular integration is also straightforward, with the result

$$\mathbf{F} = -2\sin^2(\pi\nu)n_0\gamma\mathbf{v}\,. \tag{8.1.27}$$

Adding up the contributions of different particles, we obtain

$$\mathbf{F} = -\beta T^3\gamma\mathbf{v}\,, \tag{8.1.28}$$

where

$$\beta = 2\pi^{-2}\zeta(3)\sum_a b_a \sin^2(\pi\nu_a) \tag{8.1.29}$$

with the summation taken over the spin states of light particles ($m \ll T$).

8.1.3 String dynamics with friction

Let us now see how the frictional force (8.1.28) modifies the string equations of motion [Vilenkin, 1991]. We shall begin with the strings in a flat spacetime. In the notation of chapter 6, the general form of the string equations of motion is

$$\mu x_{,a}^{\nu\;;a} = F^\nu\,, \tag{8.1.30}$$

where F^ν is a four-vector representing the frictional force and μ is the string mass per unit length. F^ν should be constructed in terms of $x_{,a}^\nu$, the temperature T and the four-velocity of the background radiation u^ν. In the local rest frame of the string, the spatial components of F^ν must then reduce to (8.1.28),

$$F^i = \beta T^3 u^i\,. \tag{8.1.31}$$

Also, the time-component of F^ν, which represents the rate of energy change in this frame, should be equal to zero,

$$F^0 = 0\,. \tag{8.1.32}$$

A covariant expression for F^ν with these properties is not difficult to find,

$$F^\nu = \beta T^3(u^\nu - x_{,a}^\nu x^{\sigma,a} u_\sigma)\,. \tag{8.1.33}$$

Substituting (8.1.33) into (8.1.30), we obtain the final form of the string equation of motion,

$$x_{,a}^{\nu\;;a} = \left(\frac{\beta T^3}{\mu}\right)(u^\nu - x_{,a}^\nu x^{\sigma,a} u_\sigma)\,. \tag{8.1.34}$$

A generalization of (8.1.34) to an arbitrary curved spacetime is

$$x^\nu_{,a}{}^{;a} + \Gamma^\nu_{\sigma\tau} x^\sigma_{,a} x^{\tau,a} = \left(\frac{\beta T^3}{\mu}\right)(u^\nu - x^\nu_{,a} x^{\sigma,a} u_\sigma),\qquad (8.1.35)$$

with the temperature T measured in the rest frame of the gas. We are particularly interested in the case of an FRW universe,

$$ds^2 = a^2(\tau)(d\tau^2 - d^2\mathbf{x}),\qquad (8.1.36)$$

with the four-velocity of radiation given by

$$u^\nu = (a^{-1}, 0, 0, 0).\qquad (8.1.37)$$

The string equations of motion (8.1.35) simplify considerably if we choose the gauge in which

$$\zeta^0 = \tau,\qquad \dot{\mathbf{x}}\cdot\mathbf{x}' = 0.\qquad (8.1.38)$$

Given this choice, the spatial components of (8.1.35) take the form

$$\ddot{\mathbf{x}} - \epsilon^{-1}\left(\frac{\mathbf{x}'}{\epsilon}\right)' + \left(2\frac{\dot{a}}{a} + \frac{\beta T^3}{\mu}a\right)(1 - \dot{\mathbf{x}}^2)\dot{\mathbf{x}} = 0,\qquad (8.1.39)$$

and the time component becomes

$$\dot{\epsilon} + \left(2\frac{\dot{a}}{a} + \frac{\beta T^3}{\mu}a\right)\dot{\mathbf{x}}^2\epsilon = 0,\qquad (8.1.40)$$

where

$$\epsilon = \left(\frac{\mathbf{x}'^2}{1 - \dot{\mathbf{x}}^2}\right)^{1/2}.\qquad (8.1.41)$$

Comparing (8.1.39–41) with the free string equations of motion in an expanding universe (6.3.15–17), we see that, somewhat surprisingly, the only difference introduced by the frictional force is the replacement of the Hubble damping factor $2(\dot{a}/a)$ by

$$\left(2\frac{\dot{a}}{a} + \frac{\beta T^3}{\mu}a\right).\qquad (8.1.42)$$

The importance of the frictional force is determined by the relative magnitude of the second term in (8.1.42). Disregarding numerical factors of order unity and using the conformal time Friedmann equation

$$\left(\frac{\dot{a}}{a^2}\right)^2 = \frac{8\pi G}{3}\rho \sim GT^4,\qquad (8.1.43)$$

we find that the friction becomes negligible when

$$T \ll G\mu\, m_{\mathrm{pl}}.\qquad (8.1.44)$$

8.2 Particle production

In this section we estimate rates of non-gravitational radiation by strings [Srednicki & Thiesen, 1987]. We shall consider an example with scalar particles described by a field χ coupled to the Higgs field ϕ of a string as before by

$$\mathcal{L}_{\text{int}} = -\lambda_\chi \phi^\dagger \phi \chi^\dagger \chi. \tag{8.2.1}$$

If the motion of a string loop involves harmonics of sufficiently high frequency (greater than the χ-particle mass), then there is a non-zero probability for particle production. This process is similar to pair production in a time-varying electromagnetic field.

We found in §8.1.1 that the effective interaction of χ-particles with the string is described by the Lagrangian (8.1.3). This Lagrangian is easily generalized to the case of a moving string:

$$\mathcal{L}_{\text{int}} = -\kappa \int d^2\zeta \sqrt{-\gamma}\, \delta^{(4)}\left(x^\mu - x^\mu(\zeta)\right) \chi^\dagger \chi. \tag{8.2.2}$$

Here, the notation is the same as in §6.1. The factor $(-\gamma)^{1/2}$ is introduced to make sure that the Lagrangian is invariant under reparametrization of the string worldsheet. In the conformal gauge (6.2.3) and with $\zeta^0 = t$, (8.2.2) takes the form

$$\mathcal{L}_{\text{int}} = -\kappa \int_0^L d\zeta\, \left|\mathbf{x}'(\zeta,t)\right|^2 \delta^{(3)}\left(\mathbf{x} - \mathbf{x}(\zeta,t)\right) \chi^\dagger \chi. \tag{8.2.3}$$

The amplitude for the production of a pair of χ-particles with momenta $\mathbf{k}_1$ and $\mathbf{k}_2$, to lowest order in κ, is

$$\begin{aligned}
\mathcal{A} &= i \int d^4x\, \langle \mathbf{k}_1, \mathbf{k}_2 | \mathcal{L}_{\text{int}}(x) | 0 \rangle \\
&= -\kappa \int_{-\infty}^{\infty} dt\, e^{i\omega t} \int_0^L d\zeta\, \left|\mathbf{x}'(\zeta,t)\right|^2 e^{-i\mathbf{k}\cdot\mathbf{x}(\zeta,t)},
\end{aligned} \tag{8.2.4}$$

where $\mathbf{k} = \mathbf{k}_1 + \mathbf{k}_2$, $\omega = \omega_1 + \omega_2$, $\omega_a = \left(\mathbf{k}_a^2 + m_\chi^2\right)^{1/2}$ and m_χ is the χ-particle mass. Using the periodicity of the loop motion, we can write

$$\int_0^L d\zeta\, \left|\mathbf{x}'(\zeta,t)\right|^2 e^{-i\mathbf{k}\cdot\mathbf{x}(\zeta,t)} = \sum_n a_n(\mathbf{k})\, e^{-i\omega_n t}, \tag{8.2.5}$$

where $\omega_n = 4\pi n/L$ and

$$a_n(\mathbf{k}) = \frac{2}{L} \int_0^{L/2} dt\, e^{i\omega_n t} \int_0^L d\zeta\, \left|\mathbf{x}'(\zeta,t)\right|^2 e^{-i\mathbf{k}\cdot\mathbf{x}(\zeta,t)}. \tag{8.2.6}$$

Substitution of (8.2.5) into (8.2.4) gives

$$A = -2\pi i \kappa \sum_n a_n(\mathbf{k})\, \delta(\omega - \omega_n). \qquad (8.2.7)$$

The total energy radiated by the loop is given by

$$PT = \int \widetilde{dk_1}\widetilde{dk_2}\, \omega |A|^2, \qquad (8.2.8)$$

where $\widetilde{dk_a} = d^3 k_a/(2\pi)^3 2\omega_a$, $T = 2\pi\delta(0)$ and $P = \dot{E}$ is the radiation power. Using (8.2.7), we can rewrite (8.2.8) as

$$P = \sum_n P_n, \qquad (8.2.9)$$

$$P_n = 4\pi\kappa^2 \int \widetilde{dk_1}\widetilde{dk_2}\, \omega\, |a_n|^2\, \delta(\omega - \omega_n). \qquad (8.2.10)$$

Since $\omega \geq 2m_\chi$, P_n is non-zero only for $n > m_\chi L/2\pi$. For macroscopic loops this means that only high frequency harmonics with $n \gg 1$ contribute to the radiation. For cuspless loops such harmonics are strongly suppressed, and particle production is negligible. For a loop with cusps, the contribution of the near-cusp region can be estimated using the expansion (6.2.32).

We shall assume that $m_\chi \ll \eta$ and estimate P_n for $\omega_n \gg m_\chi$. (We will see that the main contribution to P comes from the largest allowed values of n.) The analysis is very similar to that in §7.4.2 for gravitational radiation. Once again we choose the origin of $\mathbf{x}, t$ and ζ so that $\mathbf{x} = t = \zeta = 0$ at the cusp. It is easily shown then that the main contribution to $a_n(\mathbf{k})$ comes from the region of integration with $|\zeta|, |t| \lesssim n^{-1/3}L$. Also, $a_n(\mathbf{k})$ is substantially different from zero only if the angle between $\mathbf{k}$ and the direction of the cusp velocity is $\theta \lesssim n^{-1/3}$. This leads to the following estimates for a_n and P_n:

$$a_n \sim n^{-4/3}L, \qquad (8.2.11)$$

$$P_n \sim \kappa^2 \omega_n^4 \theta^4 |a_n|^2 \sim \kappa^2 L^{-2}. \qquad (8.2.12)$$

Since P_n in (8.2.12) is independent of n, the sum over n in (8.2.9) is divergent. However, our formalism treating strings as infinitely thin is inadequate for describing the emission of particles with $\omega \gtrsim \delta^{-1}$. Consequently, we must introduce a small-scale cut-off and terminate the sum at $n \sim L/\delta$, yielding

$$P \sim \lambda_\chi^2 \delta^3 \eta^4 L^{-1}. \qquad (8.2.13)$$

Unless the Higgs self-coupling is very small, we have $\delta \sim \eta^{-1}$ and so $P \sim \lambda_\chi^2 \eta L^{-1}$. The energy emitted during a single occurrence of a cusp is then $\Delta E \sim \lambda_\chi^2 \eta$. For loops with $L \gg \lambda_\chi^2 m_{pl}^2/\eta^3$ the power (8.2.13) is much smaller than that of gravitational radiation.

Brandenberger [1987a] has proposed an entirely different mechanism of particle production at cusps. He notes that the distance between the two branches of the string in near-cusp regions can be smaller than the string thickness δ and he suggests that the corresponding portions of antiparallel string will annihilate and disintegrate into particles. Near the cusp, the string has a seagull shape which, with an appropriate choice of axes, is described by

$$x \sim \zeta^3/L^2, \quad y \sim \zeta^2/L, \qquad z = 0. \tag{8.2.14}$$

For $\zeta \lesssim (L^2\delta)^{1/3}$ the distance between the 'strings' of the seagull is $|x| \lesssim \delta$. If the corresponding portion of the string entirely disintegrates into particles, the emitted energy is $\Delta E \sim \mu\zeta \sim \mu (L^2\delta)^{1/3}$. This obviously gives only an upper bound on the energy rate of particle production,

$$P \lesssim \mu \left(\frac{\delta}{L}\right)^{1/3}. \tag{8.2.15}$$

The actual rate can be determined by numerically solving the field equations describing the string near a cusp.

Production of high-energy cosmic rays due to this 'cusp evaporation' process has been discussed by MacGibbon & Brandenberger [1990] and by Bhattacharjee [1989]. They conclude that, even with the maximum production rate allowed by (8.2.15), the resulting flux is below present observational limits.

8.3 Global string dynamics and radiation

8.3.1 An effective action

Global strings appear when the breaking of a global symmetry gives rise to a vacuum manifold which is not simply connected. The simplest are the global $U(1)$-strings of the Goldstone model (2.1.1), which were introduced in §3.2.2. Their detailed structure and nonlinear dynamics were discussed in §4.1.3 and §4.3 from a field-theoretic viewpoint. The long-range interactions of global strings, which are due to their coupling to a massless Goldstone field, cause their dynamics to be substantially different from those of local gauge strings. Here, we wish to quantify the implications of this coupling, particularly in describing the radiative decay of closed loops. This problem becomes tractable if we employ the low-energy effective action for a global string, the Kalb–Ramond action, which was derived in §4.4.2. It is illuminating, therefore, to recall the origin of this action.

In the low-energy limit, the complex scalar field of the Goldstone model can be represented as $\phi = \eta \exp[i\vartheta(x)]$. The effective Lagrangian for (2.1.1) then simply describes a massless Goldstone field ϑ,

$$\mathcal{L} = \eta^2 \partial_\mu \vartheta \, \partial^\mu \vartheta \,. \tag{8.3.1}$$

This holds even in the presence of a global string at large distances from the core where there are no massive excitations. As we discussed in §4.4.2, there is an equivalence between a massless scalar field ϑ and an antisymmetric tensor field $B_{\mu\nu} = -B_{\nu\mu}$, that is, we can make the replacement (4.4.7),

$$\eta \partial_\mu \vartheta = \tfrac{1}{2} \epsilon_{\mu\nu\lambda\rho} \partial^\nu B^{\lambda\rho} \,. \tag{8.3.2}$$

In this tensor representation, the peculiar coupling between the Goldstone boson field and string – ϑ changes by 2π around the string – becomes a pointlike interaction between the string and $B_{\mu\nu}$ which is closely analogous to electromagnetism. Consequently, most of our analysis will be based on the corresponding action (4.4.23),

$$S = -\mu_0 \int \sqrt{-\gamma}\, d^2\zeta + \frac{1}{6} \int d^4x \sqrt{-g} H^2 + 2\pi\eta \int B_{\mu\nu} d\sigma^{\mu\nu} \,. \tag{8.3.3}$$

where the field strength tensor is

$$H_{\mu\nu\sigma} = \partial_\mu B_{\nu\sigma} + \partial_\nu B_{\sigma\mu} + \partial_\sigma B_{\mu\nu} \,, \tag{8.3.4}$$

and the surface element on the string worldsheet is given by

$$d\sigma^{\mu\nu} = \left(\dot{x}^\mu x'^\nu - \dot{x}^\nu x'^\mu \right) d^2\zeta \,. \tag{8.3.5}$$

The first term is the Nambu action (4.4.6) for the string core with bare mass per unit length $\mu_0 \sim \eta^2$, while the second and third terms effectively replace (8.3.1). The energy–momentum tensor $T^{\mu\nu}$ of the string and antisymmetric tensor fields can be found by varying the action (8.3.3) with respect to the metric $g^{\mu\nu}$:

$$\begin{aligned} T^{\mu\nu} = \frac{\mu_0}{\sqrt{-g}} &\int d^2\zeta \sqrt{-\gamma}\, \gamma^{ab}\, x^\mu_{,a} x^\nu_{,b}\, \delta^{(4)} \left[x - x(\zeta) \right] \\ &+ H^\mu_{\lambda\sigma} H^{\mu\lambda\sigma} - \tfrac{1}{6} g^{\mu\nu} H^2 \,. \end{aligned} \tag{8.3.6}$$

8.3.2 Global string dynamics

The string equations of motion in a general spacetime are found by varying the action (8.3.3) with respect to x^μ,

$$\mu_0 \left(x^\mu_{,a}{}^{;a} + \Gamma^\mu_{\nu\sigma} \gamma^{ab} x^\nu_{,a} x^\sigma_{,b} \right) = 2\pi\eta H^\mu{}_{\nu\sigma} \epsilon^{ab} x^\nu_{,a} x^\sigma_{,b} \,. \tag{8.3.7}$$

Variation with respect to $B^{\mu\nu}$ yields the field equations,

$$\partial_\mu \left(\sqrt{-g}\, H^{\mu\nu\sigma}\right) = -4\pi\eta \int d^2\zeta \epsilon^{ab} x^\mu_{,a} x^\nu_{,b} \delta^{(4)}\left[x - x(\zeta)\right]. \qquad (8.3.8)$$

Now the action (8.3.3) is invariant under arbitrary reparametrization of the worldsheet and, up to a total derivative, under the gauge transformation,

$$B_{\mu\nu} \rightarrow B_{\mu\nu} + \partial_\mu \Lambda_\nu - \partial_\nu \Lambda_\nu. \qquad (8.3.9)$$

To remove these gauge freedoms, we impose the usual conformal gauge (6.2.4–5) for the worldsheet variables,

$$\dot{x} \cdot x' = 0, \qquad \dot{x}^2 + x'^2 = 0, \qquad (8.3.10)$$

and we impose a Lorentz gauge condition on the field $B^{\mu\nu}$,

$$\partial_\mu B^{\mu\nu} = 0. \qquad (8.3.11)$$

Choosing the gauge (8.3.10–11) in flat spacetime, the equations of motion (8.3.7–8) reduce to

$$\mu_0(\ddot{x}_\mu - x''_\mu) = 4\pi\eta H_{\mu\nu\sigma}\dot{x}^\nu x'^\sigma, \qquad (8.3.12)$$

$$\partial_\sigma \partial^\sigma B^{\mu\nu} = 4\pi j^{\mu\nu}, \qquad (8.3.13)$$

where the current density $j^{\mu\nu}$ is given by

$$j^{\mu\nu} = \frac{\eta}{2} \int \delta^4[x - x(\zeta)]d\sigma^{\mu\nu}. \qquad (8.3.14)$$

For a static straight string lying along the z-axis, we can take $\zeta^0 = t$ and $\zeta^1 = z$, so (8.3.13) reduces to

$$\frac{1}{r}\frac{\partial}{\partial r}\left(r\frac{\partial}{\partial r}B^{03}\right) = 2\pi\eta\delta^{(2)}(x, y), \qquad (8.3.15)$$

where $r^2 = x^2 + y^2$ and $\delta^{(2)}(x, y) = (2\pi r)^{-1}\delta(r)$ is a two-dimensional delta function. The straight string solution is then

$$x^\mu(t, z) = (t,\, 0,\, 0,\, z), \qquad (8.3.16)$$

$$B^{03} = \eta \ln\left(\frac{r}{\delta}\right), \qquad (8.3.17)$$

with all other components zero and the cut-off δ is provided by the string thickness.

The field (8.3.17) is the static Coulomb or self-field of the string. This gives rise to a logarithmically divergent self-energy which is analogous to the divergence in electrodynamics arising from the self-energy of the electron. It can be absorbed with a generalization of Dirac's prescription for renormalizing the bare electron mass. There is a significant difference, however, because the logarithmic divergence must be cut off at both large and small scales. This problem has been considered by Lund & Regge [1976] and more recently by Dabholkar & Quashnock [1990] and Copeland, Haws & Hindmarsh [1990].

The right-hand side of the string equation of motion (8.3.12) includes both the external field and the self-field of the string. To calculate and absorb the divergent self-force of the string, one expands the retarded solution of (8.3.13) for $B^{\mu\nu}$ in the vicinity of the string worldsheet. As for string self-gravity discussed in §7.7.1, the infinite self-force takes the form,

$$F^\alpha_{\text{inf}} = -\delta\mu(\ddot{x}^\alpha - x''^\alpha) , \qquad (8.3.18)$$

where $\delta\mu$ is a logarithmically divergent integral. This is naturally cut off by the curvature radius R of the string at large scales and by the string thickness δ at small scales:

$$\delta\mu \approx 2\pi\eta^2 \ln\left(\frac{R}{\delta}\right) , \qquad (8.3.19)$$

The right-hand side of (8.3.18) can now be transferred across in (8.3.12) to yield

$$\mu(\ddot{x}^\nu - x''^\nu) = F^\nu_{\text{ren}} , \qquad (8.3.20)$$

where $\mu = \mu_0 + \delta\mu$ and F^μ_{ren} includes the finite effects of radiation back-reaction and any external fields. It would be superfluous to replicate these renormalization arguments for gravitational, antisymmetric and electromagnetic fields, so an explicit example is left until §8.4.3 for the simplest electromagnetic case. Here, however, it is important to note that, for cosmological length scales, the bare mass density μ_0 is always very small relative to the string energy residing in the Goldstone field.

Significant deviations from Nambu string behaviour are determined by the relative magnitude of the tension $\mathbf{F_t} = \mu\mathbf{x}''$ and the long-range self-interaction force $\mathbf{F_{\text{ren}}}$. For a loop of characteristic radius R, F_{ren} can be roughly estimated from (8.3.12) and (8.3.17): $F_{\text{ren}} \sim 4\pi\eta^2/R$. On the other hand,

$$F_t \sim \frac{\mu}{R} \sim \frac{2\pi\eta^2}{R} \ln\left(\frac{R}{\delta}\right) , \qquad (8.3.21)$$

and thus corrections to the motion of macroscopic loops are suppressed by a large logarithmic factor, $[\ln(R/\delta)]^{-1} \ll 1$.

Previously, we had alluded to the close connection between global strings and superfluid vortices. While the string configurations are very similar, there is, of course, an essential difference between the backgrounds in which they are immersed. We have discussed the global string

(a)   (b)

Fig. 8.1. (*a*) Collapse of a global string loop in vacuum, showing a constant contour surface for $|\phi|$. Note the Lorentz contraction during the relativistic collapse and the massive radiation after annihilation. (*b*) Vortex-rings are stable in the relativistic superfluid background (8.3.22). Illustrated is an oscillatory solution with two 'leapfrogging' rings [Shellard, 1992].

in the context of a Lorentz-invariant vacuum state. In contrast, the superfluid vortex moves through a background fluid with a non-zero energy density ρ. A global string will behave like a superfluid vortex if we introduce the homogeneous background [Davis & Shellard, 1989b]

$$H^{ijk} = \sqrt{\rho}\,\epsilon^{ijk}, \qquad i,j,k = 1,2,3, \tag{8.3.22}$$

which is equivalent to giving the Goldstone boson field ϑ a linear time dependence,

$$\vartheta = \omega t = \frac{\sqrt{\rho}}{\eta}t. \tag{8.3.23}$$

The originally static global string configuration (4.1.11) now 'spins' and there is a circular flow of momentum about the string. There is an interaction force between this external background (8.3.22) and a string moving with a velocity $\mathbf{v}$, given by integrating over the force density $f^\sigma = j_{\mu\nu}H^{\mu\nu\sigma}$. In the rest frame of the fluid, this can be shown to be [Davis & Shellard, 1989b; Gradwohl *et al.*, 1990]

$$\mathbf{F} = \gamma^2 \rho \mathbf{v} \times \mathbf{m}, \tag{8.3.24}$$

where the circulation vector $\mathbf{m}$ lies in the direction of the string and has a magnitude $|\mathbf{m}| = 4\pi\eta/\sqrt{\rho}$. Equation (8.3.24) is merely a relativistic version of the well-known Magnus force law, which implies that a moving vortex will experience a force acting at right angles, much like an electron in a magnetic field. One implication is the existence of stable vortex-ring solutions in which the tension and self-interaction forces are exactly balanced by the Magnus force (8.3.24) acting on the string due to its motion through the fluid. In the non-relativistic limit, a comparison of (8.3.21) and (8.3.24) gives the velocity–radius relationship for vortex-rings known from hydrodynamics,

$$v \approx \frac{\eta}{2\sqrt{\rho}R}\ln\left(\frac{R}{\delta}\right). \tag{8.3.25}$$

The relation (8.3.25) has been verified numerically in this modified Goldstone model, along with other predicted superfluid phenomena [Shellard & Davis, 1989; Shellard, 1992] (see fig. 8.1(*b*)). Other aspects of these relativistic superfluids have been studied by Davis [1989, 1990] and Gradwohl *et al.* [1990]. In a cosmological context, the energy density of a homogeneous background like (8.3.22) would be rapidly diluted by the expansion of the universe, so these superfluid effects are not expected to be of great significance.

8.3.3 Goldstone boson radiation

An accelerating global string will radiate Goldstone bosons, a process which will have a significant back-reaction on its motion. An order-of-magnitude estimate by Davis [1985a] suggested that global string loops on cosmological length scales would oscillate only ten times before losing most of their energy in radiation. Using the analogy of an oscillating point charge in electrodynamics, it was suggested that the wavelengths of the emitted radiation would closely correspond to the Fourier modes of the string trajectory. By employing the antisymmetric tensor formalism, Vilenkin & Vachaspati [1987a] calculated the radiation from some explicit loop solutions and confirmed these expectations. This tensor approach is very similar to the gravitational radiation calculations described in §7.5, so we shall only briefly outline the main results.

Given a global string loop with an oscillation period T, the power radiated per unit solid angle at a frequency $\omega_n = 2\pi n/T$ in a direction $\mathbf{k}$, $|\mathbf{k}| = \omega_n$, is given by

$$\frac{dP_n}{d\Omega} = 2\omega_n^2 \, j_{\mu\nu}^*(\mathbf{k}, w_n) \, j^{\mu\nu}(\mathbf{k}, \omega_n) \,, \tag{8.3.26}$$

where the Fourier transform of the current density is

$$j^{\mu\nu}(\mathbf{k}, w_n) = \frac{1}{T} \int_0^T dt \, \exp(i w_n t) \int d^3x \, \exp(-i\mathbf{k} \cdot \mathbf{x}) \, j^{\mu\nu}(\mathbf{x}, t) \,. \tag{8.3.27}$$

By splitting the trajectory into right- and left-moving modes $\mathbf{a}$, $\mathbf{b}$, this can be rewritten in a particularly simple form [Garfinkle & Vachaspati, 1987]

$$\frac{dP_n}{d\Omega} = \frac{\eta^2 n^2}{4\pi^2} \, |I_n(\hat{\mathbf{n}}_1) J_n(\hat{\mathbf{n}}_2) - I_n(\hat{\mathbf{n}}_2) J_n(\hat{\mathbf{n}}_1)|^2 \,, \tag{8.3.28}$$

where $\mathbf{n}_1$, $\mathbf{n}_2$ are unit vectors orthogonal to $\mathbf{k}$ and, as before in (7.5.8),

$$I_n(\hat{\mathbf{n}}) = \frac{1}{L} \int_0^L d\zeta \, \mathbf{a}'(\zeta) \cdot \hat{\mathbf{n}} \, \exp\left[-\frac{i}{2}(\omega_n \zeta + \mathbf{k} \cdot \mathbf{a})\right] \,,$$
$$J_n(\hat{\mathbf{n}}) = \frac{1}{L} \int_0^L d\zeta \, \mathbf{b}'(\zeta) \cdot \hat{\mathbf{n}} \, \exp\left[\frac{i}{2}(\omega_n \zeta - \mathbf{k} \cdot \mathbf{b})\right] \,, \tag{8.3.29}$$

Note that these expressions do not include radiation back-reaction, but they will provide a good approximation if the energy loss per oscillation remains small.

Calculations for Burden loop trajectories (6.2.38) [Vilenkin & Vachaspati, 1987a] and for kinky, cuspless loops (6.2.39) [Garfinkle & Vachaspati, 1987] give results qualitatively similar to those for gravitational radiation. The total power for any loop is given by

$$P = \Gamma_a \eta^2 , \tag{8.3.30}$$

with the parameter Γ_a dependent on the shape of the loop trajectory but not its scale. Apart from special cases with high symmetry, typical loops had $\Gamma_a \sim 100$. Again, the radiation spectrum behaved aymptotically as $P_n \propto n^{-4/3}$ for loops possessing cusps, while kinky cuspless loops gave $P_n \propto n^{-2}$.

The lifetime of a loop can be found by comparing the loop energy, $E \sim 2\pi L \eta^2 \ln(L/\delta)$, to the power (8.3.30) which yields

$$\tau \sim \frac{E}{P} \approx \frac{2\pi}{\Gamma_a} L \ln\left(\frac{L}{\delta}\right) = KL . \tag{8.3.31}$$

For cosmological length scales, $\ln(L/\delta) \sim 100$ and $K \sim 10$, implying that a loop will oscillate about 20 times before radiating most of its energy. This should be contrasted with gravitational radiation where loops with realistic parameters will oscillate more than 10^4 times before their demise. Antisymmetric tensor radiation from wiggly long strings has been studied by Sakellariadou [1991].

The rapid decay of global string loops implies that radiative back-reaction is dynamically significant. Direct numerical simulations of the full field theory, as described in §4.3, automatically incorporate these effects. Results for several such simulations are shown in figs. 8.2–3, for a perturbed string lying along the z-axis [Davis & Shellard, 1989c]. Periodic boundary conditions were imposed on the faces of the cubic grid through which the string passed, effectively mimicking a sinusoidal standing wave

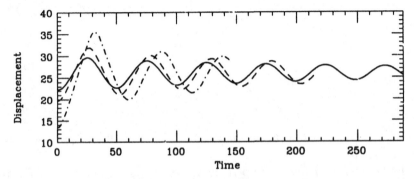

Fig. 8.2.   Damping of oscillations on a global string due to radiative back-reaction [Davis & Shellard, 1989c].

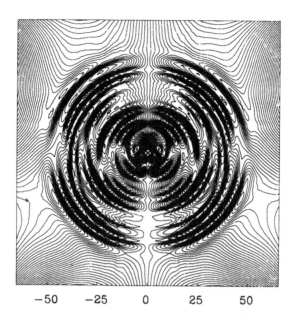

−50   −25   0   25   50

Fig. 8.3. Transverse radiation from an oscillating global string (marked with an '⊗') [Davis & Shellard, 1989c]. Contours of the Goldstone field are shown, after subtraction of the static field of the string.

on an infinite string. The other faces had absorbing boundary conditions, solving the out-going massless wave equation and thus preventing the reflection of string radiation. As shown in fig. 8.2, the damping of string oscillations is pronounced at large amplitude because the logarithm in (8.3.31) is small, $\ln(L/\delta) \lesssim 5$. However, this decay slows markedly at small amplitude.

The radiation pattern from an oscillating string is shown in fig. 8.3, after the Coulomb field of the string has been subtracted. Spectral analysis shows the predominance of the second (quadrupole) harmonic and the presence, but rapid fall-off, of higher harmonics. These numerical results have been replicated by Hagmann & Sikivie [1991], though there is an apparent disagreement over the shape of the emitted power spectrum.[*] However, their Fourier analysis of the Goldstone field ϑ was performed without any subtraction of the self-field of the string. The spectrum

---

[*] This recalls earlier controversial arguments by Harari & Sikivie [1987] suggesting that global string motion would be critically damped, giving rise to radiation with a characteristic flat $1/k$ power spectrum. However, the present authors see no substantive problems with the picture of string radiation developed above, though they are by no means impartial. These issues are addressed in considerable detail in Battye & Shellard [1993].

naturally appears flatter with the additional contribution of the string
(8.3.17).

The radiation back-reaction problem has also been studied using the
antisymmetric tensor formalism by Dabholkar & Quashnock [1990]. After
deriving the pertinent non-local back-reaction terms on the right-hand
side of (8.3.20), they were able to numerically calculate their implications
for a collapsing circular loop. The behaviour of the loop approached
the Nambu string trajectory more closely as the logarithm in (8.3.19)
increased, to the extent that the neglect of radiative effects only caused
a 1% error for cosmological strings with $\ln(R/\delta) \sim 100$.

8.4 String electrodynamics I: Formalism

8.4.1 Basic equations

In §5.1.6 we derived an effective action for a superconducting string assuming that all the relevant length-scales in the problem were much larger
than the string thickness δ. This action applies equally to both bosonic
and fermionic superconductivity, so it provides an appropriate starting
point for our analysis:

$$
S = \int d^2\zeta \sqrt{-\gamma} \left\{ -\mu + \tfrac{1}{2}\gamma^{ab}\phi_{,a}\phi_{,b} - A_\mu x^\mu_{,a} J^a \right\}
$$
$$
- \frac{1}{16\pi} \int d^4x \sqrt{-g}\, F_{\mu\nu}F^{\mu\nu} \, . \tag{8.4.1}
$$

The first term in parentheses gives the Nambu action (6.1.9), the second
term describes the inertia of the charge carriers, and the last term gives
the coupling of the current to the electromagnetic vector potential $A_\mu(x)$.
The worldsheet current J^a is given by

$$
J^a = q\tilde{\epsilon}^{ab}\phi_{,b} \, , \tag{8.4.2}
$$

where

$$
\tilde{\epsilon}^{ab} = \frac{1}{\sqrt{-\gamma}}\,\epsilon^{ab} \, . \tag{8.4.3}
$$

(Note that $\tilde{\epsilon}^{ab}$ is a tensor and so J^a transforms like a vector under worldsheet reparametrization.) This current is covariantly conserved,

$$
J^a_{;a} = \frac{1}{\sqrt{-\gamma}}\partial_a \left(\sqrt{-\gamma}\, J^a\right) = 0 \, . \tag{8.4.4}
$$

The last term in (8.4.1) is the usual electromagnetic action. Below we
shall assume for simplicity that the four-dimensional spacetime is flat,
that is, $g_{\mu\nu} = \eta_{\mu\nu}$. The electromagnetic coupling q is given by (5.1.40)
for bosonic superconductivity and (5.2.18) for fermionic. We saw in §5.1.7

that q gets renormalized by the self-inductance of the string (5.1.51) and that in both cases $q_{\text{ren}}^2 \sim 10^{-2}$ for realistic parameter values. Below, we shall take q to be the renormalized coupling, a choice which will be justified in §8.4.3.

The basic equations of string electrodynamics can be derived from the action (8.4.1) by varying it with respect to $A_\mu(x)$, $\phi(\zeta)$ and $x^\mu(\zeta)$. A variation with respect to A_μ gives

$$F^{\mu\nu}{}_{,\nu} = -4\pi j^\mu , \qquad (8.4.5)$$

where the four-dimensional current density is given by

$$j^\mu(x) = \int d^2\zeta \sqrt{-\gamma}\, J^a x^\mu_{,a}\, \delta^{(4)}\left(x - x(\zeta)\right) . \qquad (8.4.6)$$

The conservation of this current can be verified directly using (8.4.4).

The variation of (8.4.1) with respect to $\phi(\zeta)$ yields

$$\phi_{,a}{}^{;a} = -\tfrac{1}{2}q\tilde{\epsilon}^{ab}F_{ab} , \qquad (8.4.7)$$

where

$$F_{ab} = \partial_a A_b - \partial_b A_a = F_{\mu\nu}x^\mu_{,a}x^\nu_{,b} \qquad (8.4.8)$$

and

$$A_a = A_\mu x^\mu_{,a} . \qquad (8.4.9)$$

As we shall see, (8.4.7) is a relativistic generalization of (5.1.45) for current growth in an electric field.

The variation of the action (8.4.1) with respect to $x^\mu(\zeta)$ is conveniently performed using the metric definition,

$$\gamma_{ab} = x^\mu_{,a}x_{\mu,b} , \qquad (8.4.10)$$

along with the identities

$$d\gamma = \gamma\,\gamma^{ab}d\gamma_{ab} , \qquad (8.4.11)$$
$$d\gamma^{ab} = -\gamma^{ac}\gamma^{bd}d\gamma_{cd} . \qquad (8.4.12)$$

(Note that the last identity is obtained by differentiating $\gamma_{ab}\gamma^{bc} = \delta^c_a$.) The resulting equation of motion can be written in the form

$$\left[\left(\mu\gamma^{ab} + \theta^{ab}\right) x^\nu_{,a}\right]_{;b} = -F^\nu{}_\sigma x^\sigma_{,a} J^a , \qquad (8.4.13)$$

where θ_{ab} is the worldsheet energy–momentum tensor of the charge carriers,

$$\theta_{ab} = \phi_{,a}\phi_{,b} - \tfrac{1}{2}\gamma_{ab}\phi_{,c}\phi^{,c} . \qquad (8.4.14)$$

This tensor is traceless and it is covariantly conserved in the absence of an electromagnetic field,

$$\theta^b{}_{a;b} = \phi_{,b}{}^{;b}\phi_{,a} = -\tfrac{1}{2}q\tilde{\epsilon}^{bc}F_{bc}\phi_{,a}\,. \qquad (8.4.15)$$

The right-hand side of (8.4.13) is simply the Lorentz force acting on the string.

The four-dimensional energy–momentum tensor is found by varying (8.4.1) with respect to $g_{\mu\nu}(x)$. This gives

$$T^{\mu\nu}(x) = \int d\zeta \sqrt{-\gamma}\left(\mu\gamma^{ab} + \theta^{ab}\right) x^\mu_{,a}x^\nu_{,b}\,\delta^{(4)}\left[x - x(\zeta)\right] + T^{\mu\nu}_{(\mathrm{em})}(x)\,, \qquad (8.4.16)$$

where $T^{\mu\nu}_{(\mathrm{em})}$ is the usual energy–momentum tensor of the electromagnetic field.

8.4.2 Gauge conditions

The action (8.4.1) is invariant under arbitrary worldsheet reparametrizations and under electromagnetic gauge transformations. As for a free (non-superconducting) string, the most convenient choice of gauge on the worldsheet is the conformal gauge, which is specified by the conditions

$$\dot{x}^\mu x'_\mu = 0\,, \qquad \dot{x}^2 + x'^2 = 0\,. \qquad (8.4.17)$$

In this gauge $\gamma^{ab}\sqrt{-\gamma} = \eta^{ab}$ and (8.4.6–7) and (8.4.13) take the form

$$j^\mu(x) = q\epsilon^{ab}\int d^2\zeta\, x^\mu_{,a}\phi_{,b}\,\delta^{(4)}\left(x - x(\zeta)\right)\,, \qquad (8.4.18)$$

$$\Box_2\phi = -\tfrac{1}{2}q\epsilon^{ab}F_{ab}\,, \qquad (8.4.19)$$

where

$$\Box_2 = \eta^{ab}\partial_a\partial_b\,. \qquad (8.4.20)$$

We note that the metric tensor γ_{ab} has completely dropped out of the equation (8.4.19) for ϕ. This is not an accident; a massless scalar field in two dimensions is conformally invariant, so the field equation for ϕ in the conformal gauge is the same as in Minkowski space.

In the remainder of this chapter we shall be mainly interested in the case when the Lorentz force and the inertia of the charge carriers are both negligible compared to the string tension. The string equation of motion then reduces to that for a free string,

$$\Box_2 x^\mu = 0\,, \qquad (8.4.21)$$

and we can again identify the parameter ζ^0 with time t,

$$\zeta^0 = t\,. \qquad (8.4.22)$$

Note that, for the general case, this identification is inconsistent with the string equations of motion.

To fix the electromagnetic gauge, we impose the Lorentz gauge condition

$$\partial_\mu A^\mu = 0.$$

(8.4.23)

Maxwell's equations (8.4.5) then take the form

$$\Box_4 A^\mu = 4\pi j^\mu,$$

(8.4.24)

where the d'Alembertian is $\Box_4 = \eta^{\mu\nu}\partial_\mu\partial_\nu$.

8.4.3 Charge renormalization

For a static string in a constant electric field, we saw in §5.1.7 that the effect of self-inductance leads to a renormalization of the electromagnetic coupling q. Here we will show that the same renormalization of q applies for a moving string in an arbitrary electromagnetic field.

We begin with (8.4.19) which describes how the string current is affected by the electromagnetic field $F_{\mu\nu}$. This field includes both the external field and the field produced by the string itself. The solution of (8.4.24) for the field due to the string can be written as

$$A^\mu(x) = \int d^4x' \, G_{\text{ret}}(x - x') \, j^\mu(x')$$

(8.4.25)

where the retarded Green's function is

$$G_{\text{ret}}(x - x') = \frac{\delta\left(t - t' - |\mathbf{x} - \mathbf{x}'|\right)}{|\mathbf{x} - \mathbf{x}'|}.$$

(8.4.26)

To evaluate the right-hand side of (8.4.19), we need to know A^μ on the string worldsheet. When x is on the worldsheet, the integral in (8.4.25) is dominated by the vicinity of the point ζ such that $x(\zeta) = x$. Hence, using (8.4.18), we can write

$$A^\mu(x) \approx \frac{q\epsilon^{ab}}{\sqrt{-\gamma}} x^\mu_{,a} \phi_{,b} \int d^2\zeta \sqrt{-\gamma} \, G_{\text{ret}}\left(x - x(\zeta)\right).$$

(8.4.27)

This integral is Lorentz- and gauge-invariant, and we can evaluate it in the local rest frame of the string at the point x. Using the conformal gauge related to that frame, we have $\gamma = 1$, and the integral in (8.4.27) reduces to

$$I \approx \int \frac{d\zeta}{|\mathbf{x} - \mathbf{x}(\zeta)|}.$$

(8.4.28)

Near the point of interest, ζ is the length parameter along the string. The integral (8.4.28) is logarithmically divergent, but this divergence arises because we have approximated the string by a line of zero thickness. For a physical string, (8.4.28) is replaced by a large but finite integral

$$I \approx \ln\left(R/\delta\right) . \tag{8.4.29}$$

Here, δ is the string thickness and the upper integration cut-off is given by the curvature radius of the string or by the characteristic wavelength of the field ϕ describing the current. In situations of cosmological interest the logarithm in (8.4.29) is ~ 100 and can be treated approximately as a constant. Combining (8.4.27) with (8.4.29) and using the conformal gauge relation $\gamma_{ab} = \sqrt{-\gamma}\,\eta_{ab}$ and the identity $\eta_{ab}\epsilon^{ac}\epsilon^{bd} = -\eta^{cd}$, we obtain

$$\epsilon^{ab}F_{ab} = 2q\ln\left(R/\delta\right)\Box_2\phi. \tag{8.4.30}$$

With (8.4.30) taken into account we can rewrite (8.4.19) as

$$\Box_2\phi_{\text{ren}} = -\tfrac{1}{2}q_{\text{ren}}^2\epsilon^{ab}F_{ab}^{\text{ext}}, \tag{8.4.31}$$

where

$$q_{\text{ren}}^2 = \frac{q^2}{1+q^2\ln\left(R/\delta\right)}, \tag{8.4.32}$$

$$\phi_{\text{ren}} = \frac{q}{q_{\text{ren}}}\,\phi, \tag{8.4.33}$$

and the electromagnetic field F_{ab}^{ext} includes only the external field.

In the expression for the current (8.4.18), the field ϕ appears multiplied by q, and it is clear from (8.4.33) that (8.4.18) is left unchanged by the renormalization. Hence, the effect of the self-interaction can be taken into account if we replace q by q_{ren} and ϕ by ϕ_{ren} in (8.4.18–19) and interpret F_{ab} in (8.4.19) as the external field. The renormalization of q in (8.4.32) is identical to that due to the self-inductance which was derived in §5.1.7.

We emphasize that (8.4.27) which led us to (8.4.31) is an approximation including only the contribution to A_μ from the immediate vicinity of the point where the current is calculated. The non-local electromagnetic interaction between the currents in more distant parts of the string can be described by an integro-differential equation [Spergel, Press & Scherrer, 1989].

8.5 String electrodynamics II: Applications

8.5.1 Free superconducting strings

In the absence of external electromagnetic fields, the equation (8.4.19) for the scalar field $\phi(\zeta,t)$ reduces to a two-dimensional wave equation,

$$\Box_2\phi = 0. \tag{8.5.1}$$

For an oscillating loop of string the current $J^a = q\tilde{\epsilon}^{ab}\phi_{,b}$ should be a periodic function of the string parameter ζ,

$$J^a(\zeta + L, t) = J^a(\zeta, t), \tag{8.5.2}$$

where L is the invariant length of the loop. The general solution of (8.5.1) having this property is

$$\phi(\zeta, t) = q^{-1}(\lambda_0\zeta - J_0 t) + \phi_1(\zeta - t) + \phi_2(\zeta + t), \tag{8.5.3}$$

where ϕ_1 and ϕ_2 are arbitrary periodic functions. Here, λ_0 and J_0 are, respectively, a constant charge density and a constant current along the string, and the last two terms describe charge and current oscillations with zero mean. Note that for $\lambda_0 \neq 0$ the scalar field $\phi(\zeta, t)$ is not single-valued: it changes by $\lambda_0 L/q$ as we go around the loop. This does not cause any problems, since only the derivatives of ϕ are physically observable. It is clear from (8.5.3) that the worldsheet current J^a is periodic in time with the same period as the motion of the loop, $T = L/2$.

In the general case, the electromagnetic field produced by the string current is rather complicated. An interesting simple example in which the equations of string electrodynamics can be solved exactly is given by current–charge oscillations on a static straight string [Aryal, Vachaspati & Vilenkin, 1987]. Choosing the x-axis to lie along the string, we can set the string parameter ζ equal to x. The solution (8.4.25) for the field ϕ describing current and charge waves propagating along the string can be written as

$$\phi = \phi_1(x - t) + \phi_2(x + t). \tag{8.5.4}$$

The corresponding current density is

$$j^\mu = \delta(y)\delta(z)J^\mu(x, t), \tag{8.5.5}$$

where

$$J^\mu = q\left(\phi', -\dot{\phi}, 0, 0\right). \tag{8.5.6}$$

Now it is easily verified that

$$A^\mu = 2J^\mu(x, t)\ln r \tag{8.5.7}$$

is a solution of Maxwell's equations (8.4.24) with the current (8.5.5), where $r = (y^2 + z^2)^{1/2}$ is the distance from the string. The electric and magnetic fields of the string are given by

$$\begin{aligned} \mathbf{E} &= -2q\phi' r^{-2}(0, y, z), \\ \mathbf{B} &= 2q\dot{\phi}r^{-2}(0, z, -y). \end{aligned} \tag{8.5.8}$$

Clearly, there is no energy flux in directions perpendicular to the string, and thus the string does not emit electromagnetic waves. The back-reaction of the fields on the current also vanishes, since the right-hand side of (8.4.19) is $-qE_x = 0$. This means that current and charge oscillations on a straight string persist indefinitely without dissipation. (This conclusion is modified when pair production near the string is taken into account, see §8.5.5.)

8.5.2 Strings in an external field

We shall now consider the currents induced in strings by external electric and magnetic fields. The current can be found by solving (8.4.19), which can be written as

$$\Box_2 \phi = -q\mathbf{x}' \cdot (\mathbf{E} + \dot{\mathbf{x}} \times \mathbf{B}) . \tag{8.5.9}$$

For a static straight string in a uniform electric field,

$$\Box_2 \phi = -q\,\hat{\mathbf{n}} \cdot \mathbf{E} , \tag{8.5.10}$$

where $\hat{\mathbf{n}} = \mathbf{x}'$ is a unit vector along the string. If the current is uniform along the string, then $J = -q\dot{\phi}$, and we recover (5.1.47) for the current growth in an electric field,

$$\frac{dJ}{dt} = q^2\,\hat{\mathbf{n}} \cdot \mathbf{E} . \tag{8.5.11}$$

The general solution of (8.5.10), in addition to a linearly growing uniform current (8.5.11), includes charge and current oscillations as in (8.5.3).

For a closed loop, integration of (8.5.9) around the loop gives

$$\frac{d}{dt} \oint J \, d\zeta = q^2 \oint d\mathbf{x} \cdot (\mathbf{E} + \dot{\mathbf{x}} \times \mathbf{B}) = -q^2 \frac{d\Phi}{dt} , \tag{8.5.12}$$

where the magnetic flux through the loop is

$$\Phi = \int \mathbf{B} \cdot d\mathbf{S} . \tag{8.5.13}$$

For a loop oscillating in a static magnetic field, (8.5.12) implies that the average current oscillates periodically, as in an ac generator. The amplitude of the current can be estimated to be $J \sim q^2 BA/L$, where A is the loop area. Typically $A \sim 0.1L^2$, where the coefficient 0.1 is due to Lorentz and geometric factors, and thus

$$J \sim 0.1 q^2 BL . \tag{8.5.14}$$

(The numerical factor of 0.1 will be disregarded in the remainder of this chapter, but it will be reintroduced in chapter 12 for the quantitative estimates of the astrophysical implications of superconducting strings.)

The behaviour of inhomogeneous current modes is somewhat more surprising. Consider first a loop oscillating in constant homogeneous fields **E** and **B**. We shall assume that the fields are sufficiently weak for the loop dynamics to be described by the free string equations (8.4.21). We can then write

$$\mathbf{x}(\zeta, t) = \tfrac{1}{2} \left[\mathbf{a}(\zeta - t) + \mathbf{b}(\zeta + t) \right], \tag{8.5.15}$$

and (8.5.9) takes the form

$$\Box_2 \phi = -\tfrac{1}{2} q \left[(\mathbf{a}' + \mathbf{b}') \cdot \mathbf{E} + (\mathbf{a}' \times \mathbf{b}') \cdot \mathbf{B} \right]. \tag{8.5.16}$$

This equation can be solved exactly,

$$\phi = \tfrac{1}{4} q \left[(\mathbf{a} - \mathbf{b}) \cdot \mathbf{E} t + (\mathbf{a} \times \mathbf{b}) \cdot \mathbf{B} \right] + \phi_0, \tag{8.5.17}$$

where ϕ_0 given by (8.5.3) is the general solution of the homogeneous equation. If the centre of mass of the loop does not move, then **a** and **b** are periodic functions. From the solution (8.5.17), we see in this case that current and charge oscillations induced by the magnetic field have constant amplitudes, while those induced by the electric field grow linearly with time [Spergel, Piran & Goodman, 1987]. A loop exposed to an electric field for a time period Δt develops a current of amplitude

$$J \sim q^2 E \Delta t. \tag{8.5.18}$$

This amplitude growth can be understood if we note that (8.5.16) is similar to that for a forced harmonic oscillator with the period of the force equal to the period of current oscillations. The same effect occurs for a loop which moves with some velocity v in a magnetic field B. An estimate of the current in this case can be obtained by replacing E by vB in (8.5.18),

$$J \sim q^2 v B \Delta t. \tag{8.5.19}$$

Note that $\mathbf{E} = \mathbf{v} \times \mathbf{B}$ is the electric field 'seen' by the loop in its rest frame.

The effect of current growth also occurs in a loop oscillating in an inhomogeneous magnetic field $\mathbf{B}(\mathbf{x})$ [Aryal *et al.*, 1987]. If the gradient of the magnetic field is approximately constant on the scale of the loop L, then the amplitude of the current generated in a time interval Δt can be estimated to be

$$J \sim q^2 L \left| \partial_i B_j \right| \Delta t. \tag{8.5.20}$$

8.5.3 Electromagnetic radiation from loops

The radiation power from an oscillating loop of string can be calculated using the general formalism for radiation from a periodic source which was introduced in §7.5. The total power P can be expressed as a sum over harmonics

$$P = \sum_n P_n = \sum_n \int d\Omega \frac{dP_n}{d\Omega} , \qquad (8.5.21)$$

$$\frac{dP_n}{d\Omega} = -\frac{\omega_n^2}{2\pi} j_\mu^*(\omega_n, \mathbf{k}) j^\mu(\omega_n, \mathbf{k}) . \qquad (8.5.22)$$

Here, $\omega_n = 2\pi n/T$, T is the period of oscillation, $|\mathbf{k}| = \omega_n$ and

$$j^\mu(\omega_n, \mathbf{k}) = \frac{1}{T} \int_0^T dt \int d^3x \, \exp\left(i\omega_n t - i\mathbf{k} \cdot \mathbf{x}\right) j^\mu(\mathbf{x}, t) \qquad (8.5.23)$$

is the Fourier transform of the current density. Substituting the string current density (8.4.18) into (8.5.23), we obtain

$$j^\mu(\omega_n, \mathbf{k}) = \frac{q}{T} \int_0^T dt \int_0^L d\zeta \, \exp\left[i\omega_n t - i\mathbf{k} \cdot \mathbf{x}(\zeta, t)\right] \\ \times \epsilon^{ab} x^\mu_{,a}(\zeta, t) \phi_{,b}(\zeta, t) . \qquad (8.5.24)$$

One might expect that the radiation rate from a loop with a current J can be estimated using the magnetic dipole radiation formula

$$P_{\text{dipole}} = \tfrac{2}{3} \ddot{\mathbf{m}}^2 . \qquad (8.5.25)$$

The magnetic moment of the loop is $m \sim JL^2$, and (8.5.25) suggests

$$P_{\text{dipole}} \sim \omega^4 m^2 \sim J^2 , \qquad (8.5.26)$$

where $\omega \sim L^{-1}$ is the typical frequency of oscillation. However, the dipole formula is reliable only for non-relativistic sources, and the estimate (8.5.26) gives the right order of magnitude only if the loop has no cusps or kinks. Otherwise, the relativistic character of string motion near cusps and kinks drastically changes the result.

The electromagnetic radiation emitted from cusps can be analyzed along the same lines as gravitational radiation in §7.5 [Spergel et al., 1987; Vilenkin & Vachaspati, 1987b]. Since the scale of a near-cusp region involved in ultra-relativistic motion is much smaller than the overall scale of the loop L, its contribution to the radiation comes mainly in the form of higher harmonics with $n \gg 1$. As in §7.5, one finds that the dominant contribution to the integral comes from the integration region where

$$|\zeta|, |t| \lesssim n^{-1/3} L, \qquad (8.5.27)$$

with $\zeta = t = 0$ at the cusp. Evaluating this integral in the limit of large n for a loop with $\phi = -(J/q)t$, $J = $ const., one finds the following estimate for P_n,

$$P_n \approx \kappa J^2 n^{-2/3}, \qquad (n \gg 1). \qquad (8.5.28)$$

The constant coefficient κ has been calculated numerically for some simple loop trajectories, with a typical result $\kappa \sim 10$.

The most surprising feature of (8.5.28), which distinguishes electromagnetic radiation from gravitational or Goldstone-boson radiation, is that it gives a divergent total power P. We expect, of course, that with the inertia of charge carriers and electromagnetic back-reaction taken into account, the string motion near the cusp will be modified and the series in (8.5.21) will effectively be cut off at some $n \sim n_*$. The total power will then be

$$P \approx 3\kappa J^2 n_*^{1/3}. \qquad (8.5.29)$$

According to (8.5.27) and (8.5.29) an energy $\Delta E \sim n P_n T \sim \kappa J^2 T n^{1/3}$ is radiated in one oscillation period $T = L/2$ from a region with $\Delta \zeta \sim n^{-1/3}L$. The region itself has energy $\mu \Delta \zeta$, and the electromagnetic back-reaction becomes important when $\Delta E \sim \mu \Delta \zeta$. The corresponding value of n is $n_{em} \sim (\mu/J^2)^{3/2} \sim (\eta/J)^3$, where we have used the relation $\mu \sim \eta^2$ with η the symmetry breaking scale of the string. Clearly, n_* must be smaller than n_{em}, and thus

$$P \lesssim \kappa J \eta. \qquad (8.5.30)$$

A reliable estimate of the radiation power from cusps requires a better understanding of the physics of near-cusp regions.

The angular distribution of the radiation from a superconducting loop is highly asymmetric. The dominant part of the energy is emitted from the cusps in narrow beams directed along the luminal velocity. The radiation is not continuous, but comes in periodic bursts. The angular distribution of radiation averaged over one period is

$$\frac{dP}{d\Omega} \sim J^2 \theta^{-3}, \qquad (8.5.31)$$

where $\theta = 0$ is in the direction of the beam. Equation (8.5.31) applies for $n_*^{-1/3} \ll \theta \ll 1$.

The electric and magnetic fields in the propagating burst of radiation can be found in the same manner as for gravitational radiation in §7.5. If we choose the z-axis along the beam, then the electric and magnetic fields on the z-axis are perpendicular to it and their magnitudes are given by

$$E = B \sim J L z^{-1} |z - t|^{-1}. \qquad (8.5.32)$$

The behaviour of the fields at small non-zero θ can be found for $z = t$: $E = B \sim Jr^{-1}\theta^{-3}$. We expect these equations to be modified by radiative back-reaction for $|z - t| \lesssim n_*^{-1}L$ and $\theta \lesssim n_*^{-1/3}$.

The electromagnetic radiation from kinks has been analyzed in a similar manner by Garfinkle & Vachaspati [1988]. The radiation from a kink is strongly beamed in the direction of its velocity. As the kink runs around the loop, this results in a fan-like pattern of radiation. The radiation intensity near the 'fan' behaves as θ^{-1}, where θ is the angular distance from the fan surface. The total radiation power is logarithmically divergent. Assuming a cut-off at some $n \sim n_*$ we can write

$$P \sim \kappa J^2 \ln n_* . \tag{8.5.33}$$

So far we have discussed the radiation from a loop with a constant current described by a homogeneous field $\phi = -Jt/q$. In the general case, when ϕ is given by (8.5.3), the radiation power from cusps and kinks can still be estimated from (8.5.29) and (8.5.33) with J replaced by the larger of the local values of $\dot{\phi}$ and ϕ'.

8.5.4 Radiative damping and string dynamos

The non-local electromagnetic interaction between currents and charges in different parts of a string can lead to important physical effects. For a loop of size L with a current J, the typical self-field is $F \sim J/L$. At a very heuristic level, we can rewrite (8.5.11) for the current as

$$\frac{dJ}{dt} \sim kq^2 F \sim \left(\frac{kq^2}{L}\right) J , \tag{8.5.34}$$

where k is a numerical coefficient. The quantity J in (8.5.34) should be understood as the amplitude of current–charge oscillations. The solution of this equation is

$$J \propto \exp\left(\frac{kq^2t}{L}\right) . \tag{8.5.35}$$

The sign of k is, of course, of crucial importance. If $k < 0$, then (8.5.35) describes the damping of the current oscillations due to the electromagnetic back-reaction. However, for $k > 0$, (8.5.35) would correspond to oscillations being enhanced by their own electromagnetic field. In this case the energy for the exponentially growing currents has to be drawn from the energy of the moving string.

Spergel, Press & Scherrer [1989] derived an integro-differential equation describing the string self-interaction and studied its solutions numerically. Not surprisingly, most of the loop trajectories had negative values of k for all current modes. Some loops, however, had low-frequency growing modes with $k \sim +1$. The characteristic timescale of this string dynamo is

$$\tau \sim L/q^2 \,. \tag{8.5.36}$$

For the case of decaying modes, the damping time is comparable to (8.5.36) for the lowest-frequency modes and gets smaller as the frequency grows. However, we shall see in §8.6 that string electrodynamics are drastically modified in the presence of a plasma. The string currents and charges are screened by the plasma and it appears that the string dynamo cannot realistically operate in the early universe. Even in the absence of a cosmic plasma, pairs produced in the strong electric fields near the string can screen the string charges (see §5.3.1 and below), and it is conceivable that the string dynamo will not even operate in vacuum.

8.5.5 Pair production

In §5.3 we noted the large electric and magnetic fields potentially produced by heavy superconducting strings and the vacuum effects these can induce. The most dramatic of these effects is the screening of string charges by e^+e^- pairs produced in the strong electric field near the string. The screening affects portions of the string in which $\lambda > J$ or, more exactly,

$$J_a J^a \gtrsim 10^{-2} m_e^2 \,, \tag{8.5.37}$$

where λ is the charge per unit length of string and m_e is the electron mass. (In most cases of astrophysical interest, the currents involved are much greater than $0.1 m_e$ and the right-hand side of (8.5.37) can be neglected.) The rate of pair production by a string with current–charge oscillations of amplitude J and frequency ω is easy to estimate. In the course of these oscillations, string segments of length $\sim \omega^{-1}$ develop charge densities $\sim J$. These charges are screened by the created particles. The number of pairs produced per unit time per unit length of string is, therefore,

$$\frac{dN}{d\ell\, dt} \sim \omega J/e \,. \tag{8.5.38}$$

Spin-1 gauge bosons with masses $m < eJ$ will also be produced in the magnetic field of the string; their production rate is probably comparable to (8.5.38).

As pointed out in §5.3.1, the screening of fields by pair production near the string necessitates the modification of the classical equations of string electrodynamics. Notably, calculations of electromagnetic radiation and the analysis of the string dynamo will be affected. Vacuum polarization effects may also alter the electromagnetic string fields, but a quantitative analysis has yet to be given.

8.5.6 Loop lifetimes

We are now in a position to attempt to describe the typical life story of a superconducting loop. The current in a loop will generally include a dc component and a number of inhomogeneous ac modes. In the absence of external electric or magnetic fields, the ac modes are damped by radiation back-reaction and by pair production on a timescale $\tau \lesssim 100L$. The dc component of J is proportional to the total change of phase in the condensate field around the loop for bosonic superconductivity or to the total number of charge carriers for fermionic superconductivity. These quantities remain constant, as long as the current remains below critical levels. As the loop radiates away its energy by emitting electromagnetic and gravitational waves, it shrinks and the dc current in the loop grows as

$$J \propto L^{-1}. \tag{8.5.39}$$

If the loop has no cusps or kinks, then its electromagnetic radiation power is

$$\dot{E} \sim \Gamma_{em} J^2, \tag{8.5.40}$$

with $\Gamma_{em} \sim 100$. (The power is not much different for a kinky loop, but the presence of cusps can change it drastically.) The ratio of electromagnetic and gravitational power output is of the order

$$\frac{\dot{E}_{em}}{\dot{E}_g} \sim \frac{J^2}{G\mu^2}. \tag{8.5.41}$$

If the current evolves according to (8.5.39), then $\dot{E}_{em}$ gradually grows, and the net fraction of the loop's mass radiated electromagnetically during its entire lifetime is [Ostriker, Thompson & Witten, 1986]

$$f = \sqrt{f_i} \tan^{-1}(1/\sqrt{f_i}), \tag{8.5.42}$$

where f_i is the initial value of $\dot{E}_{em}/\dot{E}_g$. Eventually the current reaches the critical level $J \approx J_{max}$ discussed in §5.1.3 and §5.2.5. As the loop shrinks further, the current remains near critical and all the extra charge carriers are expelled from the string.

For a loop with cusps, the electromagnetic radiation power is dominated by bursts of radiation from near-cusp regions. The motion of the string in these regions is strongly affected by the radiation back-reaction, and the resulting power is difficult to estimate. It is expected to be much greater than the power for a cuspless loop (8.5.40). An upper bound for $\dot{E}_{em}$ is given by

$$\dot{E}_{em} \lesssim J\sqrt{\mu}. \tag{8.5.43}$$

In the vicinity of a cusp, the current tends to become super-critical. An invariant measure of the current is

$$J_a J^a = q^2 (-\gamma)^{-1/2} (\phi'^2 - \dot\phi^2) \,. \qquad (8.5.44)$$

Near a cusp $\sqrt{-\gamma} = \mathbf{x}'^2 \to 0$ and $J_a J^a \to \infty$ (unless $\dot\phi \pm \phi' = 0$). The physical origin of this effect is very simple: a moving string becomes contracted by a factor $|\mathbf{x}'|^{-1} = (1 - \dot{\mathbf{x}}^2)^{-1/2}$, and the density of charge carriers increases by the same factor. The right-hand side of (8.5.44) can be estimated using (6.2.34), $J_a J^a \sim (JL/\zeta)^2$. The current becomes super-critical in the region $|\zeta| \lesssim L J/J_{\mathrm{max}}$. As a result the loop will lose a fraction $\sim J/J_{\mathrm{max}}$ of its charge carriers.

If the motion of the loop were strictly periodic, then during the next period the current near the cusp would be exactly J_{max}. However, the new cusp could be slightly shifted due to the radiation back-reaction, and so a new portion of charge carriers would be ejected [Copeland *et al.*, 1988; Thompson, 1988b]. As a result, the loop can gradually lose a substantial fraction of its charge carriers. This process competes with current growth due to loop contraction, and it is not clear which of the two effects prevails.

A further complication arises with loop fragmentation. When the loop loses a substantial part of its energy to radiation, its configuration changes, and the loop is likely to self-intersect and fragment. The currents in daughter loops are comparable to the current in the original loop (they are not exactly equal, since part of the length of the parent loop goes into the kinetic energy of the daughter loops). The gravitational and electomagnetic radiation power from each daughter loop is also comparable to that of the parent loop. Hence, fragmentation increases the total energy output and speeds up the loop's decay.

It should be clear from this discussion that our understanding of string electrodynamics is not yet sufficiently well-developed to paint a complete picture of the demise of a superconducting loop. It should also be clear that loops can exhibit a variety of different behaviours, depending on the loop's trajectory and the magnitude of the current. Of course, this diversity is increased if one includes the effects of external fields or the presence of a plasma.

8.6 Plasma interactions

Up to this point, we have discussed the electrodynamics of strings in vacuum. However, if superconducting strings exist in nature then they will be embedded in cosmic plasmas. Since a plasma is a highly conducting medium, its presence can drastically modify the electromagnetic properties of strings, and the results obtained in previous sections for

strings in vacuum may not be valid. To elucidate the interaction of superconducting strings with plasma, let us consider an idealized situation
in which a plasma flows past a static straight string carrying a current
J [Chudnovsky, Field, Spergel & Vilenkin, 1986]. The flow velocity at
large distances from the string is $\mathbf{v} = \text{const}$. This problem is, of course,
equivalent to that of a string moving through a plasma at a constant
velocity $-\mathbf{v}$. In most situations of astrophysical interest, the resistivity
of the plasma is negligible. The magnetic field lines are then 'frozen'
into the plasma and move along with it without dissipation [Alfven &
Fälthammar, 1963]. At first sight it is not clear how the magnetic field of
the string can avoid being completely swept away by the moving plasma.
In order to have a stationary magnetic field in the vicinity of the string,
the plasma should be at rest in that region. The resolution of this apparent paradox is that the magnetic field pressure near the string is strong
enough to balance the pressure of the incoming plasma. The field is confined to a cylindrical 'cavity' surrounding the string, where the plasma is
static. (We assume for simplicity that the incoming plasma is not magnetized.) The flow of plasma around the cavity is like a flow around a blunt
body (see fig. 8.4(a)). A thin current sheet flowing through the boundary
of the cavity screens the magnetic field of the string. This boundary is
called the tangential discontinuity (TD). The gas pressure P, velocity $\mathbf{v}$
and the magnetic field $\mathbf{B}$ should satisfy the following conditions at the
tangential discontinuity [Landau, Lifshitz & Pitaevskii, 1984]

$$B_{\mathrm{n}} = 0, \qquad v_{\mathrm{n}} = 0, \qquad\qquad (8.6.1)$$

$$P_{\mathrm{out}} = P_{\mathrm{in}} + \frac{B^2}{8\pi}, \qquad\qquad (8.6.2)$$

where the subscript n indicates components normal to the tangential discontinuity. The last of these equations expresses the balance of magnetic
and gas pressure across the tangential discontinuity. The gas pressure
P_{in} depends on the amount of plasma inside the cavity which is, in turn,
determined by the manner in which the string current was generated. For
example, if a current-carrying string moves into a plasma from a vacuum
region, then the cavity will be essentially free of plasma. Alternatively, if
the string acquired the current when moving through a magnetized region
of plasma, then some of the plasma remains 'frozen into' the magnetic
field of the string.

Let us now estimate the characteristic radius of the cavity and the drag
force acting on the string. We shall first assume that the flow speed u
is well below the sound speed. The variation of the plasma density ρ is
then negligible, and we can use the Bernoulli equation for incompressible

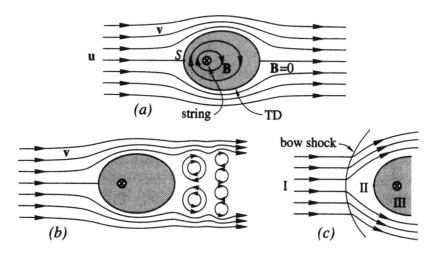

Fig. 8.4. Schematic diagram of plasma flow around a string. (a) Subsonic laminar flow: the plasma is static in the shaded region, where the magnetic field is non-zero. (b) Subsonic turbulent flow: turbulent eddies develop in the wake of the string. (c) Shock structure in a supersonic flow.

flow [Landau and Lifshitz, 1987],

$$\tfrac{1}{2}\rho v^2 + P = \text{const.} = \tfrac{1}{2}\rho u^2 + P_\infty \,. \tag{8.6.3}$$

Here, **u** and P_∞ are the gas velocity and pressure at large distances from the string. The gas velocity **v** vanishes at the stagnation point S located at the tip of the tangential discontinuity. Applying (8.6.2) and (8.6.3) at that point, we find

$$\frac{B^2}{8\pi} + P_{\text{in}} = \frac{1}{2}\rho u^2 + P_\infty \,. \tag{8.6.4}$$

If J is the current in the string and R is the characteristic radius of the cavity, then the magnetic field at the point S is $B \sim J/R$. Substituting this in (8.6.4) and assuming that $P_{\text{in}} \lesssim B^2/8\pi$, we obtain an order-of-magnitude estimate for the radius R,

$$R \sim J(\rho u^2 + P_\infty)^{-1/2} \,. \tag{8.6.5}$$

The drag force acting on a string moving through ambient plasma is roughly comparable to that on a cylinder of radius R given by (8.6.5). Using a well-known result from fluid mechanics, we can write [Landau & Lifshitz, 1987]

$$F \sim \nu \rho u \,, \tag{8.6.6}$$

where F is the force per unit length and ν is the viscosity coefficient. Since viscous drag is produced in a thin boundary layer, F is independent of the radius R. The viscosity of a neutral gas can be estimated as

$$\nu \sim v_{\mathrm{T}} \ell, \tag{8.6.7}$$

where $v_{\mathrm{T}} \sim \sqrt{T/m}$ is the thermal velocity and ℓ is the mean free path of the gas particles. The situation in plasma is more complicated because of the long-range electromagnetic interactions between the particles. Besides the mean free path ℓ, the problem also includes several other characteristic length-scales: the Debye screening length λ_{D}, the Larmor radius λ_{R} (for a magnetized plasma), and the wavelengths associated with various plasma instabilities (typically given by combinations of λ_{D} and λ_{R}). The actual mean free path of the particles should be used in (8.6.7) only if it is smaller than all other relevant length-scales. In the general case, the calculation of plasma viscosity is a rather complicated problem.

Fortunately, the problem simplifies for the case of prime physical interest. Equation (8.6.6) for the drag force applies only when the plasma flow is laminar (non-turbulent). The condition for this is that the Reynolds number, defined as

$$\mathcal{R} = \frac{uR}{\nu}, \tag{8.6.8}$$

is not large compared to unity. For $\mathcal{R} \gg 1$ the flow develops a turbulent wake behind the cavity (see fig. 8.4(b)). The drag force per unit length is determined by the impact of incoming particles,

$$F \sim \rho u^2 R. \tag{8.6.9}$$

If the plasma flow is supersonic, then the tangential discontinuity is preceeded by a bow shock, at which the flow is reduced to subsonic speed (fig. 8.4(c)). This is the situation of interest in many astrophysical applications where the strings are expected to move through the cosmic plasma at relativistic speeds. The incoming plasma in region I of fig. 8.4(c) has a uniform density ρ and velocity $\mathbf{u}$. For a strongly supersonic flow, the pressure P is negligible compared to the dynamical pressure ρu^2 and the gas temperature is $T \ll mu^2$, where m is the mean mass per particle. In the non-relativistic limit ($u \ll 1$) the density, temperature and velocity immediately behind the shock are given by [Landau & Lifshitz, 1987]

$$\rho' = 4\rho, \qquad T' = \tfrac{3}{16}mu^2, \qquad v' = \tfrac{1}{4}u. \tag{8.6.10}$$

For $u \sim 1$, these equations still give the correct order of magnitude.

To estimate the size of the cavity R, we note that the dynamical pressure of the incoming plasma is eventually balanced by the magnetic pressure in region III. Hence,

$$\rho u^2 \sim \frac{B^2}{8\pi}, \tag{8.6.11}$$

where $B \sim J/R$ is the magnetic field at the point S, and we obtain

$$R \sim \frac{J}{u\sqrt{\rho}}. \qquad (8.6.12)$$

This estimate could also be obtained from (8.6.5) if we note that in region II, $v \sim u$ and $P \sim \rho u^2$. The drag force acting per unit length of string is then

$$F \sim \rho u^2 R \sim \sqrt{\rho}\, uJ. \qquad (8.6.13)$$

So far in this section we have considered the simple case of a straight string in a non-magnetized plasma. We shall now briefly discuss the effects that can arise in more complicated situations. Equation (8.6.13) can be used to estimate the frictional force acting on a curved string moving through plasma, as long as the shock radius R is much smaller than the curvature radius of the string. For a closed loop of length L oscillating at a relativistic speed $u \sim 1$, the rate of energy loss due to shock heating of the plasma is

$$\dot{E}_{\mathrm{d}} \sim FL \sim \sqrt{\rho}JL. \qquad (8.6.14)$$

(We assume that the loop has no cusps or kinks, so that its motion is never ultra-relativistic.) If plasma dissipation is the dominant energy loss mechanism, then the lifetime of the loop is

$$\tau_{\mathrm{d}} \sim \frac{\mu}{\sqrt{\rho}J}. \qquad (8.6.15)$$

The competing energy loss mechanism is gravitational radiation with $\dot{E}_{\mathrm{g}} \sim 100 G \mu^2$. There is essentially no electromagnetic radiation from superconducting loops in plasma, since the string current is completely screened by the plasma current at the tangential discontinuity. Moreover, electromagnetic waves cannot propagate through the plasma, unless their frequency is greater than the plasma frequency,

$$\omega_{\mathrm{p}} = \left(\frac{4\pi n e^2}{m_{\mathrm{e}}} \right)^{1/2}, \qquad (8.6.16)$$

where n is the number density of electrons and m_{e} is the electron mass. For most interesting values of n and L, this frequency is much higher than the typical frequency $\omega \sim L^{-1}$ expected from loops of astronomical proportions.

The situation is, of course, different when the current is so large that the shock radius R is greater than the loop size L. In this case, the loop oscillates in a region of very low plasma density. If it is so low that $\omega_{\mathrm{p}} \sim L^{-1}$, then the loop radiates electromagnetic waves just as it

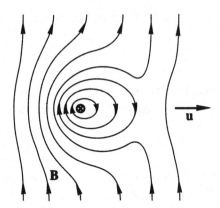

Fig. 8.5. The structure of the magnetic field in a magnetized plasma flowing past a string.

would in vacuum. The radiation pressure of these waves can sweep out the surrounding plasma, leaving the loop at the center of an expanding bubble [Ostriker *et al.*, 1986]. The bubble can also be formed for $\omega_\mathrm{p} > L^{-1}$ but now, instead of electromagnetic waves, the loop emits a magnetized relativistic plasma wind [Thompson, 1988b]. In both cases, the energy output of the loop is expected to be of the same order of magnitude, $\dot{E}_\mathrm{wind} \sim \dot{E}_\mathrm{em} \sim J^2$ (for a loop without cusps or kinks).

Finally, we consider a string moving through a magnetized plasma (see fig. 8.5). The field lines pile up in front of the string and eventually reconnect behind it. This build-up of the magnetic field around the string is consistent with the current growth in the string as it moves through a magnetized plasma. Particles move in spiral trajectories winding around the field lines, so that the plasma can move freely along the lines, but any transverse motion is suppressed. Most of the incoming plasma escapes the string by moving along the field lines downstream, but a certain fraction is trapped as the lines close up. A discussion of this process can be found in Thompson [1988b, 1990] .

9
String evolution

9.1 String formation

9.1.1 The Kibble mechanism

As we already mentioned in §3.1.4, symmetry breaking phase transitions can give rise to a stochastic network of topological defects. When the universe cools below the critical temperature T_c, the Higgs field develops an expectation value $\langle \phi \rangle$ corresponding to some point in the manifold $\mathcal{M}$ of the minima of the effective potential. Since all such points are equivalent, the choice is determined by random fluctuations and will be different in different regions of space. The values of $\langle \phi \rangle$ in two regions will be totally independent if the regions are separated by a distance greater than some correlation length ξ. If the manifold $\mathcal{M}$ has a non-trivial homotopy group, $\pi_1(\mathcal{M}) \neq I$, the stochastic field $\langle \phi(\mathbf{x}) \rangle$ will necessarily wind around $\mathcal{M}$ in a non-trivial way, and a network of strings will form with a characteristic length scale comparable to ξ. Likewise, domain walls or monopoles will be formed if $\pi_0(\mathcal{M})$ or $\pi_2(\mathcal{M})$ are non-trivial, respectively. Since correlations cannot be established at speeds greater than the speed of light, $\xi(t)$ cannot exceed the causal horizon, $d_H(t) \sim t$. Hence, we can write [Zel'dovich, Kobzarev & Okun, 1974]

$$\xi(t) \lesssim t. \tag{9.1.1}$$

The actual magnitude of ξ at the phase transition and afterwards is determined by complicated dynamical processes and can be much smaller than the upper bound (9.1.1).

Mechanisms for defect formation were first quantitively studied by Kibble [1976, 1980]. They are expected to operate in condensed matter systems such as liquid helium [Zurek, 1985a] and liquid crystals [Chuang, Durrer, Turok & Yurke, 1990; Bowick et al., 1994]. In the latter case, the formation of many defects was observed experimentally when a liquid crystal was rapidly cooled below the isotropic-nematic phase transition temperature. It should be emphasized that defect formation by the Kibble mechanism is an essentially non-equilibrium phenomenon. The equilibrium density of defects is exponentially suppressed at temperatures well below T_c. Defect networks can survive at such temperatures only because the field ϕ does not have enough time to equilibrate on scales greater than the correlation length $\xi(t)$.

As a starting point for studying the cosmological evolution of defects, we have to determine the network configuration at the time of formation. There is, however, some ambiguity as to when the defects can be considered as 'formed'. At temperatures close to T_c, thermal fluctuations taking ϕ over the potential barrier are not uncommon, and so topological knots can unwind. As a result, the defect configuration undergoes continuous change. For a second-order phase transition, the characteristic 'freeze-out' temperature is the Ginzburg temperature T_G at which the energy required to take the field ϕ over the barrier in a volume of size $\sim \xi$ is comparable to the thermal energy T (refer to §2.3.2). For the global $U(1)$-model (2.1.1–2) with $\lambda \ll 1$, T_G is very close to T_c, $T_c - T_G \sim \lambda\eta$, and the correlation length at $T = T_G$ is

$$\xi(T_G) \sim (\lambda\eta)^{-1}. \tag{9.1.2}$$

This gives the characteristic scale of the string network at formation [Kibble, 1976]. The horizon size at this temperature is $d_H \sim m_{pl}/T_G^2 \sim m_{pl}/\eta^2$, and so

$$\xi/d_H \sim \eta/\lambda m_{pl}. \tag{9.1.3}$$

This is typically a small number, unless λ is very small or η is close to the Planck scale. For a first-order phase transition, ξ is determined by the average bubble size at the time of bubble coalescence.

An additional complication with finding the initial network configuration is due to the large equilibrium concentration of defects at $T \approx T_c$. Turning again to the Goldstone model (2.1.1–2) for an example, the smallest loop of string at $T \approx T_c$ has size $r_{min} \sim m^{-1}(T) \sim [\lambda\eta(T_c - T)]^{-1/2}$ and energy $\epsilon_{min} \sim \lambda|\phi^4(T)|r_{min}^3 \sim [\eta(T_c - T)/\lambda]^{1/2}$, where $m(T)$ and $|\phi(T)|$ are given by (2.3.15) and (2.3.17) respectively. At the Ginzburg temperature, $r_{min} \sim \xi$, $\epsilon_{min} \sim \eta \sim T$, and the equilibrium density of small loops can be very high.* In fact, dynamical simulations of the phase

---

* Copeland, Haws & Rivers [1989] have pointed out that large loops and infinite strings can also appear in equilibrium at $T \lesssim T_G$. Indeed, the Hagedorn temperature for strings is $T_H \sim \epsilon_{min}$ and becomes comparable to T at $T \sim T_G$ (cf. §6.6). They argue that these equilibrium strings can have a significant effect on the dynamics of the phase transition.

transition [Hodges, 1989] indicate that such loops can initially dominate the total string length in the universe. However, as the universe expands and cools down, these equilibrium loops disappear in a few Hubble times, without having much effect on the subsequent evolution of the network – as we shall see, this evolution is not sensitive to the details of the initial state. We shall therefore ignore the equilibrium loop concentration in our discussions and analyze the general features of string networks at formation using a simple stochastic model.

9.1.2 Network configuration at formation

We first consider string formation in the global $U(1)$-model (2.1.1–2). Shortly after the phase transition, the Higgs field ϕ has the same magnitude (2.3.17) almost everywhere, while its phase θ varies on the correlation scale ξ. This situation can be simulated by randomly assigning different values of θ on a lattice with a lattice spacing comparable to ξ. The first such simulation was performed by Vachaspati & Vilenkin [1984] who used a cubic lattice and discretized the vacuum manifold by allowing θ to take only three values, 0, $2\pi/3$ and $4\pi/3$. These values were labelled 0, 1 and 2, respectively, and were randomly assigned at each lattice vertex, assuming a smooth variation between the vertices. By considering the four corner values on each cubic face, one can determine whether the plaquette is traversed by a string, an antistring, or no string at all. This is illustrated in fig. 9.1. The probability for a plaquette to be pierced by a string is 0.29. Of course, for a single cubic cell, the entry of a string through one face implies its exit through another. There is an ambiguity when two strings enter the same cell and, in this case, they are randomly assigned to join one or other of the two exiting strings. Three or more strings cannot pass through the same cell in this model. With this Monte-Carlo method applied to the whole lattice, the string segments join up to form either closed loops or else open strings which intersect with the boundaries of the cubic volume. One particular realization of the resulting lattice is illustrated in fig. 9.2.

The shapes of individual strings in the network can be characterized by the fractal dimension d, which is determined by comparing the distance R between two points on a string with the average length of string L between them, that is,

$$L = A\xi(R/\xi)^d, \qquad (9.1.4)$$

where A is a constant. Using boxes of side-length 40ξ, the values of d and A for the open strings were found to be

$$d - 2.00 \perp 0.07, \quad A = 0.97 \pm 0.03. \qquad (9.1.5)$$

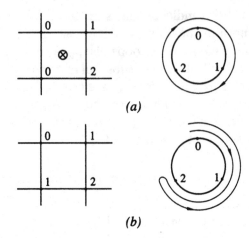

(a)

(b)

Fig. 9.1.   As we go around the face of a cubic cell, the Higgs phase describes a certain trajectory in group space (shown in the right column). A string passes through the face if that trajectory has a non-zero winding number (as in (*a*)).

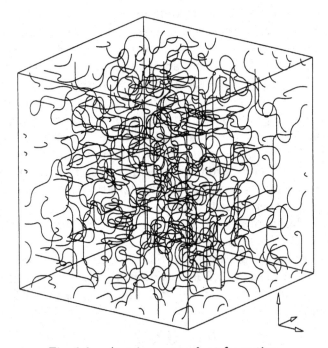

Fig. 9.2.   A string network at formation.

Hence, to good accuracy

$$L \approx R^2/\xi, \qquad (9.1.6)$$

suggesting that the strings have the shape of random walks. This could have been anticipated: since the values of θ are assigned at random, the string directions should also be uncorrelated at distances much greater than ξ.*

To define the fractal dimension of closed loops, we reinterpret R in (9.1.4) to be the 'average radius',

$$R = \Delta x_1 + \Delta x_2 + \Delta x_3, \qquad (9.1.7)$$

where the Δx_i are the extents of the loop in the three coordinate directions. The loops are then found to be nearly Brownian,

$$d = 1.90 \pm 0.05, \quad A = 0.21 \pm 0.02. \qquad (9.1.8)$$

The small deviation of d from 2 is intriguing, but its significance is not clear.

The size distribution of loops can be described by a function $n(R)$ such that $n(R)dR$ gives the number density of loops with radii in the range R to $R + dR$. On dimensional grounds, we can write

$$n(R)dR = f(R/\xi)\,R^{-4}dR. \qquad (9.1.9)$$

Numerical simulations show that, for a large range of radii, the function f is nearly a constant,

$$n(R)dR = \nu R^{-4}dR \qquad (9.1.10)$$

with $\nu = 6 \pm 2$. A remarkable property of this loop distribution is its scale invariance: $n(R)$ remains unchanged under a rescaling of ξ.

The total length of string in the form of closed loops was found to be 14%, 17% and 19% for 20ξ, 30ξ and 40ξ simulations, respectively. Larger lattices include more of the tail of the distribution (9.1.10), but the length fraction in loops appears to saturate at about 20% in the limit of infinite volume. The remaining 80% are then in the form of infinite Brownian strings. As an additional verification of this picture, simulations were performed on a lattice with periodic boundary conditions, which corresponds to a closed universe having the topology of a three-torus. In a closed universe all strings are closed, and the place of infinite strings is taken by very large loops traversing the whole universe or even running

---

* Note, however, that this argument does not constitute a proof. It is conceivable, for example, for the strings to be self-avoiding random walks, in which case the fractal dimension would be $d = 1.7$. Note also that $d = 2$ does not necessarily mean that we are dealing with a random walk, since distinct configurations can have the same fractal dimension.

around it several times. The simulations showed that periodic boundary conditions did not substantially change the loop distribution up to $L \sim 200\xi$, which is about the maximum loop length in a 40ξ simulation with non-periodic boundaries. In addition, there were a few very long loops with lengths between 800ξ and $39,000\xi$, which made up $\sim 80\%$ of the total string length as expected. A large gap in length clearly separates the loops belonging to the scale-invariant distribution (9.1.10) from the lattice-size loops playing the role of infinite strings. The prevalence of infinite strings can be attributed to the fact that a random walk on a three-dimensional lattice has a finite (and large) probability of never returning to the starting point [Scherrer & Frieman, 1986].

The results of $U(1)$-string simulations on a cubic lattice have been independently confirmed by several authors. More significantly, Scherrer & Frieman [1986] used a different type of lattice in which every cell is a tetrakaidecahedron, with three cells meeting at every edge and four edges meeting at every vertex. A value of θ between 0 and 2π was randomly assigned to each cell, assuming that the variation of θ between the neighbouring cells is along the shortest path in the group space. The string configuration was found by examining the variation of θ around the edges: whenever θ varies by $\pm 2\pi$, there is a string lying along the edge. It is easy to show that no more than one string can pass through each vertex in this construction, and thus the strings do not intersect. The results obtained by Scherrer & Frieman were entirely consistent with the exponents in (9.1.5, 8, 10), but the pre-factors differed somewhat because of the lattice geometry. For example, they found $A \approx 2$ in (9.1.4) and the proportion of length in infinite strings to be about 74%.

Kibble [1986a] used a method similar to that of Vachaspati & Vilenkin to simulate the formation of Z_2-strings in a phase transition corresponding to a complete breaking of $SO(3)$. The vacuum manifold in this case is $\mathcal{M} = SO(3)$. Kibble approximated the $SO(3)$ group by the tetrahedral group T of order 12, so that each site of a cubic lattice was assigned one out of 12 possible orientations. The probability of a string passing through a plaquette was now 3/8, so the overall string density was higher. Again, a Brownian fractal dimension and scale-invariant distribution of loops was confirmed. However, for Z_2-strings there was a substantially lower length fraction in closed loops; only about 6%.

A somewhat simpler algorithm for simulating Z_2-string formation was later employed by Aryal, Everett, Vilenkin & Vachaspati [1986] (see also Vachaspati [1991]). To imitate a continuous symmetry breaking $G \to H$, they made the simplest non-trivial choice of the discrete groups: $G = Z_4$, $H = Z_2$. The group Z_4 has four elements which can be written as I, $-I$, z, $-z$ with $z^2 = -I$. The unbroken group is $Z_2 = \{I, -I\}$. The vacuum manifold $\mathcal{M} = Z_4/Z_2 = Z_2$ consists of only two points, labelled by 0 and 1, which can be transformed into one another by either of the group elements z or $-z$. To simulate string formation, the vacuum states 0 and 1 were randomly assigned to each vertex of a cubic lattice. In order to mimic a smooth variation of the Higgs field between the vertices, a group element z or $-z$ was assigned at random to each lattice link joining adjacent vertices with different vacuum states, and the group element I to links connecting vertices in the same vacuum state. The locations of strings were determined by taking the product of the group elements associated with the four links bounding each plaquette. There is a string through a plaquette if this product is $-I$ and no string if it is I. The results obtained using this method are very similar to those of Kibble [1986a]. In particular, the fraction of length in closed loops is still about 6%.

Although the overall string density and the fraction of length in loops depend on the symmetry group and on the lattice geometry, the results of all different simulations show remarkable similarities. In all cases most of the string length (74% to 94%) is in the form of infinite Brownian strings and the remainder is in the form of (nearly) Brownian loops with a scale-invariant size distribution (9.1.10). Mitchell & Turok [1987b] made an interesting observation that the properties of a string network at formation are very similar to those of an equilibrium gas of strings above the critical density, $\rho > \rho_c$ (refer to §6.6). This similarity is not entirely surprising, since in both cases the system of strings has a large degree of randomness. An important difference between the two cases is that the equilibrium fraction of the string length in loops is determined by the ratio ρ/ρ_c, which is a free parameter, while for actual string formation this fraction is completely determined by the underlying model.

9.2 Damped epoch evolution

After formation, strings experience a significant damping force from the relatively high radiation background density. As discussed in §8.1.2, the force per unit length is (8.1.28),

$$\mathbf{F_d} = -\beta T^3 \gamma \mathbf{v}\,, \tag{9.2.1}$$

where β is a numerical factor of order unity which is defined in (8.1.29) and γ is the Lorentz factor. This can be accounted for in an expanding universe by adding a temperature-dependent term to the Hubble damping term in the string equations of motion (refer to §8.1.3). This damping term becomes negligible at temperatures

$$T \ll G\mu\, m_{\rm pl} \sim (G\mu)^{1/2} \eta\,, \tag{9.2.2}$$

where η is the string forming energy scale, $T_c \approx \eta$. The time at which damping terms become subdominant is approximately given by

$$t_* \sim (G\mu)^{-2} t_{\mathrm{pl}} \sim (G\mu)^{-1} t_{\mathrm{f}}, \tag{9.2.3}$$

where t_{pl} is the Planck time and t_{f} is the time of string formation.

By considering the damping force (9.2.1), we can develop a heuristic picture for string evolution during the damped epoch [Kibble, 1976, 1980; Vilenkin, 1985]. The kinetic energy per unit length of string is $\mathcal{E} \sim \mu v^2$, the rate of energy dissipation is $\dot{\mathcal{E}} \sim F_{\mathrm{d}} v \sim T^3 v^2$, and the characteristic damping time is of the order

$$t_{\mathrm{d}} \sim \mu/T^3. \tag{9.2.4}$$

The force due to the tension in a string of curvature radius R is $F_{\mathrm{t}} \sim \mu/R$, and the corresponding acceleration in vacuum is approximately $a \sim R^{-1}$. However, in the damped regime the string accelerates only for a time period $\sim t_{\mathrm{d}}$, after which the tension is balanced by the damping force and the string moves with a limiting velocity $v \sim t_{\mathrm{d}}/R$. The typical correlation length of the string network $L(t)$ can be expected to grow as $L \sim vt$. With $R \sim L$, this gives

$$L(t) \sim (t_{\mathrm{d}} t)^{1/2}. \tag{9.2.5}$$

Small-scale irregularities and loops of size smaller than L will be damped out in less than a Hubble time, while on scales greater than L the strings are expected to be Brownian. The correlation length L gives both the typical curvature radius of strings and the typical distance between the nearest string segments in the network. (A detailed discussion of this one-scale model will be given in the following sections.) With $t_{\mathrm{d}} \propto t^{3/2}$ from (9.2.4), $L(t)$ grows faster than the horizon scale t, $L \propto t^{5/4}$. The cross-over point roughly coincides with t_* when the undamped relativistic evolution of the network begins.

In this analysis we assumed that some particles which strongly interact with the strings are present in the thermal bath. Otherwise the coefficient β in (9.2.1) would be very small and the damping force on the strings would be negligible. We assumed also that the damping era ends before the end of the radiation era, $t_* < t_{\mathrm{eq}}$. This is justified for sufficiently heavy strings with $\eta \gtrsim 10^6 \mathrm{GeV}$. Finally, we should point out that our simple picture for the damped epoch evolution is only inferred and needs to be examined more rigorously. This epoch corresponds to the highest string densities and so may be important for baryogenesis, vorton formation and other effects.

9.3 The 'one-scale' model

9.3.1 Heuristic picture

The friction-dominated evolution of strings in the period following their formation is only transient. Strings begin to oscillate freely at the time t_* given by (9.2.3). For GUT-scale strings, the cosmological horizon expands by over 40 orders of magnitude from t_* to the present day t_0. There are two extreme cases for string evolution during this period: Either the string network, simply stretched by the expansion, can come to dominate the energy density of the universe or, alternatively, it can fragment into tiny loops and disappear completely. The entropy considerations discussed in §6.6 favour complete fragmentation, however, the rate at which the Brownian long strings can chop off loops and straighten is limited by the speed of light. The result of these competing effects is an intermediate steady-state, manifested as 'scale-invariant' network evolution, which underlies the fact that strings can be both cosmologically viable and have interesting astrophysical effects.

It is not difficult to build an intuitive picture for why a string network will evolve toward a 'scaling' regime, that is, a regime in which the characteristic scale L of the string network remains constant relative to the horizon size d_H. As the universe expands, the string energy does not grow with the scale factor because of energy losses associated with the production of small loops. These loops form when two strings collide as illustrated in fig. 9.3(a) or when a string 'curls back' on itself and self-intersects (b). The string loops then oscillate and begin to radiatively decay via the preferred channel. For strings formed from the breaking of a local symmetry this usually will be gravitational radiation, but for superconducting or global strings this may be, respectively, electromagnetic radiation or axion emission (refer to sections §7.5, §8.5 or §8.3, respectively). Provided the string network decays at a sufficient rate into an

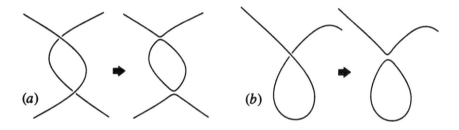

Fig. 9.3. Loop formation mechanisms: (a) collision of two strings and (b) self-intersection.

innocuous radiation background, it need not have pathological cosmological consequences.

The microphysical interaction properties of the strings discussed in chapter 4 are key to the possibility of loop formation—notably reconnection at string crossing. It is possible, however, to envisage string models where this does not happen and strings become entangled when crossing because of topological constraints (§3.3). Eventually, such strings will come to dominate the universe, but the discussion of this exceptional case is left until §9.5. Here, we shall assume a reconnection probability of order unity.

The reason the evolution of the network achieves a 'scaling' regime can be understood by considering the efficacy of loop formation mechanisms at different relative energy densities. From our discussion on string formation in §9.1, we expect the strings in the network to be Brownian random walks on the largest scales with energy E and energy density ρ_∞ given by

$$E = \frac{\mu V}{L^2}, \qquad \rho_\infty = \frac{\mu}{L^2}, \qquad (9.3.1)$$

where L is defined to be the characteristic length-scale of the network and V is a large physical volume. We can crudely estimate the energy growth rate $\dot{E}$ by assuming that the strings are merely stretched by the expansion as $L(t) \sim (a(t)/a(t_i)) L_i$ where t_i is the initial time and $L_i = L(t_i)$. (Note that we know from §6.3 that this is a significant over-estimate because subhorizon modes are significantly smoothed by the Hubble expansion.) Clearly, with these assumptions and without interactions, from (9.3.1) the string energy would grow as $E \propto a(t)$, while the radiation energy would redshift, so that strings would soon dominate the universe.

The number of loops created and lost by the network will be closely related to the number of string intersections. A string segment of length L can be expected to travel a distance L before encountering another segment in a volume L^3. This corresponds to $L^{-4}\delta t$ collisions per unit volume in a time δt, assuming relativistic string motion. Supposing, as was originally thought, that loops are typically created at sizes corresponding to the length-scale of the network, this yields a crude energy-loss rate equation

$$\dot{E} \approx \frac{\dot{a}}{a} E - \frac{1}{L^4} \cdot \mu L \cdot V , \qquad (9.3.2)$$

which becomes

$$\dot{\rho}_\infty \approx -2\frac{\dot{a}}{a}\rho_\infty - \frac{\rho_\infty}{L} . \qquad (9.3.3)$$

To determine whether this evolution will be self-similar relative to the horizon scale, we introduce a new variable $\gamma(t)$,

$$L = \gamma t. \tag{9.3.4}$$

Substituting (9.3.1) and (9.3.4) into (9.3.2) gives a dynamical equation for γ,

$$\frac{\dot{\gamma}}{\gamma} \approx -\frac{1}{2t}\left(1 - \gamma^{-1}\right), \tag{9.3.5}$$

where we have taken $a(t) \sim t^{1/2}$ in the radiation era. Clearly, $\gamma \approx 1$ is a significant fixed point: If we have a high density of strings ($\gamma \lesssim 1$), then the right-hand side becomes positive ($\dot{\gamma}/\gamma > 0$) because loop production becomes very efficient. Consequently, γ increases and the string density falls. On the other hand, if $\gamma \gtrsim 1$ for low densities, then γ will tend to fall because $\dot{\gamma}/\gamma < 0$. Hence, we might expect any initial Brownian configuration in an expanding universe to evolve toward a 'scaling' solution with constant γ yielding a correlation length comparable to the horizon size, $L \sim t$. (Note that in later chapters it will prove more convenient to use the parameter $\zeta = \gamma^{-2}$, so that $\rho_\infty = \zeta\mu/t^2$.)

The above briefly summarizes the basic principles behind the more detailed models we will now elaborate and, indeed, it provides a qualitatively accurate picture for some aspects of string evolution. Numerical simulations, to be discussed in §9.4, reveal that the actual string network evolution is a complex process, characterized by the emission of miniscule loops in vast quantities from slow-moving long strings possessing fractal-like substructure.

The notion that a string network in an expanding universe would evolve toward a 'scaling' regime, characterized by only the horizon size, was implicit in the original papers where a cosmological role for strings was first suggested [Kibble, 1976, 1980]. The intuitive picture of string evolution outlined in this section was developed by Vilenkin [1981c] and Preskill & Wise [1983] amongst others and a more quantitative analytic discussion was subsequently pursued by Kibble [1985] and Bennett [1986a,b]. The numerical simulations of Albrecht & Turok [1985, 1989], Bennett & Bouchet [1988, 1990] and Allen & Shellard [1990] have been important in delineating the successes and shortcomings of this 'one-scale' model and revealing areas where it needs to be supplemented.

We now turn to a more thorough examination of the one-scale model, describing the main dynamical processes involved and calculating the expected loop distributions. We shall then consider some further analytic developments before discussing numerical approaches to the modelling of string network evolution.

9.3.2 Strings in an expanding universe

In §6.3 we considered the dynamics of strings in a homogeneous expanding universe. Employing a conformally flat metric with conformal time τ, we obtained the equation of motion

$$\ddot{\mathbf{x}} + 2\frac{\dot{a}}{a}(1 - \dot{\mathbf{x}}^2)\dot{\mathbf{x}} = \epsilon^{-1}\left(\frac{\mathbf{x}'}{\epsilon}\right)', \tag{9.3.6}$$

where we have adopted the transverse gauge $\dot{\mathbf{x}} \cdot \mathbf{x}' = 0$ and ϵ is given by

$$\epsilon = \left(\frac{\mathbf{x}'^2}{1 - \dot{\mathbf{x}}^2}\right)^{1/2}, \tag{9.3.7}$$

with dots and primes referring, respectively, to derivatives with respect to time τ and the spacelike parameter ζ. The parameter ϵ is related to the total string energy E by

$$E = a(\tau)\mu \int d\zeta\, \epsilon. \tag{9.3.8}$$

Using (9.3.6) and (9.3.7), an additional constraint on ϵ is easily obtained

$$\dot{\epsilon} = -2\frac{\dot{a}}{a}\epsilon\dot{\mathbf{x}}^2. \tag{9.3.9}$$

If we substitute (9.3.9) in the time derivative of (9.3.8), this implies that the total rate of change of the string energy density will be given by

$$\dot{\rho} = -2\left(\frac{\dot{a}}{a}\right)\rho - 2\left(\frac{\dot{a}}{a}\right)\langle v^2\rangle\rho, \tag{9.3.10}$$

where $\langle v^2\rangle$ is the average of the string velocity squared,

$$\langle v^2\rangle = \frac{\int d\zeta\epsilon\dot{\mathbf{x}}^2}{\int d\zeta\epsilon}. \tag{9.3.11}$$

The first term in (9.3.10) merely arises from the dilution and stretching of the strings due to the expansion of the universe. The second term is a correction to our naive model (9.3.3) and describes the energy loss by the network due to the redshifting of velocities. The string energy in a comoving volume (9.3.8) will grow if $\langle v^2\rangle < 1/2$ and it will decrease otherwise.

With a flat-space value of $\langle v^2\rangle = 1/2$, the string energy would scale exactly like matter. However, in the expanding universe, Hubble damping on large scales will reduce velocities somewhat, so that we can expect $\langle v^2\rangle < 1/2$ with ρ decreasing more slowly than a^{-3}. On the other hand, the behaviour of loops on small scales will eventually approach that of loops in flat space. A non-relativistic loop of length ℓ will be affected by a damping term in (9.3.6) of order $(\ell/d_H)^2$, which becomes increasingly less significant after the loop falls inside the horizon. For $\ell \ll d_{\mathrm{H}}$, evolution will be almost indistinguishable from that in flat space and the energy of the loop will remain nearly constant. A moving loop, however, will have its translational velocity redshifted, like any particle in an expanding universe. For highly relativistic velocities, the loop energy will decay as the inverse of the scale factor, a^{-1}.

9.3.3 The 'scaling' solution

We should note that we have already drawn an implicit distinction between 'infinite' strings ρ_∞ with $\ell \gtrsim L$ and the small loops ρ_ℓ which can be expected to dominate the string energy density of a scaling solution $\rho = \rho_\infty + \rho_\ell$. Since large loops are rare, the exact length-scale ℓ_c at which we make this distinction is unimportant provided that it is not too dissimilar from L – we shall nominally take it to be comparable to the horizon, $\ell_c \sim t$. The important dynamical difference is that a large loop has a large cross-section so it is expected to interact with the long string network through reconnection, whereas a sufficiently small loop 'decouples' because it has a high probability of surviving a sufficient length of time to radiatively decay.

This distinction can be justified by considering the reconnection probability for a loop created at time t with a length,

$$\ell = \alpha t. \tag{9.3.12}$$

Assuming the long string network is in a 'scaling' regime (9.3.4) with γ constant, then the likelihood of a collision between the small loop with mean velocity $\bar{v}$ and the network in a time interval δt is approximately

$$\frac{\bar{v}\ell\delta t}{L^2} = \frac{\alpha\bar{v}}{\gamma^2}\frac{\delta t}{t}. \tag{9.3.13}$$

Hence, the overall probability of the loop surviving from its formation time t_{i} until $t = \infty$ is roughly

$$\exp\left(-\frac{\alpha\bar{v}}{\gamma^2}\right). \tag{9.3.14}$$

Provided γ is of order unity, even loops of length comparable to the correlation length-scale may survive. For very small loops, $\alpha \ll \gamma^2$, reconnection is strongly suppressed and its effect on ρ_∞ may be neglected. On the other hand, for $\gamma \ll 1$ reconnection is much more significant and might appear to preclude a scaling solution. However, statistical mechanics arguments strongly suggest that long string domination is unlikely even in this regime. As we noted in §6.6, for a given length of string, there are many more states associated with its reduction into small loops, than when it remains as a single long string. Even though the long string network is far from equilibrium, the fact that the phase space strongly favours the formation of small loops leads us to expect that reconnection effects will be subdominant under any circumstances.

Let us now describe the rate of energy loss from a 'scaling' string network in terms of a scale-invariant loop production function $f(\ell/L)$. This is defined so that $\mu f(\ell/L)d\ell/L$ gives the energy loss into loops of size between ℓ and $\ell + d\ell$ per unit time per correlation volume L^3. Using the long string energy density (9.3.1), this implies the relation

$$\dot{\rho}_\infty|_{\text{loops}} = \frac{\mu}{L^3} \int_0^{\ell_c} \frac{d\ell}{L} f(\ell/L) = c\frac{\rho_\infty}{L}, \qquad (9.3.15)$$

where the constant c is a measure of the efficiency of loop formation – recall that it merely defines the amplitude of the corresponding term in our heuristic model (9.3.3). Note also that we define $f(x)$ to be the net loop production function, which accounts for both loop creation and reconnection. We could equally extend the integration in (9.3.15) to infinity by simply assuming that $f(x) = 0$ for $x \gtrsim 1$.

Combining (9.3.10) for strings in an expanding universe with this loop energy loss (9.3.15), we obtain the long string evolution equation,

$$\dot{\rho}_\infty = -2\frac{\dot{a}}{a}\left(1 + \langle v^2\rangle\right)\rho_\infty - c\frac{\rho_\infty}{L}. \qquad (9.3.16)$$

By substituting (9.3.1) and (9.3.4), we can obtain the following simple conditions for the 'scaling' solution in the radiation- and matter-dominated eras

$$c_{\text{r}} = \left(1 - \langle v_{\text{r}}^2\rangle\right)\gamma_{\text{r}},$$

$$c_{\text{m}} = \tfrac{2}{3}\left(1 - 2\langle v_{\text{m}}^2\rangle\right)\gamma_{\text{m}}. \qquad (9.3.17)$$

Given that we can expect $\langle v^2\rangle$ to be near its flat-space value of $\langle v^2\rangle \approx 1/2$, it is evident that the loop production rate in the radiation era will be much higher than in the matter era.

The 'one-scale' model also makes testable predictions about the relaxation to scaling of an over- or under-dense initial string configuration. Assuming that $\langle v^2\rangle$ is nearly constant, the solutions for the two epochs are [Bennett & Bouchet, 1989]

$$\frac{\gamma - \gamma_{\text{r}}}{\gamma_0 - \gamma_{\text{r}}} = \left(\frac{a}{a_0}\right)^{\langle v_{\text{r}}^2\rangle - 1},$$

$$\frac{\gamma - \gamma_{\text{m}}}{\gamma_0 - \gamma_{\text{m}}} = \left(\frac{a}{a_0}\right)^{\langle v_{\text{m}}^2\rangle - 1/2}, \qquad (9.3.18)$$

where $\gamma = \gamma_0$ corresponds to the initial configuration. Related expressions can also be derived for the relaxation that occurs during the transition era.

Given the rate of loop production (9.3.15) from a 'scaling' network, we can determine the ensuing loop distribution. We define $n(\ell, t)d\ell$ to be the number density of loops in the length range ℓ to $\ell + d\ell$ at the time t, along with the corresponding loop energy density distribution $\rho_L(\ell, t)d\ell = \mu\ell\, n(\ell, t)d\ell$. From (9.3.15) and given dilution due to expansion, we obtain the rate of change of the loop energy density,

$$\dot{\rho}_L(\ell, t) = -3\left(\frac{\dot{a}}{a}\right)\rho_L(\ell, t) + g\frac{\mu}{L^4}f(\ell/L).\qquad(9.3.19)$$

Here, g is a Lorentz factor which accounts for the fact that loops are created with a non-zero centre-of-mass kinetic energy which is lost through the redshifting of velocities. As we shall see in §9.4.4, typically we have $g = (1 - v_i^2)^{1/2} \approx 1/\sqrt{2}$, where v_i is the initial loop velocity. The loop formation equation (9.3.19) can be easily integrated to yield, during the radiation epoch,

$$\rho_L(\ell, t) = \frac{g\mu}{\gamma^{5/2}t^{3/2}\ell^{3/2}}\int_{\ell/\gamma t}^{\infty}dx\sqrt{x}\,f(x).\qquad(9.3.20)$$

At late times this has the asymptotic form

$$\rho_L(\ell, t) = \frac{\mu\nu_r}{(t\ell)^{3/2}},\qquad(9.3.21)$$

where

$$\nu_r = g\gamma^{-5/2}\int_0^{\infty}\sqrt{x}\,f(x)dx.\qquad(9.3.22)$$

For loops created after matter domination, (9.3.19) the corresponding solution is

$$\rho_L(\ell, t) = \frac{\mu\nu_m}{t^2\ell},\qquad(9.3.23)$$

where

$$\nu_m = g\gamma^{-3}\int_0^{\infty}f(x)dx = gc\gamma^{-3}.\qquad(9.3.24)$$

Note that $f(x)$ need not have the same form during the matter and radiation eras; the values of g, γ and c are also expected to be different.

The one-scale model based on rate equations similar to (9.3.16) and (9.3.19) was first introduced by Kibble [1985] and further developed by Bennett [1986a,b]. These models were more detailed than the above in that they used both loop creation and reconnection functions instead of the one function $f(x)$, attempting to model these by making reasonable assumptions about loop formation mechanisms. Another version of the one-scale model was also employed by Albrecht & Turok [1989] who used it to fit their numerical results. However, the weakness in this approach is that we currently do not have convincing theoretical arguments to precisely determine these functions. As we shall see, numerical simulations have also failed to determine $f(x)$, primarily because of insufficient resolution and dynamic range.

When studying the cosmological implications of strings in the following chapters, we shall use a simple model in which all loops are assumed to be created at the same relative size (9.3.12), a fixed fraction of the horizon, $\ell = \alpha t$. The loop production function $f(x)$ is then

$$f(x) = c\,\delta\left(x - \alpha/\gamma\right). \tag{9.3.25}$$

Before determining $n(\ell, t)$ explicitly under these circumstances, we note the effect of gravitational radiation on the demise of small loops. We recall from §7.5 that loops of length ℓ decay at the rate $\dot{\ell} = -\Gamma G\mu$ where Γ depends on the loop trajectory and typically $\Gamma \approx 65$. For a loop of initial length ℓ_i formed at time t_i, $\ell = \ell_i - \Gamma G\mu(t - t_i)$. Eqns (9.3.21) and (9.3.23) give the loop densities in terms of the initial length ℓ_i. Substituting ℓ_i in terms of ℓ and t into (9.3.21) during the radiation era, assuming that $t \gg t_i$ and employing (9.3.25), we obtain

$$n(\ell, t) = \frac{\nu_r}{t^{3/2}(\ell + \Gamma G\mu t)^{5/2}}, \qquad \ell < \alpha t, \tag{9.3.26}$$

where

$$\nu_r = gc\sqrt{\alpha}\,\gamma^{-3} = g\sqrt{\alpha}\,\gamma_r^{-2}\left(1 - \langle v_r^2\rangle\right). \tag{9.3.27}$$

For the last step in (9.3.27) we have employed the scaling condition (9.3.17). For loops created during the matter-dominated era, the corresponding distribution is

$$n(\ell, t) = \frac{\nu_m}{t^2(\ell + \Gamma G\mu t)^2}, \qquad \ell < \alpha t, \tag{9.3.28}$$

where

$$\nu_m = gc\gamma^{-3} = \tfrac{2}{3}g\gamma_m^{-2}\left(1 - \langle v_m^2\rangle\right). \tag{9.3.29}$$

9.3.4 Further analytic modelling

As we shall see in §9.4, the 'one-scale' model is inadequate because it does not account for the small-scale structure that accumulates on the strings in the form of 'kinks' and 'wiggles'. This arises through string crossings and because stretching due to the expansion is insufficient to completely smooth out modes falling within the horizon. Numerical modelling has delineated the 'fractal-like' nature of this substructure but has yet to conclusively establish its 'scaling' behaviour. Although the smallest scale modes and loop creation sizes appear to grow more slowly than the horizon, there is reason to believe that the overall quantity of substructure does stabilize. With increased simulation resolution, the smallest loops may eventually be observed to scale. But in any case, we would ultimately expect a gravitational back-reaction cut-off on a length-scale $\sim \Gamma G\mu t$ which does 'scale', as was first pointed out by Bennett & Bouchet [1988].

As a first analytic approximation, the string wiggles can be described using the enhanced effective energy density $\tilde{\mu}$ which was discussed in §6.5 and §7.6. That is, viewed from the characteristic length scale L, the string can be regarded as smooth but with an equation of state which differs from that for the unperturbed string,

$$\tilde{\mu}\tilde{T} = \mu^2 \,, \tag{9.3.30}$$

where $\tilde{T}$ is the effective tension. The string motion is slowed by the presence of this small-scale structure and loops are created on scales considerably smaller than L.

The various attempts to improve on the original 'one-scale' model have generally followed the numerical simulations by introducing phenomenological terms to incorporate the dynamical mechanisms which have been identified. A significant step in this direction has been made by Kibble & Copeland [1991] in a 'two-scale' model. The characteristic length scale L was formally defined in terms of the string density (9.3.1) and for a wiggly string network it is different from the persistence length ξ of Brownian strings or the inter-string distance. To remedy this, Kibble & Copeland [1991] introduce a proper persistence length $\tilde{\xi}$, which is defined in terms of the invariant length along the string for which correlations between $\dot{\mathbf{x}}$ and $\mathbf{x}'$ persist and, when compared to the physical persistence length ξ, it provides a measure of the amount of small-scale structure. The relation between $\tilde{\xi}$ and ξ is, roughly, $\tilde{\xi} \approx (\tilde{\mu}/\mu)\xi$.

Using the two scales L and $\tilde{\xi}$, they consider the different network energy loss mechanisms, separating the loop production efficiency c into a 'constant' small loop component c_{L} and a 'variable' large loop component c_∞ (see also Copeland, Kibble & Austen [1992]). In addition, they introduce a new parameter q – the loop kinkiness – which is motivated by the relative over-density in small-scale structure which is necessary to trigger loop formation. Effectively, q is the ratio of the average small loop and long string kink densities ($q > 1$). For small loops c_{L}, the loop formation term in the long string evolution equation (9.3.16) is dependent on the amount of small-scale structure, and L^{-1} must be replaced by the persistence length $\tilde{\xi}^{-1}$. On the other hand, c_∞ is directly related to the frequency of uncorrelated long string intercommutings which depend on the average inter-string distance, a function of both L and $\tilde{\xi}$. Combining these effects in (9.3.16) gives an evolution equation for $\dot{\gamma}$ in terms of both γ and $\kappa \equiv \tilde{\xi}/t$. An auxiliary equation for $\dot{\kappa}$ is found by considering the rate of kink density growth due to long string intercommutings and loop formation.

These coupled equations always exhibit a 'scaling' solution, however, its stability is dependent on the loop kinkiness q which is required to be above a critical value q_c. Copeland *et al.* [1992] find a q_c which is probably exceeded in numerical simulations during the radiation-dominated era, but it is uncomfortably large in the matter era q_c. Whether or not the stability of the 'scaling' solution is supported by these analytic models is still in some doubt, but we note that the conclusions seem sensitive to the simplifying assumptions made regarding the nature of the small-scale structure.

A closely related approach has been adopted by Allen & Caldwell [1991a,b] who describe small-scale structure in terms of explicit 'kink-counting'. They write down a kink density rate equation which depends on the loop production rate. Like Copeland *et al.* [1992], they find a similar critical value for q_c, above which the scaling of small-scale structure is stable. For $q < q_c$, the kink density continues to grow indefinitely. However, by considering gravitational back-reaction, they show that the kink density will actually scale, but with an average kink separation $\sim \Gamma G\mu t$, which is about four orders of magnitude below the horizon for $G\mu \sim 10^{-6}$. More recent work also indicates that the introduction of gravitational back-reaction ensures the stability of the 'scaling' solution [Austen, Copeland & Kibble, 1993].

Finally, a rather different approach using a path-integral formalism has been pursued by Embacher [1992a]. By allowing more functional degrees of freedom for the small-scale structure, this has provided some insight into long string evolution in flat space, but it remains to be successfully applied to strings in an expanding universe.

9.4 Numerical modelling

Given the shortcomings of the analytic approaches to cosmic string evolution, it is fortunate that significant progress has been possible numerically. Numerical simulations, however, also have limitations related to their resolution and dynamic range, so it is hoped that the complementarity of the two approaches will aid progress by providing appropriate extrapolations.

To date, three independent groups have successfully simulated cosmic string evolution using very different numerical techniques. These are Albrecht & Turok [1985, 1989], Bennett & Bouchet [1988, 1989, 1990] and Allen & Shellard [1990]. Despite early controversy, a consensus on the nature of this evolution has emerged, particularly between the latter two groups. All the groups are agreed on the existence and stability of the scaling solution, but quantitative details from the first group have consistently been somewhat at variance. The differences remain a matter of discussion, and a disclaimer noting the partiality of one of the present authors is not inappropriate.

9.4.1 Simulation techniques

In order to successfully perform string simulations a number of serious numerical problems have had to be confronted, and these are worthy of some comment. One of the main features of an interacting string network is the presence of kinks – contact discontinuities in both $\dot{\mathbf{x}}$ and $\mathbf{x}'$ – since four kinks will result from each string reconnection. They pose the chief numerical difficulty because it is well-known that such discontinuities will be progressively smeared if standard finite-difference techniques are employed. While first-order schemes have the disadvantage of poor accuracy, higher-order schemes can develop instabilities because of 'overshoot' oscillations associated with the discontinuities. The earliest codes solving the second-order equations (9.3.6), tackled this problem by introducing artificial viscosity terms [Albrecht & Turok, 1985] or by 'pre-processing' – spreading a kink over several grid points [Bennett & Bouchet, 1988].

Later simulations solved the equations of motion in first-order form, as is more usual in numerical analysis. The simplest form for the equations of motion is in the 'light-cone' gauge with the coordinate curves u, v being the characteristics of the equation of motion, that is,

$$x_u^2 = x_v^2 = 0 \,. \tag{9.4.1}$$

Defining the unit left- and right-moving vectors,

$$\mathbf{p} = \mathbf{x}'/\epsilon - \dot{\mathbf{x}}, \qquad \mathbf{q} = \mathbf{x}'/\epsilon + \dot{\mathbf{x}}, \tag{9.4.2}$$

the string equations in an expanding universe (9.3.6) become

$$\dot{\mathbf{p}} = -\frac{\dot{a}}{a}[\mathbf{q} - \mathbf{p}(\mathbf{p}\cdot\mathbf{q})]\,,$$

$$\dot{\mathbf{q}} = -\frac{\dot{a}}{a}[\mathbf{p} - \mathbf{q}(\mathbf{q}\cdot\mathbf{p})]\,. \tag{9.4.3}$$

In these coordinates one can consider any loop to be made up of two constituent parts – a left-moving loop x_u and a right-moving loop x_v – with the coupling between them weakening as the loop becomes much smaller than the horizon.

With the equations of motion in a form similar to (9.4.3) but using the worldsheet coordinates (ζ, τ), Albrecht & Turok [1989] developed a second string simulation. This appears to tackle the kink discontinuity problem implicitly through an averaging scheme which introduces numerical viscosity. However, to preserve kink identity over long evolution periods requires considerably more sophisticated high resolution techniques. Bennett & Bouchet [1990] endeavoured to achieve this by solving (9.4.3) using the characteristic coordinates u, v. This effectively amounts to a 'shock-fronting' technique because the kinks propagate along lines of constant u or v. However, the complexity of the scheme was transferred to the underlying coordinate grids for u, v and ζ, τ which had to be evolved separately and related by interpolation. On the other hand, Allen & Shellard [1990] used artificial compression methods to prevent kink smearing. They solved (9.3.6) in a first-order form closely related to (9.4.3), but in the 'transverse' gauge using the variables

$$\mathbf{r} = \mathbf{x}' - \epsilon\dot{\mathbf{x}}, \qquad \mathbf{s} = \mathbf{x}' + \epsilon\dot{\mathbf{x}}. \tag{9.4.4}$$

In this case, the equations of motion in flat space reduce to conservation form for which the numerical analysis of shock-handling is well-developed. Analytic estimates of the numerical flux diffusion about a discontinuity can be used to apply appropriate nonlinear corrections. A positive feature of the variable redundancy that remains using (9.4.4) is that the additional gauge and energy constraints can be used to check and demonstrate the accuracy of the evolution.

Another important numerical problem is crossing detection and implementation since this proves to be very computationally intensive. The simulation is divided into segment-sized cubes, and linked lists of neighbouring segments are created. Pairs are then checked for crossing using a cascade of successively more stringent tests. This finally ends by examining the changing volume of the tetrahedron defined by the four endpoints of two nearby segments during the previous timestep. If a crossing has occurred, reconnection or the 'exchange of partners' is performed using an appropriate interpolation scheme.

Initial conditions were created by all groups using the Vachaspati–Vilenkin algorithm described in §9.1.2, with corners rounded and the addition of initial random velocities to aid the rate of relaxation. Periodic boundary conditions were employed on a box of constant comoving size. Various routines then analyzed the properties of the evolving network such as the energy density, loop production and distributions, and long string fractal dimension.

9.4.2 The 'scaling' regime

The radiation era

There is strong evidence from all numerical simulations for the 'scaling' behaviour of the long string network during the radiation-dominated era. Despite the original lack of consensus on the actual long string density, all simulations demonstrated that high or low initial string densities were driven towards a stable fixed point, $\rho_\infty t^2 = $ const., as anticipated by the arguments in §9.3. As we will discuss in some detail in §9.4.3–4, it is important to point out that not all the properties of the string network have been observed to 'scale'. Notable exceptions are the loop production function $f(x)$ defined in (9.3.15) (or the average loop size parameter α) and the distribution of small-scale structure which both appear to be affected by the numerical grid-size cut-off. However, there is good reason for believing that small-scale evolution does not substantially affect the large-scale properties of the network. Along with L and the persistence length ξ, the integrated effect of this small-scale structure appears to stabilize, which is indicated by the constancy of the effective energy density $\tilde{\mu}$ and the total relative energy loss to loops. Further evidence for 'scaling' is also provided by the constant reconnection rate between uncorrelated long strings.

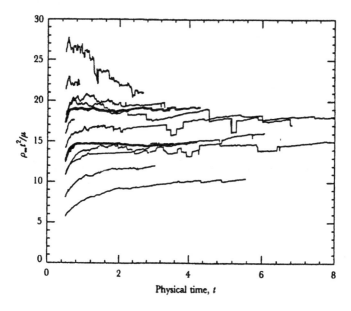

Fig. 9.4. Approach of the long string energy density toward a fixed 'scaling' value in the radiation era. All results are from simulations with over $250{,}000$ grid-points and more than $(22\xi_0)^3$ cells initially [Allen & Shellard, 1990].

The evolution of the string density is plotted in fig. 9.4 for several simulations with an initial string configuration of comoving volume $\geq (22\xi_0)^3$, where ξ_0 is the original correlation length. The rapid approach toward a single 'scaling' density is self-evident, though significant fluctuations can occur because of the limited size of the simulation. When the horizon d_H became comparable with the comoving box size, the simulation was halted, typically after d_H had grown by a factor of 30. The consensus values quoted for the long string scaling density have been: $\rho_\infty = 20(\pm 10)\mu/t^2$ [Bennett & Bouchet, 1988]; $\rho_\infty = 16(\pm 4)\mu/t^2$ [Allen & Shellard, 1990]; $\rho_\infty = 13(\pm 2.5)\mu/t^2$ [Bennett & Bouchet, 1989]. All three results are consistent within the errors and come from very different implementations, with the latter two using the robust high-resolution methods described earlier. Somewhat different from these values is $\rho_\infty = 50(\pm 25)\mu/t^2$ [Albrecht & Turok, 1989].

A striking feature of the evolved string network as shown in fig. 9.5(*a*) is the preponderance of small-scale structure – small loops, kinks and other short wavelength propagation modes or 'wiggles' – despite the much larger overall correlation length of the long strings. This qualitative observation, first noted by Bennett & Bouchet [1988], does not correspond well with the earlier discussion in §9.3.3 of a 'one-scale' model. Moreover, since loop sizes and small-scale structure are affected by the numerical resolution of the simulations it has been important to thoroughly investigate its effect. For example, the dependence of ρ_∞ on the minimum allowed loop size ℓ_{min} can be demonstrated. However, below a certain value related to the initial correlation length, $\ell_{min} \lesssim \xi_0/3$, no discernible effect on ρ_∞ was apparent. Evidently, such a cut-off did not substantially curtail the decay pathways to loops which are available to the long strings. Numerical experiments introducing additional small-scale structure in the initial configuration showed that, while the approach to scaling and loop production were altered, there was no systematic effect on the asymptotic value for ρ_∞ [Bennett & Bouchet, 1990].

The 'scaling' regime should ultimately achieve a state in which the number of reconnections in each Hubble volume per Hubble time also remains constant. With the imposition of a grid-related cut-off which falls in horizon units, this is not to be expected in numerical simulations of limited resolution. However, long string intercommutings between uncorrelated regions should stabilize to a fixed rate (this is either for crossings between different long strings or else by the same string but separated by a length greater than d_H). This is observed to be the case and a long string reconnection rate $I(t)$ per unit physical time and volume is $I(t)dt \approx 2.4(\pm 1)t^{-4}dt$ [Shellard & Allen, 1990]. Despite the presence

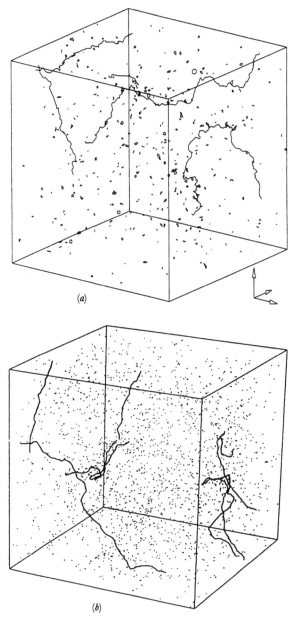

(a)

(b)

Fig. 9.5. (*a*) Portion of the numerical box showing the evolved string network during the radiation-dominated era with the box side-length $L \approx d_H/4$ after expansion by a factor $a \approx 4$ [Allen & Shellard, 1990]. Note the string substructure below the correlation length-scale and the predominance of small loops. The curved segment in the foreground, the result of a recent long string reconnection, is actively forming small loops. (*b*) Matter-dominated era string network with $L = d_H/2$ after an expansion $a = 16$ [Bennett & Bouchet, 1990]. The lower string density and reduced small-scale structure is readily apparent.

of nearly ten strings traversing a volume t^3 in the 'scaling' regime, long string reconnection is a relatively rare event because of low coherent velocities $\tilde{v}$ which, in turn, are closely related to the large effective string energy density $\tilde{\mu}$ (refer to §9.4.3).

Despite the deficiencies of the 'one-scale' model on small-scales, it does help to explain the large-scale properties of the string network. A notable success appears to be the relaxation to scaling during the radiation era, as predicted in the simplified model (9.3.18). Fig. 9.6 shows good agreement between the numerical results and an analytic fit, $\gamma - \gamma_r \sim a^{-0.57}$ using a value $\langle v_r^2 \rangle \approx 0.43$ consistent with all the simulations.

The matter era

The evolution of the long string energy density ρ_∞ during the matter-dominated era is shown in fig. 9.7. Again, the approach to a scaling solution is generally agreed, though it is somewhat slower and less definite than in the radiation era. The much lower value of $\rho_\infty = 3.5(\pm 1)\mu/t^2$ [Bennett & Bouchet, 1989] is roughly consistent with $\rho_\infty \approx 2\mu/t^2$ [Albrecht & Turok, 1990] and $\rho_\infty \approx 4\mu/t^2$ [Allen & Shellard, 1993]. It is immediately apparent from a comparison of fig. 9.5(a) and (b) – evolved networks in the radiation and matter eras – that there is considerably less small-scale structure on long strings in the matter era. We noted in §9.3.3 that a considerably lower rate of loop production is needed to maintain the matter era scaling solution, accounting for their relative scarcity in fig. 9.5(b). Dependence on a numerical loop cut-off was also noticeably

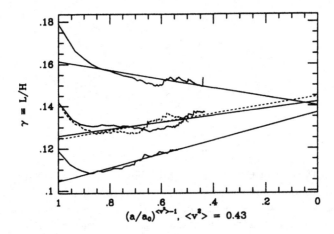

Fig. 9.6. Relaxation toward scaling: Comparisons between the numerical simulations (solid) and the predictions of the 'one-scale' model (dashed) from (9.3.18) [Bennett & Bouchet, 1990].

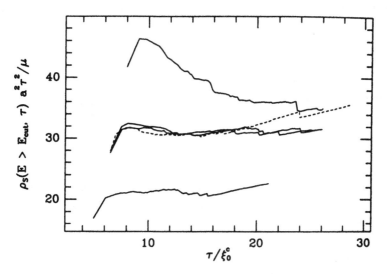

Fig. 9.7. Approach to scaling in the matter era [Bennett & Bouchet, 1990].

weaker. The average string velocity squared in the matter era was found to be $\langle v_{\rm m}^2 \rangle \approx 0.37$.

In the matter era, the 'one-scale' model appears to provide a much less satisfactory quantitative description of the large-scale evolution of the network. An attempt to fit the relaxation expression (9.3.18) shows bad disagreement, with the numerical approach to scaling considerably faster than that predicted analytically [Bennett & Bouchet, 1990]. Features of the string system have also been studied during the transition between matter and radiation domination [Albrecht & Turok, 1989; Bennett & Bouchet, 1990]. Comparison of the large-scale properties with analytic expectations from the 'one-scale' model provides reasonable agreement.

9.4.3 Long string substructure

We have already alluded to the significance of small-scale structure on long strings which originates from modes falling within the horizon and from frequent reconnections. The quantitative regulation of this substructure is not fully understood and remains to be adequately modelled analytically. There are several means by which it has been characterized numerically. The fractal structure can be employed to describe the string substructure on a variety of scales through

$$d = \frac{d \ln E}{d \ln R},$$ (9.4.5)

where E is the average string energy between string points separated by a physical distance R. On the largest scales, the strings are Brownian

($d = 2$), while on the smallest scales the strings become smooth and locally straight ($d = 1$).

An example of the string fractal structure at late times in a radiation era simulation is shown in fig. 9.8. Significant deviations from a slope of unity in this logarithmic plot begin on small scales (well above the numerical grid-size) and continue to grow in strength over two orders of magnitude, gradually attaining $d = 2$. This indicates a broad range of scales for this substructure, with the strongest power on large scales at the transition to a random walk. This fractal dimension is observed to evolve with time, spreading out between the initial correlation length ξ_0 and the growing horizon, but always decaying on any given physical scale – the strings continually become smoother on small scales. The observed slope for the fractal is in excellent agreement for the two high resolution simulations. There is reason to suppose that the intermediate decaying fractal region, (iii) in fig. 9.8, with a near-constant slope d is a slow transient caused by the artificial initial conditions and the limited dynamic range of the simulations. The true 'scaling' solution can be expected to have a more gradual transition from $d = 2$ to $d = 1$ over several orders of magnitude.

We can characterize the overall small-scale structure by measuring the effective energy density $\tilde{\mu}$ at the network correlation length-scale ξ. This simply involves comparing the invariant length of string ζ between points separated by the physical distance ξ. There are a variety of different

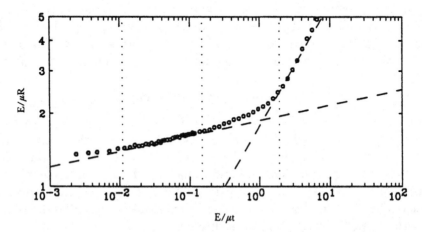

Fig. 9.8.   String fractal dimension in the radiation era after expansion by $a_\mathrm{f}/a_\mathrm{i} \approx$ 4.3 [Shellard & Allen, 1990]. Plotting the fractal profile as $\ln(E/\mu R)$ vs. $\ln(E/\mu t)$ reveals four regions: (i) the Brownian network ($d = 2$) on large scales; (ii) a transition region as d gradually falls; (iii) an intermediate fractal (perhaps a slow transient); and (iv) the smallest scales below ℓ_c where d is near unity.

definitions for ξ, ranging from L in (9.3.1) to the persistence length of Brownian strings. With ξ defined as the length-scale where the fractal dimension $d = 1.33$, the string network in the radiation era yielded [Shellard & Allen, 1990]

$$\xi(t) = 0.40(\pm 0.05)d_{\rm H}$$
$$\tilde{\mu} = 1.9(\pm 0.1)\mu. \tag{9.4.6}$$

These results were consistent with those found by Bennett & Bouchet [1990] using a slightly different ξ. The effective mass density $\tilde{\mu}$ is roughly double the unperturbed μ during the radiation era. It is interesting to note that a careful comparison with (9.3.1) indicates that the inter-string distance is roughly half the correlation length ξ. In the matter era, $\xi(t)$ remains a similar proportion of the horizon size $d_{\rm H}$, but the effective energy per unit length is considerably smaller, $\tilde{\mu} \approx 1.5\mu$. In both epochs, there is some sensitivity to whether the energy density is above or below scaling (with $\tilde{\mu}$ larger in the former case).

With up to half the string energy in small-scale structure, the discussion in §6.5 indicates that the string network dynamics will be significantly affected. Coherent velocities $v(R)$, measured on a physical length scale R, are a prime example. In fig. 9.9 the evolution of the coherent velocity profile is shown during a radiation era simulation. The approach to a stable scaling profile appears to be evident. At the correlation length $\xi(t)$, the coherent velocity has fallen to $v(\xi) \equiv \tilde{v} \approx 0.15$, in striking contrast with the much higher small-scale velocities, $\langle v_{\rm r}^2 \rangle \approx 0.43$ [Shellard & Allen, 1990]. This has important consequences for loop production mechanisms and loop fragmentation. For example, large loops ($\ell \sim L$) with such low coherent velocities become semi-static. Without sufficient angular momentum to sustain their size at formation, the loops collapse into a very small volume and their final demise is rapid and catastrophic. Small wavelength perturbations which maintain a constant amplitude throughout the collapse become relatively enhanced, the kink density per unit length rises and self-intersections eventually become unavoidable, invariably producing a multitude of tiny fragments. The implications for curved regions of string are similar, especially for regions in which uncorrelated long strings have recently reconnected. Again collapse ensues, substructure is enhanced and small loops are produced in profusion.

9.4.4 Loop production mechanisms

The original picture of cosmic string evolution assumed that the long string network lost energy to large long-lived loops with a radius comparable to the correlation length $\xi(t)$. Refinements had each loop self-

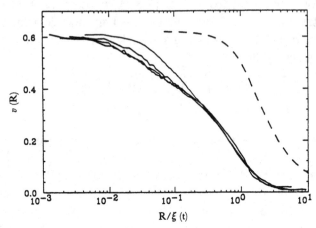

Fig. 9.9.   Long string coherent velocities evolving from initial conditions in which the persistence length of velocity correlations is equal to ξ (the dashed curve). Solid curves represent velocities after expansions $a_f/a_i = 2, 3, 4, 5$. A consistent shape and amplitude is approached on all but the smallest scales where some evolution continues.

intersecting into ten long-lived daughter loops, and simple models for large-scale structure formation were developed on this basis. The simulations, however, indicate a qualitatively different picture of severe fragmentation and direct small loop production by long strings. Net loop production is very strongly peaked about the minimum loop size allowed by the present simulations. This cut-off is a fixed physical energy scale and so loop production is not 'scaling'. In fact, the creation of long-lived non-intersecting loops over an order of magnitude above this cut-off is quite rare, and the overall loop distribution reflects this strong bias towards the smallest scales. (Recent simulations with increased resolution, however, are beginning to observe a significant spread in the loop production function.) At late times, this corresponds to the production of loops with radii peaked at a value less than one hundredth of the horizon ($\alpha \lesssim 0.005$ [Allen & Shellard, 1990] or $\alpha \lesssim 10^{-3}$ [Bennett & Bouchet, 1990]). It is conceivable that such non-scaling behaviour will persist indefinitely, even with improved resolution for the simulations. However, this possibility does not appear very likely. As the energy of the substructure is being spread over a wider and wider range of scales, one expects that the amplitude of the smallest-scale perturbations will eventually become insufficient for loop formation, and the loop production function will begin to 'scale'. At present, we can only provide an upper bound on the average loop size α. In any case, it may be that gravitational radiation back-reaction will intervene first to determine α.

We also note that the centre-of-mass loop velocity is strongly correlated with size. The smallest loops, the final products of these fragmentation processes, are generally created with highly relativistic velocities. Bennett & Bouchet [1990] find an initial average velocity for stable nonintersecting loops of $\langle v_i \rangle \approx 1/\sqrt{2}$.

As mentioned in §9.4.3, there is a close relationship between the reconnection of uncorrelated long strings and profuse loop creation – particularly evident in video animations of the evolving string network. This strong correlation between long string reconnection and subsequent small loop production has been demonstrated quantitatively [Shellard & Allen, 1990]. Cross-correlations between small loop production events yielded a $\sim r^{-2}$ correlation (consistent with a linear source). This is related to the transient $\sim r^{-2}$ two-point loop correlation function observed previously by Turok [1985] and Bennett & Bouchet [1989].

By eliminating long string reconnections – only allowing loop production on small scales – this second mechanism for loop production can be studied. Such a string network will eventually dominate the universe because the inter-string distance will not scale appropriately. In the radiation era, the long string energy density ρ_∞ grows relative to the background, so that $\rho_\infty t^2 \propto t^{1/4}$. What is remarkable, however, is that the small-scale string straightening processes maintain scale-invariant values for the correlation length $\xi(t)$ and $\tilde{\mu}$ [Shellard & Allen, 1990]. Small loops are generally created in regions of high string curvature and rarely in isolation as shown in plots by Bennett & Bouchet [1990].

In contrast, reconnections between long strings appear to have more catastrophic consequences because they generate large curvature and cause the collapse and fragmentation of much longer regions of string. Apparently, the more quiescent small-scale string straightening limits $\tilde{\mu}$ as well as $\tilde{v}$, while reconnections between long strings predominantly act to stabilize the inter-string distance relative to the horizon. In the matter and radiation eras, these processes have a different significance which may explain the variable merit of the 'one-scale' model in the different epochs.

9.5 String domination

Cosmic string evolution which approaches scale-invariance was examined in some detail in §9.3-4. There are also circumstances in which the relative energy density in strings grows and eventually dominates the universe. As discussed in §3.3.1, strings could be non-intercommuting and simply 'pass through' each other when crossing, or else they could have topological constraints causing them to become entangled and resulting in the creation of a third string stretching between them. In both of these cases, strings would eventually dominate the universe, though at different rates. Compatibility with the standard cosmology, then, constrains such strings to be relatively light.

9.5.1 Non-intercommuting strings

The classification of $U(1)$-strings in §3.3 does not distinguish between the two topologically allowed outcomes for equal winding-number strings – either 'exchanging partners' or simply passing though each other at a crossing point. A detailed study of the structure and dynamics of these strings in §4.3 was required to show that intercommuting was the actual outcome to be expected. It may be that for some other strings the alternate topological possibility is favoured or else the end result of a string interaction may depend on collision parameters, giving a very small effective intercommuting probability. The implications of such a difference are enormous. No longer can the string network actively create loops through self-intersections, so the mechanism which previously drove the evolution towards a 'scaling' solution is now absent.

We can build a simple analytic model of the evolution of non-intercommuting strings which aids intuition [Vilenkin, 1984b]. As usual, the early evolution of strings after their formation at t_0 will be friction-dominated. This will continue until $t_* \sim m_{\rm pl}^3/\eta^4$, where $m_{\rm pl}$ is the Planck mass, after which the strings will begin to oscillate freely at relativistic velocities. The persistence length of the strings $\xi(t)$ by this stage will have grown from the original value $\xi_0 = \xi(t_0)$ to $\xi \sim t$. However, the inter-string distance L will depend on the original density of strings at the phase transition because it is not being driven toward scaling by intercommutings. After t_* the strings are simply stretched on scales outside the horizon and partially straightened on smaller scales, while maintaining $\xi \sim t$. At any time, the energy of a long string between two points separated by a physical distance R is

$$E \sim \tilde{\mu} R^2/\xi, \tag{9.5.1}$$

where $\tilde{\mu}$ is the effective string energy on scale ξ. The energy density of the network is then

$$\rho_\infty(t) \sim \frac{\tilde{\mu}}{\xi_0 \xi(t)} \frac{a(t_0)}{a(t)}, \tag{9.5.2}$$

so that with $\xi \sim t$ and assuming that $\tilde{\mu}$ remains constant in time,

$$\rho_\infty(t) \propto 1/ta(t). \tag{9.5.3}$$

In the radiation era, this implies $\rho t^2 \propto t^{1/2}$ with a constant comoving mass, while in the matter era, $\rho t^2 \propto t^{1/3}$. Clearly, in either case, the strings will ultimately dominate the universe. Assuming this to be in the matter-dominated era, string domination occurs at a time t_s given by

$$\frac{\rho_\infty}{\rho}(t_s) \sim \left(\frac{t_s}{t_{eq}}\right)^{1/3} \left(\frac{t_{eq}}{t_0}\right)^{1/2} \frac{t_0}{\xi_0} G\mu \sim 1, \qquad (9.5.4)$$

where the background energy has been taken to be $\rho \sim 1/Gt^2$.

Somewhat surprisingly, numerical simulations of the evolution of non-intercommuting strings yield approximately equivalent energy density growth rates as this simple model; $\rho t^2 \propto t^{1/2}$ in the radiation era and $\rho t^2 \propto t^{1/3}$ during the matter era [Shellard & Allen, 1990]. The simulations indicate, however, that the naive model of string evolution that led to (9.5.3) is probably over-simplified. The substructure energy in the simulations appears to grow indefinitely, so that $\tilde{\mu} \propto t^{1/4}$ in the radiation era. The most notable qualitative difference in the evolution is that the strings acquire a very convoluted or 'knotted' appearance, which is generally fairly localized.

We make a slight digression here to comment on the cosmological implications of non-intercommuting strings, rather than deferring this to chapter 10. Expansion during a string dominated epoch with $\rho_\infty \propto 1/a(t)t$ will be proportional to time, $a(t) \propto t$, a fact easily observed from the evolution equation, $(\dot{a}/a)^2 \sim G\rho$. This is the same expansion rate as for a matter-filled $\Omega < 1$ universe at late redshifts with $2 + z \ll \Omega^{-1}$. Thus the evolution of a string-dominated universe with $\Omega = 1$ can mimic the behaviour of an $\Omega \sim (2+z_s)^{-1} < 1$ universe [Vilenkin, 1984b]. This could potentially solve the $\Omega = 1$ dark matter problem as well as providing a beneficial increase in the apparent age of the universe (refer to §1.3.2 and §1.3.4). However, it is difficult to envisage these strings clustering on galactic and galaxy-cluster scales, so small-scale dynamical measurements of a dark matter density would require an alternative explanation. Being light, they are also unable to provide sufficient energy density perturbations for structure formation. Indeed, if they were to dominate before t_{eq} there could be no galaxies, since there is no linear gravitational clustering in a universe with $a(t) \propto t$.

The fact that non-intercommuting strings must not dominate too early, provides a constraint on the symmetry breaking scale for their formation. Taking the apparent density of our universe in the range $0.2 \lesssim \Omega \lesssim 0.5$, constrains string domination to occur at a redshift $0 < z_s \lesssim 3$. Using the parameter values $t_{eq} \sim 10^{11}$s, $t_s \sim 10^{17}$s, and $t_0 \sim m_{pl}/\eta^2$ with $\xi_0 \lesssim t_0$, we obtain from (9.5.4) $G\mu \lesssim 10^{-20}$ and $\eta \lesssim 10^9$GeV. This constraint is appropriate for a first-order transition with $\xi_0 \sim t_0$, however, for a second-order transition with $\xi_0 \sim \eta^{-1}$ we obtain much lighter strings, $G\mu \lesssim 10^{-30}$ and $\eta \lesssim 10^4$GeV. In the latter case, the nearest string to earth would be at a distance comparable to the sun. Nevertheless, such light strings are extremely difficult to observe, unless they are superconducting.

9.5.2 A tangled network

If, at the string-forming phase transition $G \to H$, the fundamental group $\pi_1(G/H)$ is non-abelian, then strings corresponding to non-commuting elements of π_1 cannot pass through each other [Kibble, 1976]. A third string, as discussed in §3.3.1, will stretch between them after crossing. Z_N-strings will similarly form a complicated network with monopole vertices where the n strings join (refer to §4.2.2). We could envisage such a tangled network equilibrating and 'freezing' soon after formation, only to become conformally stretched by the expansion [Vilenkin, 1984b]. If this were the case, then the density of the strings for $t > t_0$ would behave simply as

$$\rho_\infty \sim \mu \xi_0^{-2} \left(\frac{a(t_0)}{a(t)} \right)^2 . \tag{9.5.5}$$

Again, these strings must eventually dominate the universe, though at an even earlier time t_s given by

$$\frac{\rho_\infty}{\rho}(t_s) \sim G\mu \frac{t_0^2}{\xi_0^2} \frac{t_{\rm eq}}{t_0} \left(\frac{t_s}{t_{\rm eq}} \right)^{2/3} \sim 1 \tag{9.5.6}$$

Repeating the previous argument for non-intercommuting strings, if $z_s \sim 1$ with $\xi_0 \lesssim t_0$ in a first-order transition, then the strings must be very light with $G\mu \lesssim 10^{-29}$ constraining $\eta \lesssim 10^4 \text{GeV}$. Such strings arising in a second-order transition ($\xi_0 \sim \eta^{-1}$) would dominate the universe very soon after formation. The discussion of their cosmological implications closely follows that given in §9.5.1.

It must be pointed out that in this picture of a 'frozen' tangled network we have ignored some dynamical degrees of freedom which may have a significant influence on the evolution. One of the questions here is whether or not the network has a stable static configuration. If it does, one expects it to be some kind of a regular lattice; for example, a hexagonal lattice could be formed by Z_3-strings in two dimensions, or a cubic lattice by Z_6-strings in three dimensions. Once the existence of an equilibrium network is established, one can ask under what circumstances will it act as an attractor for a sufficiently wide class of initial conditions. We shall see in §9.6 that numerical simulations of Z_3-strings give no evidence for a stable network formation, and it will be argued that, with some additional assumptions, the network should exhibit 'scaling' behaviour. For non-abelian strings it is similarly possible that intercommuting between strings corresponding to the same element of π_1 will be sufficiently efficient at vertex removal into closed loops to enable rapid disentanglement and scale-invariant evolution. The issue deserves closer examination.

9.6 String–monopole network evolution

In §9.5.2 the formation of a Z_N-string network was discussed. The first stage of symmetry breaking at the scale η_m will produce monopoles, followed by a string network at the second mass scale η_s,

$$ G \xrightarrow{\eta_m} K \times U(1) \xrightarrow{\eta_s} K \times Z_N . \qquad (9.6.1) $$

The monopole mass and string tension are approximately given by

$$ m \sim \frac{4\pi\eta_m}{e} , \qquad (9.6.2) $$
$$ \mu \sim \eta_s^2 , $$

with gauge coupling e. The monopoles each have N strings emanating from them and their number density n implies a typical string segment length,

$$ \xi \sim n^{-1/3} . \qquad (9.6.3) $$

The tension of each of the strings, $\sim \mu$, will induce an acceleration on the monopole, $a \sim \mu/m$, with the sum of the N contributions determining its motion. For $\xi \gg m/\mu$, monopoles can move with highly relativistic velocities.

The problem of the evolution of a string–monopole network has been studied by Vachaspati & Vilenkin [1987]. For monopoles with unconfined magnetic charge, providing a radiation damping mechanism, they argued that the network neither approached a static configuration nor disintegrated into small nets, instead exhibiting 'scaling' behaviour. Assuming there to be some scale-invariant partition between the energies in the monopoles and the strings, $\rho_m = \beta\rho$, $\rho_\infty = (1 - \beta)\rho$, we can rewrite the pressure P due to both the strings and the monopoles as

$$ P = \tfrac{1}{3}\rho \left[\beta + (1 - \beta)\left(2\langle v_s^2\rangle - 1\right)\right] . \qquad (9.6.4) $$

Depending on the value of the constant β, $P = \chi\rho$ can be either positive or negative, though it must lie in the range $-\tfrac{1}{3} < \chi < \tfrac{1}{3}$. The energy of the network will be altered by work done by this pressure in an expanding universe, as well as losses due to small loop production, monopole radiation, and monopole–antimonopole annihilation. The approriate terms for the first two mechanisms are familiar from the study of the 'one-scale' model (9.3.16), but we will attempt to estimate the magnitude of the remaining terms.

In the symmetry breaking scheme (9.6.1), the intermediate group K will be broken down to $SU(3) \times U(1)$. Typically, then, the monopoles in the model will carry $SU(3)$ or $U(1)$ magnetic charges which will be unconfined, that is, without all the flux trapped along the strings. Such monopoles with charge g will radiate when accelerated, a, by the attached strings. The electromagnetic radiation power can be estimated classically,

$$w \sim \frac{g^2}{6\pi} a^2 \sim \frac{\eta_s^4}{24\pi\eta_m^2} \,. \tag{9.6.5}$$

To have a sufficient annihilation cross-section, a monopole–antimonopole pair would be expected to have already lost most of its original energy through radiation, so that the final demise provides an inconsequential energy loss.

Combining all the appropriate terms we can write down an energy balance equation for the network evolution in an expanding universe

$$\dot\rho = -3\frac{\dot a}{a}(P + \rho) - nw - C\frac{\rho}{\xi} \,, \tag{9.6.6}$$

where C summarizes the small loop production function (here, small loops include small networks with several vertices). Now assuming that the energy partition parameter β does not change, we can rewrite the energy density as

$$\rho = b\mu n^{2/3} \sim b\mu\xi^{-2} \,, \tag{9.6.7}$$

where b is a constant. Assuming a power-law expansion $a(t) \propto t^\nu$, (9.6.6) becomes

$$3\dot\xi = \kappa\frac{\xi}{t} + \lambda \,, \tag{9.6.8}$$

where $\kappa = 9\nu(\chi + 1)/2 + 3C/2$ and $\lambda = 3w/2b\mu$. The solution of (9.6.8) takes the form

$$\xi(t) = \gamma t + (\xi_0 - \gamma t_0)\left(\frac{t}{t_0}\right)^{\kappa/3} \,, \tag{9.6.9}$$

where $\gamma = \lambda/(3 - \kappa)$, $\xi_0 = \xi(t_0)$ and $\kappa \neq 3$. (For the special case of $\kappa = 3$ the solution is $\xi = (\lambda/3)t \ln t + \text{const.}$) Causality is inconsistent with $\kappa \geq 3$, since ξ would grow faster than the horizon. For $\kappa < 3$ and sufficiently large t, the solution approaches the familiar scaling form

$$\xi(t) \approx \gamma t \,, \tag{9.6.10}$$

where $\gamma \sim \eta_s^2/24\pi\eta_m^2 \lesssim 0.01$ (for $\eta_s \lesssim \eta_m$). Compared to the standard cosmic string scenario, a considerably smaller ratio of the string correlation length to the horizon size is predicted by this simple model.

This picture of network evolution was checked with a simple numerical simulation for Z_3-strings in flat space [Vachaspati & Vilenkin, 1987]. In this simulation, transverse string degrees of freedom were eliminated and the system was allowed to evolve as a set of straight string segments connecting the monopole vertices. The direction of the motion of each monopole (assumed to be moving at $v \approx c$) was determined from the vectorial sum of the tensions of the attached strings. When a monopole–antimonopole pair approached within a fixed minimum distance, they were removed and the strings appropriately reconnected. The main outcome was the predicted self-similar evolution. The distribution of string lengths was found to be well approximated by an exponential, $\mathcal{N}(\ell) \propto \exp(-\ell/\xi(t))$, where $\mathcal{N}(\ell)$ represents the number of strings with length greater than ℓ. The evolution of this constrained network was found not to be strongly dependent on intercommuting, and the dominant 'infinite network' did not show a tendency to break up into small nets or to approach an equilibrium configuration.

Finally, we can make some brief comments about the cosmological implications of these strings. In the scenario outlined above, the main energy loss mechanism is the radiation of gauge quanta by highly relativistic monopoles. The typical frequency of radiation emitted at time t is $\omega(t) \sim \mu\epsilon/m^2 \sim (\mu/m)^2 \gamma t$, where m and $\epsilon \sim \mu\xi$ are, respectively, the monopole mass and energy. To illustrate orders of magnitude, suppose that $\eta_{\rm m} \sim \eta_{\rm s} \equiv \eta$. In this case, $\omega(t) \sim \eta^2 t$ and quanta being emitted at the present time have enormous energies, $\omega/\eta \sim 10^{41}\eta_{\rm GeV}$, where $\eta_{\rm GeV} = \eta/1{\rm GeV}$ [Vachaspati & Vilenkin, 1987].

10
Cosmological implications of strings

10.1 Summary of string properties and evolution

10.1.1 Gravitational properties of strings

Gravitational interactions of cosmic strings are characterized by a dimensionless parameter

$$G\mu \sim \left(\frac{\eta}{m_{\text{pl}}}\right)^2 ,$$

(10.1.1)

where G is Newton's constant, μ is the string mass per unit length, m_{pl} is the Planck mass and η is the string symmetry breaking scale. If η is at a grand unification scale $\eta \sim 10^{16}$GeV, then $G\mu \sim 10^{-6}$, while for an electroweak scale $\eta \sim 100$ GeV and $G\mu \sim 10^{-34}$.

We found in chapter 7 that the gravitational field outside a static straight string is given by the metric (7.1.11). It describes a conical space, which is simply a Euclidean space with a wedge removed and the two inner faces identified; the wedge has angular size

$$\Delta = 8\pi G\mu .$$

(10.1.2)

Each transverse section of this space is not a plane, but a cone with a 'deficit angle' Δ (see fig. 10.1).

The metric (7.1.11) is locally flat and can be brought to a Minkowski form in any region not surrounding the string. This implies that the presence of the string has no effect on physical processes in such a region. In particular, a test particle initially at rest relative to the string will remain at rest and will not experience any gravitational force. Although the space around the string is locally flat, its global structure is different from that of Euclidean space. One of the effects of this difference is illustrated in fig. 10.1. Two particles move towards the string along parallel paths with the same velocity **v**. As they pass the string, the particles start moving towards one another and eventually collide. The relative velocity of the particles after they pass the string is

$$\delta u = v\gamma\Delta\,,\qquad(10.1.3)$$

where the Lorentz factor $\gamma = \left(1 - v^2\right)^{-1/2}$ appears after transforming to the reference frame of one of the particles, and assuming that the line connecting the particles is perpendicular to the string direction. The effect of a string on particle motion is analogous to the Aharonov–Bohm effect; spacetime curvature is confined to the string core, but its effect is 'felt' by the particles propagating in the flat spacetime region around it.

A closed loop of string oscillates under the action of the string tension with a period comparable to its length. Consider a string segment much shorter than the local curvature radius of the loop, R. The gravitational field in the vicinity of this segment (at distances much smaller than R) is approximately described by the straight string metric (7.1.11) transformed to the appropriate Lorentz frame. However, at large distances from the loop the gravitational field averaged over the period of oscillation is just the usual Newtonian field. Hence, in contrast to a straight string, a closed loop attracts matter with the same force as any other object of equal mass.

Oscillating loops of string lose their energy by emitting gravitational waves at the rate

$$\dot{E} = \Gamma G\mu^2\,,\qquad(10.1.4)$$

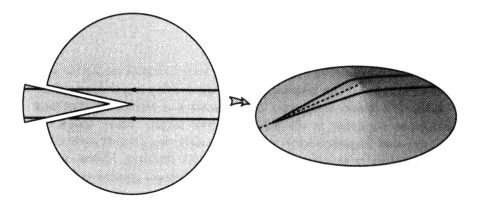

Fig. 10.1. The conical spacetime about a cosmic string is constructed by removing a wedge and identifying along its edges.

where the numerical factor Γ depends on the loop's shape and trajectory but is independent of its length L; typically, $\Gamma \sim 50$–100, with an average value $\langle \Gamma \rangle \approx 65$. As the loops lose their energy, they gradually shrink and finally disappear. The lifetime of a loop is approximately

$$\tau \sim \frac{M}{E} \sim \frac{L}{\Gamma G \mu}, \qquad (10.1.5)$$

where $M = \mu L$ is the loop's mass.[*]

10.1.2 String evolution

Strings produced at an early phase transition will form a tangled network permeating through the entire universe. These strings can be expected initially to be Brownian with a step length ξ determined by the correlation length of the symmetry breaking scalar field. There is about one string segment of length $\sim \xi$ per volume $\sim \xi^3$, and the energy density of the network is thus $\rho \sim \mu \xi^{-2}$. Since causality constrains correlations to be within the horizon, we must have $\xi < d_{\mathrm{H}}$, though typically for realistic phase transitions ξ is initially much smaller. At formation only about one fifth of the string length is in small closed loops, with the majority forming an infinite network.

Subsequently, the motion of the strings is heavily damped, because of the high background energy density, until a time

$$t_* = (G\mu)^{-2} t_{\mathrm{pl}}. \qquad (10.1.6)$$

During this damped epoch, substructure on small scales will be smoothed out and loops will rapidly shrink and decay. As discussed in §9.1, the length-scale of the network can be expected to grow as

$$\xi \sim t \left(\frac{t}{t_*} \right)^{1/4}. \qquad (10.1.7)$$

At times $t \gg t_*$, the effect of the damping force becomes negligible and, under the influence of the tension, the strings develop relativistic velocities. Oscillating strings frequently self-intersect and reconnect, chopping off convoluted segments in the form of closed loops. These loops can self-intersect and fragment further but ultimately this process ceases and a non-intersecting population of daughter loops remains. These stable loops oscillate and gradually lose their energy through gravitional radiation.

---

[*] We remind the reader that the length parameter L is defined as $L = M/\mu$. The actual length of the loop changes as the loop oscillates; its average value is $L/\sqrt{2}$.

This energy loss mechanism drives the string density down, preventing the network from dominating the universe. Indeed, there is strong evidence that the cosmological evolution of the network becomes self-similar, approaching what is termed a 'scaling' solution. In the simplest scale-invariant model, all that changes is the overall correlation length of the network ξ, which grows in proportion to the horizon (9.3.4),

$$\xi = \gamma_\xi t, \tag{10.1.8}$$

where γ_ξ is constant (not a Lorentz factor). Since the strings remain Brownian on scales above ξ the energy density of the infinite network takes the form

$$\rho_\infty = \zeta \frac{\mu}{t^2}. \tag{10.1.9}$$

While γ_ξ and ζ ($\sim \gamma_\xi^{-2}$) are closely related, the presence of small-scale structure on the strings prevents this correspondence from being exact (along with a number of other subtleties discussed in §9.3–4). Reliable measurement of these parameters has to date only been possible with numerical simulations and the values of ξ, ζ and other network parameters for the matter and radiation eras are presented in table 10.1.

Numerical simulations reveal that long strings in the 'scaling' solution are not completely smooth, possessing significant substructure over a wide range of length-scales below ξ. During the radiation era, this small-scale structure almost doubles the effective string energy density $\tilde{\mu}(\xi) \approx 1.9\mu$ averaged at the correlation length-scale ξ, while during the matter era $\tilde{\mu} \approx 1.5\mu$. Loop production acts against the build-up of these small-scale perturbations, as does string stretching due to the expansion of the uni-

Table 10.1. Scaling parameters for a long string network in the radiation and matter dominated epochs.

Epoch	Radiation:		Matter:	
Properties	BB	AS	BB	AS
ζ	13 ± 2.5	16 ± 4	3.5 ± 1	4.0 ± 2.0
$\xi(t)$	$\sim d_H/3$	$\sim 0.4 d_H$	$\sim d_H/3$	$\sim 0.4 d_H$
$\tilde{\mu}_\xi$	$\sim 1.8\mu$	$\sim 1.9\mu$	$\sim 1.4\mu$	$\sim 1.6\mu$
$v_{\rm rms}$	0.66	0.62	0.61	~ 0.58
$\langle v \rangle_\xi$	—	~ 0.15	—	~ 0.15
α	$\lesssim 10^{-3}$	$\lesssim 5 \times 10^{-3}$	$\lesssim 10^{-3}$	$\lesssim 10^{-2}$

Collated from the published results of BB [Bennett & Bouchet, 1990, 1991] and AS [Allen & Shellard, 1990; Shellard & Allen, 1990]. These parameters are all defined in §9.3.

verse. Ultimately, damping due to gravitational radiation will smooth strings on scales smaller than $\Gamma G\mu t$. One direct consequence of this substructure is that overall string motion is slowed and, although we have the expected velocity $\langle v \rangle \sim 1/\sqrt{2}$ on small scales, at the correlation length the average coherent motion is reduced to $\langle v \rangle \sim 0.15$. Relatively rare intercommutings between these turgid and wiggly long strings, however, can result in much more rapid motion by segments of long string which can be sustained for about a Hubble time and which is associated with copious loop production. We note that, unlike straight strings, strings possessing small-scale structure produce non-vanishing Newtonian fields. The effective gravitational mass per unit length is then $\mu_{\rm grav} = \tilde{\mu}(1 - \mu^2/\tilde{\mu}^2)$.

True scale invariance implies that the length of stable loops produced by the network should also remain a fixed fraction of the horizon,

$$\langle \ell \rangle = \alpha t. \qquad (10.1.10)$$

At present we do not have a reliable estimate of α but numerical simulations strongly indicate that loops are created on very small scales[*]

$$\alpha \lesssim 5 \times 10^{-3}. \qquad (10.1.11)$$

This surprisingly small relative size is a result of the small-scale structure on the long strings and, since this is cut off by gravitational back-reaction, we may reasonably expect that $\alpha \gtrsim \Gamma G\mu$. The loops are observed to form with relativistic initial velocities; the rms value is $v_{\rm rms} \approx 0.7$.

A discussion of many of the cosmological implications of strings is dependent on the density and length distribution of the loops produced by the network. Unfortunately, considerable uncertainties remain in characterizing these properties, so we shall base estimates on a simplified model whose parameters are only loosely constrained by current numerical simulations. As discussed in §9.3.3, it is not unreasonable to assume that all loops produced at a given time t_ℓ have roughly the same length, $\ell = \alpha t_\ell$. If these loops are created on scales well above the gravitational back-reaction cut-off, $\alpha \gg \Gamma G\mu t$, they will decay by a time $t \approx (\alpha/\Gamma G\mu)t_\ell \gg t_\ell$. During the radiation era, the number density of loops of length ℓ in the interval $d\ell$ is $n(\ell, t)d\ell$, where

$$n(\ell, t) = \frac{\nu}{t^{3/2}(\ell + \Gamma G\mu t)^{5/2}}, \qquad \ell < \alpha t, \qquad (10.1.12)$$

with

$$\nu \approx 0.4\zeta\alpha^{1/2}. \qquad (10.1.13)$$

---

[*] This is a conservative bound which comes from simulations by Allen & Shellard [1990]. A stronger bound, $\alpha < 10^{-3}$, is suggested by Bennett & Bouchet [1989, 1990].

Here, we have substituted numerical values from table 10.1 into the expression for the loop number density (9.3.26–27), noting also that $\zeta = \gamma_\xi^{-2}$.

For order of magnitude estimates in the following discussions, it is convenient to define the quantity

$$n_\ell(t) = \ell n(\ell, t), \tag{10.1.14}$$

which corresponds to the number density of loops with lengths $\sim \ell$ in the interval $\Delta \ell \sim \ell$. For $\Gamma G \mu t \lesssim \ell \lesssim \alpha t$, (10.1.12) yields

$$n_\ell(t) \sim \zeta \alpha^{1/2} (t\ell)^{-3/2}. \tag{10.1.15}$$

For $\ell < \Gamma G \mu t$, $n_\ell(t)$ linearly decreases with ℓ ($n_\ell \propto \ell$), and it is usually a good approximation to use (10.1.15) with a cut-off at $\ell \sim \Gamma G \mu t$.

The total mass density of loops at time $t \lesssim t_{eq}$ is then

$$\rho_L \sim \left(\frac{\alpha}{\Gamma G \mu} \right)^{1/2} \frac{\zeta \mu}{t^2}. \tag{10.1.16}$$

For $\alpha \gg \Gamma G \mu$ the loop density is much greater than that in long strings (10.1.9), simply because the loops redshift as matter. However, the possibility that $\alpha \sim \Gamma G \mu$ is not excluded by current constraints, so both densities could be comparable. In either case, the dominant contribution to ρ_L is given by the smallest loops, $\ell \sim \Gamma G \mu t$, on the verge of final decay.

For closed loops formed during the matter era we have from (9.3.28–29),

$$n(\ell, t) = \frac{\nu}{t^2 (\ell + \Gamma G \mu t)^2}, \tag{10.1.17}$$

with

$$\nu \approx 0.12\zeta. \tag{10.1.18}$$

(Note that the parameters ζ, α and ν take different values in the radiation and matter eras.) It follows from (10.1.14) and (10.1.17) that

$$n_\ell(t) \sim \nu t^{-2} \ell^{-1}, \qquad \ell \gtrsim \alpha t_{eq}, \Gamma G \mu t. \tag{10.1.19}$$

For $t_{eq} < t < (\alpha/\Gamma G \mu) t_{eq}$ some loops formed during the radiation era still survive. The length distribution of such loops is

$$n_\ell(t) \sim \frac{\zeta \alpha^{1/2}}{(t_{eq}\ell)^{3/2}} \left(\frac{t_{eq}}{t} \right)^2, \qquad \Gamma G \mu t \lesssim \ell \lesssim \alpha t_{eq}. \tag{10.1.20}$$

The total mass density of loops in the matter era can also be easily determined. At $t > (\alpha/\Gamma G \mu) t_{eq}$, when all the surviving loops were formed in the matter era, it is

$$\rho_L \sim \frac{\nu \mu}{t^2} \ln \left(\frac{\alpha}{\Gamma G \mu} \right), \tag{10.1.21}$$

which always remains comparable to the long string density (10.1.9).

We note finally that the model of string evolution described here is not directly applicable to global strings. The necessary modifications of the model and the cosmological implications of global strings will be discussed in chapter 12. Throughout this chapter we shall assume that the strings are formed as a result of a local gauge symmetry breaking.

10.2 Microwave background anisotropies

If cosmic strings exist and are sufficiently massive, their gravitational interactions can give rise to a number of observable effects. Unusual gravitational properties of strings give these a characteristic 'stringy' signature, distinguishing them from effects due to other massive objects. We begin with an analysis of the anisotropy induced by cosmic strings in the temperature of the microwave background radiation.

We mentioned in §10.1.1 that two objects initially at rest start moving towards one another when a string passes between them. If the direction of the string and its velocity $\mathbf{v}$ are both perpendicular to the line connecting the two objects, then their relative velocity is given by (10.1.3). If one of the objects is a source of radiation and the other is an observer, then the observer will detect a discontinuous change in the frequency of radiation due to the Doppler shift. In a cosmological setting, the string is backlit by a uniform black-body radiation background, and the Doppler shift results in a discontinuous change of the background temperature across the string [Kaiser & Stebbins, 1984; Gott, 1985]. The magnitude of this change is

$$\frac{\delta T}{T} = \delta u = 8\pi G\mu v\gamma . \qquad (10.2.1)$$

It is not difficult to generalize (10.2.1) for arbitrary angles between the string, its velocity, and the line of sight [Vachaspati, 1986; Vilenkin, 1986b]. Let $\hat{\mathbf{s}}$ be a unit vector pointing along the string. Since only the transverse motion of any string is observable, we can take $\mathbf{v} \cdot \hat{\mathbf{s}} = 0$. In the rest frame of the string the observer passes by the string with velocity $-\mathbf{v}$. If the observer is at rest with respect to the sources of radiation on one side of the string, then the velocity of the sources on the other side is $-\mathbf{v}+\delta\mathbf{u_0}$, where $\delta\mathbf{u_0} = 8\pi G\mu\mathbf{v}\times\hat{\mathbf{s}}$. The corresponding velocity difference in the observer's frame is $\delta\mathbf{u} = \gamma\delta\mathbf{u_0}$. The temperature discontinuity can be found from the Doppler formula $\delta T/T = \delta\mathbf{u} \cdot \hat{\mathbf{n}}$, where $\hat{\mathbf{n}}$ is a unit vector along the line of sight; thus,

$$\frac{\Delta T}{T} = 8\pi G\mu\gamma\,\hat{\mathbf{n}} \cdot (\mathbf{v} \times \hat{\mathbf{s}}) . \qquad (10.2.2)$$

The temperature distributions produced by oscillating string loops has been studied by Vachaspati [1986] and Stebbins [1988]. Not surprisingly, the discontinuity is still given by (10.2.2) with the local values of $\mathbf{v}$ and $\hat{\mathbf{s}}$. Very large temperature fluctuations are obtained near cusps, where $\gamma \to \infty$.

The pattern of the microwave background temperature fluctuations on the sky is a superposition of the contributions of strings from different redshifts between the surface of last scattering, $z = z_{ls}$, and the present, $z = 0$. It is clear that most of the strings are concentrated at high redshifts near $z = z_{ls}$. (In the matter era the projected angular length of string in the redshift interval $[z_1, z_2]$ scales as $\sqrt{z_1} - \sqrt{z_2}$.) The apparent angular size of the horizon at z_{ls} is

$$\theta_{\mathrm{H}} = z_{ls}^{-1/2}\mathrm{rad.} = 1.8° \left(\frac{z_{ls}}{10^3}\right)^{-1/2}. \tag{10.2.3}$$

With the average distance between the strings $\sim d_{\mathrm{H}}/3$ (see table 10.1), we expect the typical angular distance between the discontinuities on the sky to be of the order θ_{H}. The expected magnitude of the jumps in $\delta T/T$ is $\sim 13G\mu$, where we have used the rms string velocity in the matter era (refer to table 10.1) and averaged over directions in (10.2.2).

The picture becomes more complicated when the string substructure is taken into account. We saw in §10.1.2 that string velocities are significantly affected by small-scale structure. On small scales, the string motion is relativistic and characterized by the rms velocity (cf. table 10.1), while on the scale of the correlation length ξ the motion is typically much slower. Large temperature jumps due to rapidly oscillating wiggles will be seen only if the angular resolution of the experiment is sufficient to resolve the wiggles. Otherwise, the temperature pattern will be averaged over the width of the beam w, and the observed jump is given by (10.2.2) with μ and v replaced, respectively, by the effective mass per unit length $\tilde{\mu}$ and by the coherent string velocity on the beam-width scale w. The slow motion of strings is occasionally interrupted by periods of rapid motion triggered by string intercommutings. The rapidly moving segments have a length comparable to ξ and may give the dominant contribution to $\delta T/T$.

Besides the discontinuous Doppler shift, there are several other effects contributing to the microwave temperature fluctuations. They include: (i) temperature, velocity and potential perturbations at the surface of last scattering; (ii) the Sachs–Wolfe effect from potential perturbations between z_{ls} and us; and (iii) anisotropies from gravitational waves emitted by strings. Some of these effects may be of magnitude comparable to (10.2.2) [Brandenberger & Turok, 1986; Traschen, Turok & Brandenberger, 1986]. However, they do not have a characteristic 'stringy' signature, and so we will not discuss them here.

Bouchet, Bennett & Stebbins [1988] numerically generated a $\delta T/T$ pattern using the Bennett–Bouchet simulations of string evolution. The resulting temperature map is shown in fig. 10.2. Its angular scale is $4\theta_H$ ($\approx 7°$ for $z_{ls} = 10^3$). Curves on which the temperature changes discontinuously are clearly visible in the pattern. However, it can prove difficult to trace out an individual string, because the magnitude and sign of δT varies along the string as the string direction and velocity change. Closed loops present in the simulation are typically very small and have little effect on the overall pattern.

Analysis of this simulation shows that the rms temperature fluctuation from strings in the interval from z to $2z$ is

$$\frac{\delta_2 T}{T} \approx 6G\mu \qquad (10.2.4)$$

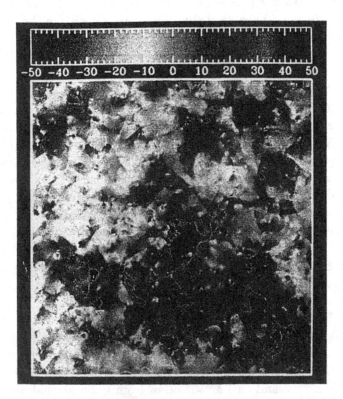

Fig. 10.2.   Temperature map of the microwave background with cosmic-string-induced anisotropy [Bouchet, Bennett & Stebbins, 1988].

and is independent of z (as it should be for a scale-invariant string evolution). The typical wavelength of these fluctuations is about half the apparent horizon size, $\theta \approx 0.5z^{-1/2}$ rad. The total rms fluctuation from all wavelengths is given by

$$\left(\frac{\delta T}{T}\right)_{\text{rms}} = \frac{\delta_2 T}{T}\left(\frac{\ln z_{\text{ls}}}{\ln 2}\right)^{1/2} = 19G\mu \qquad (10.2.5)$$

where the last equality is for $z_{\text{ls}} = 10^3$. Actual anisotropy measurements are only sensitive to a limited range of angular wavelengths and should not be compared directly with (10.2.5). The minimum wavelength that can be observed with a detector of beam-width w is $\lambda_{\text{min}} \approx 2w$. This corresponds to a maximum redshift $z_{\text{max}} \approx (4w)^{-2}$, where w is in radians. For the COBE experiment [Smoot *et al.*, 1992] $w = 0.12$ rad., and the expected value of the rms fluctuation can be obtained from (10.2.5) by replacing z_{ls} with $z_{\text{max}} \approx 4$, giving $(\delta T/T)_{\text{rms}} \approx 8.5G\mu$. Comparing this to the observed fluctuation (1.3.37) and assuming that it is due to strings, we obtain $G\mu \approx 1.3 \times 10^{-6}$. A more careful analysis of the data by Bennett, Stebbins & Bouchet [1992] gives $G\mu = (1.5 \pm 0.5) \times 10^{-6}$. A larger value of $G\mu$ would be in conflict with observations; hence,

$$G\mu \lesssim 2 \times 10^{-6}. \qquad (10.2.6)$$

A note of caution should be added here. The temperature pattern in the Bouchet *et al.* [1988] simulation was calculated using the flat-space formalism developed by Stebbins [1988], which distorts the effect of the expansion of the universe on photon propagation. This formalism gives an accurate picture of the discontinuous temperature jumps along the strings, but can otherwise lead to substantial systematic errors. Another effect neglected in the simulations is the compensation of string perturbations by matter on super-horizon scales. With these effects taken into account, the bound on $G\mu$ (10.2.6) may well change by a factor of two or more.

One possible strategy for looking for temperature jumps due to strings is to examine the temperature variations along a line, a few degrees long. Such a one-dimensional scan through the map of fig. 10.2 is shown in fig. 10.3. For comparison we also show a scan through a random-phase temperature map of the same spectrum. To mimic the finite width of the detector beam, the pattern has been smoothed with Gaussian windows of various widths. Sharp jumps of up to $40G\mu$ are clearly visible in the string temperature pattern for beamwidths $w \lesssim 2(10^3/z_{\text{ls}})^{1/2}$ arc minutes.

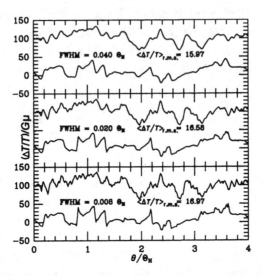

Fig. 10.3. One-dimensional scans through CMBR temperature maps contrasting (i) the sharp jumps characteristic of cosmic strings (fig. 10.2) and (ii) a random phase map with the same normalization (displaced above for comparison) [Bouchet *et al.*, 1988]. The patterns have been smoothed with Gaussian windows of width $4'(10^3/z_{\mathrm{ls}})^{1/2}$, $2'(10^3/z_{\mathrm{ls}})^{1/2}$ and $0.8'(10^3/z_{\mathrm{ls}})^{1/2}$ for the top, middle and bottom curves, respectively.

10.3 Strings as gravitational lenses

The conical nature of space around a cosmic string results in the formation of double images of galaxies or quasars located behind the string [Vilenkin, 1981b, 1984a, 1986b; Gott, 1985]. The typical angular separation between the images is comparable to the conical deficit angle, $\Delta = 8\pi G\mu$. Gravitational lensing by a string is illustrated in fig. 10.4. A conical space is obtained after the shaded wedge is discarded and its two boundaries are identified. The observer is represented by two points, O_1 and O_2, on the opposite sides of the wedge. When the observer looks above the string S, the half-space above the line ASO_1 is seen, whereas looking below the string, the half-space below the line BSO_2 is seen. It is clear that all sources located in the wedge ASB have duplicated images. For example, the light rays from a quasar Q intersect after passing on opposite sides of the string, and the observer sees two images of the quasar. The angular separation between the images, $\delta\varphi_0$, is given by the sum of the angles QO_1S and QO_2S. If ℓ and d are the distances from the string to the quasar and to the observer, respectively, and θ is the angle between the string and the line of sight, then it is easily shown that the angular separation between the images is (for $G\mu \ll 1$)

$$\delta\varphi_0 = 8\pi G\mu\ell(d+\ell)^{-1}\sin\theta. \tag{10.3.1}$$

Note that for $\theta \ne \pi/2$ the plane of the figure does not cross the string at right angles and $\angle O_1 S O_2 = \Delta\sin\theta$.

If an extended object, like a galaxy, crosses the boundary of the wedge ASB in fig. 10.4, then the observer sees one full image of the object, while the other image includes only the part lying inside ASB, and thus has a sharp edge [Paczynski, 1986].

The subscript for $\delta\varphi_0$ in (10.3.1) refers to the fact that the string was assumed to be at rest with respect to the observer. Cosmological strings are expected to move at relativistic speeds, and so (10.3.1) has to be generalized for moving strings. Let k and k' be the four-dimensional wave vectors of the light waves corresponding to the two images and then consider the invariant

$$k \cdot k' = \omega\omega'\left[1 - \cos(\delta\varphi)\right] \approx \tfrac{1}{2}\omega\omega'(\delta\varphi)^2. \tag{10.3.2}$$

In the rest frame of the string Σ_0, the two frequencies are equal, $\omega = \omega' = \omega_0$. In the observer's frame Σ, ω and ω' are slightly different because of the difference in the directions of $\mathbf{k}$ and $\mathbf{k'}$. However, this difference is comparable to $\delta\varphi$ and can be neglected to lowest order in $G\mu$, implying that $\omega\delta\varphi = \omega_0\delta\varphi_0$. The frequencies ω and ω_0 are related by $\omega = \gamma(1 - \hat{\mathbf{n}} \cdot \mathbf{v})\omega_0$, where $\hat{\mathbf{n}}$ is a unit vector in the direction from the observer to the quasar, $\mathbf{v}$ is the string velocity, and $\gamma = (1 - v^2)^{-1/2}$. Hence, we have

$$\delta\varphi = \gamma^{-1}(1 - \hat{\mathbf{n}} \cdot \mathbf{v})^{-1}\delta\varphi_0. \tag{10.3.3}$$

Depending upon the relative direction of $\mathbf{v}$ and $\hat{\mathbf{n}}$, the angular separation takes a value in the range

$$\kappa^{1/2}\delta\varphi_0 < \delta\varphi < \kappa^{-1/2}\delta\varphi_0, \tag{10.3.4}$$

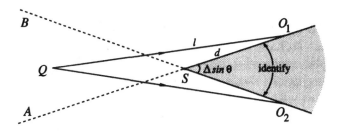

Fig. 10.4. Gravitational lensing by a cosmic string; the observer (O_1 and O_2 are identified) sees two images of the source Q.

where $\kappa = (1 - v)/(1 + v)$.

For $G\mu \sim 10^{-6}$, the typical separation of images is of the order of a few seconds of arc. Images due to a straight string are neither distorted, nor amplified. Hence, apart from a small frequency shift due to the motion of the string, one expects two identical images. Deviations from this idealized pattern can be caused by string curvature, by masses distributed near the line of sight, or by unequal time delays for the two images (a difference of a few years for $G\mu \sim 10^{-6}$). We note that more conventional gravitational lenses, such as galaxies or black holes, tend to form distorted images of unequal brightness. The number of images produced by such lenses is necessarily odd, although in practice not all of them may be resolved.

If a segment of rapidly moving string is not perpendicular to the line of sight, then the observer sees more distant parts of the string with a retardation, and the apparent direction of the string differs from its actual direction. For simplicity, we shall consider only the case when the velocity of the string is perpendicular to the line of sight. Consider a coordinate system with the y-axis along the line of sight and the z-axis in the direction of the string velocity, so that the orientation of the string is parallel to the xy-plane. The observer sees a projection of the string on the xz-plane. Two points of the string with x-coordinate difference Δx are seen with a relative time delay $\Delta t = \Delta x \cot\theta$; hence, the observed difference in the z-coordinate is $\Delta z = v\Delta t = (v \cot\theta)\Delta x$. The angle χ between the apparent and the actual direction of the string can be found from

$$\tan\chi = v \cot\theta . \tag{10.3.5}$$

If there is a double quasar due to this string, then the line connecting the two images is along the z-axis and is not, in general, perpendicular to the apparent direction of the string (which can be observed, for example, by detecting the microwave temperature discontinuity). Equation (10.3.5) only applies for a short segment of string subtending a small angle from the position of the observer. If a long straight string moves past an observer, its distant parts are seen with a retardation, and the string appears to be curved. Indeed, it can be shown that a moving straight string appears to have the shape of a hyperbola [Vilenkin, 1986b].

The angular separation between the images (10.3.1) was derived for a static string in flat spacetime. The same equation applies to a string which is at rest with respect to the Hubble flow in an expanding universe, if d and l are interpreted as comoving distances. This is easily understood if we note that the string metric in an expanding universe is the usual FRW metric with a deficit angle $\Delta = 8\pi G\mu$ (we assume $\Omega = 1$). Expressing l and $(d+l)$ in terms of the string redshift z_s and quasar redshift z_q, we obtain [Gott, 1985]

$$\delta\varphi_0 = 8\pi G\mu \sin\theta \frac{(1+z_s)^{-1/2} - (1+z_q)^{-1/2}}{1 - (1+z_q)^{-1/2}} . \qquad (10.3.6)$$

A long straight segment of string passing in front of a distant field of galaxies creates a chain of galaxy pairs whose angular separations are similar in magnitude and whose position angles are nearly identical [Vilenkin, 1984a]. Observational strategies to search for this effect have been discussed by Hogan [1987b] and Hindmarsh [1990b]. We note that the images may be misaligned because of small-scale structure on the string. A quantitative analysis of this misalignment remains to be given.

Let us now estimate the probability of the gravitational lensing of a distant object by a string. The lensing probability as a function of redshift can be calculated from numerical simulations of string evolution. However, this problem remains to be tackled, so we can give only a rough order-of-magnitude estimate. A double image is formed if the source is within a band of angular width $2\delta\varphi$ around the projection of the string on the sky. The fraction of the sky lensed by a single long string stretching across the visible universe is $\sim 10G\mu$. With about ten long strings per horizon, the lensing probability by infinite strings for a source at redshift $z \sim 1$ is

$$P_\infty(z \sim 1) \sim 100G\mu . \qquad (10.3.7)$$

At large redshifts, the projected length of string on the sky grows as $(1+z)^{1/2}$, but the angle $\delta\varphi$ decreases as $(1+z)^{-1/2}$, so the lensing probability only grows logarithmically. Lensing by closed loops can also increase the estimate (10.3.7) by a logarithmic factor $\ln(\alpha/\Gamma G\mu)$ (see (10.1.21)). With $G\mu \sim 10^{-6}$ and $\alpha \lesssim 5 \times 10^{-3}$ the correction due to this factor is less than one order of magnitude.

At small redshifts $z \ll 1$, lensing by long strings is unlikely. The fraction of the sky lensed by closed loops located at redshifts smaller than z is

$$P_L(z) \sim \frac{\Delta}{4\pi} \int_0^{3zt_0/2} dr\, 4\pi r^2 \int_0^{\alpha t_0} d\ell\, n(\ell, t_0)\frac{\ell}{r}$$

$$\sim 9\pi G\mu\nu z^2 \ln\left(\frac{\alpha}{\Gamma G\mu}\right) . \qquad (10.3.8)$$

Here we have used (10.1.17) for the loop distribution function $n(\ell, t)$ and the relation between redshift and distance, $z = 2r/3t_0$, which is valid for $z \ll 1$.

The expression (10.3.8) was derived assuming that the angular size of all loops is greater than the image separation Δ. This condition is satisfied if $z < \ell/12\pi^2 G\mu t_0$. Since most of the loops have sizes $\ell \gtrsim \Gamma G\mu t_0$

with $\Gamma \sim 100$, (10.3.8) should give the right order of magnitude for all $z \lesssim 1$. Lensing by small loops of angular size $\sim \Delta$ has been discussed by Hogan & Narayan [1984], who pointed out the possibility of a very large magnification of the image due to caustics. In this case the image configuration and luminosity can have an observable time variability.

10.4 Gravitational radiation background

Gravitational waves emitted by oscillating loops at different epochs produce a stochastic gravitational wave background. The power spectrum of this background spans a wide range of frequencies, and over much of this range has a very simple form: there is equal power in each logarithmic frequency interval [Vilenkin, 1981d; Hogan & Rees, 1984]. To clarify the origin of this result, we shall first give a simple derivation of the spectrum which uses a rather unrealistic model for the radiation spectrum of individual loops. We shall then proceed to a more sophisticated analysis, which will show, in particular, that the result is fairly insensitive to the assumed form of the loop spectrum.

10.4.1 Radiation by loops: simple derivation

A convenient measure of the intensity of the gravitational wave background is

$$\Omega_{\rm g}(\omega) = \frac{\omega}{\rho_{\rm c}} \frac{d\rho_{\rm g}}{d\omega} . \tag{10.4.1}$$

Here, $\rho_{\rm g}$ is the energy density of gravitational waves, $\rho_{\rm c}$ is the critical density and ω is the angular frequency. $\Omega_{\rm g}(\omega)$ gives the energy density in units of $\rho_{\rm c}$ per logarithmic frequency interval.

A string loop at rest emits gravitational waves at a discrete set of frequencies

$$\omega_n = 4\pi n/\ell , \tag{10.4.2}$$

where $\ell = M/\mu$ is the invariant length and M is the mass of the loop. In our first calculation we shall assume that most of the energy is emitted in the first few harmonics with $n \sim 1$. Loops decaying at a time $\sim t_1$ during the radiation era ($t_1 < t_{\rm eq}$) have length $\ell \sim \Gamma G \mu t_1$ and energy density comparable to the total energy density of the loops (10.1.16),

$$\rho \approx \left(\frac{\alpha}{\Gamma G \mu} \right)^{1/2} \frac{\zeta \mu}{t_1^2} . \tag{10.4.3}$$

Within about a Hubble time all this energy is transformed into gravitational waves of frequency

$$\omega_1 \approx 4\pi/\Gamma G \mu t_1 , \tag{10.4.4}$$

with $\Delta\omega \sim \omega$. Using the expression for the critical density, $\rho_c \approx 1/30Gt^2$, we obtain $\Omega_g(\omega) \sim 30\zeta(\alpha G\mu/\Gamma)^{1/2}$. At later times, the frequency of the waves is redshifted with the scale factor a as $\omega \propto a^{-1}(t)$ and their energy density as $\rho_g \propto a^{-4}(t)$.

For a 'comoving' frequency $\omega(t) \propto a^{-1}(t)$, the time-dependence of $\Omega_g(\omega)$ is the same as that of $\Omega_r\mathcal{N}^{1/3}$, where $\Omega_r = \rho_r/\rho_c$, ρ_r is the radiation energy density (including the energy of relativisitic particles), and $\mathcal{N}(t)$ is the effective number of spin degrees of freedom (refer to (1.3.25)). The factor $\mathcal{N}^{1/3}$ is introduced to account for the fact that the radiation energy density changes every time the temperature falls through a particle mass threshold. Hence, we can write

$$\Omega_g(\omega) \sim 30\zeta \left(\frac{\alpha G\mu}{\Gamma}\right)^{1/2} \left(\frac{\mathcal{N}}{\mathcal{N}_\omega}\right)^{1/3} \Omega_r, \qquad (10.4.5)$$

where $\mathcal{N}_\omega$ is the value of $\mathcal{N}(t)$ at the time when the gravity waves were emitted. Equation (10.4.5) applies in the frequency range corresponding to the waves emitted before t_{eq}, but later than t_*, where $t_* \sim 10^{-43}(G\mu)^{-2}$s from (10.1.6). Note that for $t \ll t_*$, string motion is heavily damped and gravitational radiation is strongly suppressed. The relevant frequency range is easily found using (10.4.4),

$$10^{-16}(G\mu)^{-1}h^2\text{s}^{-1} \lesssim \omega \lesssim 10^{-1}(G\mu)^{-2}\text{s}^{-1}. \qquad (10.4.6)$$

The right-hand side of (10.4.5) depends on ω only through $\mathcal{N}_\omega$, which is a very slowly-varying function of frequency. Over the whole frequency range (10.4.6), it is not expected to change by more than two to three orders of magnitude. The corresponding variation of $\Omega_g(\omega)$ should not exceed a single order of magnitude. After positron annihilation at $t \approx 1$s, $\mathcal{N}(t)$ remains constant to the present day.

10.4.2 Radiation by loops: improved derivation

As we already mentioned, the assumption that loops radiate only in a few lowest harmonics is rather unrealistic. For less exotic sources, the power radiated in the nth harmonic P_n is an exponentially decreasing function at large n. However, for loops having cusps or kinks P_n is a power law, $P_n \approx n^{-4/3}$ and $P_n \approx n^{-2}$ for cusps and kinks, respectively (refer to §7.5). The contribution of very high harmonics can still be neglected, but one has to include harmonics with n up to say 10 or 100. We shall now derive an improved version of (10.4.5) without making any specific assumptions about the form of P_n. In this derivation we shall still disregard the effect of loop velocities on the radiation spectrum. This is a reasonable approximation if the lifetime of the loop is greater than the Hubble time, $\alpha > \Gamma G\mu$. The initial velocity of the loop is then redshifted in the first few Hubble times, and most of the energy is radiated when the loop is essentially at rest. For $\alpha \sim \Gamma G\mu$, our results should still be useful for order-of-magnitude estimates.

Quite generally, we can write the radiation spectrum from a loop of length ℓ, averaged over various loop configurations, in the form

$$\frac{dP_\ell(\omega)}{d\omega} = G\mu^2 \ell \, g(\omega\ell) \, . \tag{10.4.7}$$

The function $g(x)$ is normalized by

$$\int_0^\infty g(x)dx = \Gamma \, , \tag{10.4.8}$$

with $\Gamma \approx 65$ as before. For a loop at rest, the spectrum is discrete, $g(x)$ is a sum of δ-functions, and the integral in (10.4.8) is really a sum.

The energy density of gravitational waves emitted at time t_1 in the interval dt_1 with frequencies from ω_1 to $\omega_1 + d\omega_1$ is

$$d\rho_g(t_1) = dt_1 d\omega_1 G\mu^2 \int_0^\infty d\ell \, n(\ell, t_1) \ell \, g(\omega_1 \ell) \, , \tag{10.4.9}$$

where $n(\ell, t)$ is the loop distribution function. To find the spectral density $d\rho_g/d\omega$ at time t, we integrate over $t_1 < t$ and take into acount the frequency redshift,

$$\omega = \frac{a(t_1)}{a(t)}\omega_1 \, , \tag{10.4.10}$$

and the energy density redshift, $\rho_g \propto a^{-4}$. This yields

$$\frac{d\rho_g}{d\omega}(t) = G\mu^2 \int_{t_*}^t dt_1 \left(\frac{a(t_1)}{a(t)}\right)^3 \int_0^{\alpha t_1} d\ell \, n(\ell, t_1)\ell \, g\left(\frac{a(t)}{a(t_1)}\omega\ell\right) \, .$$
$$\tag{10.4.11}$$

This equation can be substantially simplified in the range of frequencies corresponding to the waves emitted during the radiation era [Vachaspati & Vilenkin, 1985; Bouchet & Bennett, 1990]. For $t_1 < t < t_{eq}$, we can use $a(t) \propto t^{1/2}$ and (10.1.12) for $n(\ell, t)$. We assume that the frequency ω is sufficiently low, so that waves emitted before positron annihilation do not contribute and factors like $\mathcal{N}^{1/3}$ in (10.4.5) can be ignored. The lower limit of t_1-integration can then be replaced by $t_1 = 0$. We now go through the following chain of manipulations: (i) change the integration variables t_1, ℓ to $x = \ell/t_1$ and $z = \omega x \sqrt{tt_1}$; (ii) change the order of integration; and (iii) integrate by parts over x. The result is

$$
\frac{d\rho_{\rm g}}{d\omega}(t) = \frac{4G\mu^2\nu}{3\omega t^2}\left[(\Gamma G\mu)^{-3/2}\int_0^{\alpha\omega t} dz g(z)\left(1+\frac{z}{\Gamma G\mu\omega t}\right)^{-3/2}\right.
$$
$$
\left. -(\alpha+\Gamma G\mu)^{-3/2}\int_0^{\alpha\omega t} dz g(z)\right].
$$

(10.4.12)

The dominant contribution to the integral of $g(z)$ comes from the range $4\pi \leq z \lesssim 4\pi n_*$, where $n_* \lesssim 100$ is the mode number beyond which the radiation spectrum of loops can be truncated. Assuming $4\pi n_* \ll \Gamma G\mu\omega t$ and using the normalization condition (10.4.8), we obtain from (10.4.12)

$$
\Omega_{\rm g}(\omega) = \frac{128\pi}{9}\nu\left(\frac{G\mu}{\Gamma}\right)^{1/2}\Omega_{\rm r}\left[1-\left(1+\frac{\alpha}{\Gamma G\mu}\right)^{-3/2}\right].
$$

(10.4.13)

where we have used $\rho_{\rm r} \approx 3/32\pi Gt^2$ for the radiation density. At the present time, $\Omega_{\rm r} \approx 4 \times 10^{-5}h^{-2}$, and with $\nu \sim \zeta\alpha^{1/2}$, $\zeta \approx 13$ and $\alpha \gtrsim \Gamma G\mu$, (10.4.13) gives a lower bound on $\Omega_{\rm g}$,

$$
\Omega_{\rm g} \gtrsim 10^{-2}G\mu\, h^{-2}
$$

(10.4.14)

Equation (10.4.13) is applicable in the frequency range

$$
10^{-16}(G\mu)^{-1}n_*h^2{\rm s}^{-1} < \omega < 3 \times 10^{-11}(G\mu)^{-1}{\rm s}^{-1}.
$$

(10.4.15)

Turning now to the contribution of loops radiating during the matter era, we have to assume a specific form of the spectral function $g(z)$. The discussion in §7.5 suggests that a simple power law,

$$
g(z) = A\theta(z-4\pi)z^{-\beta}
$$

(10.4.16)

with $A = (\beta-1)(4\pi)^{\beta-1}\Gamma$ and $4/3 \leq \beta \leq 2$ should give a reasonably good approximation. With $a(t) \propto t^{2/3}$, $g(z)$ from (10.4.16) and $\beta < 2$ it is not difficult to show that the integral in (10.4.11) is dominated by $t_1 \approx t$ and $\ell \sim \Gamma G\mu t$, that is, by the loops decaying at the present time.* For $\alpha/\Gamma G\mu < z_{\rm eq}^{3/2}$, these loops were formed during the matter era, and using the loop distribution function (10.1.17) we can rewrite (10.4.11) as

$$
\frac{d\rho_{\rm g}}{d\omega}(t_0) \approx \frac{3G\mu^2\nu}{t_0}\int_0^\alpha dx\,\frac{x}{(x+\Gamma G\mu)^2}\int_0^1 dy\, y^2\, g(xy\omega t_0).
$$

(10.4.17)

---

* This was first pointed out by Caldwell & Allen [1992] on the basis of their numerical calculations.

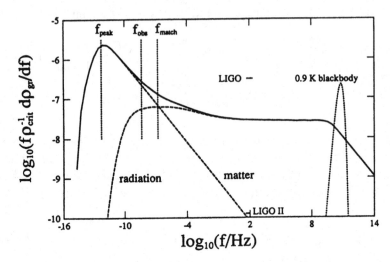

Fig. 10.5. Gravitational background spectrum for the parameters $G\mu = 10^{-6}$, $\alpha = 10^{-3}$, $\beta = 4/3$ (from Caldwell & Allen [1992]). The contributions from the radiation and matter eras are indicated by dotted lines. The frequency f on the horizontal axis is defined as $f = \omega/2\pi$.

Here, $x = \ell/t_1$, $y = (t_1/t_0)^3$, t_0 is the present time and we have assumed for simplicity that $\alpha \gg \Gamma G\mu$, so that the upper limit of x-integration can be changed to ∞. The θ-function in (10.4.16) is unimportant for $\omega > 4\pi/\Gamma G\mu t_0$ and β not very close to 2. The x- and y-integrations are then easily performed, and we obtain

$$\Omega_{\mathrm{g}}(\omega) = \frac{18\pi^2 (\beta - 1)^2 \nu\, G\mu}{(3 - \beta)\, \sin[(2 - \beta)\pi]} \left(\frac{4\pi}{\Gamma\mu\,\omega t_0}\right)^{\beta - 1} \tag{10.4.18}$$

The peak of the spectrum is reached at $\omega_{\max} \sim 4\pi/\Gamma G\mu t_0$ corresponding to $\Omega_{\mathrm{g}}(\omega_{\max}) \sim G\mu$.

The full spectrum of the gravitational background for several values of the parameters $G\mu$, α and β has been calculated numerically by Caldwell & Allen [1992]. They carefully treated the transition from the radiation to the matter era and used a discrete spectral function with $P_n \propto n^{-\beta}$ instead of (10.4.16). A sample spectrum is shown in fig. 10.5. It should be noted that the Caldwell & Allen estimates are based on the numerical simulations of Allen & Shellard [1992] which do not include gravitational back-reaction. The additional smoothing from radiation damping may act to reduce the high-frequency contributions to the radiation spectrum (refer to §7.7.1).

10.4.3 Radiation by infinite strings

The contribution of infinite strings to the gravitational radiation background can be estimated using (7.6.23) for the radiation power per unit length of a string possessing small-scale structure. The characteristic wavelength of the wiggles is comparable to the typical size of closed loops, $\langle \ell \rangle \sim \alpha t$, and the average frequency of the radiation can be estimated, $\bar{\omega} \sim 2\pi/\alpha t$. The effective string mass per unit length in the radiation era is $\tilde{\mu} \approx 2\mu$, and so (7.6.23) gives

$$\frac{dP}{d\ell} \sim \frac{4\pi^2 G\mu^2}{\alpha t}. \tag{10.4.19}$$

Multiplying this by the length of string per unit volume, $\rho_\infty/\mu \sim \zeta t^{-2}$, we obtain the radiation rate

$$\left(\frac{d\rho_g}{dt}\right)_\infty \sim \frac{4\pi^2 \zeta G\mu^2}{\alpha t^3}. \tag{10.4.20}$$

The contribution of infinite strings to Ω_g can now be obtained in a similar manner to the estimate (10.4.5). One finds that the ratio of the contributions due to infinite strings and due to loops is

$$\frac{\Omega_g(\omega)_\infty}{\Omega_g(\omega)_L} \sim \frac{4\pi^2}{\Gamma} \left(\frac{\Gamma G\mu}{\alpha}\right)^{3/2}. \tag{10.4.21}$$

Radiation from loops dominates if loops live much longer than a Hubble time ($\alpha \gg \Gamma G\mu$), but for $\alpha \sim \Gamma G\mu$ the two contributions can be comparable. Under most conditions, however, (10.4.5) and (10.4.13) would give a reasonable order-of-magnitude estimate for Ω_g. Gravitational radiation from a long string network has been studied numerically by Allen & Shellard [1992].

10.4.4 Observational bounds

Remarkably good limits on low-frequency gravitational waves have been obtained through the timing of pulsar signals. The Earth and a distant pulsar can be thought of as free masses whose positions respond to changes in the local spacetime metric. Passing gravitational waves perturb the metric and produce fluctuations in the pulse arrival times. A low level of noise in the timing data leads to a stringent bound on the gravitational wave intensity. The highest accuracy is achieved for waves with periods comparable to the total time of observation. As of 1993, the best bound is [Stinebring *et al.*, 1990]

$$\Omega_g < 4 \times 10^{-7} h^{-2} \tag{10.4.22}$$

for waves with periods near seven years. Comparing this with (10.4.13) we obtain an upper bound on $G\mu$,

$$G\mu \lesssim 4 \times 10^{-5}. \qquad (10.4.23)$$

Stronger bounds have been quoted in the literature, but these have assumed large loop sizes with $\alpha \gg \Gamma G\mu$ [Sanchez & Signore, 1988; Krauss & Accetta, 1989; Albrecht & Turok,1989] or else a specific form of the loop radiation spectrum [Caldwell & Allen, 1992].

A more stringent constraint on $G\mu$ can be obtained by considering the predictions of primordial nucleosynthesis [Davis, 1985b; Bennett, 1986b; Bennett & Bouchet, 1991]. For the standard nucleosynthesis scenario to work, the energy density in gravitational waves at $t \sim$ 1s should be less than 5.4% of the radiation density [Olive et al., 1990],

$$\Omega_{g,tot} = \frac{\rho_g}{\rho_r} < 0.054. \qquad (10.4.24)$$

The total energy density in gravitational waves at the time of nucleosynthesis can be found by integrating $\Omega_g(\omega)$,

$$\Omega_{g,tot} = \int \frac{d\omega}{\omega} \Omega_g(\omega). \qquad (10.4.25)$$

Using (10.4.11) and (10.4.13), while including the correction factors of $\mathcal{N}^{1/3}$, we find

$$\Omega_{g,tot} \approx \frac{64\pi}{9} \zeta \left(\frac{\alpha G\mu}{\Gamma}\right)^{1/2} \sum_i \left(\frac{\mathcal{N}_{nucl}}{\mathcal{N}_i}\right)^{1/3} \ln\left[\frac{t_i}{t_{i-1}}\right]. \qquad (10.4.26)$$

Here, $\mathcal{N}_{nucl} = 10.75$ is the number of relativistic spin degrees of freedom at nucleosynthesis, t_i is the time at which the i-th particle species 'freezes out', and $\mathcal{N}_i$ is the value of $\mathcal{N}$ in the interval between t_{i-1} and t_i. After calculating the sum in (10.4.26), a bound on $G\mu$ can be obtained using (10.4.24) and $\alpha \gtrsim \Gamma G\mu$. Since the particle spectrum above 1 TeV is currently unknown, one has to make some assumption about the behaviour of $\mathcal{N}(t)$ before the electroweak phase transition. Bennett & Bouchet [1991] performed the calculation for two extreme examples: First, a minimal GUT model was assumed, with no particles between the electroweak and GUT scales, and then, secondly, a 'maximal' GUT model was considered, with a very large number of particles with masses above 1 TeV such that $\mathcal{N}_i(T > 1\,\text{TeV}) \to \infty$. The resulting bounds for the two models were

$$\begin{aligned} \text{Minimal GUT}: \quad & G\mu < 6 \times 10^{-6}, \\ \text{Maximal GUT}: \quad & G\mu < 1 \times 10^{-5}. \end{aligned} \qquad (10.4.27)$$

The nucleosynthesis constraint (10.4.24) is somewhat weakened in models with inhomogeneous nucleosynthesis. According to Kurki-Suonio *et al.* [1990], the allowed excess energy density can be raised from 5.4% to 9.5%. This weakens the bounds (10.4.27) by approximately a factor of two. In all cases, the nucleosynthesis bound is stronger than the pulsar timing constraint (10.4.23), but weaker than the microwave background constraint (10.2.6).

Finally, we point out some ways of avoiding the gravitational radiation bounds (10.4.23) and (10.4.27). In the first place, the bounds were derived assuming that gravitational radiation is the dominant energy loss mechanism for strings. This does not apply to global strings which predominantly radiate Goldstone bosons. The second possibility arises in models in which strings are formed during inflation (refer to chapter 16). After the phase transition, inflation continues for some time, and the characteristic length-scale for the strings ξ grows by a very large factor. The strings begin radiating when the comoving scale $\xi(t)$ comes within the horizon. As a result, gravitational radiation at wavelengths much smaller than $\xi(t)$ will be absent from the spectrum (10.4.13) [Yokoyama, 1988, 1989]. With a suitable parameter choice, $\xi(t)$ will cross the horizon after nucleosynthesis, thus eliminating the nucleosynthesis constraint on $G\mu$. If the present-day comoving scale $\xi(t_0)$ is much greater than ten light years, then the pulsar timing constraint may also be circumvented.

10.5 Black hole formation

Oscillating string loops are unlikely to collapse to form black holes. The Schwarzschild radius of a loop of invariant length $\ell = M/\mu$ is $r_g = 2G\mu\ell$ and thus, in order to form a black hole, the loop must shrink by a factor of order $(G\mu)^{-1}$ at some time during its period of oscillation. However, even though the fraction of loops collapsing to black holes may be extremely small, they could be cosmologically significant [Hawking, 1989; Polnarev & Zembowicz, 1991]. As the universe expands, the energy density in black holes redshifts like matter, and black holes formed early in the radiation era can eventually come to dominate the universe. Another constraint comes from observations of the γ-ray background [Page & Hawking, 1976]. Black holes with masses around 10^{15}g have lifetimes comparable to the present age of the universe. Hawking radiation from such black holes contributes to the γ-ray background at energies $\sim 100\,\mathrm{MeV}$. To avoid conflict with observations, the mass density of 10^{15}g black holes must satisfy

$$\Omega_{\mathrm{bh}} \lesssim 10^{-8}. \tag{10.5.1}$$

Loops of mass M were formed at a time $t_\mathrm{f} \sim M/\mu\alpha$. At t_f, the energy density in these loops is approximately $\Omega_\mathrm{L} \sim 30\zeta G\mu$. If $\mathcal{P}$ is the probability that a loop will collapse to form a black hole, then (10.5.1) yields

$$30\zeta G\mu \mathcal{P}(\alpha\mu t_{\mathrm{eq}}/M)^{1/2} \lesssim 10^{-8}\,, \qquad (10.5.2)$$

where we have assumed that t_{f} is in the radiation era. With $M \sim 10^{15}\mathrm{g}$, $\alpha \gtrsim \Gamma G\mu$, $\Gamma \sim 100$ and $\zeta \sim 13$, this reduces to

$$(G\mu)^2\mathcal{P} \lesssim 10^{-29}\,. \qquad (10.5.3)$$

An upper bound on the probability $\mathcal{P}$ can be obtained by the following simple argument [Rees, 1990]. A loop of mass M will typically have an angular momentum $J = j\mu\ell^2 = jM^2/\mu$ with $j \sim 1$. More precisely, numerical simulations suggest that $j \sim 0.01$ for non-intersecting loops [Scherrer, 1990]. A loop can collapse to a black hole only if all three components of its angular momentum are smaller than $J_{\mathrm{max}} = GM^2$, the maximum angular momentum allowed for a black hole. Hence the probability of black hole formation is bounded by

$$\mathcal{P} \lesssim (J_{\mathrm{max}}/J)^3 \sim (G\mu/j)^3 \sim 10^6(G\mu)^3\,. \qquad (10.5.4)$$

If the actual value of $\mathcal{P}$ were close to this upper bound, then substitution into (10.5.3) would result in the stringent bound $G\mu \lesssim 10^{-7}$, a constraint which indicates the potential importance of black hole formation.

An attempt to give an estimate, rather than an upper bound for $\mathcal{P}$, has been made by Hawking [1989] who considered a class of loops constructed from a few straight segments. He found

$$\mathcal{P} \sim (G\mu)^{2n-4}\,, \qquad (10.5.5)$$

where n is the number of segments. A similar result has been obtained by Polnarev & Zembowicz [1991], who assumed that the Fourier expansion of a loop trajectory in harmonics included only the first couple of terms.

To understand the origin of (10.5.5), we recall that the motion of a loop can be described by two vector functions $\mathbf{a}$ and $\mathbf{b}$ (see (6.2.15)),

$$\mathbf{x}(\zeta,t) = \tfrac{1}{2}[\mathbf{a}(\zeta - t) + \mathbf{b}(\zeta + t)]\,. \qquad (10.5.6)$$

As ζ varies from 0 to ℓ, the vectors $\mathbf{a}(\zeta)$ and $\mathbf{b}(\zeta)$ describe closed curves of length ℓ. To form a black hole, the loop has to shrink to a size $\sim G\mu\ell$. This means that, with an appropriate choice of origin for ζ, we should have $\mathbf{b}(\zeta) \approx -\mathbf{a}(\zeta)$ with an accuracy $\sim G\mu$. If the curves $-\mathbf{a}$ and $\mathbf{b}$ are characterized by only a few parameters, these parameters must be equal within a similar accuracy. Hence, the probability $\mathcal{P}$ is given by a small power of $G\mu$.

It is not clear how many parameters are necessary for an adequate representation of a loop. Since high-frequency harmonics are efficiently damped by gravitational radiation, it is quite possible that only a few parameters are required. Even if we could delineate the appropriate parameters, there is an additional problem in defining a probability measure on the parameter space. Given these uncertainties, the estimate (10.5.5) cannot be used with any confidence to place a cosmological bound on $G\mu$.*

10.6 Baryon asymmetry

Baryon-number-violating processes discussed at the end of §8.1.1 can result in the generation of a net baryon number in a baryon-symmetric universe, or else they can erase baryon number generated at earlier times. It can be shown [Sakharov, 1967; Weinberg, 1979] that the baryon number B must vanish if B-violating interactions are in thermal equilibrium. Essentially, the reason is that, in the absence of a conservation law, B is determined by minimizing the thermodynamic potential. This leads to a vanishing chemical potential for baryons and thus to $B = 0$. B-violating scatterings of the form (8.1.19) provide a channel through which baryons and leptons can equilibrate and erase their primordial abundances.

To estimate the efficiency of this process, we find the characteristic collision time,

$$\tau \sim (\sigma \xi^{-2})^{-1} . \tag{10.6.1}$$

Here, σ is the scattering cross-section per unit length of string, ξ is the correlation length of the string network, and we have assumed that the relevant particles are relativistic. The B-violating cross-section σ is sensitive to the specific particle physics model, and its maximum value is comparable to the incident particle wavelength,

$$\sigma_{\max} \sim T^{-1} . \tag{10.6.2}$$

Baryon number destruction is efficient if τ is smaller than the Hubble time t. With $\xi \sim \gamma_\xi t$ and σ from (10.6.2),

$$\tau/t \sim \gamma_\xi^2 tT \sim \gamma_\xi^2 m_{\rm pl}/T . \tag{10.6.3}$$

If we use the value of $\gamma_\xi^2 \sim 0.1$ suggested by string simulations, then $\tau \gtrsim t$ for all temperatures well below the Planck scale. However, during the epoch when the strings are dominated by friction, γ_ξ can be greatly reduced, and with a suitable choice of parameters we can have $\tau \lesssim t$ [Brandenberger, Davis & Matheson, 1988b]. In this case, all the previously generated baryon number will be erased.

---

* The probability of black hole formation can be more reliably estimated in the case of nearly circular loops nucleating during inflation, with deviations from circular shape being caused by quantum fluctuations [Garriga & Vilenkin, 1993] (refer to §10.5).

If B-violating scattering (8.1.19) is also CP-violating, it can result in the generation of baryon asymmetry at the time when the baryon–string interaction goes out of equilibrium. This process has not as yet been studied quantitatively. An alternative possibility is related to the cusp evaporation conjecture discussed in §8.2. Kawasaki & Maeda [1988] have pointed out that heavy particles emitted at cusps can decay in a manner which violates B and CP. Assuming the maximum particle production rate allowed by (8.2.15), they found that, with a suitable particle physics model, this process can account for the observed baryon asymmetry of the universe. This mechanism is similar to the earlier suggestion by Bhattacharjee, Kibble & Turok [1982] who studied baryon number generation by loops collapsing to double lines. Such a collapse, however, requires a very special choice of initial conditions.

Baryon asymmetry can also be generated at the final stages of loop decay. When a loop shrinks to a size comparable to its thickness, it disintegrates into heavy particles which can then undergo B- and CP-violating decays. Brandenberger, Davis & Hindmarsh [1991] estimated the resulting baryon asymmetry and found that it may be comparable to that observed. The largest contribution to the baryon number of the universe again results from the earliest times, when the strings are overdamped.

11

Structure formation with strings

11.1 Defects and structure formation

The distribution of matter in the universe is far from homogeneous. Three-dimensional galaxy distribution surveys reveal a sponge-like pattern, with galaxies concentrated along filaments and sheets separated by voids typically of size $\sim 25\text{–}50\,h^{-1}\mathrm{Mpc}$. On the other hand, the isotropy of the cosmic background radiation testifies to the fact that the early universe was very nearly homogeneous. If small density fluctuations were initially present, they would grow by gravitational instability and, given the appropriate amplitude, could have evolved into currently observed structures. However, the character and origin of these initial fluctuations is unknown and remains a major cosmological enigma.

Of course, one could simply postulate a spectrum of initial fluctuations and take the view that explanations for initial conditions lie outside the realm of physics. However, it would be more satisfactory if, like most other observed phenomena, the origin of these fluctuations could be attributed to some physical mechanism. Indeed, density fluctuations in the early universe are not difficult to generate. Any cosmological phase transition, for example, produces some density inhomogeneities. However, the difficulty lies in finding a mechanism capable of generating fluctuations on sufficiently large scales. The problem is essentially due to causality; no process operating in the early universe can move matter over scales larger than the causal horizon. As a result, no appreciable fluctuations can be produced on super-horizon scales. Larger-scale fluctuations do not strictly vanish, but they are due to surface effects and decay rapidly with distance. (For a quantitative discussion, see Peebles [1980], Press & Vishniac [1980] and Zel'dovich & Novikov [1983].) Cosmological phase transitions are expected to occur early in the radiation era, say before $t < 10^{-4}$s, and can only affect comoving scales up to a few parsecs. This is clearly insufficient to explain fluctuations on scales $\sim 50\,h^{-1}\mathrm{Mpc}$.

Two ways of avoiding this difficulty have been proposed. In the inflationary universe scenario, all of the currently observed universe came out of a region which was initially smaller than the horizon, thus circumventing the causality constraint. Density fluctuations in this scenario are due to quantum fluctuations of the 'inflaton' scalar field (refer to §2.4). The predicted fluctuations are gaussian, with a nearly scale-invariant spectrum; the amplitude of fluctuations is the same on each scale at the time when that scale came within the horizon,

$$\left(\frac{\delta\rho}{\rho}\right)_{\mathrm{H}} \approx \mathrm{const.} \qquad (11.1.1)$$

The value of the constant in (11.1.1) depends sensitively on the shape of the inflaton potential. Inflation gives perhaps the most economical explanation for density fluctuations, since at the same time it helps to explain the homogeneity, isotropy and flatness of the observed universe. However, it is not clear whether or not a gaussian spectrum of the form (11.1.1) is compatible with the observed large-scale structure. The answer depends partially on the type of dark matter which dominates the universe. In particular, this spectrum appears to be ruled out if the universe is dominated by light neutrinos (see, for example, Peebles [1993]).

An alternative possibility is that density fluctuations are generated by topological defects, such as cosmic strings [Zel'dovich, 1980; Vilenkin, 1981a]. Oscillating loops of string can serve as isolated seeds for accretion, while rapidly moving long strings can create sheet-like over-densities in their wakes. When strings are produced at a phase transition, their constituent energy is taken from the surrounding plasma. As a result, on scales greater than the horizon, the density fluctuations due to strings are balanced by the corresponding variations in the matter and radiation density. However, relativistically moving strings produce density fluctuations on scales smaller than the horizon and, as the horizon grows with time, this fluctuation-generating process continues, extending to larger and larger scales.

The string mass per unit length μ required in this scenario can be easily estimated. Strings evolve in a scale-invariant manner, so that at any particular time there will be a few strings stretching across each horizon volume. The density contrast due to strings on the horizon scale is $\delta\rho \sim \mu t/t^3 \sim \mu t^{-2}$, and thus the fractional density fluctuation produced on each scale at horizon crossing is

$$\left(\frac{\delta\rho}{\rho}\right)_{\mathrm{H}} \sim 30 G\mu, \qquad (11.1.2)$$

where we have used the expression $\rho \sim (30Gt^2)^{-1}$ for the average background density. Reasonable galaxy formation scenarios are obtained for

$(\delta\rho/\rho)_{\mathrm{H}} \sim 10^{-4}$–$10^{-5}$ on galactic and cluster scales. Hence, we require $G\mu \sim 10^{-6}$. This corresponds to symmetry breaking scales in the grand unification range: $\eta \sim 10^{16}\mathrm{GeV}$ for gauge strings and $\eta \sim 10^{15}\mathrm{GeV}$ for global strings.

Apart from strings, cosmologically interesting fluctuations can be produced by light domain walls, global monopoles and global textures. The corresponding scenarios will be discussed, respectively, in chapters 13, 14 & 15. Here we shall only mention some general features common to all structure formation scenarios based on topological defects. In contrast to inflationary models, density fluctuations produced by defects are strongly non-gaussian. In the vicinity of a defect, $\delta\rho/\rho$ develops strong peaks that would be extremely unlikely in a gaussian density field. This has several important consequences. The first is that galaxies and quasars can form early. This is particularly important in a neutrino-dominated universe. With gaussian fluctuations, neutrino free-streaming smooths out the density field on galactic scales, and galaxies can only form through the fragmentation of larger objects. As a result, galaxy formation begins at $z \lesssim 1$, and it is difficult to account for the existence of quasars with $z \gtrsim 4$.

Radiation from early objects formed at $z \gtrsim 100$ can reionize the universe; as a result, anisotropies in the temperature of the microwave background can be erased on angular scales below a few degrees. On larger scales, the defects introduce characteristic features in the temperature pattern; a search for such features can be used to test the corresponding scenarios. Finally, the non-gaussian nature of these density fluctuations may provide the explanation for the sheet-like and filamentary features in the observed distribution of galaxies.

In this chapter we shall examine the generation and growth of density fluctuations in the string scenario and discuss how the predictions compare with the observed large-scale distribution of galaxies. The details of the accretion process depend on the type of dark matter dominating the universe, and we shall consider both cold dark matter, consisting of particles with very small velocities, and the opposite case with hot dark matter for which the velocity dispersion is large. In both cases we shall assume for simplicity that the universe has the critical density, $\Omega = 1$. An additional motivation for $\Omega = 1$ is that this value is predicted by inflationary scenarios (assuming that the cosmological constant is zero). The possibility that all the dark matter in the universe is baryonic will also be discussed. Nucleosynthesis considerations require that the baryon density should be bounded by $\Omega_{\mathrm{B}} \lesssim 0.16$ and in our discussion of the baryon-dominated universe we shall keep Ω as a free parameter.

11.2 Accretion onto loops

11.2.1 The Zel'dovich approximation

We shall first analyze the accretion of cold dark matter onto an isolated loop of string. Dark matter particles will be assumed to be very weakly interacting, so that non-gravitational forces on the particles can be ignored. Length scales of interest to us will be much smaller than the horizon and much greater than the loop size. On such scales the gravitational field of the loop can be approximated by that of a point mass and the accretion process can be adequately described by Newtonian physics.

We thus have a Newtonian accretion problem in an expanding universe with a point mass seed. This problem is most conveniently studied using the linear perturbation methods developed by Zel'dovich [1970]. We consider a cold-dark-matter universe in the matter-dominated era $t > t_{eq}$, with scale factor $a(t) \propto t^{2/3}$, and average density given by

$$\rho_{av} = \frac{1}{6\pi G t^2} \, . \tag{11.2.1}$$

The dark matter particle trajectories can be written as

$$\mathbf{r}(\mathbf{x}, t) = a(t) \left[\mathbf{x} + \boldsymbol{\psi}(\mathbf{x}, t) \right] \, . \tag{11.2.2}$$

Here, $\mathbf{x}$ is the unperturbed comoving position of the particle and $\boldsymbol{\psi}$ is the comoving displacement from that position. We shall call $\mathbf{x}$ the Lagrangian coordinates of the particle to distinguish them from the physical coordinates, $\mathbf{r}$. The Lagrangian coordinates can be thought of as labels distinguishing one particle from another.

The equation of motion for a dark matter particle is

$$\ddot{\mathbf{r}} = -\nabla_r \Phi \, , \tag{11.2.3}$$

where the gravitational potential $\Phi(\mathbf{r}, t)$ satisfies the Poisson equation

$$\nabla_r^2 \Phi = 4\pi G (\rho + \rho_s) \, . \tag{11.2.4}$$

Here, $\rho(\mathbf{r}, t)$ is the dark matter density and $\rho_s(\mathbf{r}, t)$ is the perturbation due to the loop seed,

$$\rho_s(\mathbf{r}, t) = m \, \delta[\mathbf{r} - \mathbf{r}_s(t)] \, , \tag{11.2.5}$$

with loop mass m and trajectory $\mathbf{r} = \mathbf{r}_s(t)$. Mass conservation implies $\rho_{av} a^3 d^3 x = \rho(\mathbf{r}, t) d^3 r$, or

$$\rho(\mathbf{r}, t) = \frac{\rho_{av}(t) a^3(t)}{|\det(\partial \mathbf{r}/\partial \mathbf{x})|} \, . \tag{11.2.6}$$

For small displacements, the Jacobian in (11.2.6) can be expanded in powers of ψ. Using (11.2.2) and the relation $\det(1+A) = 1 + \operatorname{tr} A + \mathcal{O}(A^2)$, we have, to linear order in ψ,

$$\rho(\mathbf{r}, t) = \rho_{\text{av}}(t) \left[1 - \nabla_x \cdot \psi(\mathbf{x}, t) \right] . \tag{11.2.7}$$

The fractional density perturbation is thus given by

$$\frac{\delta\rho}{\rho} = -\nabla_x \cdot \psi . \tag{11.2.8}$$

It follows from (11.2.4–5) and (11.2.7) that

$$\nabla_r \Phi = \frac{4\pi}{3} G \rho_{\text{av}} (\mathbf{r} - 3a\psi) + \frac{Gm(\mathbf{r} - \mathbf{r}_s)}{|\mathbf{r} - \mathbf{r}_s|^3} . \tag{11.2.9}$$

Combining this with (11.2.1–3) we obtain the linearized equation of motion for the displacement,

$$\ddot{\psi} + \frac{4}{3t}\dot{\psi} - \frac{2}{3t^2}\psi = -\frac{Gm(\mathbf{x} - \mathbf{x}_s)}{a^3 |\mathbf{x} - \mathbf{x}_s|^3} , \tag{11.2.10}$$

where $\mathbf{x}_s(t) = \mathbf{r}_s(t)/a(t)$ is the comoving position of the loop.

Taking the divergence of (11.2.10) and using (11.2.8) we obtain an equation for the density perturbation,

$$\left(\frac{\partial^2}{\partial t^2} + \frac{4}{3t}\frac{\partial}{\partial t} - \frac{2}{3t^2} \right) \frac{\delta\rho}{\rho}(\mathbf{x}, t) = \frac{4\pi Gm}{a^3(t)} \delta[\mathbf{x} - \mathbf{x}_s(t)] . \tag{11.2.11}$$

In the absence of seeds, $m = 0$, the general solution of this equation is

$$\frac{\delta\rho}{\rho} = A(\mathbf{x}) t^{2/3} + B(\mathbf{x}) t^{-1} , \tag{11.2.12}$$

where A and B are arbitrary functions. This decomposition into growing and decaying modes is a well-known result from linear perturbation theory (1.3.39). It also follows from (11.2.11) that if a loop seed is introduced into a homogeneous medium with $\delta\rho/\rho = 0$, the density perturbation will remain zero everywhere outside the loop. In linear perturbation theory, a point mass remains a point mass, only its mass grows as $t^{2/3}$.

Linear perturbation theory will break down when the displacement $|\psi(\mathbf{x}, t)|$ becomes comparable to $|\mathbf{x}|$. However, Zel'dovich [1970] has argued that large errors will not be made if the solution (11.2.2) is extrapolated somewhat into the nonlinear regime. This approach has been tested against exact solutions and N-body simulations and has been shown to give satisfactory results [Efstathiou & Silk, 1983; Bertschinger, 1985]. For scales on which the perturbation becomes nonlinear, particles stop expanding with the Hubble flow and start falling back towards the loop. The turnaround surface where the velocity of particles with respect to the loop is equal to zero, can be used to roughly delineate the boundary of the nonlinear region. An important advantage of the Zel'dovich approach is that is can be used outside this region even when the evolution close to the loop is strongly nonlinear.* In the following sections, the Zel'dovich approximation will be applied to determine the location of the turnaround surface.

11.2.2 Static loops

We consider an idealized situation in which the loop is formed at time $t_i > t_{eq}$ in an initially unperturbed universe. The perturbation caused by the loop at $t > t_i$ can then be found by solving (11.2.10) with the initial conditions

$$\psi(\mathbf{x}, t_i) = \dot{\psi}(\mathbf{x}, t_i) = 0. \tag{11.2.13}$$

It will be convenient to normalize the scale factor so that $a(t_i) = 1$. For a static loop, we can also choose $\mathbf{x}_s = 0$. The solution of (11.2.10) with (11.2.13) is then

$$\psi(\mathbf{x}, t) = \frac{3Gm\mathbf{x}t_i^2}{2|\mathbf{x}|^3} \left[1 - \frac{2}{5}\frac{t_i}{t} - \frac{3}{5}\left(\frac{t}{t_i}\right)^{2/3} \right]. \tag{11.2.14}$$

The late-time behaviour at $t \gg t_i$ is

$$\psi \approx -\frac{9}{10}\frac{Gm\mathbf{x}t_i^2}{|\mathbf{x}|^3}\left(\frac{t}{t_i}\right)^{2/3}. \tag{11.2.15}$$

The turnaround surface can now be found from the condition $\dot{r} = 0$. This is equivalent to $\mathbf{x} + 2\psi = 0$ or, using (11.2.12) and (11.2.15), to

$$1 - \frac{9}{5}\frac{Gmt_i^2}{|\mathbf{x}|^3}\left(\frac{t}{t_i}\right)^{2/3} = 0. \tag{11.2.16}$$

The mass enclosed by a spherical shell of initial radius $r_i = |\mathbf{x}|$ is $M = (4\pi/3)\rho_i r_i^3$, and thus the mass that has turned around at time $t \gg t_i$ is given by

$$M = \frac{2}{5}m\left(\frac{t}{t_i}\right)^{2/3}. \tag{11.2.17}$$

---

* An improvement to the Zel'dovich approximation prevents shell-crossing in nonlinear regions with an additional viscosity term, an approach known as the adhesion approximation (see, for example, Weinberg & Gunn [1990]). This has been applied successfully to the study of cosmic-string-seeded fluctuations [Avelino & Shellard, 1993].

An exact solution for the spherical accretion model [Peebles, 1980] gives the same result, with the coefficient 2/5 replaced by $(4/3\pi)^{2/3}$ – a difference by a factor of about 1.5.

When a mass shell turns around and recollapses, it is virialized by gravitational interactions between the particles in a few collapse times. The radius of the resulting virialized lump is $r \approx r_{\max}/2$, where (see, for example, Gott & Rees [1975])

$$r_{\max} = \frac{5}{4}\frac{M}{m}r_i = \frac{5}{4}\left(\frac{3}{4\pi}\frac{m}{\rho_i}\right)^{1/3}\left(\frac{M}{m}\right)^{4/3} \qquad (11.2.18)$$

is the turnaround radius for a shell of mass M, and we have used (11.2.16) and (11.2.17). The density profile of the bound object formed around the loop can now be found from (11.2.18) and $dM/dr = 4\pi\rho r^2$. This gives [Gott, 1975; Gunn, 1977]

$$\rho(r) \propto r^{-9/4} . \qquad (11.2.19)$$

For a loop formed at $t_i \ll t_{eq}$, no appreciable accretion takes place until $t \sim t_{eq}$. At later times, the mass accreted by the loop can be estimated from (11.2.17) with t_i replaced by t_{eq}. Within a Hubble time of t_{eq}, the accreted mass is comparable to the mass of the loop, and so the subsequent decay of the loop does not significantly affect the accretion pattern.

Throughout this discussion we have ignored the finite size of the loop, $L \sim m/\mu$. The mass of dark matter in a region of initial size $\sim L$ is $M_L \sim \rho_i L^3$, and our treatment applies for $M > M_L$ or $r > L^3/G\mu t_i^2$. The accretion pattern on smaller scales has been discussed by Sato [1986a,b] and Stebbins [1986] using a simple model in which the loop was represented by a massive sphere. On scales greater than the Hubble radius, strings can only induce pressure fluctuations, and we must take account of the underdensity in the background matter which compensates the loop perturbation. This is discussed for idealized loops in Shellard *et al.* [1987] and Brandenberger *et al.* [1987], but a more general treatment of compensated perturbations can be found in Veeraraghavan & Stebbins [1990].

11.2.3 Moving loops

Suppose now that the loop is formed with an initial velocity $\mathbf{v}(t_i) = \mathbf{v}_i$. For $t_i > t_{eq}$, the loop trajectory is given by

$$\mathbf{x}_s(t) = 3\mathbf{v}_i t_i (1 - a^{-1/2}) \equiv d(1 - a^{-1/2})\hat{\mathbf{e}}_z , \qquad (11.2.20)$$

where $d = 3v_i t_i$, $a(t) = (t/t_i)^{2/3}$, and the loop has been taken to move in the z-direction. Rapidly moving loops are expected to produce elongated accretion wakes of comoving length comparable to d.

The solution of the Zel'dovich equation (11.2.10) with initial conditions (11.2.13) and $\mathbf{x}_s$ from (11.2.20) has been found by Bertschinger [1987a]. The accretion pattern has an azimuthal symmetry around the z-axis, and it is sufficient to consider accretion only in the xz-plane. At $t \gg t_i$ the displacement is given by

$$\psi(\mathbf{x}, t) = -b(t)d\left[\hat{\mathbf{e}}_x f_x(\mathbf{x}) + \hat{\mathbf{e}}_z f_z(\mathbf{x})\right] , \tag{11.2.21}$$

where

$$b(t) = \frac{1}{5} \frac{Gm}{v_i^2 d} a(t) , \tag{11.2.22}$$

$$f_x = \frac{R_f - R_i}{x} + \frac{zd}{xR_i} , \tag{11.2.23}$$

$$f_z = \ln\left(\frac{R_i + z}{R_f + z - d}\right) - \frac{d}{R_i} , \tag{11.2.24}$$

with $R_i = \left(x^2 + z^2\right)^{1/2}$ and $R_f = \left(x^2 + (z-d)^2\right)^{1/2}$ which are, respectively, the comoving distances from the point $\mathbf{x}$ to the beginning and the end of the loop trajectory. It is easily verified that at large distances from the loop, $|\mathbf{x}| \gg d$, the displacement approaches the spherically-symmetric form (11.2.15) appropriate for a static loop.

The turnaround surface for a moving loop can be defined by the condition $\dot{r}_x = 0$, requiring that particles stop expanding away from the accretion wake, while expansion may continue in the z-direction. With the aid of (11.2.2) and (11.2.21–23) this condition can be rewritten as $x + 2\psi_x = 0$, or

$$x = 2d\,b(t)f_x(x, z) . \tag{11.2.25}$$

The shapes of turnaround surfaces (in the Lagrangian coordinates x, z) are illustrated at several moments in time in fig. 11.1.

At late times, when $b(t) \gg 1$, the surfaces are nearly spherical, and the accreted mass can be found from (11.2.17). At early times, when $b(t) \ll 1$, the accreted matter resides in a narrow wake trailing behind the loop. The shape of the turnaround surface in this case can be found by expanding (11.2.25) for $x^2 \ll z^2$, $(d-z)^2$. This yields

$$x^2 = 4b(t)d(d-z) , \tag{11.2.26}$$

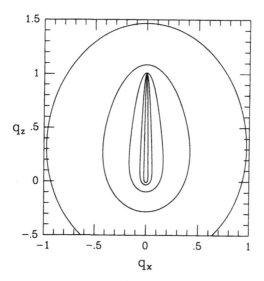

Fig. 11.1. Turnaround surfaces about a moving loop seed [Bertschinger, 1987a].

with the accreted mass given by

$$M \approx \rho_i \int_0^d \pi x^2 \, dz = \frac{3}{5} m \left(\frac{t}{t_i} \right)^{2/3} \qquad (11.2.27)$$

Somewhat surprisingly, the mass accreted in the wake of a rapidly moving loop is independent of the loop velocity and is not much different from the mass accreted by a static loop [Bertschinger, 1987a]. Hoffman & Zurek [1988] generalized this analysis for loops of finite size. As the moving loop shrinks in comoving coordinates, the accretion wake can acquire a trumpet-like shape (see also Turok [1988b]).

So far we have assumed that the loop moves as a free particle, (11.2.20). This can be modified by both dynamical friction and the gravitational radiation rocket effect. Dynamical friction is the retarding force exerted by the wake on the loop. This force slows the loop down on a timescale [Chandrasekhar, 1943]

$$t_{gd} \sim \frac{v^3 t^2}{Gm} \ln \left(\frac{v^3 t}{Gm} \right) . \qquad (11.2.28)$$

With $m = \mu L$ and $v = v_i (t_i/t)^{2/3}$, we find that this drag becomes important ($t_{gd} \lesssim t$) at times $t \sim v_i^3 t_i^2 / G\mu L$. However, most of the loops are formed with high initial velocities ($v_i \sim 1$) and small sizes ($L \sim \alpha t_i$, $\alpha \lesssim 5 \times 10^{-3}$), and decay before being affected by drag.

The gravitational rocket effect is loop self-acceleration due to gravitational radiation recoil (see §7.5.3). The loop equation of motion with the rocket effect taken into account is

$$\dot{\mathbf{v}} + 2\mathbf{v}/3t = \Gamma_{\mathrm{P}} G\mu\hat{\mathbf{n}}/L, \qquad (11.2.29)$$

where $\hat{\mathbf{n}}$ is the direction of the recoil and $\Gamma_{\mathrm{P}} \sim 10$. The solution of this equation is

$$\mathbf{v} = \mathbf{v}_{\mathrm{i}}(t_{\mathrm{i}}/t)^{2/3} + (3\Gamma_{\mathrm{P}} G\mu/5L)\hat{\mathbf{n}}t. \qquad (11.2.30)$$

Initially the velocity is redshifted, as in the case of a free particle, but then the rocket effect takes over and accelerates the loop to relativistic speeds on a timescale comparable to the loop's lifetime (10.1.5). The directions of $\hat{\mathbf{n}}$ and $\mathbf{v}_{\mathrm{i}}$ are generally different, and the resulting accretion wakes are expected to be curved.

11.2.4 Effects of hot dark matter and baryons

Suppose instead that the universe is dominated by hot dark matter consisting of particles with a large velocity dispersion at $t \sim t_{\mathrm{eq}}$. More specifically, we shall assume that the dark matter particles are neutrinos and that only one of the three light neutrino species has an appreciable mass. The neutrino mass required to close the universe is $m_\nu \approx 100\,h^2\,\mathrm{eV}$, and matter and radiation equality occurs at a redshift $z_{\mathrm{eq}} \approx 3 \times 10^4 h^2$. The rms velocity of neutrinos at a time $t \gtrsim t_{\mathrm{eq}}$ is $v_\nu = v_{\mathrm{eq}}(t_{\mathrm{eq}}/t)^{2/3}$ with $v_{\mathrm{eq}} \approx 0.2$.

The accretion of neutrinos onto a loop of string has been studied by Brandenberger, Kaiser, Schramm & Turok [1987], Brandenberger, Kaiser & Turok [1987], and Bertschinger & Watts [1987]. Their approach was to solve the linearized Liouville equation for the neutrino phase space density in the gravitational field of a loop. N-body simulations of neutrino accretion have been performed by Dobyns [1988]. Here we shall take a simple analytic approach, due to Perivolaropoulos, Brandenberger & Stebbins [1990], which is an adaptation of the Zel'dovich approximation to hot dark matter.

The typical comoving distance travelled by neutrinos in one Hubble time is

$$\lambda_\nu(t) \approx v_\nu t/a(t) \approx v_{\mathrm{eq}} t_{\mathrm{eq}} \left(t_{\mathrm{i}}^2/t\, t_{\mathrm{eq}}\right)^{1/3}, \qquad (11.2.31)$$

where again $t_{\mathrm{i}} > t_{\mathrm{eq}}$ is the time of loop formation, and the scale factor is normalized so that $a(t_{\mathrm{i}}) = 1$. On scales smaller than λ_ν, neutrinos stream out of the potential well of the loop and there is practically no accretion, while on larger scales perturbations grow in the same way as for cold dark matter. Density perturbations at a comoving distance $|\mathbf{x}|$

from the loop start growing at the time $t_s(\mathbf{x})$ when λ_ν becomes smaller than $|\mathbf{x}|$,

$$t_s(\mathbf{x}) = v_{eq}^3 t_{eq}^2 t_i^2 / |\mathbf{x}|^3 , \qquad (11.2.32)$$

or immediately at time t_i if $t_s(\mathbf{x}) < t_i$. Hence, the accretion process can be roughly described by solving the Zel'dovich equation (11.2.10) with the modified initial conditions,

$$\psi(\mathbf{x}, t_\mathbf{x}) = \dot{\psi}(\mathbf{x}, t_\mathbf{x}) = 0 , \qquad (11.2.33)$$

where

$$t_\mathbf{x} = \max\{t_s(\mathbf{x}), t_i\} . \qquad (11.2.34)$$

Clearly, this method will only have an order-of-magnitude accuracy. For an improved treatment, one has to employ the Liouville equation.

The solution of (11.2.10) and (11.2.33), for $t \gg t_\mathbf{x}$, is

$$\psi \approx -\frac{9}{10} \frac{Gm\mathbf{x}t_i^2}{|\mathbf{x}|^3} \left(\frac{t}{t_\mathbf{x}}\right)^{2/3} , \qquad (11.2.35)$$

while, for $t < t_\mathbf{x}$, $\psi = 0$. The turnaround surface and the mass M which has turned nonlinear can now be found as before. First, consider values of $|\mathbf{x}|$ such that $t_i < t_s(\mathbf{x}) < t$. Then $t_\mathbf{x} = t_s(\mathbf{x})$ and the accreted mass is given by

$$M \sim \frac{m^3}{M_{eq}^2} \left(\frac{t}{t_{eq}}\right)^2 , \qquad (11.2.36)$$

where

$$M_{eq} = v_{eq}^3 t_{eq}/G = 5 \times 10^{13} h^{-4} M_\odot \qquad (11.2.37)$$

is the mass contained in a volume of size $\sim v_{eq} t_{eq}$ at time t_{eq}.

The condition $t_s(\mathbf{x}) < t$ implies that $M > M_{eq} t_{eq}/t$. Combining this with (11.2.36) we see that only loops with

$$m \gtrsim M_{eq} t_{eq}/t \qquad (11.2.38)$$

can form nonlinear objects by a time t. The first mass scale to collapse around a loop of mass m is $M \sim m$. The condition $t_s(\mathbf{x}) > t_i$ implies that $M < M_{eq} t_{eq}/t_i$. On larger mass scales, accretion proceeds as for cold dark matter and the accreted mass is given by (11.2.17).

For the regime (11.2.36), the maximum expansion radius for a shell of mass M is $r_{max} \propto M^{2/3}$, and the density profile predicted by the simple turnaround model is $\rho \propto r^{-3/2}$. However, this model can be trusted only when it gives $\rho \propto r^{-k}$ with $k \geq 2$. For $k < 2$, analytic arguments [Fillmore & Goldreich, 1984; Hoffman & Shakem, 1985] and numerical simulations [Frenk *et al.*, 1985] show that the resulting density distribution will become

$$\rho \propto r^{-2}.$$

(11.2.39)

This distribution applies for $m \lesssim M \lesssim M_{\mathrm{eq}} t_{\mathrm{eq}} / t_{\mathrm{i}}$. On larger scales it is replaced by the cold-dark-matter distribution (11.2.19).

The accretion picture described above can be modified by loop decay, by baryon infall, and by the motion of the loop. A loop of mass m decays by gravitational radiation at $t_{\mathrm{d}} \sim m/\Gamma G\mu^2$, and the perturbation (11.2.35) is generated only on scales where $t_{\mathrm{s}}(\mathbf{x}) < t_{\mathrm{d}}$, or

$$M > M_{\mathrm{eq}} t_{\mathrm{eq}} / t_{\mathrm{d}} \equiv M_{\mathrm{d}}.$$

(11.2.40)

Collapse on smaller scales is delayed until the scale M_{d} goes nonlinear.

Baryon accretion does not begin until the decoupling of matter and radiation, $t \sim t_{\mathrm{dec}}$, when the Jeans length of baryons drops dramatically and quickly becomes smaller than the free-streaming scale of neutrinos, λ_ν. For loops that decay before t_{dec}, baryons simply fall into the potential wells created by neutrinos, and the accretion process is not significantly affected. For larger loops that decay after t_{dec}, baryons collapse in the gravitational field of the loop, forming a nonlinear clump at the core of a much wider neutrino distribution [Bertschinger, 1988]. This baryonic core survives the loop decay and can make the effect of loop decay on neutrinos less pronounced. Note that neutrino clustering in the core is suppressed by the phase space constraint [Tremaine & Gunn, 1979] and so the core can be almost entirely baryonic.

Neutrino accretion by a moving loop has been studied numerically by Bertschinger [1990]. In contrast to the cold dark matter scenario, a rapidly moving loop accretes considerably less mass than a stationary loop. This can be understood by noting that thin accretion wakes created by moving loops are washed out by neutrino free streaming whenever the comoving thickness of wakes is smaller than λ_ν.

11.3 Loops as seeds for galaxies and clusters

Oscillating loops of string can serve as localized seeds for the formation of cosmic structure [Vilenkin, 1981a]. The loops radiate away their energy and eventually disappear, but self-gravitating clumps of matter that are left behind can later evolve into galaxies and clusters of galaxies.

Most of the early work on this scenario [Vilenkin & Shafi, 1983; Turok, 1983b, 1985; Sato, 1986; Turok & Brandenberger, 1986; and many papers that followed in the period 1986–1988] was based on the 'old picture' of string evolution. It was thought that strings were smooth on the horizon scale, with about one long string per horizon volume and that about one horizon-sized loop was chopped off the network per horizon volume per expansion time. In terms of the parameters ζ and α introduced in §10.1.2, this corresponds to

$$\zeta \sim \alpha \sim 1. \tag{11.3.1}$$

The initial velocities of the loops were assumed to be small (~ 0.1), and the loops were treated as a collection of stationary accretion centres. Accretion in the wakes of long strings was also usually ignored. We shall refer to the structure formation scenario based on these assumptions as the 'old string scenario'.

The string evolution picture suggested by recent numerical simulations is quite different (refer to §9.3–4 and the brief review in §10.1). Long strings exhibit significant small-scale structure and, in consequence, loops are formed on scales well below the horizon ($\alpha \lesssim 5 \times 10^{-3}$) and with large initial velocities ($v_i \sim 0.7$). As a result, loops may be less important for structure formation than the wakes of long strings, and the vast literature on the 'old string scenario' must, therefore, be thoroughly reconsidered.

It should be noted that small-scale structure on long strings can be suppressed in some models where strings are formed during inflation (chapter 16) and in the case of global strings (chapter 12). As a result, the loop parameter α may not be so small, and some of the features of the old scenario may be preserved. In this section, we shall discuss structure formation with loops, leaving both ζ and α as free parameters.

11.3.1 Cold dark matter

In a universe dominated by cold dark matter, loops formed before t_{eq} start accreting at $t \sim t_{eq}$, and the mass accreted by a loop of initial length ℓ is, at the present time t_0,

$$M \sim \mu \ell \left(\frac{t_0}{t_{eq}} \right)^{2/3} \tag{11.3.2}$$

(see (11.2.17)). For a loop formed at $t_i = \ell/\alpha > t_{eq}$, the accreted mass is

$$M \sim \mu \ell \left(\frac{\alpha t_0}{\ell} \right)^{2/3}. \tag{11.3.3}$$

Combining this with the loop distribution function (10.1.19–20), and assuming independent accretion, we can estimate the mass distribution of gravitationally bound objects. For $\Gamma G \mu \, z_{eq}^{-1/2} \tilde{M} < M < \alpha z_{eq}^{-1/2} \tilde{M}$, we have

$$n_M \sim \frac{\zeta \sqrt{\alpha}}{t_0^3} \left(\frac{\tilde{M}}{M} \right)^{3/2} z_{eq}^{3/4}, \tag{11.3.4}$$

while, for $\alpha z_{eq}^{-1/2} \tilde{M} < M < \alpha \tilde{M}$,

$$n_M \sim \frac{\zeta \alpha^2}{t_0^3} \left(\frac{\tilde{M}}{M}\right)^3 . \qquad (11.3.5)$$

Here, $n_M = M dn/dM$ is the number density of objects with masses $\sim M$ in the interval $\Delta M \sim M$, the mass $\tilde{M}$ is defined as $\tilde{M} = \mu t_0 = 3 \times 10^{22} G\mu h^{-1} M_\odot$, and we have disregarded the numerical factors in (10.1.13) and (10.1.18). The average distance between loops of mass $\sim M$ is given by

$$d_M \approx n_M^{-1/3} . \qquad (11.3.6)$$

We now briefly review the 'old string scenario' of structure formation. Given ζ and α from (11.3.1), the only free parameter in this scenario is the mass parameter of the strings, $G\mu$. To determine the value of $G\mu$ needed for structure formation, we can choose a class of astronomical objects and require that (11.3.6) gives the correct separation for the given mass of the objects. For example, for rich clusters of galaxies $M \sim 10^{15} h^{-1} M_\odot$, $d \sim 50 h^{-1}$Mpc, and (11.3.5) and (11.3.6) yield [Turok & Brandenberger, 1986]

$$G\mu \sim 10^{-6} \left(\zeta \alpha^2\right)^{-1/3} . \qquad (11.3.7)$$

The assumption of independent accretion (the one loop–one object hypothesis) is, of course, an over-simplification. It certainly fails on small scales when loops run out of matter to accrete and larger loops start accreting objects formed by smaller loops. The total fraction of mass accreted by all loops at time $t > t_{eq}$ is

$$f_L(t) = \frac{\rho_L}{\rho}(t) \sim 6\pi\zeta\sqrt{\alpha} \left(\frac{G\mu}{\Gamma}\right)^{1/2} \left(\frac{t}{t_{eq}}\right)^{2/3} . \qquad (11.3.8)$$

It is dominated by loops of length $\ell \sim \Gamma G\mu t_{eq}$, which are the smallest loops that survive until t_{eq}. Competition between different loops becomes important when $\rho_L/\rho \sim 1$. At later times, the mass distribution (11.3.4–5) gets modified on mass scales below a certain scale $M_c(t)$, which can be roughly estimated from

$$M_c n_{M_c}(t) \sim \rho(t) . \qquad (11.3.9)$$

For the values of parameters assumed in the old scenario, $\rho_L/\rho \sim 1$ at a redshift $z \sim 10$ and $M_c(t_0) \sim 10^{14} h^2 M_\odot$. Further shortcomings of this one-to-one hypothesis and the possibility that larger objects are seeded by collections of loops and through mergers has been discussed by Zurek [1988] and Shellard & Brandenberger [1988].

A numerical simulation of loop-seeded structure formation has been performed by Melott & Scherrer [1987]. They used a simple loop distribution similar to (10.1.19–20) and followed the initial evolution of perturbations using linear perturbation theory. When perturbations became nonlinear they evolved forward to the present time using an N-body gravitational clustering code. The main conclusion of this work was that the spectrum of perturbations in a model with stationary loops and cold dark matter exhibited too much power on small scales. The resulting mass-mass correlation function at the present time was too steep in the range 1–10 Mpc, and the visual appearance of the matter distribution was too 'lumpy'. Bertschinger [1990] has argued that the situation can be improved if loop velocities are taken into account. The effect of loop motion is to redistribute power from smaller to larger scales, so it appears possible that rapidly moving loops may produce a reasonable perturbation spectrum.

The most impressive, albeit short-lived, success of the old scenario was reported by Turok [1985]. He compared the loop correlation function obtained from a numerical simulation of string evolution [Albrecht & Turok, 1985] with the observed correlation function of rich clusters of galaxies [Bahcall & Soneira, 1983] and found remarkable agreement. However, this explanation of correlations between clusters is hard to reconcile with the new picture of string evolution. Large initial velocities cause loops to move a distance more than ten times their mean separation, so any initial correlations between the loops will be rapidly washed out [Bennett & Bouchet, 1989]. Even if some correlations survive, rapidly moving loops would tend to form highly elongated objects, with an appearance unlike clusters of galaxies. Finally, if clusters are identified with objects seeded by individual loops, using values of ζ and α suggested by the new simulations (refer to §10.1.2), then (11.3.7) implies that $G\mu \gtrsim 3 \times 10^{-5}$. This exceeds the upper bounds imposed by the isotropy of the cosmic background radiation and by the intensity of the gravitational radiation background (see §10.2 and §10.4).

11.3.2 Hot dark matter

In a universe dominated by light neutrinos, accretion is suppressed on scales smaller than the neutrino free-streaming scale (11.2.31). The mass of an object seeded by a loop of length ℓ is (see (11.2.36))

$$M \sim \frac{(\mu \ell z_{\text{eq}})^3}{M_{\text{eq}}^2}, \tag{11.3.10}$$

where, as before, $M_{\text{eq}} = 5 \times 10^{13} h^{-4} M_\odot$. This equation applies to mass scales for which the onset of accretion t_s given by (11.2.32) satisfies $t_s >$

t_i, t_{eq}, where $t_i = \ell/\alpha$ is the time of loop formation. In terms of the mass M, the condition is $M \lesssim M_{max}$, where M_{max} is the smaller of M_{eq} and $M' \sim 10^{14} \alpha^{3/4} \mu_6^{3/4} h^{-5/2} M_\odot$, and $\mu_6 = G\mu/10^{-6}$. For $M > M_{max}$ accretion is not affected by free-streaming, and (11.3.10) is replaced by (11.3.2) or (11.3.3).

The form of the mass distribution n_M depends on whether M_{eq} is greater or smaller than M':

(i) If $M_{eq} < M'$ or, equivalently, $\alpha \mu_6 h^2 > 0.2$, then loops formed during the matter era are not affected by free-streaming. Combining (11.3.10) with (10.1.19) we obtain

$$n_M \sim \frac{\zeta \sqrt{\alpha}}{t_0^3} \frac{\tilde{M}^{3/2}}{M_{eq} \sqrt{M}} z_{eq}^{3/4} \qquad (11.3.11)$$

for $M < M_{eq}$. On scales greater than M_{eq} the mass distribution has the same form as for cold dark matter.

(ii) If $M_{eq} > M'$, or $\alpha \mu_6 h^2 < 0.2$, then the distribution (11.3.11) applies for $M < M''$, where $M'' \sim 3 \times 10^{13} \alpha^3 \mu_6^3 h^2 M_\odot$ is the mass accreted by a loop formed at $t \sim t_{eq}$. For $M'' < M < M'$ the distribution is found by combining (11.3.10) with (10.1.20),

$$n_M \sim \frac{\zeta}{t_0^3} \frac{\tilde{M}}{M_{eq}^{2/3} M^{1/3}} z_{eq}, \qquad (11.3.12)$$

and for $M > M'$, n_M is given by (11.3.5). In all cases, the mass distribution is cut off at the scale $M_{min} \sim 10^{11} \mu_6^{3/2} h^{-5/2} M_\odot$, below which accretion is affected by loop decay.

The value of $G\mu$ required in the 'old string scenario' can be found as before by requiring that objects with the mean separation of rich galaxy clusters have accreted a mass comparable to that of a rich cluster. This gives the same result as with cold dark matter (11.3.7), because the mass function n_M has the same form for hot and cold dark matter on cluster scales.

It is easily seen from (11.3.11–12) and (11.3.4–5) that the total mass density accreted by the loops $\rho_L(t_0)$, is dominated by objects of mass $\sim M_{max}$, which were also the first objects to collapse,

$$f_L = \rho_L/\rho(t_0) \sim 6\pi G t_0^2 M_{max} n_{M_{max}} \sim \zeta \sqrt{\alpha} \mu_6^{3/2} h^3. \qquad (11.3.13)$$

This equation, which was first obtained by Perivolaropoulos et al. [1990], applies for both $M_{eq} > M'$ and $M_{eq} < M'$. For the values of parameters assumed in the old scenario, $\rho_L/\rho(t_0) \sim 1$. This suggests that the assumption of independent accretion is justified until very recent epochs.

The loop-seeded hot dark matter scenario of structure formation has several attractive features [Brandenberger et al., 1987; Bertschinger &

Watts, 1987]. The mass function (11.3.11) $n_M \propto M^{-1/2}$, is in a better agreement with observation than the cold dark matter result ($n_M \propto M^{-3/2}$). Galaxies have baryonic interior regions and dark matter halos with flat halo rotation curves (see (11.2.39)). Galaxy formation occurs much earlier than in the hot dark matter scenario with adiabatic fluctuations, because the loop seeds survive neutrino free-streaming, thus helping to explain observations of high redshift quasars and galaxies. A numerical simulation of structure formation with loops and hot dark matter was performed by Scherrer, Melott & Bertschinger [1988], who used the following values of the parameters: $\alpha = 0.03$, $\zeta = 6$, $G\mu = 5 \times 10^{-7}$. The galaxy distribution they obtained has large voids and filaments and is visually similar to observational surveys. They also found good agreement with the observed galaxy–galaxy correlation function. We note, however, that since their values of the parameters do not satisfy (11.3.7), rich clusters of galaxies in their simulation were probably seeded by collections of smaller loops rather than by large individual loops.

Many of the positive features of the loops and hot dark matter scenario, unfortunately, disappear when the relativistic motion of loops is taken into account [Bertschinger, 1990]. Instead of a spherical halo with a flat rotation curve, a rapidly moving loop accretes a long, thin wake. The mass of dark matter in the wake is much smaller than the mass accreted by a stationary loop, and a value of $G\mu \gg 10^{-6}$ is required to accrete sufficient mass to form galaxies. However, large values of $G\mu$ give excessive power on very large scales, where free-streaming is unimportant, and lead to a potential conflict with the microwave and gravitational radiation constraints (chapter 10).

11.3.3 Baryon-dominated universe

In a baryon-dominated universe, the density parameter Ω must satisfy the nucleosynthesis constraint, $0.015 \leq \Omega h^2 \leq 0.026$ (refer to §1.3.4), and thus Ω must be substantially smaller than unity. The growth of density perturbations in this case is strongly suppressed at redshifts $1 + z < \Omega^{-1}$, but is otherwise similar to that in a cold-dark-matter universe. (Note that for $\Omega h^2 < 0.026$ decoupling occurs before t_{eq}.) The mass accreted by a loop of length ℓ is obtained by a simple modification of (11.3.2–3): For $\Gamma G\mu t_{eq} < \ell < \alpha t_{eq}$,

$$M \sim \mu\ell\Omega \left(t_0/t_{eq}\right)^{2/3}, \qquad (11.3.14)$$

while, for $\ell > \alpha t_{eq}$, we have

$$M \sim \mu\ell\Omega \left(\alpha t_0/\ell\right)^{2/3}. \qquad (11.3.15)$$

The mass distribution n_M is then given by (11.3.4–5) with $\tilde{M}$ replaced by $\Omega\tilde{M}$.

In the 'old string scenario', the normalization of $G\mu$ on the scale of rich clusters yields

$$G\mu \sim 10^{-6}\Omega^{-1}\left(\zeta\alpha^2\right)^{-1/3}. \qquad (11.3.16)$$

With $\zeta\alpha^2 \lesssim 1$ and $\Omega h^2 < 0.026$ this implies $G\mu \gtrsim 4 \times 10^{-5}h^2$, which is inconsistent with observational bounds on $G\mu$.

11.4 Wakes

11.4.1 Wakes due to straight strings

A distinctive feature of the string scenario is the formation of sheet-like wakes behind rapidly moving long strings [Silk & Vilenkin, 1984]. A straight segment of string moving with velocity v_s gives nearby matter a boost,

$$u_i = 4\pi G\mu v_s\gamma_s, \qquad (11.4.1)$$

in the direction of the surface swept out by the string. Here, we have the usual Lorentz factor, $\gamma_s = \left(1 - v_s^2\right)^{-1/2}$. This boost is due to the conical nature of space around the string and, in this sense, is a purely kinematic effect (see §10.1). The formation and evolution of wakes will be discussed in this section, and their potential role in producing large-scale structure in §11.5. We begin by considering cold dissipationless matter. In this case the two opposite streams of matter in the wake overlap within a wedge with an opening angle $8\pi G\mu\gamma_s$ (see fig. 11.2). The matter density within this wedge is doubled. As the universe expands, the surface area of the wake grows as $a^2(t)$, while the total mass of the wake grows by gravitational instability as $M \propto a(t)$. As a result, the surface density of the wake falls as $\sigma(t) \propto a^{-1}(t)$.

A quantitative description of accretion onto a wake can be obtained using the Zel'dovich formalism. We consider a wake formed by a long straight string at time $t_i > t_{eq}$. If the wake is perpendicular to the x-axis, then the displacement ψ has only one non-zero component, $\psi_x \equiv \psi$, which can be found by solving the equation

$$\ddot{\psi} + \frac{4}{3t}\dot{\psi} - \frac{2}{3t^2}\psi = 0. \qquad (11.4.2)$$

A dark matter particle labelled by the Lagrangian coordinate x is at a distance $r_x = a(t)\left[x + \psi(x,t)\right]$ from the symmetry plane of the wake, and we again assume the scale factor to be normalized so that $a(t_i) = 1$. The

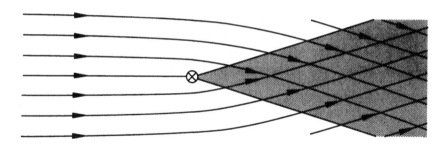

Fig. 11.2. Matter accretes onto the wake formed behind a moving string. Flow lines are shown in the rest frame of the string. The matter density is doubled in the shaded wedge.

wake begins as a velocity perturbation of magnitude (11.4.1), and thus the appropriate initial conditions are

$$\psi(x, t_i) = 0,$$
$$\dot{\psi}(x, t_i) = -u_i \epsilon(x),$$

(11.4.3)

where $\epsilon(x) = 1$ for $x > 0$ and $\epsilon(x) = -1$ for $x < 0$. The solution of (11.4.2–3) is easily found and the late-time behaviour $(t \gg t_i)$ is

$$\psi(x, t) = -\frac{3}{5} u_i t_i \left(\frac{t}{t_i}\right)^{2/3} \epsilon(x).$$

(11.4.4)

The turnaround surfaces, where particles stop expanding with the Hubble flow in the x-direction and begin falling back towards the wake, can be found from the condition $\dot{r}_x = 0$ or, equivalently, $x + 2\psi(x, t) = 0$. This yields

$$x(t) = \pm \frac{6}{5} u_i t_i \left(\frac{t}{t_i}\right)^{2/3}$$

(11.4.5)

The wake thickness d_0 and surface mass density σ_0 at the present time can then be estimated as [Vashaspati, 1986; Stebbins *et al.*, 1987]

$$d_0 \approx 2x(t_0) \left(\frac{t_0}{t_i}\right)^{2/3},$$

(11.4.6)

$$\sigma_0 \approx \rho_0 d_0 = \frac{2u_i}{5\pi G t_0} \left(\frac{t_0}{t_i}\right)^{1/3}.$$

(11.4.7)

Cold dark matter wakes can also be formed during the radiation era, $t_i < t_{eq}$, but gravitational instability* sets in only at $t \sim t_{eq}$. For $t < t_{eq}$,

---

* Here, we disregard slow logarithmic growth in the subdominant component at $t < t_{eq}$.

the wakes are simply diluted by the expansion, so that $\sigma(t) = \sigma_i t_i/t$. The initial surface density is then $\sigma_i \sim (2u_i/5\pi G t_0)(t_i/t_{eq})^{1/2}$, where the last factor is the ratio of the cold dark matter density to the total density of the universe. Hence, the surface density at the present time is

$$\sigma_0 \approx \frac{2u_i}{5\pi G t_{eq}} \left(\frac{t_i}{t_{eq}}\right)^{1/2} \left(\frac{t_{eq}}{t_0}\right)^{2/3}. \qquad (11.4.8)$$

It is clear from (11.4.7–8) that the wakes with the largest surface density are those formed at $t \sim t_{eq}$ [Vachaspati, 1986].

The evolution of the initial velocity perturbation (11.4.1) can be found by differentiating (11.4.4),

$$u(t) = |a\dot{\psi}| \approx \frac{2}{5}u_i \left(\frac{t}{t_i}\right)^{1/3}. \qquad (11.4.9)$$

The magnitude of peculiar velocities at the present time is then [Hara & Miyoshi, 1990]

$$u_0 \approx \frac{2}{5}u_i z_i^{1/2}. \qquad (11.4.10)$$

In a universe dominated by light neutrinos, wake perturbations are damped by neutrino free-streaming on all comoving scales smaller than $\lambda_\nu(t_i) = v_{eq}t_{eq}(t_i/t_{eq})^{1/3}$ (we assume that $t_i > t_{eq}$). On larger scales, $|x| > \lambda_\nu(t_i)$, the evolution of perturbations is similar to that for cold dark matter. The redshift z_{nl} at which wakes form nonlinear structures can be estimated by requiring $|x(t)| = \lambda_\nu(t_i)$ with $x(t)$ from (11.4.5). This gives

$$1 + z_{nl} \approx \frac{6}{5}\frac{u_i}{v_{eq}}z_{eq}. \qquad (11.4.11)$$

Note that z_{nl} is independent of t_i and thus all wakes formed in the matter era go nonlinear at about the same time [Stebbins et al., 1987]. In order for this to happen before the present day t_0, the string parameters must satisfy

$$u_i > 10^{-5}h^{-2}. \qquad (11.4.12)$$

If wakes do collapse before t_0, then their present thickness and surface density can be estimated using the cold dark matter equations (11.4.6–7). A more rigorous treatment of neutrino wakes, based on the linearized Liouville equation for neutrinos, leads to similar conclusions [Perivolaropoulos et al., 1990].

Baryonic wakes can start forming only after baryons decouple from radiation, $t > t_{dec}$. If the universe is dominated by cold dark matter, baryons will fall onto the cold dark matter wakes formed before t_{dec} and will flow with the dark matter in the new wakes being formed after $t_{\rm dec}$. Since the mean free path of baryons is short, there can be no counterstreaming, and a baryonic wake will be bounded by shocks on either side of the plane of symmetry. At redshifts $z < 800\,\mu_6$, the infall velocity of the wakes (11.4.1) is supersonic, and the shocks are strong. The density of the shocked gas is enhanced by a factor of about four, and its thermal velocity is comparable to u. The shocks advance outwards at a speed $\sim u$, and thus the shocked gas occupies a wedge with an opening angle $\theta \sim u/v_{\rm s}$. A detailed analysis of the shocks has been given by Stebbins *et al.* [1987], while the cooling and fragmentation of baryonic wakes has been discussed by Rees [1986] and Hara & Miyoshi [1987a].

In a neutrino-dominated universe, baryonic wakes start collapsing before neutrino wakes. However, since baryons constitute only a small fraction of the total density of the universe, the growth of these wakes by gravitational instability is strongly suppressed.

11.4.2 Wakes due to wiggly strings

As we discussed in chapter 9, numerical simulations of string evolution show that long strings possess significant amounts of small-scale structure in the form of kinks and wiggles on scales much smaller than the horizon. To an observer who cannot resolve this structure, the string will appear to be smooth, but the effective mass per unit length $\tilde{\mu}$ and tension $\tilde{T}$ will differ from their unperturbed values. The net gravitational effect of small-scale wiggles is similar to that of ordinary matter, and so a wiggly string attracts particles like a massive rod. The effective equation of state for a wiggly string is

$$\tilde{\mu}\tilde{T} = \mu^2 \,, \tag{11.4.13}$$

and the velocity boost given by a moving wiggly string to nearby matter is (refer to §7.6.2)

$$u_{\rm i} = 4\pi G \tilde{\mu} v_{\rm s} \gamma_{\rm s} + \frac{2\pi G(\tilde{\mu} - \tilde{T})}{v_{\rm s}\gamma_{\rm s}} \,. \tag{11.4.14}$$

Here, the first term is the usual velocity boost caused by the conical deficit angle and the second term is due to the Newtonian attraction of particles to the string.

Let us now consider a wiggly string moving with velocity $v_{\rm s}$ at a time $t_{\rm i}$. The wake produced by the string in one Hubble time has the shape of a strip of width $\sim v_{\rm s} t_{\rm i}$. The velocity perturbation (11.4.14) is produced at distances up to $\sim v_{\rm s} t_{\rm i}$ from the plane of the wake. For $r > v_{\rm s} t_{\rm i}$, the gravitational field of the string is like that of a stationary rod and $u_{\rm i} \sim G(\tilde{\mu} - \tilde{T})t_{\rm i}/r$. After the initial velocity perturbation is produced, the evolution of the wake is the same as it is for a straight string. All the discussion in §11.4.1 remains applicable, and so we can use (11.4.3–12) with u_i from (11.4.14). For $v_s^2 \ll 0.5$, the second term in (11.4.14) gives the dominant contribution to u_i and we can write

$$u_i \approx 2\pi G(\tilde{\mu} - \tilde{T})/v_s . \tag{11.4.15}$$

The surface density of the wake at redshift z is then

$$\sigma \approx \frac{(\tilde{\mu} - \tilde{T})}{v_s t} \left(\frac{z_i}{z}\right)^{1/2} . \tag{11.4.16}$$

Since the width of the wake is proportional to v_s, the wake mass per unit length is independent of the string velocity. If the string moves faster, the wake is wider but the surface density is decreased proportionally. (If the string moves coherently for more than one Hubble time, the resulting wake will have a variable surface density, with the denser parts being those formed at earlier times.)

The transverse dimensions of the wake at $z < z_i$ are

$$w \sim v_s t_i z_i/z , \qquad d \sim u_i t_i (z_i/z)^2 . \tag{11.4.17}$$

At $z \sim u_i z_i/v_s$, the wake thickness d becomes comparable to its width, and at later times the wake has the shape of a filament with a more or less cylindrical cross-section of diameter

$$d \sim (v_s u_i)^{1/2} t_i \left(\frac{z_i}{z}\right)^{3/2} . \tag{11.4.18}$$

The accretion of matter in a wiggly string wake has been studied numerically using the Zel'dovich approximation with the weak-field gravity string source term (7.4.3) followed, in the nonlinear regime, by an N-body code or the adhesion modification [Avelino & Shellard, 1993]. The inhomogeneous initial imprint of a typical wiggly string appears to be sufficient to cause very efficient wake fragmentation by the present day.

11.5 Wakes and large-scale structure

Wakes produced by moving long strings can provide a natural explanation for filamentary and sheet-like structures observed in the large-scale distribution of galaxies [Vachaspati, 1986]. Before we discuss the character of the predicted structures, it will be helpful to recall some relevant features and parameters of the evolving string network (§10.1). String simulations show that the coherence length of strings, beyond which directions along the string are uncorrelated, is $\xi(t) \approx t$. The inter-string separation $L(t)$ is of the same order of magnitude; in the matter era, $L(t) \approx 0.8t$. On scales smaller than ξ, the string is 'wiggly' and its effective mass per unit length and tension are $\tilde{\mu} \approx 1.5\mu$, $\tilde{T} \approx 0.7\mu$. The rms string velocity on the scale of the smallest wiggles is $\langle v^2 \rangle^{1/2} \approx 0.6$, but the coherent velocity obtained by averaging over a scale ξ is $\langle v^2 \rangle_\xi^{1/2} \approx 0.15$. (Note that these network parameters may be different for global strings and for strings formed during inflation.) Although the string velocity on the scale $\sim \xi$ is typically low, occasional intercommutings between long strings can result in much faster motion which can be sustained for about a Hubble time.

Now consider a wake formed behind a string segment of length $\sim \xi(t_i)$ moving with velocity v_s at time t_i, assuming first that the universe is dominated by cold dark matter. The distance travelled by the string in one Hubble time is $\sim v_s t_i$, and thus the initial length, width and thickness of the wake are $\ell_i \sim t_i$, $w_i \sim v_s t_i$ and $d_i \sim u_i t_i$, where u_i is given by (11.4.14). The corresponding dimensions at the present time are

$$\ell_0 \sim t_i z_i\,, \qquad w_0 \sim v_s t_i z_i\,, \qquad d_0 \sim u_i t_i z_i^2\,. \tag{11.5.1}$$

These equations apply provided they give $d_0 < w_0$ (see §11.4.2). The fraction f of the total mass of the universe accreted onto wakes formed at a time $\sim t_i$ can be estimated,

$$f \approx 2w_i d_i z_i / L^2(t_i) \approx 8\pi G (\tilde{\mu} - \tilde{T}) z_i\,. \tag{11.5.2}$$

Wakes produced by slow- and fast-moving strings differ in their spatial dimensions and surface density. For fast-moving strings ($v_s \sim 1$) they are sheet-like with dimensions $t_i z_i \times t_i z_i$, while for slow-moving strings they may have a filamentary appearance. As explained in §11.4.2, the masses of both types of wakes are comparable, but the surface density in the slow-moving string wakes is considerably higher.

Cold dark matter wakes can also form during the radiation era ($t_i < t_{eq}$), but for these gravitational instability only sets in at $t \sim t_{eq}$. For earlier t_i, the surface density of the wakes decreases as $(t_i/t_{eq})^{1/2}$, while the fraction of dark matter accreted onto all wakes formed within a Hubble time of t_i remains roughly constant. Albrecht & Stebbins [1992a] have argued that tiny wakes formed at early times can introduce excessive power on small scales. As a result, wakes formed at $t_i > t_{eq}$ are drowned in this short-wavelength noise. Cold dark matter models with string wakes, therefore, would gain little advantage in explaining large-scale structure over, say, a scale-invariant spectrum of adiabatic fluctuations due to inflation.

A more natural explanation for the observed structure may be obtained if the universe is dominated by light neutrinos [Perivolaropoulos *et al.*, 1990; Vachaspati & Vilenkin, 1991; Vollick, 1992b; Albrecht & Stebbins, 1992b]. In this case, the accretion wakes of small relativistic loops are strongly suppressed by neutrino free-streaming, so that loops play a negligible role in structure formation. The string scenario then predicts a galaxy distribution dominated by filaments and sheets with almost empty voids between them. The characteristic scale of this structure is $t_{eq}z_{eq} \sim 10\,h^{-2}$Mpc. Most of the accreted matter resides in wakes formed at $t \sim t_{eq}$, and (11.5.2) provides the following estimate for the total fraction of matter accreted into wakes,

$$f_{tot} \approx 20\,G\mu\,z_{eq} \approx 0.4\,h^2\mu_6 . \tag{11.5.3}$$

For $h \approx 0.5$ and $\mu_6 \lesssim 2$ (as required by observation – see chapter 10), most of the matter in the universe remains unclustered at the present time. This appears to concur with dynamical measurements of clusters which give values of Ω substantially smaller than unity.

The characteristic scale of the large-scale structure in this scenario is $t_{eq}z_{eq} \sim 10\,h^{-2}$Mpc. With $h \approx 0.5$, this is comparable to scales suggested by observation ($\sim 25\,h^{-1}$Mpc [de Lapparent, Geller & Huchra, 1986]). The present surface density of neutrino wakes produced after t_{eq} is smaller than that of the wakes produced at $\sim t_{eq}$, but only by a factor $\sim (t_{eq}/t_i)^{1/3}$. This means that structures on scales larger than $10\,h^{-2}$Mpc can also be prominent in this scenario. Going beyond these order-of-magnitude estimates and intuitive arguments would require numerical simulations combining a string evolution code with a gravitational clustering code. Without such simulations, a detailed quantitative comparison of the string scenario with observations will not be possible.

Neutrino wakes go nonlinear at a redshift z_{nl} which can be estimated from (11.4.11) and (11.4.14). For filamentary wakes due to slow-moving strings, this gives $1 + z_{nl} \approx 4.5\mu_6h^2$, independent of z_i, and for sheet-like wakes from fast-moving strings $1 + z_{nl} \approx 3\mu_6h^2$. Requiring that wake collapse begins at $z \gtrsim 1$, we obtain a restriction on the allowed values of μ, $\mu_6 \gtrsim 0.4\,h^{-2}$.

Baryonic wakes in a neutrino-dominated universe start collapsing after baryons decouple from radiation, $t > t_{dec}$. However, since baryons constitute only a small fraction of the total density, the growth of these wakes is strongly suppressed. Baryonic wakes could none-the-less be cosmologically significant if the energy output from primordial stars formed in the wakes were to trigger some kind of explosive amplification and lead to preferential galaxy formation along these wakes [Rees, 1986; Hara & Miyoshi, 1987a, 1990]. They could also explain the existence of quasars at redshifts $z > 3$. The scale of baryonic wakes $t_{dec}z_{dec} \sim 50\,h^{-1}$Mpc, is comparable to the largest-scale structures observed in the universe.

Finally, we mention a baryon-dominated universe. In this case, the fraction of matter accreted in the wakes is

$$f_{\rm w} \approx 20\, G\mu\, z_{\rm eq}/\Omega^{-1} \approx 0.4\, \Omega^2 h^2 \mu_6 \,. \qquad (11.5.4)$$

The parameters Ω, h and μ_6 must satisfy $\Omega h^2 \lesssim 0.026$, $\Omega \lesssim 0.16$, $\mu_6 \lesssim 2$, and thus $f_{\rm w} \lesssim 2 \times 10^{-3}$. This value is too small to account for the observed mass in clusters of galaxies, and we conclude that wakes are unable to play an important role in structure formation in a baryon-dominated universe.

We note, however, that the constraint on μ_6 (10.2.6) was obtained for a flat $\Omega = 1$ model and it may not apply in a low-density universe. At redshifts $z < \Omega^{-1}-1$, the universe expands as $a \propto t$. This rapid expansion can be expected to slow down the strings and may relax the bound on $G\mu$. The hyperbolic spatial geometry may also introduce further corrections. This scenario, as well as models with a non-zero cosmological constant, require further study.

11.6 Wakes and the origin of cosmic magnetic fields

The string model of structure formation may help to resolve another cosmological mystery – the origin of cosmic magnetic fields [Vachaspati & Vilenkin, 1991]. The flow of matter in baryonic wakes produced by wiggly strings is expected to be turbulent. The string small-scale structure induces velocity gradients in the flow, and order-of-magnitude estimates show that the corresponding Reynolds numbers are very large. Moreover, Vachaspati [1992] has argued that post-recombination wakes tend to develop strong, non-uniform shocks, which are also indicative of turbulence. The characteristic scale of turbulence at a time t is set by the size of string wiggles $\sim \Gamma G\mu t$ and is comparable to the wake thickness. The typical velocity on this scale is $\sim 4\pi G\mu$ and the corresponding vorticity is

$$\omega \sim 4\pi/\Gamma t \sim 0.1 t^{-1} \,. \qquad (11.6.1)$$

Vorticity in the baryonic flow can give rise to a primordial magnetic field through a mechanism of the kind suggested by Harrison [1970] and by Mishustin & Ruzmaikin [1971]. Electrons and protons in turbulent eddies experience friction as they move through radiation, but the effect of friction on electrons is much greater because of their small mass. As a result electrons develop a velocity relative to protons. There is an electric current and hence a magnetic field. Vorticity is necessary, because a non-vortical flow would immediately lead to charge accumulation, and the current would terminate. With ω from (11.6.1) and $t \sim t_{\rm dec}$, the resulting magnetic field can be estimated as [Harrison, 1970]

$$B(t_{\rm dec}) \sim 10^{-4} \omega \sim 10^{-17} h\, {\rm Gauss} \qquad (11.6.2)$$

on a comoving scale $\sim 5\,h^{-1}\mu_6$ kpc. This seed field can then be amplified by turbulence and by galactic dynamos.

11.7 Outlook*

At the time of writing, theories of structure formation are in a state of flux. Observational data is accumulating at an astonishing pace and theories rise and fall within the span of a few years. The string scenario of structure formation has survived its first decade, but its future remains uncertain. Since its inception in 1980–81, our view of structure formation with strings has undergone substantial revision caused by improvements in our understanding of the evolution of a cosmic string network. At present, the most promising version of the scenario appears to be that in which the universe is dominated by light neutrinos and the dominant perturbations are produced by long string wakes. The wakes provide a natural explanation for the sheets and filaments observed in the large-scale distribution of galaxies. The voids between the wakes remain essentially unperturbed, allowing us to reconcile an $\Omega = 1$ universe with dynamical measurements in clusters which give values of Ω substantially below unity. An infusion of some extra cold dark matter, as in some recent mixed dark matter models, may have the same beneficial effects found in gaussian scenarios. As we emphasized earlier, however, firmer quantitative predictions will depend on more detailed studies, probably of a numerical nature.

The 'old string scenario', which emphasized accretion onto closed loops, was superseded when string simulations demonstrated that loops are typically created with sizes considerably smaller than the horizon. However, it would be premature to discard the old scenario completely. For global strings, small-scale wiggles are efficiently damped by Goldstone boson radiation, and the resulting loop sizes may be larger than those for gauge strings. Small-scale structure on long strings can also be suppressed and loop sizes correspondingly increased in models where strings are formed during inflation (see chapter 16). In fact, inflationary models in which string loops are formed by quantum nucleation predict no infinite strings. In all these models, accretion onto closed loops may play an important role and some of the features of the old scenario may be preserved.

Like most theories of structure formation, whether the string scenario will stand or fall will depend on improvements in cosmic background radiation measurements. As we discussed in chapter 10, strings are expected to cause linear discontinuities in the radiation temperature of magnitude $\delta T/T \approx 2 \times 10^{-5}\mu_6 \approx 20\,G\mu$. For values of μ required by the standard string scenario, this prediction is likely to be ruled out or confirmed within the next few years.

---

* Please refer to section 0.3 of the paperback preface for a more recent update.

12

Cosmology of superconducting
and global strings

12.1 Superconducting strings

Strings predicted in a wide class of elementary particle theories behave like superconducting wires. Such strings can carry large electric currents and their interaction with cosmic plasmas can give rise to a variety of astrophysical effects. To discuss the astrophysical consequences of superconducting strings, we must specify the mechanism responsible for the generation of string currents. There are several possibilities: (i) The string currents could be induced by galactic magnetic fields. Obviously, this mechanism can only work after galaxies are formed. (ii) It is possible that a large-scale primordial magnetic field was present in the early universe, which was generated long before galaxy formation. We have no solid observational evidence for such a field and no compelling theories predicting why it should exist, but neither can we rule out this possibility (for a review see Rees [1987]). If there were primordial fields, then the strings could develop large currents at very early times. (iii) Finally, small-scale current and charge fluctuations are produced on strings at the time when they become superconducting. Net currents and charges due to such fluctuations can stabilize small closed loops of string. Here we shall consider all three mechanisms separately.

To remind the reader of the basic properties of superconducting strings, we begin with a brief review of string electrodynamics (for a detailed discussion see §8.4–5). In this section it will be convenient to retain $\hbar$ and c explicitly in all equations, rather than adopting the fundamental units in which they are set to unity – as in the rest of the monograph.

12.1.1 String electrodynamics: a review

The defining property of a superconducting string is its response to an applied electric field: the string develops an electric current which grows in time,

$$dJ/dt \sim (ce^2/\hbar)E\,. \qquad (12.1.1)$$

Here, E is the field component along the string and e is the elementary charge ($e^2/\hbar c \sim 10^{-2}$).*

The charge carriers in the string can be bosons or fermions. We consider first the case of fermionic superconductivity. Models of this type have fermions which are massless inside the string and have a finite mass m outside the string. Particles inside the string can be thought of as a one-dimensional Fermi gas. When an electric field is applied, the Fermi momentum grows as $\dot{p}_F = eE$, and the number of fermions per unit length, $n = p_F/2\pi\hbar$, also grows:

$$\dot{n} \sim eE/\hbar\,. \qquad (12.1.2)$$

The particles move along the string at the speed of light. The resulting current is $J = enc$, and dJ/dt is given by (12.1.1).

The current continues to grow until it reaches a critical value

$$J_c \sim emc^2/\hbar\,, \qquad (12.1.3)$$

when $p_F = mc$. At this point, particles at the Fermi level have sufficient energy to leave the string. Consequently, in this simplified picture, the growth of the current terminates at J_c and the string starts producing particles at the rate (12.1.2). The fermion mass m is model-dependent, but it does not exceed the symmetry breaking scale of the string, η. Hence,

$$J_c \lesssim J_{\max} \sim e(\mu c^3/\hbar)^{1/2}\,, \qquad (12.1.4)$$

where we have used the relation $\mu \sim \eta^2 c/\hbar$. Grand unification strings can carry enormous currents, $J_{\max} \sim 10^{31}$esu/s, while for electroweak-scale strings $J_{\max} \sim 10^{17}$esu/s. Note that the actual value of the critical current J_c is highly model-dependent and is discussed in some detail in chapter 5.

Superconducting strings can also have bosonic charge carriers. This occurs when a charged scalar or gauge field develops a vacuum expectation value inside the string. As a result the electromagnetic gauge invariance inside the string is broken, indicating superconductivity. The critical current J_c for this type of string is determined by the energy scale at which the gauge invariance is broken. It is model-dependent, but is still bounded by $J_{\max}$ from (12.1.4).

---

* In chapter 5 the electromagnetic coupling of the string current was called q and an expression was derived for q in terms of e. Here, we shall be interested only in order-of-magnitude estimates and so we shall not distinguish between q and e.

Superconducting strings can develop currents not only in electric, but also in magnetic fields. Consider a segment of string moving at a speed v in a magnetic field B. In its rest frame the string 'sees' an electric field $E \sim (v/c)B$, and so the current grows at the rate

$$dJ/dt \sim (e^2/\hbar)vB. \tag{12.1.5}$$

A closed loop of length L oscillating in a magnetic field acts as an ac generator and develops an ac current of amplitude

$$J \sim 0.1(e^2/\hbar)BL. \tag{12.1.6}$$

The factor of 0.1 appears because the area of the loop is typically of the order $A \sim 0.1L^2$.

An oscillating current-carrying loop in vacuum emits electromagnetic waves. For a loop without kinks or cusps the radiation power is

$$\dot{E}_{em} \sim \Gamma_{em} J^2/c, \tag{12.1.7}$$

where the numerical factor Γ_{em} depends on the loop's shape, but not on its length; typically, $\Gamma_{em} \sim 100$. The ratio of the power in electromagnetic waves to that in gravitational waves is

$$\frac{\dot{E}_{em}}{\dot{E}_g} \sim \frac{e^2}{\hbar c}\left(\frac{G\mu}{c^2}\right)^{-1}\left(\frac{J}{J_{max}}\right)^2, \tag{12.1.8}$$

and we see that for sufficiently large currents electromagnetic radiation can become the dominant energy loss mechanism for the loop.

For a loop with kinks, Γ_{em} in (12.1.7) has a weak logarithmic dependence on the current; its characteristic range is $10^2 \lesssim \Gamma_{em} \lesssim 10^3$. If the loop has cusps, then for $J \ll J_{max}$ the radiation power is dominated by the emission of short periodic bursts of highly directed energy from near-cusp regions. An estimate of $\dot{E}$ in this case is complicated by the fact that the string motion near the cusp is strongly affected by radiation back-reaction. Only an upper bound on this radiation power has been obtained,

$$\dot{E}_{em} \lesssim JJ_{max}/c. \tag{12.1.9}$$

The evolution of a superconducting loop as it gradually loses energy and finally decays is discussed in §8.5.6.

Interaction with plasma

String electrodynamics is substantially modified in the presence of a plasma. Consider a straight segment of string carrying a current J and moving with velocity v through a plasma of density ρ. Charged particles cannot penetrate the region of strong magnetic field near the string, and if the motion of the string is supersonic, a shock front will form at some distance r_s from the string. The physics of the shock region is rather complicated; for a detailed discussion see §8.6. The shock radius r_s can be estimated if we note that in the shock region the ram pressure of the plasma is balanced by the magnetic pressure,

$$\rho v^2 \sim B^2/8\pi \,. \qquad (12.1.10)$$

Substituting $B \sim J/cr_s$ we obtain

$$r_s \sim J/cv\sqrt{\rho}\,. \qquad (12.1.11)$$

The string current is completely screened by the currents flowing in the plasma in the vicinity of the shock, and so the magnetic field of the string is confined to a region of radius $\sim r_s$ near the string. The analysis of electromagnetic radiation from strings in vacuum is inapplicable in the presence of a plasma. The typical frequency of electromagnetic waves from a loop of length L is $\omega \sim c/L$. For macroscopic loops it is much smaller than the plasma frequency, and it is well known that such low-frequency waves do not propagate in plasma.

The damping force per unit length of string due to its interaction with a plasma is comparable to the force acting on a cylinder of radius r_s,

$$F_d \sim \rho v^2 r_s \sim J(v/c)\sqrt{\rho}\,. \qquad (12.1.12)$$

This force has to be compared with the force due to the tension,

$$F_t \sim \mu c^2/R \qquad (12.1.13)$$

where R is the curvature radius of the string. If $JR\sqrt{\rho} \ll \mu c^2$, then $F_d \ll F_t$ and the effect of the plasma on the string trajectory is negligible. In this case, $v \sim c$, the rate of energy dissipation for a loop of length L is

$$\dot{E}_d \sim F_d Lc \,, \qquad (12.1.14)$$

and the corresponding lifetime is

$$\tau_d \sim \mu c^2 L/\dot{E}_d \sim \mu c/J\sqrt{\rho}\,. \qquad (12.1.15)$$

In the opposite limit, the strings are over-damped, and the string velocity can be found from $F_d \sim F_t$:

$$v \sim \frac{\mu c^3}{RJ\sqrt{\rho}} \ll c\,. \qquad (12.1.16)$$

12.1.2 Strings within the Galaxy

In the absence of primordial magnetic fields, strings can develop appreciable currents only after galaxies form and generate their own magnetic fields in galactic dynamos. In our Galaxy, both the magnetic field and the ionized gas that supports it are concentrated in a thin disk of radius $R \approx 15\,\text{kpc}$ and thickness $h \approx 0.4\,\text{kpc}$. The large-scale galactic field is oriented along the spiral arms, with a characteristic scale in this direction of $\sim 4\,\text{kpc}$. The scale in the transverse direction is given by the thickness of the arms $\sim 0.4\,\text{kpc}$. The average strength of this large-scale field is $B \approx 2 \times 10^{-6}\,\text{Gauss}$. The galactic magnetic field also has a random component of comparable magnitude, $\delta B/B \sim 1$, with a typical scale $\sim 100\,\text{pc}$.

If a loop of string finds itself in a galaxy like our own, it develops a current

$$J \sim 2 \times 10^{20} L_{\text{pc}}\,\text{esu/s}, \tag{12.1.17}$$

where L_{pc} is the loop's length in parsecs ($1\,\text{pc} = 3 \times 10^{18}\text{cm}$). Here it is assumed that the current (12.1.17) is below critical and we shall disregard the possible build-up of the current in the course of the loop's evolution (see the discussion in §8.5.6). As we discussed in §12.1.1, a relativistically moving string shocks the surrounding plasma. The temperature near the shock reaches $T \sim m_{\text{n}}c^2 \sim 1\,\text{GeV}$, where m_{n} is the nucleon mass, and the gas becomes completely ionized. The shock radius r_{s} is found from (12.1.11), $r_{\text{s}} \sim 10^{-7}L$, where we have used the average gas density in the disk, $\rho \sim 2 \times 10^{-24}\text{g/cm}^3$. The magnetic field strength at the shock can be found from (12.1.9), that is, $B \sim 0.1\,\text{Gauss}$.

If there is a superconducting loop in our Galaxy, it may be directly observable through synchrotron emission by electrons from regions with strong magnetic fields near the shocks [Chudnovsky *et al.*, 1986]. The motion of the string may also be observable: for an object at a distance $\sim 10\,\text{kpc}$, relativistic motion corresponds to an angular displacement of a few arc seconds per year.

The intensity of synchrotron radiation has been estimated by Chudnovsky *et al.* [1986] who assumed that the region of strong magnetic field near the string is essentially free of plasma (refer to the discussion in §8.6) and that plasma diffuses into this region only in a thin layer near the shock. With this assumption, a rough estimate of the total radiation intensity from a loop of length L gives $dE/dt \sim 4 \times 10^{28} L_{\text{pc}}^{5/2}\text{erg/s}$, with a characteristic wavelength $\lambda \sim 20\,\text{cm}$. However, Thompson [1990] has pointed out that as the string moves through a magnetized plasma and develops its current, a substantial amount of plasma can be captured on the field lines in the vicinity of the string. In this case, the radiation power will be greatly increased (by several orders of magnitude) and the spectrum will extend towards much smaller wavelengths. A quantitative estimate is difficult because it depends on the details of magnetic field line reconnection behind the string.

The loop sizes that we can expect to find in a galaxy depend on whether or not the loops can be captured by galaxies. We know that loops are formed with relativistic speeds and then are slowed down by the Hubble expansion and by the gravitational drag of matter. They can be gravitationally captured by galaxies* if their present velocities are smaller than the galactic escape velocity, $v < 200\text{km/s}$. This will be true for loops formed at $z > 10^3$.

Even if the loops are captured, we can expect to find loops of string in our Galaxy only if the density of loops in the universe is greater than the density of galaxies, $n_\text{G} \sim 0.03\,\text{Mpc}^{-3}$. Using (10.1.17) for the number density of loops n_L, and requiring that $n_\text{L} > n_\text{G}$, we obtain a bound on the loop's length,

$$L < 30\,\text{pc}. \tag{12.1.18}$$

Here we have assumed that the loops were formed in the matter era, but a similar constraint applies for loops formed in the radiation era, $L < \alpha ct_\text{eq} \lesssim 50\,\text{pc}$, where we have used the bound (10.1.11) on the loop creation size, $\alpha \lesssim 5 \times 10^{-3}$.

If the loops are moving too fast to be captured, they spend only a small fraction of their time in galaxies. The volume occupied by magnetized plasma in our Galaxy is $V \sim 3 \times 10^{-7}\text{Mpc}^3$. In order to have at least one loop in this volume, we need the loop density $n > V^{-1} \sim 3 \times 10^6\text{Mpc}^{-3}$. The corresponding loop sizes are tiny (much smaller than a parsec).

Galactic loops lose their energy through gravitational radiation and by plasma dissipation. The corresponding lifetimes are given by (10.1.5) and (12.1.15),

$$\tau_\text{g} \sim 10^6 (G\mu/c^2)^{-1} L_\text{pc}\text{s}, \tag{12.1.19}$$

$$\tau_\text{d} \sim 10^{30} (G\mu/c^2) L_\text{pc}^{-1}\text{s}. \tag{12.1.20}$$

Only loops for which both of these times is greater than the age of the universe will survive in the Galaxy until the present day. The corresponding range of the parameters L_pc and $G\mu/c^2$ is shown in fig. 12.1.

---

* Loops can also be slowed down and captured by the frictional force due to their interaction with plasma. However, it can be easily shown that such loops will not survive for much longer than one galaxy crossing time. They will decay by plasma dissipation on a timescale comparable to their slow-down time. We note an exception to this rule: the loops captured by the plasma can survive and even multiply if the strings are very light, so that their motion is over-damped (see §12.1.3).

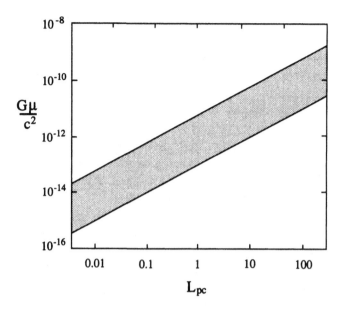

Fig. 12.1. The shaded area shows the range of parameters for which galactic loops can survive until the present day t_0.

For loops of size $L < 30\,\mathrm{pc}$, which we can expect to find in the Galaxy, $G\mu/c^2$ cannot exceed 10^{-10}. Note that this value is far below the value of $G\mu/c^2 \sim 10^{-6}$ required in the string scenario of galaxy formation. We note also that, for any given value of $G\mu/c^2$, the allowed range of L is rather narrow: it does not exceed two orders of magnitude.

The current (12.1.17) is smaller than the critical current (12.1.3) only if the energy scale of string superconductivity is

$$m > 2 \times 10^5 L_{\mathrm{pc}}\,\mathrm{GeV}\,. \tag{12.1.21}$$

Otherwise, the current saturates at $J \simeq J_\mathrm{c}$ and the string produces particles at the rate [Witten, 1985a]

$$dN/dLdt \sim eB/\hbar \sim 10^{12}\,\mathrm{cm}^{-1}\,\mathrm{s}^{-1}\,. \tag{12.1.22}$$

This effect, however, is cosmologically insignificant: the energy output of loops in the form of particles does not exceed $10^{29}\mathrm{erg/s}$.

Superconducting loops can also produce relativistic particles through several other mechanisms. (i) For a loop with cusps, the current can reach critical values in near-cusp regions even if the condition (12.1.21) is satisfied and the average current is less than J_c. In this case, the charge carriers may be ejected from these regions. (ii) Relativistic shocks and powerful bursts of electromagnetic radiation from cusps can accelerate plasma particles to ultra-relativistic energies. Although no new particles are produced, this process would contribute to the density of cosmic rays in the Galaxy. (iii) Electron–positron pairs are produced near string segments in which the electric charge density is greater than the current (see §8.5.6). Estimates of the average rate of pair production are comparable to (12.1.22).

The total energy output of loops in the form of particles depends on the size distribution of loops and on the scale of string superconductivity m. It can be shown, however, that particle production gives a negligible contribution to the density of galactic cosmic rays, while the contribution of shock acceleration can be substantial [Chudnovsky *et al.*, 1986; Hayashi, 1988; Kobayakawa & Samura, 1992].

Heavy superconducting loops with cusps can give rise to observable effects even in distant galaxies. Bursts of electromagnetic radiation from cusps will accelerate particles in the plasma, producing relativistic jets. It has even been proposed that jets observed in quasars may have this origin [Vilenkin & Field, 1987]. Quasars are present in the greatest numbers at $z \sim 3$, corresponding to a cosmic time $\sim 10^9$ years. The most numerous loops at that time have length $L \sim 3 \times 10^{10}(G\mu/c^2)\,\mathrm{pc}$. The total energy of such a loop is $E \sim 10^{78}(G\mu/c^2)^2\mathrm{erg}$, and for $(G\mu/c^2) \gtrsim 10^{-8}$ this would be sufficient to explain the energy output of quasars ($\gtrsim 10^{61}\mathrm{erg}$).

12.1.3 Light superconducting strings

In the previous section we assumed that the damping force acting on strings is always smaller than the string tension, so that their motion is relativistic. Here we shall consider the opposite case with very light superconducting strings. As these strings move through magnetized cosmic plasmas and develop electric currents, they become over-damped and are essentially 'frozen' into the plasma. Now in our own galaxy, the interstellar plasma is in a state of turbulent motion. As string loops are dragged along by the plasma, they will be stretched and twisted by turbulent eddies. Smaller loops will be chopped off them and, in turn, become stretched and twisted. The process may continue until the entire volume occupied by the plasma is filled with a stochastic network of strings [Chudnovsky & Vilenkin, 1988].

To describe this network quantitatively, let us consider a string segment having a curvature radius R. The velocity of this segment relative to the plasma is given by (12.1.16). We shall see shortly that the currents in the string network are near critical, and so (12.1.16) can be rewritten as

$$v \sim \frac{c}{kR}\left(\frac{\hbar c \mu}{e^2 \rho}\right)^{1/2}. \qquad (12.1.23)$$

Here, k is a model-dependent numerical factor given by

$$k = J_c/J_{max} \lesssim 1, \tag{12.1.24}$$

with J_{max} defined in (12.1.4).

Observations indicate that the characteristic scale of turbulent motion in the interstellar plasma is $l \sim 100\,\mathrm{pc}$ (for a review see Zel'dovich, Ruzmaikin & Sokoloff [1983]). The typical velocity on that scale is $v_l \simeq 10^6\,\mathrm{cm/s}$ and the stochastic magnetic field is $B_l \simeq 10^{-6}\mathrm{Gauss}$. On scales $R > l$, the motion of the plasma is well approximated by the Kolmogorov spectrum, $V_R \sim V_l(R/l)^{1/3}$, while for $R < l$ the spectrum is roughly that of magnetohydrodynamic turbulence. In the latter case, there is an equipartition between plasma kinetic and magnetic energies,

$$B_R^2/8\pi \sim \rho v_R^2/2, \tag{12.1.25}$$

where

$$v_R \sim v_l(R/l)^{1/4}. \tag{12.1.26}$$

Comparing the turbulent velocity v_R with the string velocity (12.1.23), we find the scale R_* at which $v \sim v_R$;

$$R_* \sim 10^{16}k^{-4/5}\mu_{-6}^{2/5}\,\mathrm{cm}. \tag{12.1.27}$$

Here, μ_{-6} is μ in units of $10^{-6}\mathrm{g/cm}$ ($\mu \sim 10^{-6}\mathrm{g/cm}$ for electroweak-scale strings with $\eta \sim 100\,\mathrm{GeV}$). Strings with curvature radii $R \ll R_*$ are unaffected by turbulence, while strings with $R \gg R_*$ are dragged along by the plasma. In the latter case, the large-scale behaviour of the strings is similar to that of the magnetic field lines which are 'frozen' into the plasma.

As long strings are stretched and twisted by turbulent eddies, they develop fractal shapes with a minimum scale $\sim R_*$. The length of string is most efficiently increased by eddies of size $\sim R_*$. Every such eddy interacting with the string stretches it by $\sim R_*$ on a time scale

$$t_* \sim R_*/v_*, \tag{12.1.28}$$

where v_* is the turbulent velocity on the scale R_*. The distance between the nearest strings in the network is also $\sim R_*$. Closed loops of size $\lesssim R_*$ will be constantly formed by string intersections at the rate of about one loop per volume R_*^3 per time t_*. Subsequently, such loops shrink under their tension and lose energy to heat and particle production. The efficiency of this dissipation mechanism is sufficient to balance the stretching of strings by the turbulence. The whole random network of strings can be generated from a single 'seed' loop of size $R > R_*$.

Although this picture is intuitively plausible, it still requires numerical verification. The process of string proliferation is similar to a turbulent magnetic dynamo, which is known to work if the turbulence has a nonzero average helicity. With a non-helical turbulence, the magnetic field drifts to the smallest turbulent scales, where it is dissipated (see, for example, Zel'dovich *et al.* [1983]). When a string loop is stretched and twisted, there is a possibility that it will self-intersect many times and break up into many loops of size smaller than R_*, which then shrink and disappear. To definitely ascertain whether or not a stochastic network will form, one has to perform a numerical simulation of strings interacting with a turbulent plasma.

Assuming that a string network does indeed form, let us now estimate typical string currents. On scales below R_*, tension causes the strings to cross the magnetic field lines, and electric currents are constantly induced in the strings. The rate of current growth is $dJ/dt \sim e^2 Bv/\hbar$, and using (12.1.23), (12.1.25), and the expression for J_{max} (12.1.4), it is easily verified that the current on the scale R_* reaches the critical value J_c in a time $t \lesssim t_*$. This justifies the earlier assumption that $J \sim J_c$.

The string energy dissipated by plasma friction and transformed into heat per volume R_*^3 per time t_* is $E_d \sim \mu c^2 R_*$ and the energy dissipated in the form of relativistic particles is $E_p \sim k\eta(eBR_*/\hbar c)R_*$ (compare with (12.1.22)). Using (12.1.23) and (12.1.25) we find $E_p/E_d \sim (B^2/\rho v_*^2)^{1/2} \sim 1$.

Although individual strings in the network are very difficult to detect, their cumulative effect can be quite significant. The energy output in the form of high-energy particles due to all the strings in the galaxy is

$$\frac{dE}{dt} \sim \frac{\mu c^2 v_*}{R_*^3} V \sim 10^{40} k^{11/5} \mu_{-6}^{-1/10} \, \text{erg/s} \,, \qquad (12.1.29)$$

where $V \simeq 10^{67} \text{cm}^3$ is the volume occupied by magnetized plasma in the galaxy. Somewhat remarkably, the value 10^{40}erg/s coincides with the estimates of the energy input necessary to sustain the observed concentration of cosmic rays in the galaxy [Simpson, 1983]. Note that the μ-dependence in (12.1.29) is extremely weak, but a small value of k can reduce this estimate.

Particles produced by the strings have energies $\sim k\eta$, but their energy can be strongly degraded by pair production and other inelastic processes in the strong magnetic field of the strings [Thompson, 1988c; Berezinsky & Rubinstein, 1989]. A distinctive feature of the string contribution to cosmic rays is that we expect the string network to produce equal numbers of particles and antiparticles.

Strings are not likely to be captured by a star, since the plasma wind from the star would not allow the string to approach it. However, strings from the network can be trapped within stars during star formation. When this occurs, 'contaminated' stars will develop a string network of their own, with a typical scale R_* depending on the plasma density inside the star and on the character of its turbulent motion. The energy output of strings can be comparable to the total stellar luminosity, and so the string network could have a significant effect on stellar evolution [Chudnovsky & Vilenkin, 1988; Wang, 1990]. However, the probability for a star to capture a string at formation can be rather small, so that the effect of strings may not be immediately apparent on the HR diagram. The energy dissipated by strings inside the star is transformed partly into heat and partly into relativistic particles. All particles except neutrinos are quickly thermalized and their energy is also turned into heat. Neutrinos can escape from the star, and string-contaminated stars are expected to be powerful sources of high-energy neutrinos and antineutrinos. The absence of a substantial flux of energetic neutrinos from the Sun indicates that it is not string-contaminated.

12.1.4 Strings and primordial magnetic fields

String currents

In this section we shall consider the astrophysical effects of strings in the presence of a primordial magnetic field. As the universe expands, the magnetic flux is conserved, and the field changes as $B \propto a^{-2}(t)$. The strength of the primordial magnetic field can be characterized by the ratio of the magnetic energy to the radiation energy,

$$\epsilon = \rho_B/\rho_\gamma, \tag{12.1.30}$$

which remains roughly constant during the course of cosmological evolution. An observational upper bound on the present strength of the intergalactic magnetic field, $B_0 \lesssim 10^{-9}$ Gauss, gives an upper bound on ϵ, that is, $\epsilon \lesssim 10^{-7}$. During the radiation era, $\rho_\gamma \sim 1/Gt^2$ and $B \sim (\epsilon/G)^{1/2}(c/t)$. If the strings are not over-damped and move relativistically, then the correlation length of strings at a time t is $\sim ct$, and the current induced by the magnetic field on that scale is

$$J \sim (e^2/\hbar)Bvt \sim (vce^2/\hbar)(\epsilon/G)^{1/2}$$
$$\sim 2 \times 10^{32}\epsilon^{1/2} \text{ esu/s}, \tag{12.1.31}$$

where we have used $v \sim 0.15c$ for the coherent velocity of strings on the scale ct. (Of course, this estimate is only valid provided $J < J_c$). Currents induced at earlier times, which have shorter wavelengths, are gradually damped by pair production and by ohmic dissipation in the plasma [Thompson, 1988b, 1990]. Short-wavelength ac contributions to the current are also given by small-scale wiggles on long strings, but their amplitude, $J_\lambda \sim (e^2/\hbar)B\lambda$, is much smaller than (12.1.31) for $\lambda \ll vt$.

Hence, we expect the current in long strings to have amplitude (12.1.31) and coherence length $\sim ct$. The assumption that strings are not over-damped is justified if the drag force due to plasma dissipation (12.1.12) is smaller than the force from the string tension (12.1.13),

$$\frac{F_d}{F_t} \sim \frac{e^2}{\hbar c}\frac{c^2}{G\mu}\frac{v^2}{c^2}\epsilon^{1/2} < 1. \tag{12.1.32}$$

In the matter era the magnetic field scales as $B \propto t^{-4/3}$, and the string current is

$$J \sim (vce^2/\hbar)(\epsilon/G)^{1/2}(t_{eq}/t)^{1/3}. \tag{12.1.33}$$

Expanding plasma shells

Closed loops are chopped off the string network with sizes much smaller than the horizon, so we expect them to carry dc currents of magnitude (12.1.31) or (12.1.33). As the loops oscillate, they lose their energy to gravitational radiation and to plasma dissipation at the rates given by (10.1.4) and (12.1.14), respectively. In the matter era, an additional energy loss mechanism can come into play and the final stages of loop decay can be rather dramatic. In the vicinity of an oscillating loop, the plasma is shock-heated up to relativistic temperatures. As the plasma expands, it leaves the loop oscillating in a region of much lower density. If this density gets so low that the plasma frequency ω_p is smaller than the oscillation frequency L^{-1}, then the loop starts radiating electromagnetic waves, just as it would in vacuum. The radiation pressure of these waves will then blow the surrounding plasma into a thin expanding shell, similar to the nebula around the Crab pulsar [Ostriker, Thompson & Witten, 1986].

Thompson [1988b, 1990] has subsequently argued that the loop's surroundings may never be cleared so efficiently that $\omega_p < L^{-1}$. Plasma trapped in the magnetic field of the string and gradually shed off as the loop loses energy may be sufficient to completely damp the low-frequency electromagnetic radiation from the loop. The loop then loses energy in the form of a magnetized, relativistic plasma wind. The rate of energy loss in the wind is expected to be comparable to the electromagnetic rate in vacuum (12.1.7). Like radiation, this plasma wind will blow the plasma into an expanding shell, and so there is little difference between the two pictures in terms of the effect of the loop on the surrounding plasma. For definiteness, we shall adopt the picture in which the plasma shell is driven by electromagnetic radiation.

When a loop is chopped off the network, its current is comparable to that in the long strings and is given by (12.1.31) or (12.1.33). However, as the loop radiates away its energy and shrinks, the dc component of the current grows as $J \propto L^{-1}$, and the electromagnetic power of the loop increases (see §8.5.6). By the end of the loop's life this power can become greater than that in gravitational waves, even if it was much smaller initially. By this stage, the loop may have lost much of its initial velocity, and its effect on the plasma will be similar to that of a stationary explosion centre. Alternatively, the loop may be accelerated by the rocket effect (§7.5.3), and if its velocity is much greater than the expansion speed of the shell, it will produce a tubular blast wave.

We found in §8.5 that the radiation pattern from an oscillating loop is rather asymmetric, and thus the plasma shells are not expected to be spherical, even for stationary loops. In the extreme case, when the loop has strong cusps, most of the radiation power is emitted in narrow beams directed along the cusp velocities, and the loop creates plasma jets rather than expanding shells. However, according to string simulations, loops with well-developed cusps are relatively rare. The asymmetries introduced by kinks are very mild by comparison and, assuming small loop velocities, we expect most of the shells to have the shape of distorted spheres, being of comparable size in all directions.

The Ostriker–Thompson–Witten scenario

Ostriker *et al.* [1986] suggested that bubbles blown by heavy superconducting loops in the ambient plasma can explain the formation of galaxies and large-scale structure. In this model, hereafter called the OTW scenario, bubble interiors are identified with voids observed in the large-scale distribution of galaxies, and galaxies are formed in the dense shells of gas at the bubble walls.

An important role in the OTW scenario is played by the interaction of the expanding plasma shells with the cosmic background radiation. At $z > 100$ the motion of the shells is efficiently damped by scattering off background photons (Compton drag). For $100 > z > 10$, the drag force is sufficient to slow down electrons, but not protons. As a result the gas is cooled, but its bulk motion is not substantially slowed down, and the shells can reach gigantic sizes. Cold gas in the shells can then fragment into galaxies. At $z < 10$, Compton cooling is inefficient, and the gas in the bubbles remains hot. Loops decaying in this epoch create a hot intergalactic medium. The largest structures are thus created by loops decaying at $z \sim 10$. The OTW scenario is, in a sense, opposite to the 'standard' string scenario of structure formation: instead of attracting matter by their gravitational fields, the strings repel matter by their electromagnetic radiation.

The characteristic size R_0 of the bubbles at the present time can be estimated from simple energy considerations. The kinetic energy of the shell, $K \sim \rho_b R_0^3 (R_0/t_0)^2 \sim \Omega_b R_0^5 / G t_0^4$, where $\rho_b = \Omega_b \rho_0$ is the baryon density, should be comparable to the electromagnetic energy released by the loop, ΔE_{em}; hence,

$$R_0 \sim (G t_0^4 \Delta E_{em} / \Omega_b)^{1/5} . \tag{12.1.34}$$

In order to explain observed structure, we must have at least $R_0 \sim 20 \, h^{-1} \mathrm{Mpc}$, that is,

$$R_0 / c t_0 \gtrsim 10^{-2} , \tag{12.1.35}$$

and the bubbles must fill a sizeable fraction of space in the universe. This implies that the comoving number density of loops responsible for the formation of the bubbles is $n_0 \sim R_0^{-3}$, and using (10.1.19) with $\nu \sim 0.5$ we find the characteristic length of the loops,

$$L \sim 5 \times 10^{-7} t_0 . \tag{12.1.36}$$

These loops were formed at

$$z_L \sim (\alpha t_0 / L)^{2/3} \lesssim 500 , \tag{12.1.37}$$

where we have used the bound on α (10.1.11). If they released the bulk of their electromagnetic energy at $z \sim 10$, their peculiar velocities at that time were $0.014 \lesssim v_L \lesssim 0.7$, and the comoving distances traversed by the loops would lie in the range $10 \, h^{-1} \mathrm{Mpc} \lesssim d_0 \lesssim 500 \, h^{-1} \mathrm{Mpc}$. At the lower end of this range, d_0 has the same order of magnitude as R_0, and the model of stationary loops should give a reasonable approximation. Following OTW, we shall assume this to be the case and proceed to use (12.1.34–37) to impose constraints on the two model parameters, $G\mu/c^2$ and ϵ.

From (12.1.34–35), the energy density released by the loops is, in terms of the average density ρ_0,

$$\delta_0 \sim \frac{\Delta E_{em}}{R_0^3 \rho_0 c^2} \sim \Omega_b \left(\frac{R_0}{c t_0} \right)^2 \gtrsim 10^{-5} . \tag{12.1.38}$$

On the other hand,

$$\delta_0 \sim f \rho_L / \rho_0 \sim 10^2 f G\mu/c^2 , \tag{12.1.39}$$

where f is the fraction of the loop's energy emitted in the form of electromagnetic waves, and ρ_L is the mass density of the loops. Combining (12.1.38) with (12.1.39) we obtain

$$f G\mu/c^2 \gtrsim 10^{-7} . \tag{12.1.40}$$

Since $f < 1$ and microwave anisotropies (10.2.5) require $G\mu/c^2 \lesssim 2 \times 10^{-6}$, this implies $f \gtrsim 0.1$ and $G\mu/c^2 \gtrsim 10^{-7}$.

As we already mentioned, the electromagnetic output of a loop grows towards the end of its life, so that f can be much greater than the initial ratio of the electromagnetic and gravitational powers,

$$f_i \sim \frac{J^2}{G\mu^2 c^2} \sim 10^{-4} \left(\frac{v}{c}\right)^2 \left(\frac{c^2}{G\mu}\right)^2 \epsilon \frac{z_L}{z_{eq}}. \qquad (12.1.41)$$

Here, z_L is given by (12.1.37), v is the characteristic speed of long strings, and we have used (12.1.33) and (12.1.36). The relation between f and f_i is given by (8.5.42). For $f_i \lesssim 1$ it reduces to $f \sim \sqrt{f_i}$, and (12.1.37), (12.1.40) and (12.1.41) yield

$$\epsilon \gtrsim 10^{-8}(c/v)^2 h^2. \qquad (12.1.42)$$

Since long string velocities are significantly smaller than the speed of light, it follows from (12.1.42) that the OTW scenario can work only with the magnetic field close to its observational upper bound, $B_0 \sim 10^{-9}$ Gauss.

Spectral distortion of the CBR and other effects

Like all structure formation models based on explosions, the OTW scenario predicts a large distortion in the spectrum of the cosmic background radiation (CBR). In fact, the predicted distortion may already exceed observational bounds. This spectral distortion is caused by the interaction of the CBR with the expanding plasma shells. As energy is transferred from the plasma to radiation, scattering processes tend to establish a Bose–Einstein distribution for the photons. However, at $z < 10^4$ the corresponding relaxation time becomes greater than the Hubble time, and the resulting spectrum has an excess of photons at high energies and a deficit at low energies [Sunyaev & Zel'dovich, 1970]. The spectral distortion can be characterized by the y-parameter, which we shall not define rigorously, but which is roughly the mean frequency shift per photon, $y \approx \delta\nu/\nu$. If the radiation energy density is increased by $\delta\rho_\gamma$, then y can be estimated as

$$y \sim \frac{1}{4}\frac{\delta\rho_\gamma}{\rho_\gamma}. \qquad (12.1.43)$$

The present observational bound on y is [Mather *et al.*, 1990]

$$y < 4 \times 10^{-4}. \qquad (12.1.44)$$

In the OTW scenario, let $\delta\rho c^2$ be the energy density transferred from the loops to plasma shells in a Hubble time, t. Then $\delta\rho/\rho \sim \delta_0$, where ρ is the total density at time t and δ_0 is given by (12.1.39). If a fraction x of this energy is dumped into radiation, then using the relation $\rho_\gamma/\rho =$

$2 \times 10^{-5} h^{-2}(1 + z)$, the y-parameter can be estimated as [Ostriker & Thompson, 1987; see also Signore & Sanchez, 1990]

$$y \sim 10^6 x f (G\mu/c^2)(1 + z)^{-1} h^2 \sim 0.1 x (1 + z)^{-1} h^2 , \qquad (12.1.45)$$

where we have used (12.1.40) to obtain the second expression.

If all the energy of the shells were transferred into radiation, $x \sim 1$, then the largest distortion would be produced at the smallest allowed redshift, $z \sim 10$. This would be the case if bubble collisions are frequent, so that the kinetic energy of colliding shells is turned into heat and then transferred into radiation. With $x \sim 1$ and $h > 0.5$, (12.1.45) gives $y \gtrsim 2 \times 10^{-3}$, which is in conflict with the bound (12.1.44). However, if bubbles do not completely fill the space and collisions are rare, the string energy that went into accelerating the bubbles at $z \lesssim 100$ can avoid ever being dumped into the CBR. At $z \gg 100$, the bulk motion of plasma is efficiently damped by the Compton drag, and thus a lower bound on y can be obtained by setting $x \sim 1$ and $z \sim 100$ in (12.1.45). This gives $y \gtrsim 2 \times 10^{-4}$, which does not contradict current observations.

Finally, we briefly mention some other observational side effects of the OTW scenario which have been discussed in the literature. (i) Charge carriers can be expelled from a loop when the loop current becomes super-critical due to the shrinking of the loop. The masses of these charge carriers can be as high as 10^{15} GeV, and their decay products can reach the Earth in the form of ultra-high-energy cosmic rays. The resulting cosmic ray spectrum has been studied by Hill, Schramm & Walker [1987]. (ii) Vilenkin [1988] has pointed out that pairs produced by current–charge oscillations in long strings can make these strings visible in the distribution of high-energy γ-rays and neutrinos over the sky.* (iii) High-energy cosmic rays can also be produced via Fermi acceleration of particles in the gigantic shocks driven by superconducting loops. The resulting proton energies can reach $\sim 10^{11}$ GeV [Thompson, 1990]. (iv) Babul, Paczynski & Spergel [1987] have suggested that bursts of energy from cusps at high redshifts ($z \sim 10^3$) can be observed as γ-ray bursts at MeV energies. (At such redshifts, $e^+ e^-$ pair production makes the universe opaque to γ-rays with energies above ~ 100 MeV.)

The potential importance of these effects is currently hard to assess. As in most of the work on astrophysical applications of superconducting strings, the uncertainties in the string evolution picture are aggravated by gaps in our understanding of superconducting string physics and by the

---

* Thompson [1988c] and Berezinsky & Rubinstein [1989] have argued that particles can be drastically degraded in energy due to pair production in the strong magnetic field of the string. Note, however, that in the case of current–charge oscillations their conclusions do not apply in regions where the charge density of the string exceeds the current, so that the electric field is greater than the magnetic field.

complexity of the astrophysical processes involved. The same applies to the OTW scenario. Our conclusion regarding this scenario is that it can conceivably explain the observed large-scale structure if the mass parameter of the superconducting strings lies in the range $10^{-7} \lesssim G\mu/c^2 \lesssim 10^{-6}$ and the present strength of the primordial magnetic field is near its observational bound, $B_0 \sim 10^{-9}$Gauss. The scenario is, however, under considerable observational pressure. We note also that the possible roles of long strings and of the high loop velocities in this scenario remain largely unexplored.

12.1.5 Springs and vortons

In §5.1.3–4 we commented upon the potential existence of superconducting string loops which were stabilized against collapse by repulsive stresses associated with trapped worldsheet currents and charges. The 'vorton' possesses both a net charge and current and is stabilized by the angular momentum of its charge carriers, while the spinless 'spring' has only a current and is stabilized by magnetic pressure. We envisage vorton formation at or soon after the string-forming phase transition, while springs would more probably be created at late times with the assistance of primordial magnetic fields. Note that throughout this section we shall generally follow the discussion of Davis & Shellard [1989a]. We also reintroduce fundamental units $\hbar = c = 1$ for the remainder of the chapter.

Some vorton stability issues were raised in §5.1.4. In particular in the bosonic superconducting string model (5.1.1–2), the 'chiral' state with equal charge and current densities is an attractor, so it implies the radial stability of a classical circular vorton. In this case, the stable vorton radius, for a given winding number N, was given by

$$R_v = N \left(\frac{2\Sigma}{\mu} \right)^{1/2} , \qquad (12.1.46)$$

where Σ is an integral over the condensate cross-section which is approximately equal to the inverse of a dimensionless coupling constant (5.1.11–12). The quantum mechanical stability of the vorton was also maximized in the 'chiral' state, however, it remained susceptible to charge carrier tunnelling. A semiclassical estimation of the vorton lifetime [Davis,1988], suggests that long-lived vortons which survive until the present day t_0 must have a winding number greater than

$$N_{\min} \sim \ln \left(t_0 \sqrt{4\pi\mu} \right) \left(\frac{4\pi\mu}{m_\sigma^2} \right)^{3/2} , \qquad (12.1.47)$$

where m_σ is the mass of the charge carriers off the string. Typically, for GUT-scale strings, we can expect $N_{\min} \gtrsim 10^3$ and $R > 10^{-25}$cm. The issue of classical vorton stability under more general non-radial perturbations has been more recently studied by Carter & Martin [1993]. A degree of model-dependence is found, ranging between stability and marginal instability, however, electromagnetic fields which are not included in their formal analysis will tend to aid stability.

Spring states, on the other hand, are only attainable for loops on much larger scales because their stability depends almost entirely on the logarithmic contribution of the magnetic pressure (5.1.29). Indeed, Peter [1993] has demonstrated that spring states cannot be achieved through worldsheet pressure alone.

Vorton formation

The formation of strings at a cosmological phase transition was discussed in §9.1. If these strings become superconducting, then loops have the potential to relax into long-lived vorton states if they possess net current and charge such that $N > N_{\min}$. The nearer to the phase transition that this criterion is satisfied, then the more the process will be assisted by frictional forces during the damped epoch. The formation of vortons when strings are moving relativistically is more questionable because charge carriers can be easily ejected from the string at cusps and kinks.

Subsequent to a string-forming transition, a Brownian network will form with about 20% of the total string length in a scale-invariant distribution of loops (9.1.10). The number density of these Brownian loops with average radius R in the range R to $R + dR$ is

$$n(R) \approx \nu R^{-4} dR, \qquad (12.1.48)$$

where $\nu \approx 6 \pm 2$ and a lower cut-off is set by the network correlation length ξ. For simplicity, we shall assume that strings become superconducting at the same time and that the correlation length of the trapped charged fields is the same. Typically, then, a Brownian loop of length $L \sim R^2/\xi$ will pass through L/ξ uncorrelated regions, thus acquiring a net charge $Q \sim e(L/\xi)^{1/2}$ and current $J \sim e(L/\xi)^{1/2}$ by the Kibble mechanism. The pertinent average value of N will be

$$N \sim \left(\frac{L}{\xi}\right)^{1/2} \sim \frac{R}{\xi}. \qquad (12.1.49)$$

Given that $\xi \sim T_c^{-1} \sim \mu^{-1/2}$, we see from (12.1.46) that any Brownian loop will have an average radius comparable to the corresponding stable vorton radius for the same N, that is, $R \sim R_v$.

String motion after the phase transition at t_c is damped for many orders of magnitude in time, particularly if the strings are light. The end of the damped era is at

$$t_* \sim (G\mu)^{-1} t_c \,. \tag{12.1.50}$$

During this period, small-scale wiggles on a loop will be gradually erased until the coherence length has grown from $\xi \ll t_c$ to $\xi \sim t_*$. By comparing the string tension with the frictional force $T^3 v$, we can see that loops of different radii will fall into three distinct regimes determined by the length-scales:

$$\begin{aligned} R_c &\sim (G\mu)^{1/4} t_c \,, & N_c &\sim (G\mu)^{-1/4} \,, \\ R_* &\sim t_* \,, & N_* &\sim (G\mu)^{-1} \,, \end{aligned} \tag{12.1.51}$$

If $R < R_c$, then throughout the entire history of the loop, its motion will be critically damped. Substructure will rapidly disappear and the loop will be likely to achieve a final vorton state before sustaining significant charge and current losses. For $R_c < R < R_*$, the loop's motion will be initially damped, but in the later stages it will have to survive a period of relativistic motion when it is only damped by radiative effects. Finally, loops with $R > R_*$ break away from the network after t_* and their motion is relativistic initially.

To make an estimate of the density of long-lived vortons, let us suppose that only loops with $R < R_*$ relax into vortons. Assuming $N_{\min} < N_*$, we have

$$\rho_v \sim \left(\frac{t_c}{t}\right)^{3/2} \int_{R_{\min}}^{R_*} (\mu R) \, \nu R^{-4} dR \sim \frac{\nu \mu^2}{N_{\min}^2} \left(\frac{t_c}{t}\right)^{3/2} \,. \tag{12.1.52}$$

Of course, this is an over-estimate because we have ignored effects such as loop fragmentation through self-intersection. Nevertheless, these arguments do suggest that if vorton solutions exist, then it may be possible for them to form if $N_{\min} < N_*$, and even more likely if $N_{\min} < N_{cr}$.

Vorton cosmology

The cosmological implications of vortons are best illustrated by comparing with the well-known monopole problem which confronts grand unified models. This will be discussed in some detail in §14.3, but here it suffices to know that monopoles and baryons would be of nearly equal abundance at the present day if initially only one monopole was formed per horizon volume ($n_M \sim t_c^{-3} \ll \xi^{-3}$). However, with $m_M \sim 10^{18}$ GeV, we obtain the approximate estimate, $\rho_M \sim 10^{18} \rho_B$, which reveals the disastrous consequences of monopole formation. This implies that the creation of even one long-lived GUT-scale vorton in 10^{18} horizon volumes would rule out the underlying superconducting string model. This is clearly a difficulty that must be addressed in models requiring large string currents comparable to the string tension, such as the OTW scenario. However, it is a model-dependent constraint with many uncertainties. For example, if we assume that vorton creation is suppressed after t_*, we only require that $N_{min} > N_*$ to save the model, that is, $m_\sigma \lesssim \sqrt{\mu}/4$. The formation of springs presents a similar problem for models with primordial magnetic fields.

On the other hand, vortons may have a benign, even beneficial, cosmological role because they are potential dark matter candidates. Vorton formation in light superconducting string models is much more likely because the damped epoch lasts a comparatively long time (12.1.50). A pertinent example is provided by electroweak-scale strings with $\eta \sim 100\,\mathrm{GeV}$ which have correspondingly larger critical winding numbers,

$$N_{cr} \sim 10^9, \qquad N_* \sim 10^{36}. \qquad (12.1.53)$$

In this case, one can easily find a natural parameter range for which $\Omega_v \approx 1$,

$$N_{min} \sim 2 \times 10^5, \qquad m_\sigma \sim 0.3\sqrt{\mu}. \qquad (12.1.54)$$

The dominant vortons are the lightest with a mass $m_v \sim 10^7\,\mathrm{GeV}$. Being highly-charged magnetic dipoles, they would face the same cosmological constraints as ultra-massive charged particles. Since the smallest vortons will be decaying, they may produce a characteristic cosmic ray signature.

Vorton physics is undoubtedly interesting, but many issues, including questions relating to stability, await further study before their potential cosmological role can be discussed with more definiteness.

12.2 Global string cosmology

12.2.1 Global string evolution

The distinguishing feature of global strings, as opposed to gauge strings, is their coupling to a Goldstone boson field which gives rise to long-range interactions. As discussed in §8.3.2, the dynamical differences are not great on cosmological scales, but for an enhanced radiative decay into Goldstone bosons. The global strings will approximately obey the Nambu equations of motion (6.2.11–13) except with a renormalized tension. For the $U(1)$-string of the Goldstone model (2.1.1),

$$\mu \approx 2\pi\eta^2 \ln\left(\frac{R}{\delta}\right), \qquad (12.2.1)$$

where η is the scalar field expectation value and the logarithmic factor is due to the Goldstone boson self-field of the string, cut off at large scales by the typical string radius of curvature R and at small scales by the string thickness δ. A global string loop with $\ln(R/\delta) \sim 100$ will radiate away most of its energy in only 20 oscillations, a result which should be contrasted with a GUT-scale gauge string loop which decays in about $(\Gamma G\mu)^{-1} \sim 10^4$ oscillations.

No serious attempt has been made to develop a quantitative model of the cosmological evolution of a global string network, so we are limited to a heuristic picture based upon the marriage of results for gauge string networks (§9.3–4) and radiation calculations for global strings (§8.3.3). Because of the dynamical similarities, we can expect to achieve 'scaling', that is, evolution which is scale-invariant with respect to the horizon size, as described in §9.3. However, the radiation back-reaction scale will be much larger, with the gravitational scale $\Gamma G\mu t$ being replaced by $K^{-1}t$, where

$$K \approx \frac{2\pi}{\Gamma_a} \ln\left(\frac{t}{\delta}\right) . \tag{12.2.2}$$

Typically, we have $\Gamma_a \sim 100$ and $K \sim 10$ for late cosmological epochs. The average size of loops produced by the network, $\langle \ell \rangle = \alpha t$, will be closely related to this back-reaction cut-off, $\alpha \sim K^{-1}$. In numerical studies of gauge strings, the large-scale properties of the network were remarkably independent of the loop-size and grid resolution cut-offs, so we might reasonably suppose that global strings will evolve toward a very similar scaling density,

$$\rho_\infty \approx \frac{\zeta\mu}{t^2} \sim \frac{2\pi\zeta\eta^2}{t^2} \ln\left(\frac{t}{\delta}\right) , \tag{12.2.3}$$

but for the additional logarithmic term. The correlation length of the network will be comparable, $\xi \sim d_H/3$, but the small-scale structure will be reduced by the additional radiative damping.

On the assumption that all loops created at any one time have roughly the same size, $\ell \approx \alpha t$, we can expect a loop density during the radiation era which is a minor modification of (10.1.12),

$$n(\ell, t) = \frac{\nu}{t^{3/2}(\ell + K^{-1}t)^{5/2}} , \tag{12.2.4}$$

with $\nu \sim \zeta\sqrt{\alpha}$. For $\alpha \sim K^{-1}$, the total density in loops is comparable to (12.2.3). Similar straightforward replacements can be made in (10.1.17) and (10.1.21) to obtain the expected global string loop densities in the matter era. We can then, as a first approximation, substitute the values of ζ from the gauge string parameters summarized in table 10.1, using $\alpha \sim K^{-1} \sim 0.1$.

Many of the cosmological implications of global strings can be taken over directly from their gauge counterparts, using the renormalized string density (12.2.1). This is largely true of the microwave background aniso-tropies produced by global strings, as well as their gravitational lensing (see §7.3), so the reader is referred back to the pertinent sections in chapter 10. Note, however, that the logarithmic correction implies that the symmetry breaking scale is about one order of magnitude smaller. For example, the microwave anisotropy constraint (10.2.6), $G\mu \lesssim 2 \times 10^{-6}$, implies that $\eta \lesssim 7 \times 10^{14}$ GeV. The diminished small-scale structure on the strings creates differences, such as increasing the orientation correlations of double images. Of course, the gravitational radiation background produced by global strings is negligible.

Large-scale structure formation with global strings is expected to be qualitatively similar to that with gauge strings, but may have important quantitative differences. The properties of wakes will be different because of the strings' reduced wiggliness and higher coherent string velocities (refer to §11.4.2). Loops created by the network should be much larger, and potentially more important than in the gauge case.

12.2.2 Goldstone boson background

The radiative decay of a global string network gives rise to a background of massless Goldstone bosons [Davis, 1985a]. Estimates of the density of this background closely follow that calculated for gravity waves in §10.4, but they are even more subject to uncertainty given our crude model for global string evolution. Goldstone boson radiation by global strings was discussed in §8.3.3 using an antisymmetric tensor formalism. A loop of invariant length ℓ radiates at a discrete set of frequencies, $\omega_n = 4\pi n/\ell$. For a rough estimate of the Goldstone boson background we shall assume, as in §10.4.1, that most of the energy is emitted in the first few harmonics with $n \sim 1$. Then, in about a Hubble time, almost all the energy in the network (12.2.3) will be converted into Goldstone bosons of frequency $\omega \sim 4\pi K/t$, and lying in the range $\Delta\omega \sim \omega$. Comparing this contribution with the radiation background $\Omega_r = \rho_r/\rho_c$ as in (10.4.5), we obtain the relative energy density in a logarithmic frequency interval,

$$\Omega(\omega) \equiv \frac{\omega}{\rho_c}\frac{d\rho}{d\omega} = 30\zeta \, G\mu_\omega \left(\frac{\mathcal{N}}{\mathcal{N}_\omega}\right)^{1/3} \Omega_r, \qquad (12.2.5)$$

where $\mathcal{N}_\omega$ and μ_ω are, respectively, the effective number of spin degrees of freedom and the string tension at the time of emission. This will apply for bosons emitted before t_{eq} and after $t_* \sim 10^{-43}(G\mu)^{-2}$s, prior to which string motion will be heavily damped. A more elaborate derivation for both radiation and matter eras can be made merely by replicating the steps in §10.4.2, but this is not warranted by current data.

The infinite string network will also directly contribute a substantial proportion of the Goldstone boson background. The infinite strings can be expected to radiate mainly at frequencies between those corresponding to the correlation length ξ and the back-reaction scale $K^{-1}t$. Given loop production at $\alpha \sim K^{-1}$, the long string contribution may well be comparable to (12.2.5).

A significant background of Goldstone bosons will increase the effective number of spin degress of freedom $\mathcal{N}$, which will in turn disrupt primordial nucleosynthesis unless $\Omega_{\text{tot}} < 5.4\%$ (refer to §10.4.4). This will give the same GUT model-dependent results as those for gravity waves (10.4.27), implying $\eta \lesssim 10^{15}\text{GeV}$ in a minimal model. After t_{eq}, these massless Goldstone bosons will redshift like radiation, so their contribution to the mass density of the Universe will become insignificant.

12.2.3 Axion constraints

The axion, a pseudo-Goldstone boson arises through the breaking of a global $U(1)$-symmetry, which is added to the standard model in an elegant solution of the strong CP-problem [Peccei & Quinn, 1977; Weinberg, 1978; Wilczek, 1978]. It also appears naturally in a number of theories of fundamental interactions, notably superstring models. Additional interest in the axion has been spurred by its potential astrophysical role as a cold dark matter candidate. Axions acquire a small mass during the quark–hadron phase transition and may come to dominate the energy density of the universe. However, axion cosmology is complicated by the potential appearance of topological defects; global strings are produced at the breaking of the global Peccei–Quinn symmetry $U_{\text{PQ}}(1)$ [Vilenkin & Everett, 1982] and, subsequently, domain walls form when the axion mass 'switches on' [Sikivie, 1982]. Here, we shall follow events in a thermal scenario in which the universe heats up to a temperature at which the Peccei–Quinn symmetry is restored, $T > \eta$, after a hypothetical period of inflation. Topological defects must inevitably appear in this scenario, though they may also arise during inflation because of quantum effects which will be discussed in chapter 16. Since domain walls will form as well as strings, we shall prefigure some discussions in chapter 13.

As the universe cools below $T \sim \eta$, a global string network will be formed with the Goldstone boson field θ winding by 2π around each string. The strings will then evolve toward 'scaling' as discussed in the §12.2.1. The crucial difference between axionic and 'regular' global strings is due to the anomalous nature of the Peccei–Quinn symmetry. Soft instanton effects induce a θ-dependent potential of the form (13.1.11), where the axion mass $m(T)$ grows as

$$m(T) \approx 0.1 m_{\text{a}} (\Lambda_{\text{QCD}}/T)^{3.7} \qquad (12.2.6)$$

at high T, until saturating near $T \sim \Lambda_{\text{QCD}} \approx 200\,\text{MeV}$ at the value

$$m_{\text{a}} = \frac{2m_\pi \eta_\pi}{\eta_{\text{a}}} \left(\frac{m_{\text{u}} m_{\text{d}}}{m_{\text{u}} + m_{\text{d}}} \right)^{1/2} \approx 6 \times 10^{-4}\text{eV} \left(\frac{10^{10}\text{GeV}}{\eta_{\text{a}}} \right) \qquad (12.2.7)$$

(see, for example, Kim [1987], Kolb & Turner [1990]). Here, m_u, m_d are the up and down quark masses and m_π, η_π are the pion mass and decay constant.

The parameter N in (13.1.11) is a model-dependent integer. For $N > 1$, the approximate $U_{PQ}(1)$-symmetry has an exact Z_N subgroup, and symmetry breaking gives rise to N degenerate vacua at $\theta = 2\pi n/N$ with $n = 0, 1, ..., N-1$. Each global string will then get attached to N domain walls (13.1.13) which interpolate between the neighbouring vacua. Although such a hybrid network may approach 'scaling', it will soon come to dominate the energy density of the universe unless η is unacceptably low (refer to §13.5).

The only viable axion models in the thermal scenario, therefore, are those with $N = 1$. In this case each wall will be attached to a single string, and reconnections of walls and strings lead to a rapid demise of the network (§13.6). Axionic domain walls can be considered 'formed' at a time t_w when the wall thickness $m^{-1}(t)$ becomes comparable to the inter-string separation, $t_w m(t_w) \sim 1$. The walls become dynamically important when $\mu/t \sim \sigma$, where the wall tension $\sigma(t)$ is given by (13.1.14). This happens at $\tilde{t} \sim 3t_w$, and using (12.2.6) for $m(t)$ we obtain*

$$\tilde{t} \sim 10^{-5} \left(\frac{m_a}{10^{-3}\text{eV}} \right)^{0.36} \text{s} \,. \tag{12.2.8}$$

At this time the walls pull the strings together, and the system quickly disintegrates into disconnected pieces of wall bounded by a string [Everett & Vilenkin, 1982; Shellard, 1986a]. The string loops bounding the walls will have typical length $\ell \sim K^{-1}\tilde{t}$ and will oscillate at a frequency $\omega \sim 4\pi K/\tilde{t} \sim 4m(\tilde{t})$. Within a Hubble time they will decay by radiating massive axions.

Let us now estimate the axion contribution to the mass density of the universe. According to (12.2.5) the axion number density will have a frequency distribution given by

$$\frac{dn_a}{d\omega} \sim \frac{\zeta \mu_\omega}{\omega^2 t^2} \,, \tag{12.2.9}$$

where μ_ω has a logarithmic dependence on ω. Clearly, the dominant contribution to n_a is given by the lowest frequencies corresponding to axions radiated at $t \sim \tilde{t}$ when the network decayed. The wall energy at $t \sim \tilde{t}$ is comparable to that of strings, and we can still use (12.2.3) to

---

* To express T in terms of t, we have used the relation (1.3.28) with ~ 60 spin degrees of freedom, which corresponds to temperatures above Λ_{QCD} and below the electroweak scale. As we shall see, this is the most interesting temperature range for network decay.

estimate the total energy density of the network. Hence, we can write

$$n_{\rm a} \sim \frac{\zeta\tilde{\mu}}{\tilde{\omega}\tilde{t}^2} \left(\frac{\tilde{t}}{t}\right)^{3/2} \sim \frac{\zeta\eta^2}{2K\tilde{t}} \ln\left(\frac{\tilde{t}}{\delta}\right)\left(\frac{\tilde{t}}{t}\right)^{3/2}, \qquad (12.2.10)$$

where a tilde indicates that the corresponding quantity is evaluated at $t = \tilde{t}$. When the temperature drops below $\Lambda_{\rm QCD}$, the axion mass density is $\rho_{\rm a} = n_{\rm a} m_{\rm a}$, and the axion contribution to the present density of the universe can be estimated as

$$\Omega_{\rm a} \sim \frac{n_{\rm a} m_{\rm a}}{\rho_{\rm r}}(t_{\rm eq}) \sim 10^{-2}\zeta h^{-2}\left(\frac{m_{\rm a}}{10^{-3}{\rm eV}}\right)^{-1.18}, \qquad (12.2.11)$$

where $\rho_{\rm r} \sim 1/30Gt^2$ is the radiation density. Requiring that $\Omega \lesssim 1$ and using $\zeta \sim 10$ we obtain cosmological constraints on the axion mass $m_{\rm a}$ and the Peccei–Quinn symmetry breaking scale $\eta_{\rm a}$,

$$m_{\rm a} \gtrsim 10^{-4}{\rm eV}, \qquad \eta_{\rm a} \lesssim 10^{11}{\rm GeV}. \qquad (12.2.12)$$

These constraints are similar to those derived by Davis [1986] and Davis & Shellard [1989c] who used somewhat different network parameters and assumed the dominant axion emission to be from long strings.* We note also that (12.2.12) is an order of magnitude stronger than the bounds on a uniform axion field obtained in the inflationary scenario [Abbott & Sikivie, 1983; Preskill, Wise & Wilczek, 1983; Dine & Fischler, 1983].

The cosmological bounds (12.2.12) are close to conflicting with astrophysical bounds which derive from the acceptable levels of axion emission from red giants and Supernova 1987a. Conservative estimates with generic couplings of these astrophysical constraints imply

$$m_{\rm a} \lesssim 10^{-3}{\rm eV}, \qquad \eta_{\rm a} \gtrsim 10^{10}{\rm GeV}. \qquad (12.2.13)$$

Hence, the axion mass is allowed to take values only in the narrow window between 10^{-3} and $10^{-4}{\rm eV}$. For $m_{\rm a} \sim 10^{-4}{\rm eV}$ axions may be the dominant form of matter in the present universe. By the time they come to dominate, their velocities are redshifted to $v_{\rm a} \lesssim 10^{-8}$, and thus axions provide an example of cold dark matter (refer to §1.3.6).

There can be little doubt that all the above estimates are plagued by substantial uncertainties which demand resolution. True quantitative modelling will require an improved understanding of radiation from a string network and the complicated dynamics of domain wall formation and annihilation.

---

* We should mention the dissenting opinion of Harari & Sikivie [1987] and Sikivie [1992] who assume a very different spectrum for axion radiation by strings (refer to §0.3.3) and obtain a weaker bound on $m_{\rm a}$.

13

Domain walls

13.1 Field theory

13.1.1 Domain wall models

Domain walls arise in models with spontaneously broken discrete symmetries. The model generally involves a set of real scalar fields ϕ_i with a Lagrangian of the form

$$\mathcal{L} = \tfrac{1}{2}(\partial_\mu \phi_i)^2 - V(\phi), \tag{13.1.1}$$

where the potential $V(\phi)$ has a discrete set of degenerate minima. We shall assume that $V(\phi) = 0$ at the minima. Gauge fields typically do not affect domain wall structure and have not been included in (13.1.1).

A planar wall orthogonal to the z-axis is described by functions $\phi_i(z)$ interpolating between two different minima at $z \to \pm\infty$. This field configuration is invariant with respect to Lorentz boosts in the xy-plane, indicating that only transverse motion of the wall has physical significance. The functions $\phi_i(z)$ satisfy the equations

$$\phi_i''(z) - \partial V/\partial \phi_i = 0, \tag{13.1.2}$$

which have a first integral

$$\tfrac{1}{2}\phi'^2 - V(\phi) = 0, \tag{13.1.3}$$

where $\phi'^2 = \phi_i' \phi_i'$. With (13.1.3) taken into account, the energy–momentum tensor of the wall,

$$T_{\mu\nu} = \partial_\mu \phi_i \partial_\nu \phi_i - g_{\mu\nu}\mathcal{L}, \tag{13.1.4}$$

takes the form

$$T_\mu^\nu(z) = \phi'^2(z)\, \mathrm{diag}\,(1,1,1,0). \tag{13.1.5}$$

The fact that $T_3^3 = 0$ is a simple consequence of the conservation law, $\partial_z T_3^3 = 0$, and the fact that the remaining components of T_μ^ν are proportional to δ_μ^ν is due to the tangential Lorentz invariance of the wall. The surface energy density of the wall is given by

$$\sigma = \int dz \phi'^2(z), \qquad (13.1.6)$$

and it follows from (13.1.5) that the wall tension in the two tangential directions is also equal to σ.

Examples

(i) A prototypical model of discrete symmetry breaking is a single scalar field with a double-well potential,

$$V(\phi) = \tfrac{1}{4}\lambda(\phi^2 - \eta^2)^2. \qquad (13.1.7)$$

The Z_2 symmetry of the potential is broken when ϕ chooses between the two minima at $\phi = \pm\eta$. The corresponding domain wall solution has already been discussed in §3.2.1,

$$\phi(z) = \eta \tanh\left[(\lambda/2)^{1/2}\eta z\right]. \qquad (13.1.8)$$

The thickness of this wall is approximately

$$\delta \sim (\sqrt{\lambda}\,\eta)^{-1}, \qquad (13.1.9)$$

and its tension is

$$\sigma = \tfrac{4}{3}\sqrt{\lambda}\,\eta^3. \qquad (13.1.10)$$

(ii) We next consider a complex scalar field, $\phi = f e^{i\theta}$, with a Z_N-symmetric Lagrangian,

$$\mathcal{L} = \partial_\mu \phi^\dagger \partial^\mu \phi - \tfrac{1}{4}\lambda(\phi^\dagger \phi - \eta^2)^2 + 2(m^2\eta^2/N^2)(\cos N\theta - 1). \quad (13.1.11)$$

Here, N is an integer, and it will be assumed that $m^4 \ll \lambda\eta^4$, so that $\mathcal{L}$ has an approximate $U(1)$ symmetry which is explicitly broken by the last term. Potentials like (13.1.11) arise naturally in axion models [Peccei & Quinn, 1977; Dine, Fischler & Srednicki, 1981]. At low energies, the heavy degrees of freedom described by the variable f are not excited, and the effective Lagrangian for θ can be obtained by setting $f = \eta$ in (13.1.11),

$$\mathcal{L}_\theta = \eta^2(\partial_\mu\theta)^2 + 2(m^2\eta^2/N^2)(\cos N\theta - 1). \qquad (13.1.12)$$

Note that for $N = 1$ this is identical to the sine-Gordon model (3.1.4).

The minima of the potential in (13.1.12) are at $\theta = 2\pi n/N$ with $n = 0, 1, ..., N-1$, so there are N distinct vacua which can be separated by domain walls [Sikivie, 1982]. The domain wall solution interpolating between two neighbouring minima is

$$\theta = \frac{2\pi n}{N} + \frac{4}{N} \tan^{-1} \exp(mz), \qquad (13.1.13)$$

the wall thickness is $\delta_w \sim m^{-1}$ and the wall tension is given by

$$\sigma = 16m\eta^2/N^2. \qquad (13.1.14)$$

It can be shown that there are no other stable wall solutions in this model (see, for example, Coleman [1985a]).

The $U(1)$-symmetric model described by the first two terms in (13.1.11) has string solutions with θ changing by 2π around a string. The effect of the symmetry-violating term is that each string gets attached to N domain walls [Vilenkin & Everett, 1982]. Note that the thickness of the string core, $\delta_s \sim (\sqrt{\lambda}\eta)^{-1}$, is much smaller than the wall thickness δ_w.

(iii) Our final example is a model with an approximate $O(N)$ symmetry [Ishihara, Kubotani & Nambu, 1992],

$$V(\phi) = V_0 + \frac{1}{4}\lambda(\phi_i\phi_i - \eta^2)^2 - \epsilon\sum_{i=1}^{N}\phi_i^4. \qquad (13.1.15)$$

It is assumed that $\epsilon \ll \lambda$, and the constant $V_0 \approx \epsilon\eta^4$ is added to ensure that $V = 0$ in the vacuum. The symmetry is explicitly broken by the last term to a non-abelian discrete group and the vacuum manifold is reduced to a discrete set of minima at $\phi_i^{(j,\pm)} \approx \pm\eta\delta_i^j$. The model has domain walls interpolating between the neighbouring vacua $\phi^{(j,\pm)}$, $\phi^{(k,\pm)}$ with $j \neq k$. By symmetry, it is clear that all these walls have the same tension. Opposite vacua $\phi^{(j,+)}$, $\phi^{(j,-)}$ are separated by a high potential barrier, and the walls connecting such vacua are unstable. The model also allows a stable configuration in which three walls at 120° to one another are joined along a line. Unlike the previous example, there is no string at the wall junction, the scalar field does not go through zero, and the energy density is comparable to that inside the wall. The wall junctions can themselves be joined at 'knots', and a stable configuration is the tetrahedral arrangement shown in fig. 13.1.

13.1.2 Walls bounded by strings

These hybrid defects can arise in models with a sequence of phase transitions [Kibble, Lazarides & Shafi, 1982b]. Consider, for example,

$$G \xrightarrow{\phi_1} H \xrightarrow{\phi_2} I, \qquad (13.1.16)$$

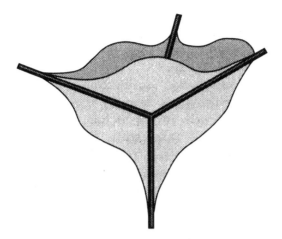

Fig. 13.1. Six walls joined at a 'knot' in an equilibrium configuration.

with $\pi_0(G) = \pi_1(G) = I$ and $\pi_0(H) \neq I$. According to the theorem of §3.3.2, $\pi_1(G/H) = \pi_0(H)$, and thus the first phase transition gives rise to strings. A closed path around a string corresponds to a path between two disconnected components of H in G, as illustrated in fig. 3.10. Since the field ϕ_1 in (13.1.16) is invariant under H, it returns to its initial value when the path is completed.

The second phase transition in (13.1.16) involves a discrete symmetry breaking and leads to the formation of domain walls. It is not difficult to see that each string remaining from the first phase transition must now get connected to a domain wall. The field ϕ_2 is not invariant under H, and without a domain wall it would have to change discontinuously as we go around a string. Since walls can terminate on strings, they are not topologically stable and can decay by quantum nucleation of holes bounded by strings (see §13.2.5). The absence of topologically stable defects in (13.1.16) is not surprising if we note that the vacuum manifold $M = G$ has trivial homotopy groups π_0 and π_1.

Both strings and walls in models like (13.1.16) are classified by the elements of $\pi_0(H)$. Some of the walls may be unstable with respect to splitting into several (N) smaller mass walls. Then the corresponding strings get attached to N domain walls. Of course, the strings can be similarly unstable and split, so that each wall gets bounded by its own string. In the rest of this chapter, the term 'walls bounded by strings' will be used only in reference to models where a single domain wall is attached to each stable string.

A grand unified symmetry breaking scheme which produces domain walls bounded by strings is given in §2.1.4. Here, the first stage (2.1.68a) breaks $Spin(10)$ to produce strings associated with a discrete Z_2^C charge conjugation symmetry. Since Z_2^C is not an observed symmetry of nature, it must be broken at a subsequent stage (2.1.68b) when domain walls will appear – attached to the strings.

Walls bounded by strings can also be formed in models with an approximate symmetry [Vilenkin & Everett, 1982]. Consider, for example, the model (13.1.11) with $N = 1$. It has a unique vacuum state at $|\phi| = \eta$, $\theta = 0$ and no exact symmetries. However, for $m^4 \ll \lambda\eta^4$ the model has an approximate, broken $U(1)$ symmetry and accompanying global string solutions. If the symmetry were exact, the phase θ would change uniformly around the string, but the last term in (13.1.11) forces θ to zero, so that all the variation of θ from 0 to 2π is confined to a domain wall centered at $\theta = \pi$, which is attached to the string. As already noted, the thickness of the wall in this type of model is much greater than the thickness of the string core.

13.2 Domain wall dynamics

13.2.1 Action for a domain wall

In most situations of cosmological interest the thickness of a domain wall can be neglected compared to its other dimensions and the wall can be treated as in infinitely thin surface. The spacetime history of the wall can then be represented by a three-dimensional worldsheet,

$$x^\mu = x^\mu(\zeta^a), \qquad a = 0, 1, 2. \qquad (13.2.1)$$

The action describing the dynamics of the wall can be derived from the action of the underlying field theory,

$$S = \int d^4x \sqrt{-g}\,\mathcal{L}, \qquad (13.2.2)$$

with $\mathcal{L}$ given in (13.1.1).

In the vicinity of the worldsheet it is convenient to use the coordinates ζ^a and z, where z is the normal distance from the surface (13.2.1). The metric in these coordinates has the form

$$ds^2 = \gamma_{ab}\,d\zeta^a d\zeta^b - dz^2, \qquad (13.2.3)$$

where γ_{ab} is the worldsheet metric

$$\gamma_{ab} = g_{\mu\nu}x^\mu_{,a}x^\nu_{,b}. \qquad (13.2.4)$$

The spacetime volume element in the new coordinates is

$$\sqrt{-g}\,d^4x = \sqrt{\gamma}\,d^3\zeta dz, \qquad (13.2.5)$$

where $\gamma = \det(\gamma_{ab})$. The next step is to note that in the thin wall limit all fields appearing in the Lagrangian $\mathcal{L}$ depend only on z. In this approximation, the fields $\phi_i(z)$ are given by the same functions as for an infinite plane wall. Integrating out the z-dependence, we obtain from (13.2.2)

$$S = -\sigma \int d^3\zeta \sqrt{\gamma}, \qquad (13.2.6)$$

where the constant σ is given by

$$\sigma = -\int \mathcal{L}\, dz. \qquad (13.2.7)$$

The action (13.2.6) is proportional to the three-volume of the wall's worldsheet. It is very similar to the Nambu action (6.1.9) for a string and to the action (6.1.10) for a relativistic particle. The meaning of the parameter σ is easily understood if we note that the energy density of a static configuration of scalar fields is $T_0^0 = -\mathcal{L}$ (see (13.1.1)). It is clear from (13.2.7), then, that σ is the mass per unit area of the wall. The expression (13.1.6) obtained previously for σ is equivalent to (13.2.7) because of the relation (13.1.3). Corrections to the wall action due to the finite width of the wall have been discussed by Gregory, Haws & Garfinkle [1990].

Variation of (13.2.6) with respect to the metric $g_{\mu\nu}$ gives the energy–momentum tensor of the wall

$$T^{\mu\nu}\sqrt{-g} = -2\frac{\delta S}{\delta g_{\mu\nu}} = \sigma \int d^3\zeta\, \delta^{(4)}[x - x(\zeta)]\sqrt{\gamma}\gamma^{ab}x^\mu_{,a}x^\nu_{,b}. \qquad (13.2.8)$$

13.2.2 Wall motion

The equations of motion of a domain wall are derived by varying the action (13.2.6) with respect to $x^\mu(\zeta)$. Using the identity

$$d\gamma = \gamma\gamma^{ab}d\gamma_{ab} \qquad (13.2.9)$$

and the definition of the metric (13.2.4) we obtain

$$\gamma^{-1/2}\partial_a(\sqrt{\gamma}\,\gamma^{ab}x^\mu_{,b}) + \Gamma^\mu_{\nu\sigma}\gamma^{ab}x^\nu_{,a}x^\sigma_{,b} = 0. \qquad (13.2.10)$$

In flat spacetime with $g_{\mu\nu} = \eta_{\mu\nu}$ this equation reduces to

$$\partial_a(\sqrt{\gamma}\,\gamma^{ab}x^\mu_{,b}) = 0. \qquad (13.2.11)$$

Since the wall action is invariant under worldsheet reparametrization, we are free to impose three arbitrary gauge conditions. However, no gauge conditions have been found that would reduce (13.2.11) to an exactly soluble linear equation. As a result, much less is known about the dynamics of domain walls than about the dynamics of strings.

A choice of gauge similar to the conformal gauge for strings is

$$\gamma_{01} = \gamma_{02} = 0, \qquad \gamma_{00} = \sqrt{\gamma}. \qquad (13.2.12)$$

With this choice the wall equations of motion (13.2.11) simplify somewhat,

$$\ddot{x}^\mu + \epsilon^{AC}\epsilon^{BD}\partial_A(x^\nu_{,C}\, x_{\nu,D}\, x^\mu_{,B}) = 0. \qquad (13.2.13)$$

Here, dots represent derivatives with respect to ζ^0, the indices A, B, C, D take values $1, 2$, and we have used the relations

$$\gamma^{AB} = {}^{(2)}\gamma^{-1}\epsilon^{AC}\epsilon^{BD}\gamma_{CD}, \qquad \gamma_{AB} = x^\nu_{,A}\, x_{\nu,B}, \qquad (13.2.14)$$

where ${}^{(2)}\gamma = \gamma_{00}$ is the determinant of γ_{AB}. Clearly, $x^0 = \zeta^0$ is a solution of (13.2.13), and so we can identify ζ^0 with time t. The motion of the wall is then described by a vector function $\mathbf{x}(\zeta^1, \zeta^2, t)$.

We do not know how to solve (13.2.13) in general, but a special family of solutions can be found in the form

$$\mathbf{x}(\zeta^1, \zeta^2, t) = \mathbf{n}\zeta^2 + \mathbf{x}_\perp(\zeta^1, t). \qquad (13.2.15)$$

Here, $\mathbf{n}$ is a unit vector and $\mathbf{x}_\perp$ lies in the plane perpendicular to $\mathbf{n}$. The wall surface is obtained by simply translating the planar curve $\mathbf{x}_\perp$ in the direction of $\mathbf{n}$. Interestingly enough, a substitution of (13.2.15) into (13.2.12) and (13.2.13) gives equations identical to the Nambu equations of motion for a planar string,

$$\ddot{\mathbf{x}}_\perp - \mathbf{x}_\perp'' = 0, \qquad (13.2.16)$$

$$\dot{\mathbf{x}}_\perp \cdot \mathbf{x}_\perp' = 0, \qquad \dot{\mathbf{x}}_\perp^2 + \mathbf{x}_\perp'^2 = 1, \qquad (13.2.17)$$

where primes stand for derivatives with respect to ζ^1. This means that any planar string solution $\mathbf{x}_\perp(\zeta, t)$ can be turned into a domain wall solution by simply translating it in the direction perpendicular to the plane, as in (13.2.15).

If $x_0^\mu(\zeta^a)$ is a known solution of the wall equations of motion, one can study the behaviour of linearized perturbations about this solution. Since only motion transverse to the wall is observable, the perturbed wall trajectory can be represented as

$$x^\mu = x_0^\mu + \xi n_0^\mu, \qquad (13.2.18)$$

where $n_0^\mu(\zeta^a)$ is a unit normal to the unperturbed worldsheet. The perturbation $\xi(\zeta^a)$ can be thought of as a scalar field in a three-dimensional curved spacetime. For a domain wall moving in Minkowski space, it can be shown that the field ξ satisfies the equation [Garriga & Vilenkin, 1991]

$$(\Box + R)\xi = 0, \qquad (13.2.19)$$

where the d'Alembertian and the scalar curvature R are calculated using the unperturbed worldsheet metric.

Small perturbations on a collapsing spherical wall have been studied by Widrow [1989b] in the gauge $\zeta^0 = t$, $\zeta^1 = \theta$, $\zeta^2 = \phi$. The wall trajectory in this gauge is described by the function $r(\theta, \phi, t)$. Representing r as $r_0(t) + \delta r$ and linearizing in δr, Widrow has shown that the relative perturbation $\delta r/r$ has growing modes with $\delta r/r \propto r^{-1}$ as $r \to 0$. Hence, nearly spherical walls can become highly asymmetric in the course of their collapse.

13.2.3 Walls in an expanding universe

We now consider the behaviour of small perturbations on walls in an expanding universe

$$ds^2 = a^2(\tau)(d\tau^2 - d\mathbf{x}^2). \tag{13.2.20}$$

Assuming that the unperturbed wall lies in the xy-plane, we choose the gauge

$$\zeta^0 = \tau, \quad \zeta^1 = x, \quad \zeta^2 = y. \tag{13.2.21}$$

The motion of the wall is then described by a single function $z(x, y, \tau)$ and the action (13.2.6) takes the form

$$S = -\sigma \int \int \int d\tau dx dy \, a^3(\tau) \sqrt{\gamma_0}, \tag{13.2.22}$$

where

$$\gamma_0 = 1 - \dot{z}^2 + z_x'^2 + z_y'^2, \tag{13.2.23}$$

$z_x' = \partial z/\partial x$ and $\dot{z} = \partial z/\partial \tau$. The corresponding equation of motion is

$$\left(\frac{\partial}{\partial \tau} + 3\frac{\dot{a}}{a}\right)\left(\gamma_0^{-1/2}\dot{z}\right) - \frac{\partial}{\partial x}\left(\gamma_0^{-1/2}z_x'\right) - \frac{\partial}{\partial y}\left(\gamma_0^{-1/2}z_y'\right) = 0. \tag{13.2.24}$$

Assuming a power-law expansion

$$a(\tau) \propto \tau^\alpha, \tag{13.2.25}$$

we can write the linearized equation for small perturbations as

$$\ddot{z} + 3\alpha\tau^{-1}\dot{z} - z_{xx}'' - z_{yy}'' = 0. \tag{13.2.26}$$

The solution of this equation, which is well-behaved as $\tau \to 0$ is a superposition of waves of the form

$$z = A\tau^{-\nu}J_\nu(k\tau)\exp(ik_x x + ik_y y), \tag{13.2.27}$$

where $k^2 = k_x^2 + k_y^2$ and $\nu = (3\alpha - 1)/2$.

The properties of the solutions (13.2.27) are similar to those of perturbations on straight strings discussed in chapter 6. The physical wavelength of the waves, $\lambda = 2\pi a(\tau)/k$, always grows proportionally to the scale factor, while the behaviour of the amplitude depends on the magnitude of $k\tau$. To order of magnitude, the quantity $k\tau$ is equal to the ratio of the horizon size to the wavelength, $k\tau \sim t/\lambda$. As $k\tau \to 0$, the time-dependent factor in (13.2.27) approaches a constant. This means that waves with $\lambda \gg t$ are conformally stretched: both the amplitude and the wavelength grow as $a(\tau)$, with the shape of the wall remaining unchanged. In the opposite limiting case $k\tau \gg 1$, $\lambda \ll t$, the physical amplitude of the wave, $a(\tau)z$, decreases as $a^{-1/2}$ and the ratio of the amplitude to the wavelength is proportional to $a^{-3/2}$. Hence, perturbations of a domain wall on scales smaller than the horizon are rapidly smoothed by the expansion of the universe.

13.2.4 Dynamics of thick domain walls

The thin-wall approximation is justified as long as the wall thickness δ is negligible compared to all the other dimensions in the problem. It breaks down when the curvature radius of the wall becomes comparable to δ or when the wall intersects itself (or other walls). To study the wall dynamics in the general case, one has to solve the scalar field equation

$$\Box\phi + V'(\phi) = 0. \tag{13.2.28}$$

A family of exact solutions of this equation has been found by Vachaspati, Everett & Vilenkin [1984]. Suppose that $\phi = \phi_0(z)$ is the solution of (13.2.28) describing a static wall lying in the xy-plane (one example is the solution (13.1.8) for the $\lambda\phi^4$ theory). Then it is easily verified that

$$\phi = \phi_0[z - f(k_a x^a)], \tag{13.2.29}$$

with $a = 0, 1, 2$ and an arbitrary function f, is also a solution provided that

$$k_a k^a = 0. \tag{13.2.30}$$

Solutions of the form (13.2.29) describe plane waves of arbitrary shape propagating along the wall at the speed of light. It is clear that such waves propagate without dissipation even when their wavelength is comparable to or smaller than the wall thickness. Another example of this phenomenon is for strings where solutions describing travelling waves on strings can be obtained from static string solutions [Vachaspati & Vachaspati, 1990].

The internal structure of the walls has an important effect on what happens at wall intersections. When domain walls collide, we expect scalar particle production and topology-changing processes similar to string intercommuting. This has been confirmed numerically in two and three dimensions; in the $N = 1$ model (13.1.11), intersecting walls were observed to 'exchange partners' and reconnect [Shellard, 1986a,b]. This is implicit in the numerical studies of domain walls in an expanding universe to be discussed in §13.5.2. Widrow [1989a] studied the collapse of spherical domain walls and their subsequent annihilation into massive particles.

13.2.5 Decay of metastable domain walls ·

We have already mentioned in §13.1.2 that in models with walls bounded by strings, the walls are metastable and can decay by quantum nucleation of holes bounded by strings. Here, we shall estimate the corresponding decay rate for a planar wall [Kibble *et al.*, 1982b]. Semiclassically, the hole nucleation is described by an instanton, which is a solution of Euclidean (imaginary time, $t = i\tau$) field equations, approaching the unperturbed wall solution at $\tau \to \pm\infty$. The decay probability can be expressed as

$$\mathcal{P} = Ae^{-B} , \qquad (13.2.31)$$

where B is the difference between the Euclidean actions of the instanton (S) and of the unperturbed wall (S_0), that is,

$$B = S - S_0 . \qquad (13.2.32)$$

The pre-factor A can be calculated by analyzing small perturbations about the instanton [Coleman, 1985a].

If the radius of the nucleating hole is much greater than the wall thickness,

$$R \gg \delta_{\rm w} , \qquad (13.2.33)$$

we can use the thin-string and thin-wall approximation, in which the actions for the string and for the wall are proportional to the corresponding worldsheet areas,

$$S = \mu \int dS_2 + \sigma \int dS_3 . \qquad (13.2.34)$$

Here, μ and σ are the string and wall tensions, respectively. The worldsheet of a static wall lying in the xy-plane is the three-dimensional hyperplane $z = 0$. In the instanton solution, this hyperplane has a 'hole' which is bounded by the closed worldsheet of the string.

A planar wall is invariant with respect to longitudinal Lorentz boosts, which turn into Euclidean rotations after the change $t \to i\tau$. In order to preserve this symmetry, the string worldsheet should be a sphere,

$$x^2 + y^2 + \tau^2 = R^2 . \qquad (13.2.35)$$

The corresponding tunnelling action is

$$B = 4\pi R^2 \mu - \frac{4\pi}{3} R^3 \sigma \,. \tag{13.2.36}$$

Minimizing this with respect to R we find

$$R = \frac{2\mu}{\sigma} \,, \tag{13.2.37}$$

$$B = \frac{16\pi\mu^3}{3\sigma^2} \,. \tag{13.2.38}$$

The Lorentzian evolution of the hole after nucleation can be found by making the replacement $\tau \to -it$ in (13.2.35), giving

$$x^2 + y^2 = R^2 + t^2 \,, \tag{13.2.39}$$

The hole expands, rapidly approaching the speed of light.

If η_s and η_w are, respectively, the symmetry breaking scales of strings and walls, then it follows from (13.1.9–10), (13.2.33) and (3.2.10) that the thin-defect approximation is justified provided $\eta_s \gg \eta_w$. In this limit $B \gg 1$, and the nucleation probability can be extremely small.

13.3 Gravitational effects of domain walls

13.3.1 Gravitational field of planar and spherical walls

The gravitational field of a domain wall is rather unusual and is very different from that of an ordinary massive plane. As for strings, the origin of the difference is due to the large wall tension. From (13.2.8), the energy–momentum tensor of a wall lying in the yz-plane is

$$T_\mu^\nu = \sigma\delta(x)\,\text{diag}\,(1,0,1,1)\,. \tag{13.3.1}$$

The wall tension σ in the y- and z-directions is equal to the surface energy density of the wall. To preview the bizarre gravitational properties of domain walls, we note that the Newtonian limit of the Einstein equations for a static distribution of matter is

$$\nabla^2\Phi = 4\pi G(\rho - T_i^i)\,, \tag{13.3.2}$$

where $\rho = T_0^0$ and Φ is the gravitational potential. For the energy–momentum tensor (13.3.1) this yields

$$\nabla^2\Phi = -4\pi G\rho\,, \tag{13.3.3}$$

indicating that the gravitational field of the wall is repulsive. The general-relativistic analysis of the problem confirms this conclusion, but it holds some other surprises.

The first surprise is that the Einstein equations with the source term (13.3.1) have no static solutions having planar and reflectional symmetry. The only static solution (up to a coordinate transformation), with $T^{\nu}_{\mu} = 0$ everywhere except in the yz-plane and having the required symmetry, is [Taub, 1956]

$$ds^2 = (1 - K|x|)^{-1/2}(dt^2 - dx^2) - (1 - K|x|)(dy^2 + dz^2). \quad (13.3.4)$$

The metric is continuous at $x = 0$, but its x-derivatives are discontinuous, resulting in a δ-function energy–momentum tensor

$$T^{\nu}_{\mu} = \frac{K}{8\pi G}\delta(x)\,\text{diag}\left(1, 0, \tfrac{1}{4}, \tfrac{1}{4}\right). \quad (13.3.5)$$

This is not a form consistent with (13.3.1) and thus the gravitational field of a planar wall must be non-static.

A time-dependent solution of Einstein's equations for a planar domain wall with energy–momentum tensor (13.3.1) has been found by Vilenkin [1983c] and by Ipser & Sikivie [1984]:

$$ds^2 = (1 - \kappa|x|)^2 dt^2 - dx^2 - (1 - \kappa|x|)^2 e^{2\kappa t}(dy^2 + dz^2), \quad (13.3.6)$$

where $\kappa = 2\pi G\sigma$. This metric has some very interesting properties. The (x, t)-part of (13.3.6) is a (1+1)-dimensional Rindler metric describing a flat space in the frame of reference of a uniformly accelerated observer. An observer at $x = 0$ will see test particles moving away from the wall with an acceleration $a = \kappa = 2\pi G\sigma$, in agreement with the Newtonian analysis. On hypersurfaces $x = $ const., the metric is that of a (2+1)-dimensional de Sitter space,

$$ds_3^2 = dt^2 - e^{2\kappa t}(dy^2 + dz^2). \quad (13.3.7)$$

The metric (13.3.6) has an event horizon: an observer at $x = 0$ never sees particles or light cross the surfaces $x = \pm\kappa^{-1}$. On the other hand, it takes a finite proper time for a particle to reach $|x| = \kappa^{-1}$. In addition, the usual de Sitter horizon of radius κ^{-1} exists in the yz-plane. The singularity at $|x| = \kappa^{-1}$ is not a true singularity of the metric. In fact, it can be shown that the metric (13.3.6) is locally flat everywhere except on the wall itself. A coordinate transformation can be found which transforms the metric to the Minkowski form on one side of the wall,

$$ds^2 = dt^{*2} - dx^{*2} - dy^{*2} - dz^{*2}. \quad (13.3.8)$$

In these coordinates the location of the wall ($x = 0$) is given by [Ipser & Sikivie, 1984]

$$x^{*2} + y^{*2} + z^{*2} = \kappa^{-2} + t^{*2}. \quad (13.3.9)$$

Hence, from the point of view of an inertial observer, the wall has the shape of a sphere! It comes in from large distances, halts its collapse at radius κ^{-1} and re-expands, always moving with a constant outward acceleration $a = \kappa$. By reflection symmetry, it is clear that observers on the other side of the wall will also see themselves enclosed by an outwardly accelerating sphere.

How can a spherically shaped wall in (13.3.9) be reconciled with the flat wall in (13.3.6)? As often happens in general relativity, the apparent discrepancy is due to a different choice of coordinates. It is well-known that the spatially-flat coordinates (13.3.7) cover only half of de Sitter space, while the full space is covered by the metric (see, for example, Hawking & Ellis [1973])

$$ds_3^2 = dt^2 - \kappa^{-2}\cosh^2(\kappa t)(dr^2 + \sin^2 r d\theta^2)\,. \qquad (13.3.10)$$

This suggests that the metric (13.3.6) covers only part of the domain wall spacetime and that the full worldsheet metric is given by (13.3.10), which has spherical spatial sections.

In the general case, solutions of the Einstein equations describing the gravitational field of a domain wall can be found by matching vacuum solutions on three-dimensional timelike hypersurfaces. The matching condition [Israel, 1966] relates the difference of extrinsic curvatures on the two sides of the hypersurface to the energy–momentum tensor of the wall. Ipser & Sikivie [1984] used this formalism to study the gravitational field of collapsing spherical walls in asymptotically-flat space. Not surprisingly, the space outside the wall is Schwarzschild and the walls always collapse to form black holes.

The gravitational field of 'thick' domain walls has been discussed by Widrow [1989c] who studied plane-symmetric solutions of the combined Einstein and scalar field equations. As expected, at large distances from the wall the metric approaches (13.3.6).

13.3.2 A wall-dominated universe

To study the effect of domain walls on the expansion of the universe, we shall first find the effective equation of state of the wall system. Using (13.2.8) for the energy–momentum tensor and the gauge conditions (13.2.12) with $\zeta^0 = t$, we can write (in flat spacetime)

$$\rho = T_0^0 = \sigma \int d^2\zeta\, \delta^{(3)}[\mathbf{x} - \mathbf{x}(\zeta,t)]\,, \qquad (13.3.11)$$

$$T_i^i = \sigma \int d^2\zeta\,(2 - 3\dot{\mathbf{x}}^2)\,\delta^{(3)}[\mathbf{x} - \mathbf{x}(\zeta,t)]\,. \qquad (13.3.12)$$

For an isotropic 'gas' of domain walls the pressure is $P = -\frac{1}{3}T_i^i$, and the effective equation of state is [Zel'dovich, Kobzarev & Okun', 1975]

$$P = (\langle\dot{\mathbf{x}}^2\rangle - \tfrac{2}{3})\rho\,. \qquad (13.3.13)$$

If the walls are highly relativistic, $\langle \dot{\mathbf{x}}^2 \rangle \to 1$ and (13.3.13) gives the usual equation of state for a relativistic gas, $P = \frac{1}{3}\rho$. In the opposite limiting case of non-relativistic walls, $\langle \dot{\mathbf{x}}^2 \rangle \ll 1$ and

$$P = -\tfrac{2}{3}\rho\,. \tag{13.3.14}$$

We shall see in the following sections that, once formed, domain walls eventually dominate the energy density of the universe. Here, we shall find the expansion rate of the universe during the wall-dominated epoch. If the universe is homogeneous on scales much greater than the wall separation, it can be approximately described by an FRW metric

$$ds^2 = dt^2 - a^2(t)d\mathbf{x}^2\,. \tag{13.3.15}$$

With the equation of state $P = \alpha\rho$ the scale factor is given by

$$a(t) \propto t^{2/3(\alpha+1)}\,. \tag{13.3.16}$$

In our case, the coefficient α depends on the mean square velocity of the walls $\langle \dot{\mathbf{x}}^2 \rangle$. The value suggested by numerical simulations of walls in a matter-dominated universe ($a \propto t^{2/3}$) is $\langle \dot{\mathbf{x}}^2 \rangle \sim 0.16$ [Press, Ryden & Spergel, 1989]. As we shall see shortly, the expansion of a wall-dominated universe is much faster than $t^{2/3}$, so we can expect $\langle \dot{\mathbf{x}}^2 \rangle$ to be even smaller. To a good approximation, then, we can use the equation of state (13.3.14) for non-relativistic walls, and (13.3.16) implies [Kobzarev *et al.*, 1974]

$$a(t) \propto t^2\,. \tag{13.3.17}$$

In a universe expanding as t^2, the average distance between the walls also grows as t^2 and rapidly becomes greater than the horizon. This happens at $t \sim (G\sigma)^{-1}$. Equation (13.3.17) may still apply in this case to describe the universe on very large scales, but of course the situation seen by a local observer will be quite different. An inertial observer will see domain walls moving away towards the horizon. As the walls fade away with increasingly high redshifts, the spacetime around the observer will asymptotically approach Minkowski space.

13.4 Interaction with particles

13.4.1 Reflection probability

Particle scattering by a planar domain wall reduces to a one-dimensional scattering problem. The essential features are revealed for the simplest case of scalar particles interacting with a wall in the ϕ^4-model (13.1.7). Suppose the particles are described by a field χ which is coupled to the field of the wall ϕ through

$$\mathcal{L}_{\text{int}} = -\tfrac{1}{2}\tilde{\lambda}\phi^2\chi^2\,, \tag{13.4.1}$$

where a tilde has been placed over λ to distinguish it from the self-coupling of ϕ. With a plane wave ansatz, $\chi = \chi(z)\exp(-i\omega t + ik_x x + ik_y y)$, the field equation for $\chi(z)$ can be written as

$$\chi'' + k_z^2 \chi - U(z)\chi = 0, \tag{13.4.2}$$

where $k_z^2 = \omega^2 - k_x^2 - k_y^2 - m_\chi^2$ and m_χ is the mass of χ far away from the wall ($m_\chi = \tilde{\lambda}^{1/2}\eta$ if χ obtains all its mass from the interaction with ϕ). The potential $U(z)$ is given by

$$U(z) = \tilde{\lambda}[\phi^2(z) - \eta^2] = -\tilde{\lambda}\eta^2 \cosh^{-2}[(\lambda/2)^{1/2}\eta z], \tag{13.4.3}$$

where $\phi(z)$ is the domain wall solution (13.1.8).

Equation (13.4.2) can be solved exactly [Landau & Lifshitz, 1977]. The reflection coefficient for the χ-particles has the form

$$R = \frac{\cos^2 \beta}{\sinh^2 \alpha + \cos^2 \beta}, \tag{13.4.4}$$

with

$$\alpha = \left(\frac{2}{\lambda}\right)^{1/2}\frac{\pi k_z}{\eta}, \qquad \beta = \frac{\pi}{2}\left(1 + \frac{8\tilde{\lambda}}{\lambda}\right)^{1/2}. \tag{13.4.5}$$

If the particle wavelength is much smaller than the thickness of the wall, $k_z \gg \delta^{-1} \sim \sqrt{\lambda}\,\eta$, then $\alpha \gg 1$, and the reflection coefficient is exponentially small, while in the opposite limit with $k_z \to 0$, $R \approx 1$ and almost all particles are reflected.

In the long-wavelength limit, the wall potential (13.4.3) can be adequately approximated by a δ-function,

$$U(z) \approx -\nu\delta(z), \tag{13.4.6}$$

where

$$\nu = -\int_{-\infty}^{\infty} U(z)dz = 2(2/\lambda)^{1/2}\tilde{\lambda}\eta. \tag{13.4.7}$$

The scattering problem with a δ-function potential is easily solved; the reflection coefficient is

$$R = \frac{\nu^2}{4k_z^2 + \nu^2}. \tag{13.4.8}$$

The thin-wall approximation (13.4.6) is justified if the particle wavelength is much greater than δ both inside and outside the wall. For small k_z, the wavelength inside the wall is $[U(0)]^{-1/2} \sim \tilde{\lambda}^{-1/2}\eta^{-1}$, while the wavelength outside is $\sim k_z^{-1}$; thus we must require

$$\tilde{\lambda} \ll \lambda, \qquad k_z \delta \ll 1. \tag{13.4.9}$$

It is easily verified with the above conditions that (13.4.8) is in agreement with (13.4.4).

An interesting special case is the scattering of ϕ-particles off a ϕ-wall. The equation for linear perturbations about the wall solution has the form (13.4.2) with

$$U(z) = -3\lambda\eta^2 \cosh^{-2}\left[(\lambda/2)^{1/2}\eta z\right], \qquad (13.4.10)$$

and the reflection coefficient can be found by replacing $\tilde{\lambda}$ with 3λ in (13.4.4–5). Surprisingly, this gives $\beta = 5\pi/2$ and $R = 0$, so there is no reflection at all! (see, for example, Jackiw [1977]). Similarly, it can be shown that θ-particles are not reflected from the sine-Gordon wall (13.1.13).

A more general case when ϕ and χ are multi-component fields has been studied by Everett [1974]. He assumed that the VEV of ϕ gives masses $m_\chi \sim \tilde{\lambda}^{1/2}\eta$ to some members of the χ-multiplet, while other members remain massless. In general, different components of χ will acquire masses on the two sides of the wall (since the VEVs of ϕ are different). Suppose that a massless χ-particle is moving towards the wall with momentum **k**. Everett has shown that, as in our simple model, the outcome of scattering depends on the relative size of the particle wavelength $2\pi/k_z$ and the wall thickness δ. A long-wavelength particle is reflected from the wall with a very high probability, while in the short-wavelength limit the massless χ-state on one side of the wall is adiabatically rotated to a massless state on the other side, and the reflection probability is nearly zero. The transition between the two regimes is at $k_z \sim \nu$, where

$$\nu \sim \min\left(\frac{m_\chi^2}{m_\phi}, m_\phi\right) \qquad (13.4.11)$$

and $m_\phi \sim \sqrt{\lambda}\eta$ is the ϕ-particle mass. The same conclusions apply if the light particles are not strictly massless, but have a small mass $m \ll m_\chi$.

For walls bounded by strings, the particle masses must be the same on the two sides of the wall, since particles can get from one side to the other by going around a string. But particle masses will generally vary inside the wall, with light members of the multiplet acquiring masses $\sim m_\chi$. The reflection coefficient will then behave in the same manner as when the masses on the two sides of the wall are different.

13.4.2 Frictional force

We can now estimate the frictional force acting on a domain wall moving with velocity v relative to the radiation background. For simplicity we shall assume that the wall is not ultra-relativistic, so that $(1-v^2)^{-1/2} \sim 1$.

At low temperatures when $T \ll \nu$, with ν given in (13.4.11), almost all particles are reflected from the wall. The density of particles per spin degree of freedom is $n \sim T^3$, the momentum transfer per collision is $\Delta p \sim Tv$ (averaged over the two sides of the wall), and we can express the force per unit area on the wall as [Kibble 1976]

$$F \sim N_{\mathrm{w}} n T v \sim N_{\mathrm{w}} T^4 v, \qquad (13.4.12)$$

where N_{w} is the number of light particle states changing their mass across the wall.

In the high-temperature regime $T \gg \nu$, only particles with momentum $k_z \lesssim \nu$ are reflected, and the average momentum transfer is $\sim \nu v$. The resulting force per unit area is

$$F \sim N_{\mathrm{w}} \nu^2 T^2 v. \qquad (13.4.13)$$

If the wall is described by a singlet field ϕ, it will typically interact only with massive particle states. The frictional force is then given by (13.4.12–13) at $T > m_\chi$, but for $T \ll m_\chi$ the χ-particles are not present in the thermal bath and the force is exponentially suppressed. For axionic walls, there can be a large non-thermal density of axions at low temperatures (refer to §12.2.3). But the frictional force still vanishes, since axions pass through axion walls without reflection.

13.5 Cosmological evolution

13.5.1 Domain wall formation

Much of what has been said in §9.1 about the formation of strings at a phase transition is equally applicable to domain walls. The initial shape of the walls after the phase transition is determined by the random variation of the scalar VEV, $\langle \phi \rangle$. Hence, one expects the walls to be very irregular, random surfaces with a typical curvature radius $\sim \xi$, where ξ is the correlation length of $\langle \phi \rangle$. Although the details depend on the complicated dynamics of the phase transition, the essential features of the resulting wall system can be seen in a simple Monte Carlo simulation [Harvey, Kolb, Reiss & Wolfram, 1982; Vachaspati & Vilenkin, 1984].

For Z_2-walls, a cubic volume of size $N\xi$ is divided into N^3 cubic cells and to each cell a number $+1$ or -1 is assigned at random with equal probability (these values correspond to $\langle \phi \rangle = \pm \eta$ in the model (13.1.7)). These are called plus-cells and minus-cells, respectively. The walls lie on the boundaries between cells of opposite sign.

To characterize the system of domain walls, one can look for the size distribution of clusters of connected plus-cells. (Two cells are connected if they have a common face, and the size of a cluster is the number of cells in the cluster.) Of course, the distribution of minus-cell clusters will be similar. This is a typical problem of percolation theory,* which is concerned with the statistical properties of systems like ours for various types of lattices at different concentrations of plus-cells, p. The central concept of percolation theory is the critical concentration p_c at which an infinite plus-cluster first appears (in an infinite lattice). The value of p_c is different for different lattices, but in all three-dimensional cases it is smaller than 0.5. (For a cubic lattice, $p_c = 0.31$.) In our case, $p = 0.5$, and so the system is above the percolation threshold.

The following properties of percolating systems are known for $p > p_c$: (i) There is only *one* infinite plus-cluster (and one infinite minus-cluster if $p < 1 - p_c$). (ii) The number density of finite clusters, n_s, decreases exponentially with their size s,

$$n_s \propto s^{-\tau} \exp(-\alpha s^{2/3}), \qquad (13.5.1)$$

where the numerical coefficients τ and α depend on p.

The implications of these results for a system of domain walls are straightforward. The system is dominated by *one* infinite wall of very complicated topology [Kibble, 1976]. In addition, there are some finite closed walls, mostly of spatial extent $R \sim \xi$. The probability of finding closed walls with $R \gg \xi$ exponentially decreases with R, $\ln n \propto -R^2$. In a typical simulation on a lattice of size $N = 40$, approximately 98% of all plus-cells and 87% of the total wall area belong to the 'infinite' cluster.

Models with more than two degenerate vacua can give rise to cellular structures with linear junctions where several walls meet and with 'knots' connecting several junctions. Depending on the model, the linear junctions may or may not contain strings. For example, the non-abelian model (iii) of §13.1.1 has three-wall junctions without strings and four-junction knots, while the Z_N-model (ii) has N-wall junctions with strings and no knots. As in the Z_2 case, cellular wall structures are expected to be dominated by a single, infinite, connected wall-cluster.

13.5.2 Evolution

Z_2-walls

Once domain walls are formed, we now consider how the system will evolve. The tension σ in convoluted walls produces a force per unit area $f \sim \sigma/R$, where R is the mean curvature radius. In vacuum the walls would rapidly develop relativistic speeds, but in the early universe their motion is resisted by a damping force F_d due to particle scattering. This force is given by (13.4.12) or (13.4.13), depending on the relative

---

* For a review of percolation see Stauffer [1979].

magnitude of T and ν, where ν is defined in (13.4.11). If the self-coupling of the ϕ-field describing the wall and its couplings to other fields are not very small, then $\nu \sim \eta$, and $F_{\rm d}$ is given by (13.4.12) at all times after wall formation. Omitting numerical factors, and using $T^4 \sim \rho \sim 1/Gt^2$, we can write $F_{\rm d} \sim T^4 v \sim v/Gt^2$, with wall velocity v.

At very early times the walls are over-damped and their typical velocity v is determined by the balance between tension and friction, $f \sim F_{\rm d}$:

$$v \sim G\sigma t^2/R. \tag{13.5.2}$$

Small-scale irregularities on the walls are damped out, and the typical curvature radius of walls at time t is $R(t) \sim vt$. Substituting this in (13.5.2) we find [Kibble, 1976]

$$R(t) \sim (G\sigma)^{1/2}t^{3/2}, \tag{13.5.3}$$
$$v(t) \sim (G\sigma)^{1/2}t^{1/2}. \tag{13.5.4}$$

The walls reach relativistic speeds at the time

$$t_* \sim (G\sigma)^{-1} \tag{13.5.5}$$

when their characteristic length-scale becomes comparable to the horizon.

The contribution of domain walls to the energy density of the universe is

$$\rho_{\rm w} \sim \sigma R^2/R^3 \sim \sigma/R, \tag{13.5.6}$$

so that, for $t < t_*$,

$$\rho_{\rm w}/\rho \sim (t/t_*)^{1/2}. \tag{13.5.7}$$

At $t \sim t_*$, $\rho_{\rm w} \sim \rho$ and the universe becomes dominated by domain walls.

Suppose instead that the field ϕ of the wall is very weakly coupled, so that $\nu \ll \eta$. In this case, for $\nu \lesssim T \lesssim \eta$, we have $F_{\rm d} \sim \nu^2 m_{\rm pl} v/t$, and (13.5.3) is replaced by $R \sim vt$ with

$$v \sim \left(\frac{\sigma}{\nu^2 m_{\rm pl}}\right)^{1/2} \tag{13.5.8}$$

or $v \sim 1$, whichever is smaller. For $\sigma/\nu^2 m_{\rm pl} \ll 1$, we have $v \ll 1$, and (13.5.8) turns smoothly into (13.5.4) at $T \sim \nu$. In the opposite limit with $\sigma/\nu^2 m_{\rm pl} \gg 1$, the walls move relativistically from the outset. In this case, the main energy loss mechanism is through wall collisions and reconnections, which are accompanied by the copious production of ϕ-particles. The expansion of the universe will smooth out sub-horizon irregularities on the walls. However, the actual efficiency of this process is unknown and it is possible that walls – like strings – will develop significant small-scale structure. By analogy with string evolution, we expect the evolution of domain walls to be scale-invariant, with the typical scale of the large walls comparable to the horizon. The energy density of the walls is then given by

$$\rho_{\mathrm{w}} \sim \sigma/t, \tag{13.5.9}$$

$$\rho_{\mathrm{w}}/\rho \sim G\sigma t. \tag{13.5.10}$$

Once again, the universe becomes wall-dominated at $t \sim t_*$. We note that the characteristic length of the walls cannot become greater than the horizon. Otherwise correlations in the field ϕ would be able to establish on scales greater than t, in contradiction with causality.

In studying cosmic string evolution in chapter 9, we paid much attention to the evolution of closed loops. Loops are of interest because of the existence of long-lived non-intersecting loop trajectories. No such trajectories are known for closed domain walls, and it is intuitively clear that a rather special configuration is required to make a closed wall non-intersecting. For example, one can try to stabilize a toroidal wall by setting up travelling waves in the directions of large and small circles. However, even if non-intersecting walls exist, they will inevitably be destroyed by collisions with horizon-crossing walls.

Numerical simulations of domain wall evolution have been performed by Press, Ryden & Spergel [1989]. Rather than idealize the walls as infinitely thin sheets (the approach taken by Kawano [1990]) they studied the evolution of the scalar field throughout space. This gives a much more realistic treatment of wall reconnections and particle production. However, for computational reasons they had to modify the field equation (13.2.28) so that the wall thickness was time-dependent growing in proportion to the scale factor.* A potential danger of this modification is that it can erase small-scale structure on the walls, also destroying small closed walls when the wall thickness becomes comparable to their size.

Domain wall simulations were performed in two and three dimensions. The initial values of ϕ were chosen randomly to lie between $-\eta$ and $+\eta$ and the initial value of $\dot{\phi}$ was set equal to zero everywhere. Although not particularly realistic, these initial conditions do serve the purpose of pushing the field ϕ into the different minima of the potential in different regions of space. The late-time behaviour of the system was not sensitive to the choice of initial state. The main conclusion from the simulations is that the wall annihilation and reconnection processes are very efficient and that at any time there is only about one large wall per horizon (see

---

* Without this modification the wall thickness decreases as a^{-1} in comoving coordinates, rapidly becoming smaller than the grid-size resolution of the simulation.

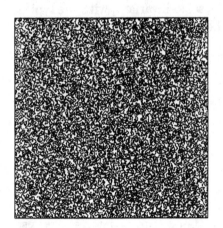 

Fig. 13.2.   A two-dimensional domain wall network shown when the horizon is 1/100 and 1/10 of the box-size. Slices through three-dimensional simulations have a very similar appearance (from Press *et al.* [1989]).

fig. 13.2). In a matter-dominated universe ($a \propto t^{2/3}$), the mean square velocity of the walls is $v \approx 0.4$ and is roughly independent of time, while the time evolution of the wall energy density is given by

$$\rho_{\mathrm{w}} \propto t^{-\nu} \qquad\qquad (13.5.11)$$

with $\nu \approx 0.96$. The deviation from $\nu = 1$, as in (13.5.9), is intriguing, but given the limited dynamical range of the simulations, its significance is not clear.

Z_N and non-abelian walls

Ryden, Press & Spergel [1990] also performed simulations of Z_N-string–wall structures in which each string is attached to N domain walls (as in Example (ii) in §13.1.1). One might expect that under the action of the wall tension, the strings could reach equilibrium positions, so that the whole network would 'freeze', with walls being stretched between the strings by the cosmological expansion. For example, an equilibrium hexagonal structure might be formed in a two-dimensional model with $N = 3$. However, simulations indicate that the string and wall annihilation processes in Z_N-models are still rather efficient, and no tendency towards 'freezing' has been observed. A snapshot of a two-dimensional simulation for $N = 5$ is shown in fig. 13.3. The evolution of the wall energy density in a Z_N-network can still be fitted by (13.5.11) with ν slowly decreasing as N is increased. This agrees with the intuitive expectation

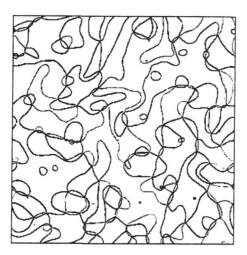

Fig. 13.3. Two-dimensional Z_N-string–wall system for $N = 5$ at the time when the horizon is $1/8$ times the size of the box. Different types of walls are shown using different shades of grey (from Ryden *et al.* [1990]).

that for higher values of N the energy of the system should be dissipated less efficiently.

Simulations of non-abelian wall structures arising in models like Example (iii) in §13.1.1 have been performed by Kubotani [1992]. In this case for sufficiently large values of N, the energy dissipation rate is much smaller than in Z_N-models, and there is even some evidence for the for-

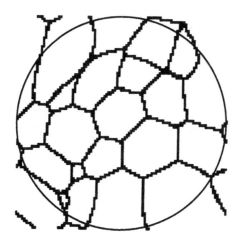

Fig. 13.4. A non-abelian wall structure in a model with approximate $O(7)$ symmetry (from Kubotani [1992]).

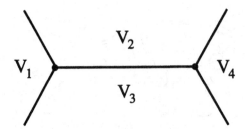

Fig. 13.5. These triple wall junctions can annihilate only if the vacuum V_4 is the same as V_1.

mation of equilibrium hexagonal structures (see fig. 13.4). The difference is due to the large number of different types of wall junctions in the non-abelian case (in contrast to just two types corresponding to strings and antistrings in the Z_N case). As a result, nearby junctions seldom annihilate, even when they are pulled together by the walls. For example, the junctions in fig. 13.5 will be able to annihilate only if the vacuum V_4 is the same as V_1. In a model with $2N$ degenerate vacua the corresponding probability is $[2(N-2)]^{-1}$.

For all types of domain walls, we conclude that the energy density of the wall system decreases no faster than (13.5.9), and wall domination is predicted no later than the time t_* given by (13.5.5). If there are domain walls in our universe, then the universe would already be wall-dominated, unless

$$(G\sigma)^{-1} > t_0 , \qquad (13.5.12)$$

where t_0 is the present time. With $\sigma \sim \eta^3$, where η is the symmetry breaking scale of the walls, this implies $\eta < 100\,\mathrm{MeV}$.

13.5.3 Observational bounds

A stringent constraint on the wall tension σ can be derived from the isotropy of the microwave background. Consider first Z_2-walls: If the interaction of walls with matter is negligible, then there will be one or only a few walls stretching across the present horizon. They introduce a density fluctuation

$$\frac{\delta\rho}{\rho} \sim G\sigma t_0 \sim 10^{60}(\eta/m_{\mathrm{pl}})^3 \qquad (13.5.13)$$

and a comparable fluctuation in the temperature of the microwave background. Observations constrain $\delta T/T \lesssim 10^{-5}$, and thus models predicting topologically stable domain walls with $\eta \gtrsim 1\,\mathrm{MeV}$ should be ruled out [Zel'dovich et al., 1975]. Background temperature fluctuations produced by the walls on sub-horizon scales have been discussed by Stebbins & Turner [1989] and Turner, Watkins & Widrow [1991].

The stringent bound (13.5.13) apparently rules out any scenarios [Hill, Schramm & Fry, 1989] in which light domain walls generate cosmologically significant density fluctuations [Stebbins & Turner, 1989; Press *et al.*, 1989]. The density fluctuations on a comoving scale of present size l were produced by the walls at time

$$t_l \sim l^3/t_0^2, \tag{13.5.14}$$

when that scale crossed the horizon. Initially, at the time t_l,

$$\delta\rho/\rho(t_l) \sim G\sigma t_l, \tag{13.5.15}$$

and subsequently $\delta\rho/\rho$ will grow as $t^{2/3}$ by (1.3.39). At present, therefore, we can expect

$$\frac{\delta\rho}{\rho} \sim G\sigma t_l(t_0/t_l)^{2/3} \sim G\sigma l. \tag{13.5.16}$$

From the microwave isotropy bound $G\sigma t_0 < 10^{-5}$, and so $\delta\rho/\rho < 10^{-5}$ on all scales in the visible universe. This is obviously very unsatisfactory, because to form some bound objects, like galaxies, we need to have $\delta\rho/\rho \gtrsim 1$ on the corresponding scales.

Massarotti [1991a] has argued that this negative verdict for the light-domain-wall scenario can be avoided if the walls strongly interact with surrounding matter. In this case, the motion of the walls is severely damped and the characteristic wall separation can be much smaller than the horizon. Another way out may be to invoke non-abelian walls which can form equilibrium cellular structures [Kubotani, 1992]. The microwave background distortion in this case can be much smaller than that for relativistically moving walls [Nambu *et al.*, 1991].

13.5.4 Avoiding wall domination

There are some exceptions to the cosmological constraint that the energy scale of domain walls cannot exceed 1 MeV. If the formation of walls is followed by inflation, the walls can be inflated away far beyond the present Hubble radius.

Wall domination can also be avoided in models where a discrete symmetry is broken and then restored at a lower temperature. As we saw in §3.2.5, the properties of topological defects in this type of models are somewhat different from the usual case. Most importantly, the wall tension σ is temperature-dependent. If the discrete symmetry is broken at a temperature T_1 and restored at T_2, then for $T_2 \ll T \ll T_1$,

$$\sigma \sim \lambda^{-1}T^3, \tag{13.5.17}$$

where λ is the quartic scalar field coupling. Combining (13.5.17), (13.5.5) and (13.5.7) and omitting numerical factors and coupling constants, we find that the walls are always over-damped and that

$$\rho_{\rm w}/\rho \sim (t_{\rm pl}/t)^{1/4} \ll 1 \qquad (13.5.18)$$

for $t \gg t_{\rm pl}$, where $t_{\rm pl}$ is the Planck time. Hence, in models of this type the universe never becomes wall-dominated.

Yet another possibility is to allow a small bias, so that one of the two vacuum states separated by the walls has a slightly smaller energy density than the other: $\Delta\rho_{\rm v} = \epsilon \neq 0$. This is only an approximate discrete symmetry. Regions of the higher density vacuum tend to shrink, the corresponding force per unit area on the walls is $\sim \epsilon$. The energy difference ϵ becomes dynamically important when this force becomes comparable to the tension, $f \sim \sigma/R$. To prevent wall domination, this has to happen when $R < (G\sigma)^{-1}$. This requirement gives a lower bound for the asymmetry ϵ [Vilenkin, 1981b; Gelmini, Gleiser & Kolb, 1989]

$$\epsilon > G\sigma^2 . \qquad (13.5.19)$$

What happens, then, if $\epsilon < G\sigma^2$? At $t \sim t_*$ the universe enters a period of wall domination. The vacuum pressure on the walls becomes dynamically important when the distances between the walls are $R \sim \sigma/\epsilon$. Subsequent events look different from regions where $\rho_{\rm v} = 0$ and where $\rho_{\rm v} = \epsilon$. The regions with $\rho_{\rm v} = \epsilon$ are dominated by the false vacuum energy. The corresponding de Sitter horizon is

$$H^{-1} \sim (G\epsilon)^{-1/2} < R . \qquad (13.5.20)$$

Since the Hubble radius is smaller than the inter-wall separation, observers in a false vacuum region will find themselves in an eternally expanding de Sitter space. On the other hand, observers in a region with $\rho_{\rm v} = 0$ will see the false vacuum regions bounded by the walls collapse to form black holes. Soon afterwards, the universe will become black hole dominated. Black holes whose interiors are inflationary universes have been discussed by Sato, Sasaki, Kodama & Maeda [1981], Farhi, Guth & Guven [1990] and others.

Finally, wall domination can be avoided if the discrete symmetry responsible for the walls is embedded in a continuous symmetry group. Models of this kind give rise to domain walls bounded by strings. As we shall see in §13.6, these hybrid defects decay well before they can dominate the energy density of the universe.

13.6 Evolution of walls bounded by strings

Walls bounded by strings are formed in two steps. Strings form at an earlier phase transition and approach scale-invariant evolution as discussed in chapter 9. A later phase transition gives rise to walls which are bounded by the strings, together with some closed walls. A Monte Carlo simulation of these phase transitions was performed by Vachaspati & Vilenkin [1984] for the case when the walls are formed soon after the strings, so that the two correlation lengths are comparable. The results suggest that the system is dominated by one infinite cluster comprising about 90% of the total wall area and string perimeters. This cluster has a complicated topology and is very 'holey', so that its intersection with a plane gives a large number of short pieces. Some finite walls bounded by strings are also formed in numbers rapidly decreasing as their size increases. Closed domain walls are very rare. If the walls are formed much later than the strings, then the properties of the system are expected to change on scales smaller than the horizon and to remain essentially unchanged on larger scales.

The reconnection properties of domain walls and strings will determine the subsequent evolution of the hybrid network. These interactions have been studied numerically for the axion model (13.1.11) which possesses both global $U(1)$-strings and sine-Gordon soliton walls [Shellard, 1986a]. We have already noted that $U(1)$-strings inevitably reconnect (§4.3.3), as do domain walls (§13.2.4). The key interaction here is that between a string and a domain wall which is illustrated in fig. 13.6(a–c). The vortex-string (with a forming domain wall attached) approaches the wall (a) and 'slices' it in half (b). The string 'exchanges partners' with one wall section and is drawn away by the wall tension (c). The remaining wall sections join together. The generic nature of the interaction is predetermined by the string twist direction and the phase gradient of the wall and it has been observed at all velocities (up to $v \approx 0.9$).

Turning now to the cosmological evolution of the string–wall network, we have to assess the relative importance of the various forces acting on strings and walls. The tension in a string of curvature radius R, $F \sim \mu/R$, is smaller than the wall tension σ for $R > \mu/\sigma$. Since the typical scale of strings $R(t)$ grows with time, the walls eventually dominate the dynamics of the system.

The importance of damping forces acting on strings and walls depends on the relative magnitude of the temperature T, the symmetry breaking scales $\eta_{\rm s}$ and $\eta_{\rm w}$, and the Planck mass $m_{\rm pl}$, as well as on the strength with which the field ϕ of the wall is coupled to itself and to other fields (refer to §13.4.2). The number of possible cases is somewhat intimidating, and we shall only consider the situation when both damping forces are negligible. This case arises, for instance, if ϕ is a strongly coupled field interacting only with heavy particles of mass $\sim \eta_{\rm w}$. The wall tension

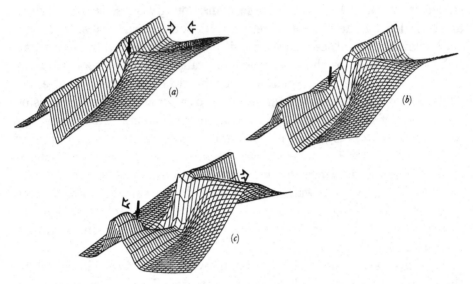

Fig. 13.6. Reconnection during the collision of a domain wall and a string (with a domain wall attached) [Shellard, 1986a]. The real part of the scalar field ϕ in (13.1.11) is plotted and the location of the string zero is marked with an arrow.

is then $\sigma \sim \eta_{\rm w}^3$, and since the walls are formed at $T \sim \eta_{\rm w}$, it is always greater than the string friction, $F_{\rm s} \sim T^3 v$. The wall friction is also very small, since particles coupled to the walls disappear from the thermal bath at $T < \eta_{\rm w}$. Another example is very weakly coupled walls with $\sigma/\nu^2 m_{\rm pl} \gg 1$ (see (13.5.8)). The wall friction is then negligible, and the string friction is also negligible, provided the walls are formed after the end of the string 'damped epoch', $t > (G\mu)^{-2} t_{\rm pl}$ from (9.2.3). The condition for this is $\eta_{\rm w} m_{\rm pl}/\eta_{\rm s}{}^2 < 1$. Such a string–wall system will rapidly disappear [Vilenkin & Everett, 1982].

Let $R_{\rm i}$ be the initial value of the string correlation length $R(t)$ at the time of wall formation. If $R_{\rm i} \ll \mu/\sigma$, then initially the walls have little effect on string dynamics. Since we know that friction is even less important, the strings will move relativistically with $R(t) \sim t$. Similarly, in the absence of friction, the motion of the domain walls will also be relativistic, and their characteristic scale will be comparable to the horizon. The walls become dynamically important at a time $t_{\rm w} \sim \mu/\sigma$. At later times, the walls will tend to shrink pulling the strings towards one another, and the holes in the walls will grow in size. The strings will frequently intersect and intercommute with both walls and strings, as demonstrated numerically (see fig. 13.6). As a result, the walls connecting the strings will be rapidly cut into pieces of size hardly exceeding $t_{\rm w}$.

The motion of wall pieces bounded by strings is not periodic, and we do not expect any non-self-intersecting trajectories. Consequently, these pieces will further fragment by multiple self-intersections. When the size of the pieces becomes much smaller than μ/σ, their dynamics will be determined mainly by the strings, and non-intersecting trajectories may be possible. The lifetime due to gravitational radiation of a piece of size R is (see (10.1.5))

$$\tau \sim R/G\mu \ll (G\sigma)^{-1}. \tag{13.6.1}$$

Other energy loss mechanisms, such as particle production or friction, can reduce this estimate, but in any case (13.6.1) gives an upper bound on τ.

In conclusion, the string–wall network will completely decay well before the time $t_* \sim (G\sigma)^{-1}$. The ratio of the mass density of the network to the total density of the universe will never exceed

$$\left.\frac{\rho_{\mathrm{ws}}}{\rho}\right|_{\max} \sim \left(\frac{\tau}{t_{\mathrm{w}}}\right)^{1/2} G\sigma t_{\mathrm{w}} \ll \left(\frac{t_{\mathrm{w}}}{t_*}\right)^{1/2} \ll 1, \tag{13.6.2}$$

and thus these vacuum structures will never dominate the universe.

If $R_{\mathrm{i}} > \mu/\sigma$, the walls become dynamically important immediately after their formation, and the system breaks up into pieces of size $\sim R_{\mathrm{i}}$. Note that in this case R_{i} is not necessarily comparable to the horizon, since the walls can form during the period when the strings are over-damped. Once again, it can be easily verified that the pieces decay before they dominate the energy density. An extension of this analysis to the case when the wall friction is significant has been discussed with qualitatively similar conclusions by Everett & Vilenkin [1982] and Kibble, Lazarides & Shafi [1982b].

Shellard [1986a, 1990] and Ryden, Press & Spergel [1990] have performed two-dimensional numerical simulations of domain walls bounded by strings for the model (13.1.11) with $N = 1$. The coupling of strings and walls to gravity and to thermal particles was not included, and the dominant decay mechanism of oscillating wall pieces was particle production. As expected, all defects rapidly disintegrated into particles. A snapshot of the simulation by Ryden *et al.* [1990] is shown in fig. 13.7.

Although the hybrid defects are not expected to produce any direct observational effects, their decay can give rise to an observable spectrum of gravitational waves. If the constituent fields of strings and walls have baryon-number-violating decays, it may also be possible to construct scenarios in which they account for the observed baryon asymmetry of the universe.

Fig. 13.7. Decaying wall–string network in two dimensions [Ryden *et al.*, 1990].

The evolution of walls bounded by strings can be modified if there is a period of inflation between the two phase transitions. Inflation can push strings out to arbitrarily large scales; then the evolution of walls will proceed as discussed in §13.5. The only difference from topologically stable Z_2-walls is that the walls can now decay by the quantum nucleation of holes bounded by strings. Hole nucleation, however, is a tunnelling process and is typically suppressed by a large exponential factor (see §13.2.5). The corresponding decay time can be much larger than the time $t_* \sim (G\sigma)^{-1}$ at which the walls come to dominate the universe.

14

Monopoles

14.1 Field theory

Localized defects, or monopoles, arise if the manifold $\mathcal{M}$ of degenerate vacua contains non-contractible two-surfaces. For any field configuration of finite energy, the Higgs field at spatial infinity takes values in the vacuum manifold, and thus the two-sphere of infinite radius is mapped into a two-dimensional surface in $\mathcal{M}$. If this surface cannot be smoothly contracted to a point, a monopole must be present and the Higgs field will depart from the vacuum manifold in the monopole core.

Like strings, monopoles which have formed as a result of gauge symmetry breaking will carry a unit of magnetic flux. However, for strings the flux corresponds to one of the broken symmetry generators and is confined to the string core, while monopoles have long-range radial magnetic fields corresponding to unbroken symmetry generators. Hence, the monopoles are in fact magnetic monopoles whose possible existence was suggested by Dirac [1931] long before the appearance of spontaneously broken gauge theories.

This section gives an overview of the basic properties of monopoles. The literature on monopoles is very extensive and the reader who is interested in further details and references is referred to excellent reviews by Coleman [1983] and Preskill [1985].

14.1.1 The 't Hooft–Polyakov monopole

The existence of monopole solutions was first demonstrated by 't Hooft [1974] and Polyakov [1974, 1975] in a gauge model possessing $SO(3)$ symmetry with a Higgs field ϕ in a triplet representation (see *Example 2* in §2.1.3). The potential $V(\phi)$ is chosen so that ϕ develops a vacuum expectation value with $\phi^a \phi^a = \eta^2$, for example,

$$V(\phi) = \tfrac{1}{4}\lambda(\phi^a \phi^a - \eta^2)^2 \, . \tag{14.1.1}$$

The $SO(3)$ symmetry is then spontaneously broken to $U(1)$, and the vacuum manifold is $\mathcal{M} \cong S^2$. The unbroken $U(1)$ is the group of rotations about ϕ^a; it can be identified with the gauge group of electromagnetism.

The monopole corresponds to the simplest topologically non-trivial field configuration in which the Higgs field points in the radial direction,

$$\phi^a = \eta h(r) \frac{x^a}{r}, \qquad (14.1.2)$$

and the gauge field is

$$A_i^a = -[1 - K(r)]\epsilon^{aij} \frac{x^j}{er^2}, \qquad A_0^a = 0. \qquad (14.1.3)$$

At spatial infinity ϕ should approach the vacuum value and the covariant derivative of ϕ should vanish,

$$D_\mu \phi^a = (\partial_\mu \phi^a - e\epsilon^{abc} A_\mu^b \phi^c) \to 0, \qquad r \to \infty. \qquad (14.1.4)$$

Hence, the asymptotic behaviour of the functions $h(r)$ and $K(r)$ should be

$$h(r) \to 1, \qquad K(r) \to 0, \qquad r \to \infty. \qquad (14.1.5)$$

Note that in the absence of gauge fields the energy of the Higgs field (14.1.2) would be linearly divergent at large distances. At the origin, regularity requires that

$$h(0) = 0, \qquad K(0) = 1. \qquad (14.1.6)$$

Unlike the case of a string, the vector field (14.1.3) is not pure gauge, even in the asymptotic region. With $h(r) \approx 1$, the field tensor is given by

$$F_{ij}^a \approx \frac{x^a}{r} \epsilon_{ijk} \frac{x^k}{er^3}, \qquad F_{0i}^a = 0. \qquad (14.1.7)$$

Note that $F_{\mu\nu}^a$ is proportional to ϕ^a, indicating that the field (14.1.7) corresponds to the unbroken symmetry generator. To make this correspondence more precise, 't Hooft [1974] suggested the following definition for the electromagnetic field tensor,

$$\mathcal{F}_{\mu\nu} = \frac{\phi^a}{|\phi|} F_{\mu\nu}^a + \frac{1}{e|\phi|^3} \epsilon^{abc} \phi^a (D_\mu \phi^b)(D_\nu \phi^c). \qquad (14.1.8)$$

In the gauge where ϕ points in the same direction everywhere,

$$\phi^a = \delta^{a3}|\phi|, \qquad (14.1.9)$$

the terms quadratic in A_μ^a drop out, and the definition (14.1.8) reduces to

$$\mathcal{F}_{\mu\nu} = \partial_\mu A_\nu^3 - \partial_\nu A_\mu^3. \qquad (14.1.10)$$

It can also be verified that $\mathcal{F}_{\mu\nu}$ satisfies the ordinary Maxwell equations at all points where $\phi \neq 0$.

In the presence of a monopole, the insertion of (14.1.2) with $h(r) \approx 1$ and (14.1.7) into (14.1.8) yields

$$\mathcal{F}_{ij} \approx \epsilon_{ijk}\frac{x^k}{er^3}, \quad \mathcal{F}_{0i} = 0. \tag{14.1.11}$$

The magnetic field is

$$\mathbf{B} = \frac{g\mathbf{r}}{4\pi r^3}, \tag{14.1.12}$$

with a magnetic charge,

$$g = -\frac{4\pi}{e}. \tag{14.1.13}$$

14.1.2 The charge quantization condition

A relation similar to (14.1.13) between the magnetic charge g and the elementary electric charge e was first derived by Dirac [1931] who realized that the quantum mechanics of a charged particle in the presence of a monopole cannot be consistently formulated for arbitrary values of g and e. The action for a charged particle in a magnetic field is

$$S = S_0 + e\int_1^2 \mathbf{A} \cdot d\mathbf{x}, \tag{14.1.14}$$

where S_0 is the action for a free particle. The amplitude $\mathcal{A}$ for the particle to go around a closed path Γ is proportional to

$$\mathcal{A} \propto \exp\left[ie\oint_\Gamma \mathbf{A} \cdot d\mathbf{x}\right] = \exp\left[ie\int_\Sigma \mathbf{B} \cdot d\mathbf{S}\right], \tag{14.1.15}$$

where the surface Σ is bounded by Γ. The surface Σ can be chosen to lie either to the left or to the right of the monopole (see fig. 14.1) and, in order for $\mathcal{A}$ to be single-valued, the magnetic flux through a surface surrounding the monopole should satisfy

$$e\oint \mathbf{B} \cdot d\mathbf{S} = 2\pi n, \tag{14.1.16}$$

where n is an integer. With $\mathbf{B}$ from (14.1.12) this gives

$$eg = 2\pi n. \tag{14.1.17}$$

The smallest allowed magnetic charge is $g = 2\pi/e$. For a 't Hooft–Polyakov monopole the condition (14.1.17) is satisfied with $n = 2$.

Later in this chapter, when discussing electromagnetic interactions of monopoles in an astrophysical setting, we shall switch from the Heaviside

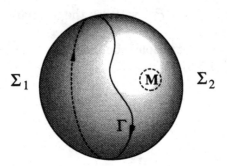

Σ_1 Σ_2

Fig. 14.1. The phase factor (14.1.15) should not depend on the choice of the surface Σ_1 or Σ_2.

units employed here to cgs units which are traditionally used in astrophysics. In these units, the fine structure constant is $\alpha = e^2$ (rather than $\alpha = e^2/4\pi$), and (14.1.12) and (14.1.17) are replaced by

$$\mathbf{B} = \frac{g\mathbf{r}}{r^3}, \tag{14.1.18}$$

$$eg = \frac{n}{2}. \tag{14.1.19}$$

14.1.3 Monopole mass and structure

The monopole solution (14.1.2–3) has two characteristic length-scales, r_s and r_v. These are, respectively, the radii of the regions in which the scalar and vector fields depart significantly from their asymptotic behaviour (14.1.5). The scales r_s and r_v and the monopole mass m can be estimated using a crude variational argument [Preskill, 1985] (compare with the discussion of the vortex in §4.1.1). The energy of a static monopole is given by

$$m = \frac{1}{2} \int d^3x \left[\mathbf{B}^a \cdot \mathbf{B}^a + \mathbf{D}\phi^a \cdot \mathbf{D}\phi^a + \frac{1}{2}\lambda(\phi^a\phi^a - \eta^2)^2 \right], \tag{14.1.20}$$

where

$$B^{ai} = \frac{1}{2}\epsilon^{ijk}F^a_{jk}. \tag{14.1.21}$$

To order of magnitude, the three terms in (14.1.20) are

$$m \sim 4\pi \left(\frac{1}{e^2 r_v} + \eta^2(r_v - r_s) + \lambda\eta^4 r_s^3 \right), \tag{14.1.22}$$

where we have assumed that $r_v > r_s$. The first term is the energy of a magnetically charged sphere of radius r_v. The second is due to the gradient energy of the scalar field in the region $r_s < r < r_v$; this term is absent when $r_s > r_v$. Finally, the last term is the contribution of the vacuum energy in the region $r < r_s$. The energy is minimized by

$$r_s \sim (\sqrt{\lambda}\eta)^{-1} = m_s{}^{-1}, \qquad r_v \sim (e\eta)^{-1} = m_v{}^{-1}, \qquad (14.1.23)$$

where m_s and m_v are the masses of scalar and vector particles, respectively. Substitution of (14.1.23) into (14.1.22) yields

$$m \sim \frac{4\pi}{e^2} m_v. \qquad (14.1.24)$$

From (14.1.23), it is clear that the assumption $r_v > r_s$ is justified only when $m_v < m_s$ or, equivalently, when $e^2 < \lambda$. In the opposite case $e^2 > \lambda$, the situation is more complicated because of the presence of a long-range scalar field, $(h - 1) \propto r^{-1}$, in the range $m_v{}^{-1} < r < m_s{}^{-1}$. It can be shown, however, that the estimate (14.1.24) still applies. In the next subsection we shall see that in the limit $e^2 \gg \lambda$ the Higgs and gauge fields and the monopole mass can be calculated exactly.

It follows from (14.1.24) that the Compton wavelength of the monopole is much smaller than the size of its core,

$$\lambda_C \sim \frac{e^2}{4\pi} m_v^{-1} \ll r_v. \qquad (14.1.25)$$

Hence, to high accuracy the monopole can be treated as a classical object.

14.1.4 The Bogomol'nyi bound and the Prasad–Sommerfield limit

For arbitrary values of e and λ, the monopole mass is bounded from below by [Bogomol'nyi, 1976]

$$m \geq \frac{4\pi}{e^2} m_v. \qquad (14.1.26)$$

It follows from (14.1.20) that

$$m \geq \frac{1}{2} \int d^3x \, (\mathbf{B}^a \cdot \mathbf{B}^a + \mathbf{D}\phi^a \cdot \mathbf{D}\phi^a)$$

$$= \frac{1}{2} \int d^3x \, (\mathbf{B}^a \pm \mathbf{D}\phi^a)^2 \mp \int d^3x \, \mathbf{B}^a \cdot \mathbf{D}\phi^a \qquad (14.1.27)$$

$$\geq \left| \int d^3x \, \mathbf{B}^a \cdot \mathbf{D}\phi^a \right|.$$

Now, since $\mathbf{D} \cdot \mathbf{B}^a = 0$, we have

$$\left| \int d^3x \, \mathbf{B}^a \cdot \mathbf{D}\phi^a \right| - \left| \oint_{r \to \infty} d\mathcal{S} \, \mathbf{n} \cdot \mathbf{B}^a \psi^a \right| = \frac{4\pi}{e^2} m_v, \qquad (14.1.28)$$

where in the last step we have used the asymptotic forms of $\mathbf{B}^a$ and ϕ^a. The bound (14.1.26) follows immediately from (14.1.27) and (14.1.28).

The Bogomol'nyi bound (14.1.26) is actually saturated in the limit $\lambda/e^2 \to 0$. In this limit, the first inequality in (14.1.27) becomes an equality, and the energy is minimized by requiring

$$\mathbf{B}^a \pm \mathbf{D}\phi^a = 0, \tag{14.1.29}$$

where the sign depends on the monopole charge. The first-order differential equations (14.1.29) can be solved in closed form,

$$h = \coth\xi - \xi^{-1}, \quad K = \xi/\sinh\xi, \tag{14.1.30}$$

where h and K are from (14.1.2–3) and we have introduced a dimensionless variable $\xi = m_v r$. The solution (14.1.30) was first found by Prasad & Sommerfield [1975] by solving the field equations in the limit $\lambda/e^2 \to 0$. The monopole mass in this limit is $m = (4\pi/e^2)m_v$. An interesting property of the solution (14.1.30) is that the force between two equally charged monopoles is equal to zero. The magnetic repulsion is balanced by the attractive force due to the long-range scalar field. Exact solutions describing several static monopoles have also been found [Forgacs, Horvath & Palla, 1981b].

14.1.5 Monopoles and grand unification

The general condition for the existence of monopoles in a model with a symmetry breaking $G \to H$ is that the vacuum manifold $\mathcal{M} = G/H$ should contain non-contractible two-surfaces or, equivalently, $\pi_2(G/H) \neq I$ (refer to §3.3.3). In grand unified theories, a semi-simple group G is broken, in several stages down to $H = SU(3) \times U(1)$. Since $\pi_2(G/H) \cong \pi_1(H)$, we have

$$\pi_2(G/H) \cong \pi_1(SU(3) \times U(1)) \cong Z, \tag{14.1.31}$$

and thus the existence of monopole solutions is an inevitable prediction of GUTs [Preskill, 1979]. This conclusion is quite general and does not depend upon the choice of the group G or the intermediate stages of the symmetry breaking.

In the general case, monopoles can have magnetic charges corresponding to several unbroken symmetry generators. For example, in realistic GUT models the lowest-mass, stable monopoles typically carry both electromagnetic and colour magnetic charges. The colour magnetic field, however, is screened at the characteristic QCD length-scale.

14.2 Interaction with particles

14.2.1 The drag force

A monopole moving through a plasma in the early universe experiences a drag force due to its interaction with charged particles. A rough estimate of this force can be obtained from the equation

$$\mathbf{F} \sim n \sigma v_{\mathrm{T}} \Delta \mathbf{p} , \qquad (14.2.1)$$

where $n \sim \mathcal{N}_c T^3$ is the particle density, $\mathcal{N}_c$ is the effective number of light charged particle species, T is the temperature, $v_{\mathrm{T}} \sim 1$ is the thermal velocity of the particles, σ is the scattering cross-section and $\Delta \mathbf{p}$ is the average momentum transfer per collision. For a monopole moving with a non-relativistic velocity $\mathbf{v}$, $\Delta \mathbf{p} \sim -T\mathbf{v}$. On dimensional grounds we expect $\sigma \sim (eg)^2 T^{-2}$, and so (14.2.1) yields

$$\mathbf{F} \sim -\mathcal{N}_c T^2 \mathbf{v} . \qquad (14.2.2)$$

To obtain a better estimate, we consider the scattering of a charged particle in the magnetic field of a monopole. As we shall see, the main contribution to the drag force comes from small scattering angles and large impact parameters, where the particle can be treated classically. In the rest frame of the monopole, the magnetic field is given by (14.1.18) and the equation of motion for a particle of charge e is

$$\frac{d\mathbf{p}}{dt} = eg \frac{\mathbf{v} \times \mathbf{r}}{r^3} . \qquad (14.2.3)$$

The momentum transfer in small-angle scattering of a relativistic particle $(v \approx 1)$ is given by (see fig. 14.2)

$$\begin{aligned}
\Delta p &= eg \int_{-\infty}^{\infty} dt \, \frac{\sin \phi}{r^2} \\
&= eg\rho \int_{-\infty}^{\infty} \frac{dt}{(\rho^2 + t^2)^{3/2}} = \frac{2eg}{\rho} ,
\end{aligned} \qquad (14.2.4)$$

where ϕ is the angle between $\mathbf{r}$ and $\mathbf{v}$, ρ is the impact parameter, and the integration is performed over the unperturbed particle trajectory. The scattering angle is

$$\theta = \frac{\Delta p}{p} = \frac{2eg}{\rho p} , \qquad (14.2.5)$$

and the differential cross-section can be found from

$$\frac{d\sigma}{d\theta} = 2\pi\rho \left| \frac{d\rho}{d\theta} \right| = \frac{8\pi(eg)^2}{p^2\theta^3} . \qquad (14.2.6)$$

The drag force can now be calculated following the same steps outlined in §8.1. The result is

$$\mathbf{F} = \beta T^2 f(u)\mathbf{u},$$

(14.2.7)

where $\mathbf{u}$ is the plasma velocity in the rest frame of the monopole, $f(u)$ is a slowly-varying function with $f(0) = 1$ and $f(1) = 3/2$,

$$f(u) = \frac{3}{2u^2}\left[1 + \frac{1-u^2}{2u}\ln\left(\frac{1-u}{1+u}\right)\right],$$

(14.2.8)

and

$$\beta = \frac{2\pi}{9}Cg^2\sum_a b_a e_a^2.$$

(14.2.9)

The summation in (14.2.9) is over the spin states of light charged particles ($m \ll T$), $b = 1$ for bosons and $b = 1/2$ for fermions, and

$$C = \int_{\theta_{\min}}^{\theta_{\max}}\frac{d\theta}{\theta}$$

(14.2.10)

In the non-relativistic limit, $u \ll 1$, (14.2.7) was obtained by Goldman, Kolb & Toussaint [1981].

The integration cut-offs in (14.2.9) are determined from the following considerations. The upper cut-off should be set at $\theta_{\max} \sim 1$, where the small-angle scattering approximation breaks down. Small scattering angles correspond to large impact parameters. Since particles scatter off each other and not only off the monopole, it is clear that impact parameters much greater than the mean free path of the particle in plasma should not contribute to the drag force. The cross-section for light particle interactions is $\sigma \sim e^4/T^2$, and the mean free path is $l \sim (n\sigma)^{-1} \sim (\mathcal{N}e^4 T)^{-1}$. Setting $\rho \sim l, p \sim T$ in (14.2.5) we obtain $\theta_{\min} \sim \mathcal{N}e^4$ and $C \approx \ln(\mathcal{N}e^4)^{-1}$. For $1 \lesssim \mathcal{N} \lesssim 100$, $C \sim 5$–10, and depending on the parameters, $\beta \sim (1$–$5)\mathcal{N}_c$, in agreement with (14.2.2).

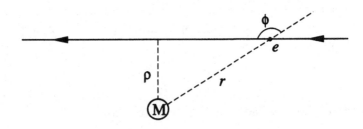

Fig. 14.2.   Small-angle scattering of a charged particle off a monopole, M.

The equation of motion for a non-relativistic monopole with friction taken into account is

$$m\dot{\mathbf{v}} = -\beta T^2 \mathbf{v}. \tag{14.2.11}$$

Its solution (disregarding cosmological expansion) is

$$\mathbf{v} = \mathbf{v}_0 e^{-t/\tau}, \tag{14.2.12}$$

where

$$\tau = m/\beta T^2. \tag{14.2.13}$$

The characteristic timescale τ is the time in which the monopole 'forgets' its initial velocity; essentially, it plays the role of a mean free time for the monopole. The mean free path of a monopole moving with a thermal velocity $v_T \sim (T/m)^{1/2}$ is then

$$l \sim (\beta T)^{-1}(m/T)^{1/2}. \tag{14.2.14}$$

Finally, we obtain a covariant generalization of (14.2.7). In the covariant formalism, the four-vector of force F^μ should be expressed in terms of the velocity four-vectors of the monopole v^μ and of the plasma u^μ. In the local rest frame of the monopole, the spatial components of F^μ should be given by (14.2.7), while the time-component of F^μ representing the rate of energy change should be equal to zero, $F^0 = 0$. A covariant expression for F^μ satisfying these requirements is easily found,

$$F^\mu = \beta T^2 f(v_{\text{rel}})(u \cdot v)^{-1}[u^\mu - (u \cdot v)v^\mu], \tag{14.2.15}$$

where v_{rel} and T are the monopole velocity and the temperature measured in the rest frame of the plasma. The velocity v_{rel} can be expressed in terms of the scalar product $u \cdot v$ as $v_{\text{rel}}^2 = 1 - (u \cdot v)^{-2}$.

14.2.2 Baryon decay catalysis

In a typical grand unified theory, the heavy gauge fields present at the monopole core have baryon-number-violating couplings. Consequently, light fermions scattering off monopoles can change their baryon number, for example

$$M + p \longrightarrow M + \pi^0 + e^+. \tag{14.2.16}$$

This reaction is essentially proton decay with the monopole playing the role of catalyst. The size of the core is $\sim m_v^{-1}$, where m_v is the gauge boson mass, and one might expect that the cross-section for B-violating scattering is no greater than the geometrical cross-section, $\sigma \sim m_v^{-2}$. For $m_v \sim 10^{14}$GeV this gives $\sigma \sim 10^{-56}$cm^2. Such a small cross section would be of little astrophysical interest. However, Rubakov [1981, 1982] and Callan [1982a,b] have shown that the s-wave component of the fermion

wave function is greatly enhanced near the monopole core, making the
B-violating cross-section comparable to a typical strong-interaction cross-
section. In the non-relativistic limit,

$$\sigma \sim \sigma_0 v^{-1}, \qquad (14.2.17)$$

where $\sigma_0 \sim 10^{-28} \mathrm{cm}^2$ and v is the relative fermion–monopole velocity.[*]
The mathematical treatment of the Callan–Rubakov effect is rather subtle
and will not be discussed here. We refer the reader to the original papers
and to the review article by Preskill [1985]. Here, we shall only give a
heuristic argument, which makes the fermionic s-wave enhancement somewhat less mysterious [Gregory, Perkins, Davis & Brandenberger, 1990].
We first note that only s-waves can effectively interact with the core, since
wave functions with non-zero angular momenta are strongly suppressed
near the origin. Semiclassically, the s-wave corresponds to radial motion,
and the Lorentz force on the particle vanishes (see (14.2.3)) However,
there is also a force,

$$\mathbf{F} = \nabla(\mu \cdot \mathbf{B}), \qquad (14.2.18)$$

due to the interaction of the magnetic moment μ of the particle with the
magnetic field of the monopole. With the right choice of polarization, the
s-wave magnetic moment μ is radial, and the force (14.2.18) is attractive.
This long-range attraction results in the s-wave enhancement near the
core.

14.3 Formation and evolution of monopoles

14.3.1 Formation

The initial density of monopoles at the time of formation can be estimated
using the standard Kibble argument of §3.1.4, that is, in this case

$$n_M(t_i) \sim p\xi^{-3} \qquad (14.3.1)$$

Here, ξ is the Higgs field correlation length at the phase transition, and
the numerical factor p depends on the geometry of the vacuum manifold
$\mathcal{M}$. The value of p can be estimated by discretizing $\mathcal{M}$ and physical
space, as was done for strings in §9.1. The model (14.1.1) in which $\mathcal{M}$
is S_2 yields $p \sim 0.1$ [Kibble, 1976]. The magnitude of ξ depends on the
dynamics of the phase transition, but causality constrains it to be smaller
than the horizon, $\xi \lesssim t_i$, and thus [Einhorn, Stein & Toussaint, 1980]

$$n_M(t_i) \gtrsim p t_i^{-3}. \qquad (14.3.2)$$

---

[*] For a composite object like a proton, the velocity dependence in (14.2.17) is a deli-
cate issue, since it is not clear whether we should use the velocity of the proton or
the velocities of quarks inside the proton. For a discussion see Cragie [1986].

One might expect the distribution of monopoles and antimonopoles at formation to be random, with the usual $\sqrt{N}$ fluctuations of the magnetic charge,

$$\delta N \sim \sqrt{N} \sim (L/\xi)^{3/2}. \tag{14.3.3}$$

Here, $N = N_+ + N_- \sim (L/\xi)^3$ is the total number of monopoles and antimonopoles in a volume L^3, and $\delta N = N_+ - N_-$ is the magnetic charge fluctuation. However, this expectation is wrong: positions of monopoles and antimonopoles are strongly correlated [Einhorn *et al.*, 1980]. To see this, note that the total magnetic charge inside a volume can be expressed as a surface integral over the boundary of the volume, where the integrand depends on the direction of the Higgs field at the boundary. For example, in the $SO(3)$ model (14.1.1),

$$\delta N = \frac{1}{8\pi} \oint dS^{ij} |\phi|^{-3} \epsilon_{abc} \phi^a \partial_i \phi^b \partial_j \phi^c. \tag{14.3.4}$$

The integrand is of the order ξ^{-2} and varies randomly on scales $\sim \xi$; hence

$$\delta N \sim L/\xi. \tag{14.3.5}$$

We see that δN depends on the square root of the surface area, not of the volume. In this sense, the magnetic charge fluctuation (14.3.5) is just a surface effect, and there are no real volume charge fluctuations in the system.

Turning now to the cosmological evolution of monopoles, it is convenient to introduce the monopole-to-entropy ratio $n_{\rm M}/s$, where $s \sim \mathcal{N}T^3$ is the entropy density and $\mathcal{N}$ is the effective number of helicity states for particles with $m < T$. This ratio is not affected by the expansion of the universe, as long as it is adiabatic, and changes only through monopole-antimonopole (M$\bar{\rm M}$) annihilation. In terms of this ratio, the causality bound (14.3.2) can be rewritten as

$$\frac{n_{\rm M}}{s}(t_{\rm i}) \gtrsim p\sqrt{\mathcal{N}} \left(\frac{T_{\rm i}}{m_{\rm pl}}\right)^3, \tag{14.3.6}$$

where $T_{\rm i}$ is the temperature of the phase transition at which the monopoles were formed. For a second-order or weakly first-order phase transition, $T_{\rm i} \sim \eta \sim em$, where η is the monopole symmetry breaking scale.

14.3.2 Annihilation mechanisms

Mechanisms for M$\bar{\rm M}$ annihilation in the early universe were first studied by Zel'dovich & Khlopov [1978] and by Preskill [1979]. We begin by considering a diffusive capture mechanism which operates at early times, when the plasma is dense and the monopole mean free path is short.

The motion of monopoles in plasma is like the Brownian motion of heavy dust particles in a gas or liquid. Monopoles move with thermal velocities, $v_T \sim (T/m)^{1/2}$, and the typical step of the Brownian walk is given by the mean free path l in (14.2.14). The attractive force between monopoles and antimonopoles introduces a slight bias in their random walks, so that they gradually drift towards each other. The average drift velocity is determined by the balance of electromagnetic attraction and the drag force (14.2.2),

$$v(r) \sim g^2/\beta T^2 r^2. \tag{14.3.7}$$

The drag force actually aids the annihilation process, because it dissipates the energy of the pair, allowing a bound state to be formed. Capture occurs when $g^2/r \sim T$, hence the capture radius is

$$r_c \sim g^2/T. \tag{14.3.8}$$

Once a bound M$\bar{\text{M}}$ pair is formed, the monopole and antimonopole spiral around one another, losing energy to plasma drag and to radiation, and eventually annihilate into Higgs particles and gauge bosons.[*] If $d \sim n_M^{-1/3}$ is the average distance between the monopoles, then the characteristic timescale of M$\bar{\text{M}}$ annihilation is

$$\tau_a \sim \frac{d}{v(d)} \sim \frac{\beta T^2}{g^2 n_M}. \tag{14.3.9}$$

A comparison with the Hubble time, $t \sim (\mathcal{N}G)^{-1/2}T^{-2}$, gives

$$\frac{\tau_a}{t} \sim \frac{\beta s T}{\sqrt{\mathcal{N}} g^2 n_M m_{\text{pl}}}. \tag{14.3.10}$$

If $\tau_a > t$, then annihilation is unimportant and the monopole-to-entropy ratio remains constant. However, with $n_M/s \sim$ const., (14.3.10) shows that τ_a/t decreases with time, until it becomes of order unity. At later times, annihilations reduce the monopole density to the value at which $\tau_a \sim t$, that is,

$$\frac{n_M}{s} \sim \frac{\beta}{g^2\sqrt{\mathcal{N}}} \frac{T}{m_{\text{pl}}}. \tag{14.3.11}$$

---

[*] Plasma dissipation reduces the orbital radius of the pair on a timescale which is much shorter than the Hubble time, but probably becomes unimportant when the radius gets smaller than the thermal particle wavelength, $r \ll T^{-1}$. The rate of radiative energy loss is $\dot{\mathcal{E}} \sim -g^6/m^2 r^4$, and the corresponding lifetime of the pair is $\tau \sim m^2 r^3/g^4$. With $r \sim T^{-1}$, we obtain $\tau \sim m^2/g^4 T^3$. In some models annihilation can occur much later than capture. However, since monopoles of a bound pair cannot escape annihilation, we shall not distinguish between annihilation and capture in the following discussion.

The diffusive capture process is effective as long as the mean free path l of monopoles in plasma is smaller than the capture radius r_c or, equivalently, as long as the temperature is higher than

$$T_d \sim m/g^4 \beta^2 . \tag{14.3.12}$$

For $T \sim T_d$, the monopole-to-entropy ratio can still be estimated from (14.3.11),

$$\frac{n_M}{s} \sim \frac{1}{g^6 \beta \sqrt{\mathcal{N}}} \frac{m}{m_{pl}} . \tag{14.3.13}$$

At lower temperatures, monopoles and antimonopoles can capture each other only by emitting radiation. For a pair with thermal incident velocities, $v_T \sim (T/m)^{1/2}$, the classical radiative capture cross-section is [Zel'dovich & Khlopov, 1978; see also Dicus, Page & Teplitz, 1982]

$$\sigma_c \sim (g^2/T)^2 (T/m)^{3/5} . \tag{14.3.14}$$

The corresponding annihilation time, $\tau_a \sim (n_M \sigma_c v_T)^{-1}$, is always greater than the expansion time,

$$\frac{\tau_a}{t} \sim g^2 \beta \left(\frac{m}{T} \right)^{1/10} \gg 1 , \tag{14.3.15}$$

where we have used n_M/s from (14.3.13). Hence, for $T < T_d$ annihilation cannot keep pace with expansion, and the monopole-to-entropy ratio 'freezes' at the value (14.3.13) [Preskill, 1979]. If this 'freeze-out' occurs before monopole annihilation becomes efficient, then instead of (14.3.13), n_M/s retains its initial value. In summary, if the initial value of n_M/s is greater than (14.3.13), then monopole annihilation processes reduce it to (14.3.13), while in the opposite case n_M/s remains essentially unchanged from the time of monopole formation.

As we have explained in §14.1.5, the existence of magnetic monopoles is a generic prediction of grand unification. For a typical grand unification scale, $\eta \sim 10^{15}$GeV, monopoles are 'superheavy', with mass $m \sim \eta/e \sim 10^{16}$GeV. From (14.3.6), the initial value of n_M/s cannot be less than $\sim 10^{-12} m_{16}^3$, where $m_{16} = m/10^{16}$GeV. For $m_{16} \gtrsim 1$, (14.3.13) predicts the final value

$$n_M/s \sim 10^{-12} m_{16} , \tag{14.3.16}$$

while for $m_{16} < 1$ the causality constraint (14.3.6) gives

$$n_M/s \gtrsim 10^{-12} m_{16}^3 , \tag{14.3.17}$$

where we have used $g^2 \sim \beta \sim N \sim 100$. The problem with the predictions (14.3.16–17) is that they are in a drastic conflict with observation.

The simplest way to derive a bound on the present monopole abundance is to require that the mass density of monopoles does not exceed the critical density,

$$mn_M \lesssim \rho_c, \tag{14.3.18}$$

which yields

$$n_M \lesssim 10^{-21} m_{16}^{-1} \mathrm{cm}^{-3}. \tag{14.3.19}$$

Bounds on monopole abundance are more often expressed in terms of a potentially observable quantity, the monopole flux, $F_M = n_M v$, where v is the average velocity of monopoles. In the early universe monopoles are very slow-moving, but after galaxy formation they can be accelerated by gravitational and magnetic fields. The gravitational field of a galaxy like our own will accelerate a monopole up to $v_G \sim 10^{-3}$. The magnetic field in our galaxy has strength $B \sim 3 \times 10^{-6}$Gauss and a coherence length $L \sim 10^{21}$cm. A monopole crossing one coherence length is accelerated to

$$v_B \sim (gBL/m)^{1/2} \sim 3 \times 10^{-3} m_{16}^{-1/2}. \tag{14.3.20}$$

For $m < 10^{17}$GeV, v_B is greater than v_G and the monopoles are ejected from the galaxy, and thus do not cluster in our galactic halo. In fact, it can be shown that for $m < 10^{19}$GeV, monopoles 'evaporate' from the halo in a time comparable to the age of the galaxy [Turner, Parker & Bogdan, 1982]. For any value of m, we expect the present velocities of the monopoles to be greater than or comparable to $v_G \sim 10^{-3}$. Returning now to the bound (14.3.19), we can rewrite it as

$$F_M = n_M v \lesssim 3 \times 10^{-14} m_{16}^{-1} v_{-3} \mathrm{cm}^{-2} \mathrm{s}^{-1}, \tag{14.3.21}$$

where $v_{-3} = v/10^{-3}$. However, if monopoles are clustered in galaxies, then the local density and flux can be significantly greater.

As monopoles are accelerated by the galactic magnetic field, the magnetic energy density $U = B^2/8\pi$ is dissipated at the rate

$$dU/dt = -gn_M \langle \mathbf{v} \cdot \mathbf{B} \rangle, \tag{14.3.22}$$

where angular brackets indicate averaging. For $m \lesssim 10^{17}$GeV, the main monopole acceleration mechanism is magnetic, $\langle \mathbf{v} \cdot \mathbf{B} \rangle \sim vB$, and the characteristic dissipation timescale is $\tau_d \sim B/8\pi gn_M v$. The time needed to regenerate the magnetic field is comparable to the galactic rotation period, $\tau \sim 10^8$ years. Requiring that $\tau < \tau_d$, we obtain the limit [Parker, 1970]

$$F_M \lesssim 10^{-15} \mathrm{cm}^{-2} \mathrm{s}^{-1}. \tag{14.3.23}$$

For $m \gg 10^{17}$ GeV, monopole velocities are $v \sim v_G \gg v_B$, and they are only slightly deflected by the galactic magnetic field. In this case some monopoles will gain energy and others will lose energy, depending on the direction of their motion relative to $\mathbf{B}$ and on the sign of their magnetic charge. To linear order in $\mathbf{B}$, when the deflection of monopoles by the magnetic field is neglected, the average energy gain vanishes. However, at second order in B there is a non-zero effect; the corresponding flux limit is [Turner *et al.*, 1982]

$$F_M \lesssim 10^{-16} m_{16} \, \text{cm}^{-2} \text{s}^{-1} \,, \qquad m > 10^{17} \text{GeV} \,. \qquad (14.3.24)$$

An even stronger bound on the monopole abundance can be obtained by considering the astrophysical consequences of baryon catalysis. As we explained in §14.2.2, monopoles can catalyze nucleon decay reactions such as $p \to \pi^0 + e^+$ or $n \to \pi^- + e^+$ with a typical strong-interaction cross-section (14.2.17). Since practically all the rest energy of the nucleon is released in the process, monopole catalysis has tremendous potential for liberating energy. One can easily obtain limits on the number of monopoles that could be captured in different types of stars, so that their luminosities would not be strongly affected. If ρ_N is the mass density of nucleons in a star and v is their average velocity, then the energy output due to baryon decay is

$$\dot{\mathcal{E}} \sim N_M \rho_N \sigma v \sim 3 \times 10^3 N_M (\rho_N / g \, \text{cm}^{-3}) \text{erg/s} \,. \qquad 14.3.25$$

Here, N_M is the total number of monopoles in the star and we have used (14.2.17) with $\sigma_0 \sim 10^{-28} \text{cm}^2$. Neutron stars have densities $\rho_N \sim 10^{15} \text{g/cm}^3$ and luminosities as low as $3 \times 10^{30} \text{erg/s}$. Hence, the number of monopoles captured in such stars cannot exceed $N_M \sim 10^{12}$. Similarly one finds $N_M \lesssim 10^{28}$ for the Sun and $N_M \lesssim 10^{18}$ for white dwarfs.

In order to translate these limits into bounds on the monopole flux, one needs to estimate how many monopoles are captured by different types of stars during their lifetimes. Omitting the details, we shall only quote the resulting bounds for F_M. The strongest bound follows from considering neutron stars [Kolb, Colgate & Harvey, 1982; Dimopolous, Preskill & Wilczek, 1982; Freese, Turner & Schramm, 1983],

$$F_M \lesssim 10^{-20} \sigma_{-28}^{-1} \text{cm}^{-2} \text{s}^{-1} \,, \qquad (14.3.26)$$

where $\sigma_{-28} = \sigma_0 / 10^{-28} \text{cm}^2$. This is five orders of magnitude stronger than the Parker limit (14.3.23). In fact, the bound can be improved by yet another seven orders of magnitude if one takes into account monopoles captured by the star when it was on the main sequence, before it became a neutron star [Frieman, Freese & Turner, 1988]. Analysis of baryon decay catalysis in white dwarfs yields a bound three orders of magnitude weaker than (14.3.26) [Freese, 1984].

The observational bounds on the monopole flux must now be compared with the theoretical prediction (14.3.16). The present value of the entropy density is $s \sim 10^3 \text{cm}^{-3}$, and we can rewrite (14.3.16) in terms of the flux F_M as

$$F_M^{\text{pred.}} \sim 3 \times 10^{-2} m_{16} v_{-3} \text{cm}^{-2} \text{s}^{-1}. \tag{14.3.27}$$

Comparison with the previous bounds reveals a discrepancy of many orders of magnitude. Thus, one is forced to conclude that grand unified theories (at least the simplest of them) are incompatible with the standard cosmology; this discrepancy is known as *the monopole problem*. In the next section we review some of the attempts to explain the failure of monopoles to appear at predicted densities.

14.3.4 Solutions to the monopole problem

The most attractive solution suggested to date is the inflationary universe scenario [Guth, 1981]. In this scenario the universe goes through a period of exponential expansion, and all of the presently observed universe arises from a tiny region which was initially smaller than the causal horizon. Monopoles, or any other defects formed before inflation, are diluted by an enormous factor and are never seen again. After inflation the universe thermalizes at some temperature T, and some monopole–antimonopole pairs are produced by thermal fluctuations. Their density is suppressed by the Boltzmann factor $\exp(-2m/T)$ and is typically very small [Turner, 1982; Lazarides, Shafi & Trower, 1982]. However, it is possible in some models for the monopole abundance produced at thermalization to be acceptably small, but still potentially detectable [Preskill, 1983].

We now mention some other possibilities. Langacker & Pi [1980] suggested a model in which the universe goes through a phase with the $U(1)$ symmetry of electromagnetism spontaneously broken. During this phase monopoles and antimonopoles are connected by flux tubes (strings), as they would be in a superconductor. As the monopole–antimonopole pairs are pulled together by strings, the annihilation can be very efficient. A detailed discussion of this scenario is given in §14.4.2. The main criticism of the Langacker–Pi model is that it depends sensitively on the details of the Higgs structure of a specific model.

A more radical proposal would be to give up grand unification altogether. If the electromagnetic $U(1)$ is not embedded in a spontaneously broken semi-simple group, then there are no magnetic monopoles and there is no monopole problem.

We finally mention an unsuccessful, but interesting attempt to resolve the monopole problem. If the initial monopole density is indeed as high as suggested by (14.3.16), then the universe rapidly becomes monopole-dominated. Due to gravitational instability, monopoles form clumps, and their annihilation rate can be greatly enhanced. A detailed analysis of this scenario shows that either the enhancement is not sufficient to solve the monopole problem, or, if the annihilation rate is efficient, the resulting entropy production is too large, and the baryon-to-entropy ratio is reduced below an acceptable level [Goldman, Kolb & Toussaint, 1981; Fry, 1981; Dicus *et al.*, 1982].

14.4 Monopoles connected by strings

Monopoles formed at one phase transition can become connected to strings at a subsequent phase transition [Langacker & Pi, 1980; Lazarides, Shafi & Walsh, 1982; Vilenkin, 1982c]. As discussed in §3.2.5, a typical sequence of phase transitions leading to this series of events is

$$G \to K \times U(1) \to K.$$

$$(14.4.1)$$

The first phase transition gives monopoles carrying a magnetic charge of the $U(1)$ gauge field. At the second transition the magnetic field is squeezed into flux tubes connecting monopoles and antimonopoles. Closed loops of string can also be formed.

14.4.1 Physical properties

If the string length is much greater than its thickness, the dynamics of a monopole–antimonopole (MM̄) pair connected by a string can be described by the action

$$S = -m \int ds_1 - m \int ds_2 - \mu \int dS.$$

$$(14.4.2)$$

Here, the first two integrations are over the monopole and antimonopole worldlines and the third is over the string worldsheet. If η_M and η_s are the monopole and string symmetry breaking scales, respectively, then the monopole mass m and the string tension μ can be estimated as

$$m \sim \eta_M/e,$$

$$(14.4.3)$$

$$\mu \sim \eta_s^2.$$

$$(14.4.4)$$

A monopole attached to a string moves with a proper acceleration

$$a = \frac{\mu}{m}.$$

$$(14.4.5)$$

For the case of a straight string, the motion is hyperbolic,

$$x^2 - t^2 = r_0^2, \quad y = z = 0,$$

$$(14.4.6)$$

where $r_0 = m/\mu$ and the string is assumed to lie along the x-axis. An MM̄ pair connected by a sufficiently long straight string, $l \gg \delta_s/2G\mu$, collapses to a black hole before the distance between the monopoles becomes comparable to the string thickness δ_s. (Note that, as the string shortens, its energy goes into the kinetic energy of the monopoles.)

In a cosmological setting, strings are not expected to be straight. The acceleration of the monopoles will change direction as the monopoles move along the curved string. The string itself will wiggle under the action of its tension, and the system will undergo a complicated motion in which the energy is constantly exchanged between the monopole kinetic energy and the kinetic and potential energy of the string. If the string is not particularly wiggly, with its curvature radius comparable to its length, the average energies of the monopoles and the string will also be comparable. For a long string with $l \gg m/\mu$, the motion of the monopoles will be ultra-relativistic.

In the simplest models all the magnetic flux of the monopoles is confined into the strings. However, in the general case, and in most realistic models, stable monopoles have unconfined, typically non-abelian, magnetic charges (for example, a colour magnetic charge [Lazarides & Shafi, 1980; t'Hooft, 1976a]). In this case, accelerating monopoles can lose energy by radiating gauge quanta. Since the radiation frequency is typically very high, we can ignore the complications associated with the long-distance behaviour of non-abelian fields and make estimates analogous to the electromagnetic case. In this case, the rate of energy loss by a relativistic monopole is [Landau & Lifshitz, 1975]

$$\dot{\mathcal{E}}_{\mathrm{M}} \sim -g^2 \frac{dv^\mu}{ds}\frac{dv_\mu}{ds} \sim -\left(\frac{g\mu}{m}\right)^2 . \tag{14.4.7}$$

A competing energy loss mechanism is gravitational radiation. We can roughly estimate its power using the quadrupole formula (7.5.1), as we did for string loops. A naive application yields[*]

$$\dot{\mathcal{E}}_{\mathrm{g}} \sim -G\mu^2 . \tag{14.4.8}$$

A comparison with (14.4.7) suggests that the radiation of gauge quanta dominates for $\eta_{\mathrm{M}} < m_{\mathrm{pl}}$.

In models where monopoles get connected by strings, the strings are not topologically stable and can break producing monopoles and anti-monopoles at the free ends. The breaking of a string is a tunnelling process and its probability is typically very small. The corresponding

---

[*] We note that this formula should be used with even greater caution than a similar estimate for string loops, (7.5.2–3). The reason is that the motion of monopoles can be ultra-relativistic, and the validity of the quadrupole formula in this case is highly dubious. Here we use (14.4.8) in lieu of a more reliable estimate.

instanton is obtained by changing $t \to it$ in (14.4.6), giving

$$x^2 + t^2 = r_0^2, \quad y = z = 0. \tag{14.4.9}$$

The Euclidean worldsheet of the string is the xt-plane with a circular hole of radius r_0, and the monopole–antimonopole worldline is the circle bounding the hole. The semiclassical tunnelling probability is proportional to $\exp(-B)$, where

$$B = S_{\mathrm{E}} - S_{\mathrm{E}}^0 = 2\pi r_0 m - \pi r_0^2 \mu = \pi m^2/\mu. \tag{14.4.10}$$

Here, S_{E} is the instanton action and S_{E}^0 is the action for a straight string without a monopole (the corresponding worldsheet is the xy-plane without a hole excised). Hence, the probability for a string to break per unit length per unit time is [Vilenkin, 1982c]

$$\mathcal{P} \propto \exp(-\pi m^2/\mu). \tag{14.4.11}$$

From (14.4.3) and (14.4.4), we see that $\pi m^2/\mu \sim (\pi/e^2)(\eta_{\mathrm{M}}^2/\eta_{\mathrm{s}}^2)$. This suggests that such strings are essentially stable for $\eta_{\mathrm{M}} > \eta_{\mathrm{s}}$.

We finally consider the case of transient strings with a temperature-dependent tension (refer to §3.2.5),

$$\mu \propto T^2. \tag{14.4.12}$$

In an $M\bar{M}$ pair connected by such a string, the string tension decreases slightly during each oscillation period, and in the absence of dissipation, the monopoles would get further and further apart with each oscillation. The evolution of the oscillation amplitude d with the temperature can be found from the adiabatic invariant

$$\oint p\,dq \propto \mathcal{E}^{3/2}/\mu, \tag{14.4.13}$$

where $\mathcal{E} = \mu d$ is the energy of the pair and we have assumed non-relativistic motion. This gives

$$\mathcal{E} \propto T^{4/3}, \quad d \propto T^{-2/3}. \tag{14.4.14}$$

This behaviour can, of course, be strongly modified by dissipation (see §14.4.3).

14.4.2 Formation and evolution

We consider a model of the type (14.4.1) in which monopoles formed at the first phase transition $(T = T_{\mathrm{M}})$ become connected by strings at the second phase transition $(T = T_{\mathrm{s}} < T_{\mathrm{M}})$. Until string formation, the monopoles evolve in the manner discussed in §14.3. Depending on the details of the model, the Higgs field correlation length ξ_{s} at the second phase transition can be of the order or less than the average distance

d between the monopoles at $T = T_s$. If $\xi_s \sim d$, then most M$\bar{\text{M}}$ pairs will be connected by the shortest possible strings of length $l \sim d$. Some longer strings and closed loops would also be present, and the length distribution of strings can be found using a Monte Carlo simulation. One finds that open strings and closed loops of length much greater than d are exponentially suppressed. This result was first obtained by simulating a phase transition in a two-dimensional version of the system, in which vortices played the role of monopoles and domain walls the role of strings [Sikivie, 1983; Everett, Vachaspati & Vilenkin, 1985; Lee & Weinberg, 1990]. A full three-dimensional simulation was performed by Copeland, Haws, Kibble, Mitchell & Turok [1986]. The exponential character of the distribution can be understood by noting that, as we travel along a string, at each step there is a certain probability of obtaining the combination of phases corresponding to a monopole which will terminate the string [Sikivie, 1983]. Mitchell & Turok [1987b] studied the distribution of open and closed strings using arguments from statistical mechanics. They showed that, for a given monopole density, the entropy is maximised when

$$n_{\text{open}}(l) \propto \exp(-l/\bar{l})\,, \tag{14.4.15}$$

$$n_{\text{closed}}(l) \propto l^{-5/2}\exp(-l/\bar{l})\,, \tag{14.4.16}$$

with $\bar{l} \sim d$. Here, $n(l)dl$ is the number density of strings with lengths from l to $l + dl$. As for topologically stable strings, the configuration of monopoles and strings at formation is expected to resemble the equilibrium configuration, and indeed the numerical simulations of Copeland *et al.* [1986] are well fitted by (14.4.15) and (14.4.16).

In their work, Copeland *et al.* [1986] used the same lattice spacing to simulate both phase transitions in (14.4.1). This amounts to the assumption that $\xi_s \sim d$. For $\xi_s \ll d$, strings are expected to be Brownian and have average length $\bar{l} \sim d^2/\xi_s$. We expect that open and closed strings with $l \gg \bar{l}$ will be exponentially suppressed for the same reason as before. In the following discussion of string evolution we shall first assume the strings to be nearly straight and then consider the case of Brownian strings.

The lifetime of an M$\bar{\text{M}}$ pair connected by a string is determined essentially by the time it takes to dissipate the energy stored in the string. (Even if the string is almost straight, the monopoles are unlikely to collide head-on and annihilate, since for $\eta_s \ll \eta_M$ the string thickness is much greater than the size of the monopole core.) From the causality constraint (14.3.2), the monopole separation d at temperature T is bounded by

$$d \lesssim (tt_M)^{1/2} \sim m_{\text{pl}}/TT_M\,. \tag{14.4.17}$$

At the time of string formation $T = T_s \sim \eta_s$, and for a pair connected by a nearly straight string, the energy to be dissipated is

$$\mathcal{E} \sim \mu d \lesssim \frac{m_{\rm pl} T_s}{T_M} . \tag{14.4.18}$$

Monopoles are pulled by the strings with a force $\mu \sim \eta_s^2$, while the frictional force acting on a monopole is $F \sim T^2 v$ from (14.2.7). The corresponding rate of energy loss is given by

$$\dot{\mathcal{E}} \sim -T^2 v^2 . \tag{14.4.19}$$

The characteristic monopole velocity is

$$v \sim \left(\frac{\mu d}{m}\right)^{1/2} , \tag{14.4.20}$$

provided that this gives $v < 1$; otherwise it is $v \sim 1$. For $v < 1$ the lifetime of a pair can be estimated as

$$\tau \sim \frac{\mu d}{T^2 v^2} \sim \frac{m}{T^2} , \tag{14.4.21}$$

while for $v \sim 1$ we have

$$\tau \sim \frac{\mu d}{T^2} \lesssim \frac{T_s}{T_M} \frac{m_{\rm pl}}{T^2} , \tag{14.4.22}$$

where the last inequality follows from (14.4.18). We see that in both cases monopoles annihilate on a timescale much shorter than the Hubble time, $t \sim m_{\rm pl}/T^2$ [Holman, Kibble & Rey, 1992].[*]

If monopoles have unconfined magnetic charges, they will also lose energy through the radiation of gauge quanta at the rate $\dot{\mathcal{E}} \sim -(g\mu/m)^2$ (see (14.4.7)). However, the corresponding lifetime of a pair is $\tau \sim m_{\rm pl} T_M/T_s^2 T$ and becomes smaller than the Hubble time only at $T \sim T_s^2/T_M \ll T_s$. Clearly, the lifetime estimates (14.4.21) and (14.4.22) are unaffected by gauge radiation.

For the case of Brownian strings, there is one more energy loss mechanism to consider: friction due to the interaction of moving strings with the background plasma. The small-scale string wiggles are damped by this frictional force and the coherence length of the strings $\xi(t)$ grows according to (9.2.4–5) as

$$\xi \sim \left(\frac{\mu m_{\rm pl}}{T^5}\right)^{1/2} . \tag{14.4.23}$$

---

[*] Our treatment differs from Holman *et al.* [1992] in that they assume efficient capture and annihilation of monopoles on their first encounter inside the strings. The conclusions, however, are not significantly different.

By the time that ξ becomes comparable to the average monopole separation, the strings connecting the monopoles are essentially straight. For $T_s < T_M^2/m_{pl}$ this happens in less than a Hubble time. Otherwise, ξ can be found from (14.4.23), the typical string length is $\ell \sim d^2/\xi$, and the monopole velocity is $v \sim (\mu\xi/m)^{1/2}$ or $v \sim 1$, whichever is smaller. Estimating the dissipation rate from (14.4.19), one finds that the lifetime of a pair can exceed one Hubble time only if $T_s > T_M(T_M/m_{pl})^{1/4}$. For heavy monopoles with $m \gtrsim 10^{15}$ GeV this means the string-forming phase transition must follow soon after the monopoles are formed.

We conclude that monopoles connected by strings annihilate, typically in less than a Hubble time after string formation. Closed loops of string lose their energy to friction and to gravitational radiation, and thus the whole system of monopoles and strings rapidly decays. As an interesting by-product of this decay, baryon-number-violating processes involved in monopole annihilation can produce some net baryon number, and it is conceivable that they could even account for the entire baryon asymmetry of the universe [Dixit & Sher, 1992; Farris, Klephart, Weiler & Yuan, 1992].

The evolution of monopoles connected by strings is different if there is a period of inflation between the two phase transitions (or if the monopole-forming phase transition is itself inflationary). For $\mu \ll m^2$, the string decay probability (14.4.11) is negligible, and if the monopoles are pushed beyond the present horizon, the evolution of strings is identical to that of topologically stable strings (see chapter 9).

14.4.3 Langacker–Pi model

An interesting model of the type (14.4.1) has been suggested by Langacker & Pi [1980] as a possible solution to the monopole problem. The model is based upon an unusual pattern of symmetry breaking,

$$SU(5) \xrightarrow{T_M} SU(3) \times SU(2) \times U(1) \xrightarrow{T_1} SU(3) \xrightarrow{T_2} SU(3) \times U(1)_{em}$$
$$(14.4.24)$$

Monopoles formed at the first phase transition get connected by strings at the second phase transition. But at the third transition the $U(1)$ symmetry is restored and the strings disappear. The tension of such transient strings is temperature-dependent, $\mu \sim \kappa T^2$, where κ is a combination of coupling constants (see §3.2.5). The discussion in the previous section suggests that the Langacker–Pi mechanism should suppress the monopole density very efficiently, provided that T_1 and T_2 are separated by more than a Hubble time. However, a more careful analysis shows that there are some exceptions to this rule.

The frictional energy loss for an MM̄ pair connected by a straight string is governed by the equation

$$T^{4/3}\frac{d}{dt}(T^{-4/3}\mathcal{E}) = -2\beta T^2\langle v^2\rangle = -\frac{2\beta}{3m}T^2\mathcal{E}. \qquad (14.4.25)$$

Here, the factors of $T^{4/3}$ account for the temperature-dependence of the string tension (see (14.4.12)), angular brackets stand for averaging over a period, and β is given by (14.2.9). The solution of this equation is $\mathcal{E} \propto t^{-p}$, where

$$p = \frac{2}{3} + \frac{0.2\beta m_{\rm pl}}{\sqrt{\mathcal{N}}m}, \qquad (14.4.26)$$

$\mathcal{N}$ is the number of light particle species at temperature T, and we have used (1.3.28) relating t and T in the radiation era. The monopole separation varies as $d \sim t^{1-p}$. With $\mathcal{N} \sim 100$ and $\beta \sim 1$, we obtain $p \sim 2 \times 10^{-2}m_{\rm pl}/m$.

For $m \ll 10^{17}$GeV, we have $p \gg 1$ and dissipation is very efficient. However, for superheavy monopoles with $m \gtrsim 10^{17}$GeV the decline of the monopole separation can be very slow, and it can take much longer than a Hubble time to dissipate the energy of a pair [Gates, Krauss & Terning, 1992]. In models where monopoles come to dominate the energy density of the universe, it can be shown that the dissipation is even less efficient. In general, the density of monopoles surviving after the last phase transition in (14.4.24) is model-dependent, and a more reliable estimate would require detailed analysis.

14.4.4 Causality considerations

Some doubts have been raised as to whether efficient monopole annihilation is consistent with causality [E. Weinberg, 1983; Lee & Weinberg, 1984]. The argument is that the magnetic charge fluctuations (14.3.5) on a scale L cannot be erased on a timescale shorter than L. This would be true if, in order to erase the fluctuations, the monopoles had to travel a distance $\sim L$. However, as we observed in §14.3.1, the fluctuation (14.3.5) is just a surface effect which can be erased by a slight reshuffling of the monopoles near the surface. As long as monopole annihilation does not require superluminal velocities, its efficiency is a dynamical question and it is not constrained by any causality principles [Vilenkin, 1984c; Everett *et al.*, 1985; Copeland *et al.*, 1986; Lee & Weinberg, 1990; Kibble & Weinberg, 1991].

For monopoles connected by strings, the strings help monopoles and antimonopoles to find each other, and the exponential length distribution of strings at formation ensures rapid annihilation of the monopoles. Even

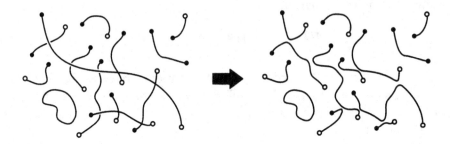

Fig. 14.3. Long strings are cut into small pieces by intercommuting with much more numerous shorter strings.

with a different initial state, an exponential length distribution would be rapidly established by intercommuting processes [Vilenkin, 1982c]. Long strings would be chopped into short pieces by intercommutings with the much more numerous shorter strings (see fig. 14.3). The probability for a string of length ℓ to avoid intercommuting (per unit time) is an exponentially decreasing function of ℓ, and thus very long strings are exponentially suppressed. Even in the worst possible scenario, when the intercommuting probability is negligible and strings can freely pass through one another, an exponential length distribution would eventually be established by the spontaneous breaking of strings through the nucleation of $M\bar{M}$ pairs.* These arguments demonstrate the existence of dynamical mechanisms which can decrease the monopole density at a rate much faster than suggested by the causality 'constraint'.

Similar arguments apply to domain walls bounded by strings, leading to the conclusion that causality imposes no interesting constraints on the rate of their decay. As we discussed in chapter 13, a wall–string network is expected to break into pieces and disintegrate. This expectation is confirmed by dynamical simulations in two and three dimensions [Shellard, 1986a, 1990; Ryden *et al.*, 1990].

14.4.5 Metastable monopoles?

Since domain walls and strings can be metastable (refer to §13.2.5 and §14.4.1), one could expect the same to be true for monopoles. What one

---

* The characteristic timescale for the string breaking is $t_b \sim \mathcal{P}^{-1/2}$, where $\mathcal{P}$ is the pair nucleation probability per unit length of string per unit time. This probability is typically very small, and t_b can be much greater than the present age of the universe. However, causality constraints, if correct, should be valid at any time.

needs is a model with a sequence of symmetry breakings

$$G \to H_1 \to H_2 \,, \qquad (14.4.27)$$

with

$$\pi_2(H_1/H_2) \neq I \,, \qquad \pi_2(G/H_2) = I \,. \qquad (14.4.28)$$

The first of these conditions indicates that monopoles will be formed at the second phase transition in (14.4.27), and the second condition ensures that the model has no absolutely stable monopoles, and thus the monopoles are only metastable. Somewhat surprisingly, it can be shown that models satisfying (14.4.28) do not exist [Preskill & Vilenkin, 1993].

According to the second fundamental theorem (3.3.10), monopoles arising from a symmetry breaking $G \to H$ are classified by non-contractible loops in H which can be contracted to a point in G. However, in a sequential symmetry breaking (14.4.27) all loops that can be contracted to a point in H_1 can also be contracted in G, and the impossibility of (14.4.28) follows immediately. The difference between domain walls and strings on the one hand and monopoles on the other is related to the fact that $\pi_2(G) = I$ for all compact Lie groups, while $\pi_0(G)$ and $\pi_1(G)$ may well be non-trivial.

14.5 Global monopoles

14.5.1 Physical properties

Monopoles can be formed as a result of local, as well as global symmetry breaking. In the latter case we shall call them global monopoles. The topological conditions for the formation of monopoles are the same in both cases, but their physical properties are very different [Barriola & Vilenkin, 1989; Linde, 1990].

The simplest model that gives rise to global monopoles is described by the Lagrangian

$$\mathcal{L} = \tfrac{1}{2}\partial_\mu \phi^a \partial^\mu \phi^a - \tfrac{1}{4}\lambda(\phi^a\phi^a - \eta^2)^2, \qquad (14.5.1)$$

where ϕ^a is a scalar field triplet, $a = 1, 2, 3$. The model has a global $SO(3)$ symmetry, which is spontaneously broken to $SO(2)$. This is just a global version of the model discussed in §14.1.1. Monopoles are characterized by the topological charge (14.3.4),

$$N = \frac{1}{8\pi} \oint dS^{ij} |\phi|^{-3} \epsilon_{abc} \phi^a \partial_i \phi^b \partial_j \phi^c \,, \qquad (14.5.2)$$

and the simplest $N = 1$ solution is the spherically-symmetric configuration

$$\phi^a = \eta h(r)\frac{x^a}{r} \,. \qquad (14.5.3)$$

The function $h(r)$ vanishes at $r = 0$ and approaches 1 at $r \gg \delta$, where $\delta \sim (\sqrt{\lambda}\eta)^{-1}$ is the size of the monopole core. Outside the core, where $h(r) \approx 1$, the dynamics of the scalar field can be described by a nonlinear σ-model, in which the triplet ϕ^a is confined to lie on a sphere $\phi^a \phi^a = \eta^2$. (For a discussion of σ-models and their properties, refer to §15.1.)

For $r \gg \delta$, the energy–momentum tensor of the monopole is (in spherical coordinates)

$$T_t^t = T_r^r = \eta^2/r^2, \qquad T_\theta^\theta = T_\varphi^\varphi = 0. \qquad (14.5.4)$$

The total monopole energy or mass is given by

$$m = 4\pi \int_0^R T_t^t \, r^2 dr \approx 4\pi\eta^2 R, \qquad (14.5.5)$$

where R is the cut-off radius. This is linearly divergent as $R \to \infty$ but, in practice, the integral in (14.5.5) is cut off at roughly the distance to the nearest antimonopole. The energy of an $M\bar{M}$ pair separated by a distance R is $\mathcal{E} \sim 4\pi\eta^2 R$, and the attractive force acting on M and $\bar{M}$ is

$$F = \partial\mathcal{E}/\partial R \sim 4\pi\eta^2. \qquad (14.5.6)$$

Note that the force is independent of the distance, as for monopoles connected by strings.

The stability of the spherically-symmetric monopole solution (14.5.3) has been a matter of some debate [Goldhaber, 1989; Rhie & Bennett, 1991; see also Perivolaropoulos, 1992]. Goldhaber [1989] has shown that for cylindrically-symmetric (φ-independent) distortions of the monopole field, the energy of a static monopole can be expressed as

$$\mathcal{E} = \int \rho \, dr \, dy \, d\varphi, \qquad (14.5.7)$$

where $y = \ln[\tan(\theta/2)]$,

$$\rho = \frac{|\phi|^2}{2} \left[\sin^2\tilde{\theta} + \left(\frac{\partial\tilde{\theta}}{\partial y}\right)^2 + \frac{r^2}{\cosh^2 y} \left(\frac{\partial\tilde{\theta}}{\partial r}\right)^2 \right]$$
$$+ \frac{1}{2}\left(\frac{\partial|\phi|}{\partial y}\right)^2 + \frac{r^2\lambda}{4\cosh^2 y}\left(|\phi|^2 - \eta^2\right)^2, \qquad (14.5.8)$$

and $\tilde{\theta}$ is the polar angle in ϕ-space ($\tilde{\theta} = \theta$ in the spherically-symmetric case). Goldhaber then noted that for $|\phi| = \eta$ and $\partial\tilde{\theta}/\partial r = 0$, (14.5.8) reduces to the energy density for the sine-Gordon model (3.1.4–5). The energy is minimized by a sine-Gordon soliton,

$$\tilde{\theta} = 2\tan^{-1}\exp\left(y - y_0\right), \qquad (14.5.9)$$

where y_0 is an arbitrary constant. From the translational invariance of the sine-Gordon model it follows immediately that the monopole energy is independent of y_0. The solution with $y_0 = 0$ corresponds to a spherically-symmetric monopole, while in the limiting cases of $y_0 \to +\infty$ and $y_0 \to -\infty$ all the gradient energy is concentrated in thin 'strings' pointing, respectively, in the $\theta = \pi$ and $\theta = 0$ directions.

The collapse of the spherically-symmetric field (14.5.3) into a 'string' at no energy cost suggests a possible instability when deviations from $|\phi| = \eta$ and $\partial\tilde{\theta}/\partial r = 0$ are allowed. However, Rhie & Bennett [1991] have argued that such an instability will not arise. When a spherical monopole is distorted so that some energy is transported, say, from the southern to the northern hemisphere, an effective tension develops pulling the monopole core to the north, towards another spherically-symmetric configuration. An instability can arise only if the monopole core is artificially held fixed. This picture is supported by numerical simulations of global monopoles [Bennett & Rhie, 1990] (refer to §14.5.3).

14.5.2 Gravitational field

The large energy in the scalar field surrounding global monopoles suggests that they can produce strong gravitational fields. For an isolated, static monopole the metric must have spherical symmetry and can be written as

$$ds^2 = B(r)dt^2 - A(r)dr^2 - r^2(d\theta^2 + \sin^2\theta d\varphi^2). \qquad (14.5.10)$$

The scalar field ϕ^a is given by (14.5.3) with the usual relation between the spherical coordinates r, θ, φ and the Cartesian coordinates x^i. The functions $A(r)$, $B(r)$ and the scalar field magnitude $h(r)$ can be found by solving the combined system of the Einstein and scalar field equations.

The problem substantially simplifies outside the monopole core, where we can use the σ-model approximation and set $h(r) = 1$. In this approximation, the energy–momentum tensor is given by (14.5.4), and the general solution of Einstein's equations is [Barriola & Vilenkin, 1989]

$$B = A^{-1} = 1 - 8\pi G\eta^2 - 2Gm_c/r. \qquad (14.5.11)$$

The mass parameter m_c appears in (14.5.11) as a constant of integration. To determine its value, one has to analyze the behaviour of the scalar field at $r \lesssim \delta$, but an order-of-magnitude estimate is easy to obtain; it yields $m_c \sim \lambda^{-1/2}\eta$. Harari & Lousto [1990] studied the problem numerically and found that m_c is in fact negative,

$$m_c \approx -20\lambda^{-1/2}\eta, \qquad (14.5.12)$$

in the whole range $0 \le 8\pi G\eta^2 < 1$.

For reasonable values of η and λ, the mass m_c is totally negligible on astrophysical scales. Neglecting the mass term in (14.5.11) and rescaling the r and t variables, we can rewrite the monopole metric as

$$ds^2 = dt^2 - dr^2 - (1 - 8\pi G\eta^2)r^2(d\theta^2 + \sin^2\theta d\varphi^2). \qquad (14.5.13)$$

This metric describes a space with a deficit solid angle; the area of a sphere of radius r is not $4\pi r^2$, but a little smaller. The surface $\theta = \pi/2$ has the geometry of a cone with a deficit angle

$$\Delta = 8\pi^2 G\eta^2. \qquad (14.5.14)$$

By symmetry, all surfaces passing through the origin and cutting space into two symmetric parts have the same geometry.

A striking feature of the metric (14.5.13) is that the monopole exerts no gravitational force on the matter around it (apart from the tiny repulsive effect of the core). This can be understood qualitatively if we note that the gravitational mass density, $\rho_g = T_0^0 - T_i^i$, vanishes for the energy–momentum tensor (14.5.4). Some properties of the global monopole metric are similar to those of the metric (7.1.11) for a gauge cosmic string, in which all surfaces normal to the string are cones with a deficit angle $\Delta = 8\pi G\mu$, μ being the string mass per unit length. An important difference, however, is that the monopole metric is not locally flat.

We now turn to light propagation in the gravitational field of a global monopole. Consider a light signal propagating from a source S to an observer O. Without loss of generality, we can assume that both S and O lie on the surface $\theta = \pi/2$. It is clear from symmetry that the whole trajectory must lie on that surface. Moreover, it is easily verified that the geodesic equation describing light propagation on the surface $\theta = \pi/2$ of the metric (14.5.13) is identical to that on the surface $z = $ const. of the cosmic string metric (7.1.11) with the same deficit angle. The problem is thus reduced to the problem of light propagation in the metric of a cosmic string (§7.2). If S, O, and the monopole M are perfectly aligned, then the image will appear in the form of a ring of angular diameter

$$\delta\varphi_0 = 8\pi^2 G\eta^2 l(d+l)^{-1}, \qquad (14.5.15)$$

where d and l are the distances from the monopole to the observer and to the source, respectively. If SM and OM are misaligned by a small angle $\alpha < \Delta$, then the observer will see two point images separated by an angle $\delta\varphi_0$ given by (14.5.15). Note that $\delta\varphi_0$ is independent of α.

The subscript of $\delta\varphi_0$ refers to the fact that the monopole was assumed to be at rest with respect to the observer. As we shall explain shortly, cosmological monopoles are expected to move at relativistic speeds, and so (14.5.15) has to be generalized to the case of a moving monopole. The same argument as we used for a string in §10.3 yields

$$\delta\varphi = \gamma^{-1}(1 - \mathbf{n} \cdot \mathbf{v})^{-1}\delta\varphi_0, \qquad (14.5.16)$$

where $\mathbf{n}$ is a unit vector along the line of sight, $\mathbf{v}$ is the monopole velocity, and $\gamma = (1 - v^2)^{-1/2}$.

For a typical grand unification scale $\eta \sim 10^{16}\mathrm{GeV}$, the angular separation (14.5.15) is of the order of 10 arcsec and is certainly within an observable range. The crucial issue is the expected density of global monopoles which we address in the next section.

14.5.3 Evolution

The large attractive force between global monopoles and antimonopoles suggests that $M\bar{M}$ annihilation is very efficient and that the monopole over-production problem does not exist. In fact, it could be so efficient that not a single monopole would be left within the observable universe. One can draw the analogy between global monopoles and monopoles connected by strings, for which we know that large $M\bar{M}$ separations are exponentially suppressed, and so no $M\bar{M}$ pairs are expected to survive in the observable universe (see §14.4.2). However, there is an important distinction. Unlike monopoles connected by strings, a global monopole is not paired with any particular antimonopole, and so it is not clear how efficiently it can find a partner.

It appears that the only reliable way to understand the fate of global monopoles is to perform numerical simulations of their evolution. Such a simulation has been performed by Bennett & Rhie [1990]. They used the σ-model approximation, which gives an accurate description of scalar field dynamics everywhere except near the monopole core. Since the dominant part of the monopole energy is in the field outside the core, this does not seem to be a serious problem. The initial field configuration was chosen by randomly assigning the field direction at every fourth grid point and then using an interpolation scheme for the intermediate points. The simulations showed that monopole evolution rapidly settles into a scale-invariant regime with only a few monopoles per horizon volume at all times. More precisely, the monopole density at scaling is given by

$$n_M = (4 \pm 1.5)\, d_H^{-3}, \qquad (14.5.17)$$

in both radiation and matter eras, where d_H is the horizon length.

To get a better insight into the mechanism of $M\bar{M}$ annihilation, Bennett & Rhie [1990] simulated the evolution of isolated $M\bar{M}$ pairs. With the positions of the poles constrained to remain fixed, they found that the field energy collapsed into a narrow 'string' with width comparable to the grid spacing. However, when the simulation was run without artificially fixing the positions of M and M̄, the poles moved relativistically and annihilated before the field between them could collapse into a 'string'.

By analogy with gauge monopoles, one could expect that, unless the collision is head-on, the monopoles will form a bound state and gradually spiral down as they lose energy by Goldstone boson radiation. However, as the following argument shows, the radiative lifetime of the pair is very short [Barriola & Vilenkin, 1989]. The energy flux carried away by Goldstone bosons is of the order $T_i^0 \sim \partial_t \phi^a \partial_i \phi^a$. At a distance R from the pair comparable to the monopole separation, $\partial_\mu \phi^a \sim \eta/R$ and $T_i^0 \sim \eta^2/R^2$. Hence, the energy loss rate is $\dot{\mathcal{E}} \sim T_i^0 R^2 \sim \eta^2$, and the lifetime of the pair is $\tau \sim R$. Thus Goldstone boson radiation is a very efficient energy loss mechanism: the MM̄ separation decreases at a speed comparable to the speed of light. This picture is in a qualitative agreement with the simulations. Although the radiative lifetime of a pair has not been studied quantitatively, no long-lived spiraling pairs have been observed.

14.5.4 Cosmological implications

The gravitational fields of global monopoles can cause clustering in matter and they can induce anisotropies in the microwave background. From (14.5.17), the average number of monopoles per horizon is $N_{\mathrm{H}} \sim 4$, and the amplitude of density fluctuations on the horizon scale can be roughly estimated as $\delta\rho_{\mathrm{H}} \sim N_{\mathrm{H}}^{1/2} m d_{\mathrm{H}}^{-3}$, where $m \sim 4\pi\eta^2 n_{\mathrm{M}}^{-1/3}$ is the typical energy per monopole. This gives a scale-invariant spectrum of fluctuations [Bennett & Rhie, 1990]

$$\left(\frac{\delta\rho}{\rho}\right)_{\mathrm{H}} \sim 30 G\eta^2 . \tag{14.5.18}$$

The expected microwave background anisotropy is $\delta T/T \sim (\delta\rho/\rho)_{\mathrm{H}}$.

Bennett & Rhie [1993] numerically generated a number of $\delta T/T$ patterns using their simulations of monopole evolution. To imitate the finite width of the microwave detector beam, the patterns were smoothed over $10°$ angular scales. The resulting temperature maps have a gaussian appearance, without any prominent features. The rms value of $\delta T/T$ is

$$\left(\frac{\delta T}{T}\right)_{\mathrm{rms}} = (18.7 \pm 2.1) G\eta^2 , \tag{14.5.19}$$

and the power spectrum is essentially identical to that due to a Harrison–Zel'dovich spectrum of primordial density fluctuations (see §1.3.6). Comparing (14.5.19) to the observed rms fluctuation (1.3.37), and assuming that it is due to global monopoles, we have

$$G\eta^2 = (5.9 \pm 1.2) \times 10^{-7} . \tag{14.5.20}$$

Higher values of $G\eta^2$ would be in conflict with observations.

Bennett & Rhie have pointed out that density fluctuations generated by global monopoles can later evolve into galaxies and clusters. Beside the gaussian fluctuations due to the variation of the scalar field on the horizon scale, one could expect that relativistically moving monopoles will produce filamentary structures, in a manner similar to the wakes formed behind rapidly moving strings (refer to §11.4.1). Accretion of matter in monopole wakes has been discussed by Perivolaropoulos [1992] assuming that the field of a moving monopole can be obtained by Lorentz-boosting the spherically-symmetric solution (14.5.3). However, since most of the monopole energy is distributed far from the core, it is not clear how the core can enforce itself as a symmetry centre. With large deviations from spherical symmetry, the accretion pattern can be strongly modified.

The spectrum of density fluctuations induced by monopoles has been numerically calculated by Bennett, Rhie & Weinberg [1993]. If the spectrum is normalized at $8h^{-1}$Mpc, as in (1.3.47), the required value of $G\eta^2$ is greater than (14.5.20) by a factor between 2.5 and 4 (depending on the Hubble parameter h). This conflict with the microwave background measurements can be avoided if we assume that galaxies are clustered more strongly than the overall mass density. The necessary 'biasing' factor on the scale of $8h^{-1}$Mpc is $b_8 = 2.5$ for $h = 1$ and $b_8 = 4$ for $h = 0.5$, where b_8 is defined in (1.3.49).

14.5.5 Global monopoles connected by strings

Global monopoles and antimonopoles can become connected by global strings in the same manner as discussed for gauge monopoles in §14.4. However, the long-range interactions between the M$\bar{\text{M}}$ pairs in the global case substantially change the properties and the evolution of these hybrid defects. Let η_M and η_s be the symmetry breaking scales of monopoles and strings, respectively. The tension in a global string is (see (3.2.15))

$$F_s \approx 2\pi\eta_s^2 \ln(R/\delta_s)\,, \tag{14.5.21}$$

where δ_s is the thickness of the string core and the upper cut-off R is set by the typical monopole separation. From (14.5.6), the long-range force between the monopoles is $F_M \sim 4\pi\eta_M^2$. For $\eta_M \gg \eta_s$, this force can dominate the string tension for a considerable period of time. Monopole evolution will then be almost unaffected by the strings, with the monopole separation growing as $R \sim t$. The strings will become dynamically important at

$$t \sim \delta_s \exp(2\eta_M^2/\eta_s^2)\,. \tag{14.5.22}$$

At this stage, they will pull the monopoles and antimonopoles together, and the resulting pairs will decay in a few Hubble times. We note, however, that for $\eta_M \gg \eta_s$ the decay time (14.5.22) can easily exceed the present age of the universe.

If the monopoles are diluted away by inflation, then the second phase transition can give rise to metastable global strings. As in the case of gauge strings, the global strings can break producing $M\bar{M}$ pairs, but the corresponding tunnelling action is proportional to $\exp(k\eta_M^2/\eta_s^2)$, with numerical coefficient $k \sim 1$ [Preskill & Vilenkin, 1993]. The tunnelling probability depends exponentially on the action and, for $\eta_M > \eta_s$, this probability can be extremely small, so that metastable global strings decay at an even slower rate than their gauge counterparts.

15

Textures

Textures arise in models in which the vacuum manifold has a non-trivial third homotopy group, $\pi_3(\mathcal{M}) \neq I$. Unlike other defects, in a texture the scalar field remains in the vacuum manifold everywhere. Consequently, the energy of the texture comes entirely from the gradient energy of the field. For a local gauge symmetry breaking, the energy of the gauge field can also contribute. However, here we shall be primarily interested in the properties and cosmological effects of global textures, deferring the discussion of gauge textures until the end of the chapter.

15.1 Effective action

To derive the low-energy effective action for a global texture, we start with a general scalar field Lagrangian,

$$\mathcal{L} = \tfrac{1}{2}\partial_\mu \Phi_i \partial^\mu \Phi_i - V(\Phi), \qquad (15.1.1)$$

with $i = 1, ..., n$. We assume that the symmetry group G of $\mathcal{L}$ is spontaneously broken and that the minima of $V(\Phi)$ form a vacuum manifold $\mathcal{M}$ with $\pi_3(\mathcal{M}) \neq I$. As we discussed in chapter 2, the dimension d of $\mathcal{M}$ is equal to the number of broken generators of G.

At low energies, the massive degrees of freedom are not excited, and the field Φ is constrained to stay in $\mathcal{M}$. In this case, it can be parametrized by d fields ϕ^A, which can be regarded as coordinates on the manifold $\mathcal{M}$,

$$\Phi_i = \Phi_i(\phi^A), \quad A = 1, ..., d. \qquad (15.1.2)$$

These fields represent the Goldstone bosons resulting from the global symmetry breaking. Substituting (15.1.2) into (15.1.1) we can write the effective action for $\phi^A(x)$ in the form

$$S = \frac{1}{2}\int d^4x \sqrt{-g}\, G_{AB}(\phi)\partial_\mu \phi^A \partial^\mu \phi^B, \qquad (15.1.3)$$

where

$$G_{AB}(\phi) = \frac{\partial \Phi_i}{\partial \phi^A} \frac{\partial \Phi_i}{\partial \phi^B} \tag{15.1.4}$$

is the metric on the manifold $\mathcal{M}$ and g is the determinant of the spacetime metric. The action (15.1.3) is that of a nonlinear σ-model.

The field equations are obtained by varying (15.1.3) with respect to ϕ^A,

$$\Box \phi^A + \Gamma^A_{BC} \phi^B{}_{,\mu} \phi^{C,\mu} = 0. \tag{15.1.5}$$

Here, $\Box$ is the covariant Laplacian in the metric $g_{\mu\nu}$ and Γ^A_{BC} are the Christoffel symbols in the metric G_{AB},

$$\Gamma^A_{BC} = \tfrac{1}{2} G^{AD} \left(\frac{\partial G_{BD}}{\partial \phi^C} + \frac{\partial G_{CD}}{\partial \phi^B} - \frac{\partial G_{BC}}{\partial \phi^D} \right). \tag{15.1.6}$$

The energy–momentum tensor for textures can be obtained from (15.1.3) by varying with respect to $g^{\mu\nu}$,

$$T_{\mu\nu} = G_{AB}(\phi) \left(\phi^A{}_{,\mu} \phi^B{}_{,\nu} - \tfrac{1}{2} g_{\mu\nu} \phi^A{}_{,\sigma} \phi^{B,\sigma} \right). \tag{15.1.7}$$

As an example of a model which gives rise to textures, consider the Lagrangian (15.1.1) with $n = 4$ and

$$V(\Phi) = \tfrac{1}{4} \lambda (\Phi_i \Phi_i - \eta^2)^2. \tag{15.1.8}$$

This model has an $SO(4)$ symmetry which is spontaneously broken to $SO(3)$. The vacuum manifold, $\Phi_i \Phi_i = \eta^2$, is a three-sphere S^3 which can be parametrized by three angular variables χ, $\tilde{\theta}$ and $\tilde{\varphi}$ as

$$\Phi_i = \eta(\sin \chi \sin \tilde{\theta} \cos \tilde{\varphi}, \ \sin \chi \sin \tilde{\theta} \sin \tilde{\varphi}, \ \sin \chi \cos \tilde{\theta}, \ \cos \chi) \tag{15.1.9}$$

where tildes have been placed over θ and φ in order to avoid confusion with spatial angular variables. The effective action for this model is then

$$S = \frac{1}{2} \eta^2 \int d^4 x \sqrt{-g} \left\{ (\partial_\mu \chi)^2 + \sin^2 \chi \left[(\partial_\mu \tilde{\theta})^2 + \sin^2 \tilde{\theta} (\partial_\mu \tilde{\varphi})^2 \right] \right\} \tag{15.1.10}$$

A simple example of a texture is a spherically-symmetric configuration

$$\chi = \chi(t, r), \qquad \tilde{\theta} = \theta, \qquad \tilde{\varphi} = \varphi, \tag{15.1.11}$$

with χ interpolating between $\chi = 0$ at $r = 0$ and $\chi = \pi$ as $r \to \infty$. If most of the variation of χ between 0 and π occurs on scales $r \lesssim R$, we shall call R the characteristic size of the texture. The effective action (15.1.10) can be used as long as $R \gg m_\Phi^{-1} \sim (\sqrt{\lambda}\eta)^{-1}$.

The 'winding' of Φ around the three-sphere can be quantified using the topological current density [Skyrme, 1961]

$$j^\mu \sqrt{-g} = \frac{1}{12\pi^2\eta^4} \epsilon^{\mu\nu\sigma\rho} \epsilon^{ijkl} \Phi_i \partial_\nu \Phi_j \partial_\sigma \Phi_k \partial_\rho \Phi_l . \qquad (15.1.12)$$

If Φ_i is constrained to lie on the three-sphere, the Jacobian $\det(\partial\Phi/\partial x)$ must vanish (since the four components of Φ are not independent) and

$$\partial_\mu(\sqrt{-g}\, j^\mu) = 0 . \qquad (15.1.13)$$

The corresponding topological charge (or winding number) is given by

$$Q = \int d^3x \sqrt{-g}\, j^0 . \qquad (15.1.14)$$

For a texture configuration of the form (15.1.11), the winding number is

$$Q = \frac{1}{\pi} \left[\chi(r=0) - \chi(r \to \infty) \right] . \qquad (15.1.15)$$

An important difference from other topological defects is that textures in a three-dimensional Euclidean space do not necessarily have integral charge Q.

15.2 Field equations with a constraint

Instead of introducing a new set of independent variables ϕ^A, one can work with the original variables Φ_i and impose the condition that Φ_i should stay in the vacuum manifold as a constraint. For example, in the $SO(N)$-symmetric model (15.1.8) the constraint would be

$$\Phi_i\Phi_i - \eta^2 = 0 . \qquad (15.2.1)$$

Introducing a Lagrange multiplier α, we can write the corresponding Lagrangian as

$$\mathcal{L} = \tfrac{1}{2}\partial_\mu\Phi_i\partial^\mu\Phi_i - \tfrac{1}{2}\alpha(\Phi_i\Phi_i - \eta^2) . \qquad (15.2.2)$$

Variation with respect to Φ_i gives

$$\Box\Phi_i + \alpha\Phi_i = 0 , \qquad (15.2.3)$$

and variation with respect to α recovers the constraint (15.2.1).

The parameter α can be excluded from (15.2.3) by acting with a projection operator,

$$(\delta_{ij} - \eta^{-2}\Phi_i\Phi_j)\Box\Phi_j = 0 . \qquad (15.2.4)$$

If Φ_i and $\dot\Phi_i$ are given on a hypersurface $t = $ const., then (15.2.4) determines the transverse (orthogonal to Φ_i) components of $\ddot\Phi_i$, while the longitudinal component has to be found with the aid of (15.2.1). Bennett

& Rhie [1990] used equations (15.2.4) and (15.2.1) in numerical simulations of global monopoles and textures.

To derive another convenient form of the field equations for Φ_i, we
multiply (15.2.3) by Φ_i and use (15.2.1), giving

$$\alpha = -\eta^{-2}\Phi_i \Box \Phi_i \,. \tag{15.2.5}$$

Now, acting with the $\Box$ operator on (15.2.1) we have

$$\Phi_i \Box \Phi_i + \partial_\mu \Phi_i \partial^\mu \Phi_i = 0\,, \tag{15.2.6}$$

and combining this with (15.2.3) and (15.2.5) yields

$$\Box \Phi_i + \eta^{-2}(\partial_\mu \Phi_j \partial^\mu \Phi_j)\Phi_i = 0\,. \tag{15.2.7}$$

It can be easily shown that the constraint (15.2.1) is satisfied by solutions
of (15.2.7), provided that it is satisfied by the initial conditions.

15.3 Texture collapse

The scaling argument used in the proof of Derrick's theorem (§3.1.2)
shows immediately that textures are unstable to collapse. Under a rescal-
ing $\mathbf{r} \to \alpha \mathbf{r}$ the energy of a texture changes as $E \to \alpha^{-1}E$, and there is
nothing to stablilize it against shrinking. When the texture shrinks to a
size $R \lesssim m_\Phi^{-1}$, the field gradients are strong enough to pull Φ over the
potential barrier to unwind the topological knot.

This argument tacitly assumes that the texture is in a flat spacetime.
A texture in an expanding universe is stretched by the expansion if its
size is greater than the horizon and starts collapsing only when it comes
within the horizon. Stable textures can exist in a closed universe with the
topology of a three-sphere. One example is the $SO(4)$ model (15.1.9) with
χ, $\tilde\theta$ and $\tilde\phi$ set equal to the three angular variables on the three-sphere in
a closed FRW model [Davis, 1987a].

A simple analytic solution describing the collapse and unwinding of a
texture in the model (15.1.10) in flat spacetime has been found by Turok
& Spergel [1990]. With a spherically-symmetric ansatz (15.1.11), the
action (15.1.10) takes the form

$$S = 2\pi\eta^2 \int dt\, dr\, r^2 \left(\dot\chi^2 - \chi'^2 - 2r^{-2}\sin^2\chi\right)\,. \tag{15.3.1}$$

The field equation for χ becomes

$$\ddot\chi - \chi'' - \frac{2}{r}\chi' + \frac{\sin 2\chi}{r^2} = 0\,, \tag{15.3.2}$$

which has as a solution

$$\chi = 2\cot^{-1}(-t/r)\,, \quad t < 0\,. \tag{15.3.3}$$

This describes a texture of unit winding number with χ ranging from 0 at the origin to π at infinity. The texture collapses towards the origin at relativistic velocities, so that its size at time t is $R \sim |t|$. At $t = 0$ the field χ is singular at the origin and is equal to π everywhere outside the origin. This singularity signals a breakdown of the effective action approximation at $R \lesssim m_\Phi^{-1}$.

Because of the singularity at $r = t = 0$, the continuation of the solution (15.3.3) to positive values of t is not unique; Turok & Spergel [1990] suggest

$$\chi = 2 \cot^{-1}(r/t), \qquad t > 0, \quad r < t, \qquad (15.3.4a)$$
$$= 2 \cot^{-1}(t/r), \qquad t > 0, \quad r > t. \qquad (15.3.4b)$$

This solution describes an expanding shell of Goldstone boson radiation of radius $R \sim t$ which has $\chi = \pi$ both at $r = 0$ and as $r \to \infty$ (see fig. 15.1).

Numerical simulations of spherically-symmetric texture collapse show that the actual behaviour of $\chi(r, t)$ at $r < t$ is more complicated [Barriola, 1993]. In addition, as the field Φ leaves the vacuum manifold during the texture unwinding, some of the energy gets converted into the massive modes of Φ. This energy loss, however, is roughly independent of texture size and is negligible for textures of astronomical dimensions.

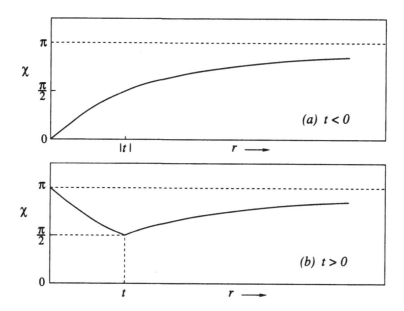

Fig. 15.1. Texture solution showing (a) texture collapse and (b) an expanding shell of Goldstone boson radiation in the aftermath.

Simulations of texture collapse from non-spherically-symmetric initial configurations suggest that the spherically-symmetric solution (15.3.3) is approached in the final stages of the collapse [Spergel, Turok, Press & Ryden, 1991; Prokopec, Sornborger & Brandenberger, 1992]. The simulations also demonstrated that it is not necessary for Φ to wind all the way around the three-sphere: winding more than half-way is sufficient for collapse. If Q is the winding number (15.1.14), then a rough criterion for collapse is $|Q| > 0.5$.

15.4 Gravitational field of a collapsing texture

Texture collapse can lead to high energy densities and strong gravitational fields. The relativistic nature of textures makes these fields rather different from those due to the collapse of ordinary matter. The linearized metric perturbation resulting from spherically-symmetric collapse (15.3.3) has been found by Turok & Spergel [1990], Durrer [1990] and Notzold [1991]. Properties of the exact solution of coupled Einstein and σ-model equations have been studied by Barriola & Vachaspati [1991a,b].

To find the gravitational field of a collapsing texture we first note that the solution (15.3.3) for χ depends only on the ratio r/t. This means that the texture evolves in a self-similar way: its size changes as $R \sim t$, but the 'shape' remains the same. Assuming that the gravitational field of the texture is also self-similar, we can write the metric in the form

$$ds^2 = e^\mu dt^2 + e^\nu dtdr - e^\lambda dr^2 - e^\omega r^2 d\Omega^2, \qquad (15.4.1)$$

where μ, ν, λ and ω are functions of $u = r/t$. The functions μ and ν can be eliminated by a coordinate transformation of the form

$$\tilde{t} = tf(u), \qquad \tilde{r} = rg(u). \qquad (15.4.2)$$

The only non-vanishing components of $S_{\mu\nu} = T_{\mu\nu} - \frac{1}{2}g_{\mu\nu}T$ are then given by

$$S_{tt} = \frac{\eta^2}{r^2}u^4\chi'^2, \qquad S_{rr} = \frac{\eta^2}{r^2}u^2\chi'^2,$$

$$S_{\theta\theta} = \eta^2 \sin^2 \chi, \qquad S_{\varphi\varphi} = S_{\theta\theta} \sin^2 \theta, \qquad (15.4.3)$$

$$S_{tr} = -\frac{\eta^2}{r^2}u^3\chi'^2,$$

where primes stand for derivatives with respect to u. Note that, surprisingly, $S_{\mu\nu}$ is independent of the metric coefficients λ and ω. When (15.4.1) and (15.4.3) are substituted in Einstein's equations and in the field equation for χ, one finds that the equations are consistent only if $\lambda' = 0$. The system reduces to two independent equations for ω and χ

[Barriola & Vachaspati, 1991a],

$$\omega'' + \frac{2}{u}\omega' + \frac{1}{2}\omega'^2 = -\epsilon\chi'^2 , \tag{15.4.4}$$

$$\chi'' + \frac{2}{u}\chi' + \omega'\chi' + e^{-\omega}\frac{\sin 2\chi}{u^2(u^2 - 1)} = 0 , \tag{15.4.5}$$

where

$$\epsilon = 8\pi G\eta^2 . \tag{15.4.6}$$

At this point we shall assume that the metric perturbation due to the texture is small, $|\omega| \ll 1$. This is justified if $\eta \ll 10^{18}$ GeV, so that $\epsilon \ll 1$. To lowest order in ω, we can use the flat-space solution (15.3.3) for χ, and (15.4.4) then takes the form

$$\omega'' + \frac{2}{u}\omega' = -\frac{4\epsilon}{(1 + u^2)^2} . \tag{15.4.7}$$

It is easily verified that this equation is also valid at $t > 0$, if one uses (15.3.4) to describe the unwinding of a texture. The general solution of (15.4.7) is

$$\omega = 2\epsilon u^{-1}\tan^{-1} u + A + Bu^{-1}. \tag{15.4.8}$$

The constants A and B are fixed by requiring that $\omega \to 0$ as $t \to -\infty$, and we can write the final form of the metric as [Notzold, 1991]

$$ds^2 = dt^2 - dr^2 - r^2(1 + \omega)d\Omega^2 , \tag{15.4.9}$$

where

$$\omega = -2\epsilon\left[1 + \frac{t}{r}\cot^{-1}(-t/r)\right] . \tag{15.4.10}$$

The trigonometric function in (15.4.10) has been chosen so that it applies at both positive and negative values of t.

Turning now to the properties of the metric (15.4.9), we first note that the lines of constant r, θ and ϕ are geodesics. This can be immediately seen from the geodesic equations

$$\ddot{t} = -\tfrac{1}{2}r^2\dot{\theta}^2\partial_t\omega ,$$
$$\ddot{r} = \tfrac{1}{2}\dot{\theta}^2\partial_r[r^2(1 + \omega)] , \tag{15.4.11}$$
$$[\dot{\theta}r^2(1 + \omega)]\dot{} = 0 ,$$

where dots represent derivatives with respect to the affine parameter along the geodesic. Hence, the coordinate system in the metric (15.4.9) is comoving with the particles which were initially at rest at $t \to -\infty$. Since the particles stay at constant r, one might think that they do not move with respect to the origin. (Note that Δr is the proper distance in the metric (15.4.9).) However, this is not so, because the origin is in fact not at $r = 0$. At late times, $t \gg r$, the asymptotic form of ω is $\omega \approx -2\pi\epsilon t/r$. The corresponding metric is flat and can be obtained from the Minkowski metric by a coordinate transformation $R = r - vt$, $T = t - vr$ with $v = \pi\epsilon$ (R and T are the Minkowski coordinates). The weak-field approximation breaks down at $r \sim \epsilon t$, but this is only a consequence of a coordinate singularity at $R = 0$, and the flat-space metric

$$ds^2 = dt^2 - dr^2 - (r - \pi\epsilon t)^2 d\Omega^2 , \qquad (15.4.12)$$

can be used at small values of R. The range of the coordinate r is determined by the condition $R > 0$, and we conclude that the particles fall towards the centre at a speed $v = \pi\epsilon$.

The behaviour of test particles in the field of a texture can be qualitatively understood in the following way. The gravitational mass of a texture within a radius r is

$$
\begin{aligned}
M(r) &= \int_0^r (T_t^t - T_r^r - T_\theta^\theta - T_\phi^\phi)\, 4\pi r^2 dr \\
&= 16\pi\eta^2 \int_0^r \frac{r^4 dr}{(r^2 + t^2)^2} .
\end{aligned}
\qquad (15.4.13)
$$

This mass is negligible when the size of the texture, t, is large compared to r, while for $|t| \lesssim r$ the gravitational acceleration of a particle is $GM(r)/r^2 \sim \epsilon/r$. A particle at a distance r from the origin experiences this acceleration for a time interval $\Delta t \sim r$ and reaches the velocity $v \sim \epsilon$, independent of r.

15.5 Cosmological evolution

The initial texture configuration is set at the phase transition when the scalar field Φ develops a vacuum expectation value. Φ takes values in the vacuum manifold $\mathcal{M}$ and winds around $\mathcal{M}$ in a non-trivial way on scales greater than the correlation length, $\xi \lesssim t$. Subsequent evolution is determined by the nonlinear dynamics of the field Φ. Topological texture 'knots' collapse at relativistic speeds and, when the gradient energy becomes sufficient to lift Φ over the potential barrier, they unwind themselves and dissipate into Goldstone bosons. As a result, the correlation length of Φ quickly grows to about the horizon scale, d_H, and then keeps growing with it [Turok, 1989b]. In this scaling regime texture knots collapse and unwind at a fixed rate per horizon volume per Hubble time,

$$\frac{dn}{dt} \approx \kappa d_H^{-4} , \qquad (15.5.1)$$

where $\kappa = $ const.

The coefficient κ can be interpreted as the probability for a non-trivial winding to occur in a correlation-length-size volume. This probability can be estimated by discretizing the vacuum manifold and the physical space, as was done for strings in §9.1.2. There, the vacuum manifold S^1 was approximated by the three vertices of an equilateral triangle. More generally, a d-sphere can be approximated by the vertices of a $(d+1)$-dimensional tetrahedron. A three-sphere, which is the vacuum manifold in this texture model, is then represented by a five-point set. The physical three-space R^3 can be triangulated by a diamond lattice, so that each point is connected to four others and the lattice spacing is $\sim \xi$. For the mapping from the vacuum manifold to R^3 to be non-trivial around some point of the lattice, that point and the four neighbouring points should all have different values of Φ. The probability of this occurring [Turok, 1989b] is

$$\kappa \sim 5 \times 4 \times 3 \times 2 \times 1/5^5 \approx 0.04 . \qquad (15.5.2)$$

Similar estimates have been obtained by Borrill, Copeland & Liddle [1991a] and by Leese & Prokopec [1992] using more sophisticated techniques.

In the radiation era, the texture field with a coherence length $d_{\rm H} \sim 2t$ has energy density $\rho_{\rm T} \sim \eta^2/d_{\rm H}^2$, which remains a fixed fraction of the total density,

$$\frac{\rho_{\rm T}}{\rho} \sim 8G\eta^2 . \qquad (15.5.3)$$

Most of this energy is transformed into massless Goldstone bosons in each Hubble time. The Goldstone bosons redshift like radiation, and their energy density $\rho_{\rm GB}$ obeys the equation

$$\dot{\rho}_{\rm GB} + 4\frac{\dot{a}}{a}\rho_{\rm GB} = k\frac{\eta^2}{t^3} , \qquad (15.5.4)$$

where $k \sim 1$ is a constant coefficient which at present must be determined numerically. During the radiation era $a \propto t^{1/2}$, so the solution of (15.5.4) can be written as

$$\frac{\rho_{\rm GB}}{\rho} \approx 30kG\eta^2 \ln(t/t_*) , \qquad (15.5.5)$$

where t_* is the time of the phase transition when the evolution began. In the matter-dominated era $\rho_{\rm GB}$ redshifts faster than the background density, and the dominant contribution to ρ_{GB} is given by the Goldstone bosons produced in the last Hubble time. Hence, for $t \gg t_{\rm eq}$, we have $\rho_{\rm GB}/\rho \approx 20kG\eta^2$. Interactions of Goldstone bosons are suppressed by inverse powers of the symmetry breaking scale η, and for very large η the Goldstone boson background can be practically unobservable.

Numerical simulations of texture evolution have been performed by
Spergel *et al.* [1991]. Rather than using the nonlinear σ-model (15.1.3),
they evolved the full $SO(4)$ model (15.1.1) with the potential (15.1.8).
The advantage of this approach is that it gives a realistic treatment of
texture unwinding. The chief difficulty is that in comoving coordinates
the unwinding scale, $m_\Phi^{-1} \sim (\sqrt{\lambda}\,\eta)^{-1}$, changes inversely with the scale
factor a and rapidly becomes smaller than the resolution of the simula-
tion. To circumvent this problem, Spergel *et al.* [1991] made the coupling
λ in the potential (15.1.8) time-dependent, so as to keep m_Φ^{-1}/a a constant. This procedure is justified by the fact that the field Φ remains in
the vacuum manifold during most of its evolution, and one expects that
the dynamics of Φ will not be strongly affected by the time variation of
the potential, as long as the vacuum manifold remains fixed.

Spergel *et al.* [1991] evolved the field Φ in an FRW universe, starting
with a random initial configuration. They observed texture knots collapsing and unwinding on the horizon scale. The rate of texture collapse
observed in the simulations is well fitted by (15.5.1) with $\kappa \approx 0.04$. The
seemingly perfect agreement with (15.5.2) is, of course, fortuitous, since
(15.5.2) can be expected to give only an order-of-magnitude estimate of
κ. The energy density of Goldstone bosons found in the simulations is
also in good agreement with (15.5.5) for $k \approx 0.2 \pm 0.1$.

15.6 Cosmological effects of texture

15.6.1 Microwave background anisotropy

The gravitational fields of textures can significantly distort the isotropy
of the cosmic microwave background. The coherence length of the texture
field is $\sim d_{\rm H}$, and the density fluctuation induced by textures on any scale
at horizon crossing can be roughly estimated from (15.5.3),

$$(\delta\rho/\rho)_{\rm H} \sim 8G\eta^2 . \tag{15.6.1}$$

The microwave temperature fluctuations are expected to be of a comparable magnitude,

$$\delta T/T \sim (\delta\rho/\rho)_{\rm H} \sim 8G\eta^2 . \tag{15.6.2}$$

Apart from the gaussian fluctuations due to a random variation of Φ
over horizon scales, the temperature distribution will have characteristic
features produced by collapsing texture knots. Turok & Spergel [1990],
Durrer [1990] and Notzold [1991] calculated the photon frequency shift
in the gravitational field of a collapsing spherical texture (15.4.9). The
result is

$$\frac{\delta T}{T} = \frac{\pi\epsilon t_0}{(t_0^2 + 2r_0^2)^{1/2}} , \tag{15.6.3}$$

where t_0 and r_0 are the time and radius of closest approach of the photon to the centre of collapse, which is assumed to be at $t = r = 0$. Photons reaching the closest approach before the collapse ($t_0 < 0$) are redshifted, and those reaching it after the collapse ($t_0 > 0$) are blueshifted. This can be illustrated by the spacetime diagram in fig. 15.2. The former photons climb out of the potential well of the collapsing texture, while the latter fall into the expanding shell of Goldstone boson radiation.

Consequently, collapsing textures are seen as cold or hot spots on the microwave sky. To estimate the number of such spots, it is convenient to use comoving coordinates and the conformal time τ. Using the relation $d_H = a(\tau)\tau$, we can rewrite the texture collapse rate (15.5.1) as

$$\frac{dn_c}{d\tau} \approx \kappa \tau^{-4}, \qquad (15.6.4)$$

where n_c is the number density of knots per comoving volume. The number of spots N is found by integrating (15.6.4) along the photon's wavefront back to some redshift z. Assuming that only textures collapsing within a fraction of the horizon distance, $\beta\tau$, from the wavefront have an effect, we obtain [Spergel et al., 1991]

$$N(z) = \int_{\tau_z}^{\tau_0} d\tau \, 4\pi(\tau_0 - \tau)^2 \beta\tau \left(\frac{dn_c}{d\tau}\right) \approx 2\pi\kappa\beta \left(\frac{\tau_0}{\tau_z}\right)^2 \approx 0.2\beta z. \quad (15.6.5)$$

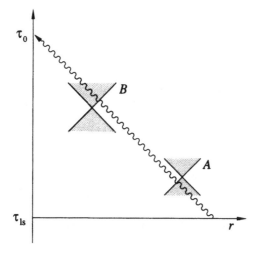

Fig. 15.2. Spacetime diagram showing a photon propagating towards us, $\tau = \tau_0$, from the surface of last scattering, $\tau = \eta_s$. The photon is redshifted as it comes out of the collapsing knot at A and is blue-shifted as it falls into the expanding shell of radiation at B.

Here, τ_0 is the present conformal time and we have assumed that $z \gg 1$. The parameter $\beta < 1$ accounts for the fact that the spherical collapse model (15.3.3) and the expression (15.6.3) for the temperature fluctuation are expected to apply only towards the end of the collapse. The typical angular size of cold and hot spots due to textures collapsing at a redshift z is $\beta\tau_z/\tau_0 \approx \beta z^{-1/2}$ radians. The maximum temperature fluctuation at the centre of the spots is $\delta T/T \approx \pi\epsilon$. The spots with a distribution (15.6.5) should occur at all redshifts up to the last scattering surface, $z = z_{\text{ls}}$. For example, with $z \sim 40$, (15.6.5) implies that there should be about 10β hot (and cold) spots of angular size $\gtrsim \beta \times 10°$.

Bennett & Rhie [1993] numerically generated a number of $\delta T/T$ patterns using a σ-model simulation of texture evolution. To mimic the finite width of the microwave detector beam, the patterns were smoothed over $10°$ angular scales. The resulting temperature maps have a gaussian appearance, without any prominent features. This indicates that a smaller beam width would be required to resolve the hot and cold spots from collapsing texture knots. The rms value of $\delta T/T$ is

$$(\delta T/T)_{\text{rms}} = (10 \pm 1)G\eta^2, \qquad (15.6.6)$$

and the power spectrum is essentially identical to that due to a Harrison–Zel'dovich spectrum of primordial density fluctuations (see §1.3.6). If the observed fluctuations (1.3.37) are due to textures, then we must have $G\eta^2 = (1.0 \pm 0.2) \times 10^{-6}$ or, equivalently,

$$\epsilon = (2.5 \pm 0.5) \times 10^{-5}. \qquad (15.6.7)$$

15.6.2 Structure formation

As already mentioned, an inhomogeneous texture field of coherence length $\sim d_{\text{H}}$ induces matter density fluctuations on the horizon scale. The fractional amplitude of the fluctuations at horizon crossing is scale-independent and is given by (15.6.1). On scales greater than the horizon, the texture density fluctuations are balanced by fluctuations in the matter density. This is a consequence of energy conservation at texture formation – these are isocurvature fluctuations. On the horizon scale, as texture modes of wavelength $\sim t$ start to evolve at relativistic speeds, the energy balance is destroyed, and matter density fluctuations are left behind. That a spectrum of the form (15.6.1) can be produced as a result of a global symmetry breaking was pointed out by Vilenkin [1982b] (see also Press [1980]). In the context of cosmic texture this idea was revived by Turok [1989b] who argued that density fluctuations due to texture could be responsible for the formation of structure in the universe.

Although the fluctuation spectrum (15.6.1) is of Harrison–Zel'dovich form, the nonlinear dynamics of textures introduce strongly non-gaussian features. In particular, the collapse of texture knots gives rise to isolated spikes in the density field. We shall now proceed to find the matter density profile in such spikes, assuming that the universe is dominated by cold dark matter. Recall from §15.4 that for a spherically-symmetric texture collapse, test particles develop a velocity $v = \pi\epsilon$ towards the centre, where $\epsilon = 8\pi G\eta^2$. If ρ_0 is the initial density of particles, then the mass of matter within a sphere of comoving radius r is $M = 4\pi\rho_0 r^3/3$. The physical radius of the sphere is $R = r - \pi\epsilon t$, where t is the time elapsed after the knot collapse. The density profile $\rho(R)$ can be found from

$$\rho(R) = \frac{1}{4\pi R^2}\frac{dM}{dR} = \rho_0 \left(1 + \frac{\pi\epsilon t}{R}\right)^2.$$

(15.6.8)

Hence, the fractional density fluctuation is [Turok & Spergel, 1990; Notzold, 1991]

$$\frac{\delta\rho}{\rho_0} = \frac{2\pi\epsilon t}{R}.$$

(15.6.9)

The mass distribution of bound objects formed by collapsing knots has been calculated by Gooding, Spergel & Turok [1991] using the texture collapse rate (15.6.4) and (15.6.9). Subsequent analysis has shown, however, that distributed field gradients give a larger contribution to $\delta\rho/\rho$ than the isolated knots, and thus it appears that the only reliable way to study the texture-seeded structure formation is through numerical simulations. Such simulations have been performed by Park, Spergel & Turok [1991] who combined the texture evolution code of Spergel *et al.* [1991] with an N-body clustering code in a cold dark matter dominated universe. Hydrodynamical and dissipative effects have been included by Cen, Ostriker, Spergel & Turok [1991]. Although this work captured the qualitative features of the texture scenario, the size of the simulation box was apparently too small, and the numerical results are therefore unreliable.

More recently, Bennett, Rhie & Weinberg [1993] studied the linear evolution of texture-seeded density fluctuations using a σ-model numerical code. The results are very similar to those discussed in §14.5.4 for global monopoles. With ϵ from (15.6.7), the density fluctuation on the scale of $8h^{-1}$Mpc at the present time lies in the range 0.25 to 0.4, depending on the Hubble parameter h. These values of $\delta\rho/\rho$ can be reconciled with $\delta\rho/\rho \approx 1$ suggested by galaxy counts if galaxies are actually more strongly clustered than the mass distribution. The 'biasing' factors required to account for this difference are $b \approx 2.5$ for $h = 1$ and $b \approx 4$ for

$h = 0.5$. (For the definition of b refer to (1.3.49).*) We were informed that similar results have been more recently obtained by Pen, Spergel & Turok [1993].

15.7 Gauge texture

For a local gauge symmetry breaking, the scalar field gradients can be cancelled by the gauge potential, and a non-trivial winding of the scalar field Φ does not necessarily indicate the presence of a topological defect. Consider, for example,

$$\Phi(x) = D(g)\Phi_0, \qquad A_\mu = ie^{-1}g^{-1}\partial_\mu g, \qquad (15.7.1)$$

where $g(x)$ is a position- and time-dependent element of the symmetry group G, $D(g)$ is the transformation matrix of the Higgs field Φ, and Φ_0 is some point in the vacuum manifold $\mathcal{M}$. It is clear from (15.7.1) that

$$\mathcal{D}_\mu\Phi = 0, \qquad \mathcal{F}_{\mu\nu} = 0, \qquad (15.7.2)$$

and that the field Φ remains in $\mathcal{M}$ everywhere. Hence, (15.7.1) is a vacuum configuration of zero total energy. If $\pi_3(\mathcal{M})$ is non-trivial, then configurations with different windings of the field Φ describe topologically distinct vacua which can tunnel into one another. For a discussion of this tunnelling and of the gauge theory vacuum structure to which it leads, the reader is referred to a review by Coleman [1985a]. The important point for us here is that the vacuum configurations (15.7.1) have no classical dynamics – the Goldstone bosons of the global texture have been 'eaten' by the Higgs mechanism.

This is not to say, however, that the classical dynamics of a gauge texture is always trivial. Consider, for example, a non-vacuum field configuration in which $\Phi(\mathbf{x})$ has a non-trivial winding and $A_\mu = 0$. The dynamical evolution of this configuration is governed by the tendency of the gauge field to compensate the gradients of Φ and the competing tendency for the scalar field to unwind. The outcome is determined by the characteristic size of the texture L [Turok & Zadrozny, 1990]. In order to cancel the scalar gradients, the gauge potential must reach a magnitude $A_\mu \sim (eL)^{-1}$, where e is the gauge coupling. The field equation for A_μ can be symbolically written as $\Box A_\mu \sim j_\mu$, where the current is given by $j \sim e\eta^2/L$ and η is the magnitude of Φ. These equations suggest that the characteristic timescale for A_μ to reach the required magnitude is

---

* A large value of the biasing factor in the texture scenario ($b \sim 3$–4) was also suggested by Park et al. [1991] and by Bennett et al. [1993] in order to avoid unreasonably large velocity dispersion in clusters of galaxies. In this sense the microwave background measurements do not give much of an additional constraint, although it is not clear what physics can provide the needed 'biasing'.

$\tau_A \sim (e\eta)^{-1} \sim m_A^{-1}$, where m_A is the gauge boson mass. On the other hand, the timescale for texture collapse is $\tau_\Phi \sim L$. Hence, we expect that textures of size $L \ll m_A^{-1}$ will unwind, while textures with $L \gg m_A^{-1}$ will not. If the scalar field mass m_Φ is much greater than m_A, then in the range $m_\Phi^{-1} \ll L \ll m_A^{-1}$ the behaviour of gauge textures will be similar to that of global textures. This is not surprising since in the limit $e \to 0$, when m_A becomes small, a gauge texture should turn into a global texture.

In a cosmological context, gauge textures are not expected to exhibit the scaling behaviour described in §15.5. They can do so only for a brief period of time, $m_\Phi^{-1} < t < m_A^{-1}$. By the end of this period, the fields Φ and $\mathcal{A}_\mu$ evolve into a vacuum configuration of the form (15.7.1) with all the excess energy transferred into propagating gauge and Higgs bosons.

When the gauge field $\mathcal{A}_\mu$ adjusts itself to cancel the Higgs field gradients, it also develops a non-trivial winding. This change of winding can induce a change in certain fermionic charges through the axial current anomalies ['t Hooft, 1976b]. For example, in electroweak theory, baryon and lepton numbers can change due to a variation in the winding number of the $SU(2)$ gauge field. This baryon-number-violating process plays the central role in a recently proposed mechanism for baryon number generation at the electroweak phase transition [Shaposhnikov, 1987, 1988; Turok & Zadrozny, 1990; McLerran *et al.*, 1991]. This topic, however, is beyond the scope of this monograph and the reader is referred to the original literature.

16

Topological defects and inflation

Inflation increases the size of the universe by an enormous factor, so that all of the presently observed universe arises from one tiny initial region (refer to §2.4). Topological defects which formed before inflation or during its early stages would have been drastically diluted by the expansion and never seen again. The only defects we can hope to observe now are those formed after or near the end of inflation. In this chapter we shall discuss various defect formation mechanisms and the constraints, imposed by inflation, on the corresponding symmetry breaking scales. As before, we shall use the term 'inflaton' for the scalar field whose vacuum energy drives the inflation.

16.1 Formation of defects after inflation

As discussed in §2.4, microwave background measurements impose a bound on the thermalization temperature after inflation,

$$T_{\text{th}} \lesssim 2 \times 10^{16} \text{GeV} \,. \tag{16.1.1}$$

The critical temperature of any post-inflationary phase transition must satisfy $T_{\text{c}} < T_{\text{th}}$, and since the symmetry breaking scale η is usually comparable to T_{c}, one expects that the only defects which can be formed after inflation will have

$$\eta \lesssim T_{\text{th}} \,. \tag{16.1.2}$$

The string scenario of structure formation, as well as scenarios based on global monopoles and textures, require $\eta \sim 10^{16} \text{GeV}$. However, (16.1.1) and (16.1.2) suggest that production of such defects after inflation is only marginally possible [Pollock, 1987; Vishniac, Olive & Seckel, 1987; Lyth, 1987]. The situation is slightly improved for global strings, for which the required value of η is $\sim 10^{15} \text{GeV}$. But realistic models are still difficult to construct, because the thermalization temperature T_{th} is typically much smaller than the maximum value (16.1.1).

The formation of superheavy defects is relatively easy to arrange in models based on Brans–Dicke-type gravity [Copeland, Kolb & Liddle, 1990]. In such models, the thermalization of the false vacuum can occur through bubble nucleation and can be very efficient. In this case, the thermalization temperature can therefore be made very close to the value (16.1.1).* Moreover, the expectation values of the inflaton field in different bubbles are uncorrelated, so defects can be formed at the inflationary phase transition itself. This is in contrast to slow-rollover inflation, when the inflaton field is expected to be very homogeneous on all scales in the observable universe. It should be noted that the bound (16.1.2) is not a rigid one. The 'soft link' in the argument is the assumption that $\eta \sim T_c$. In many models the coupling constants can be adjusted so that $T_c \ll \eta$, and defects with $\eta > T_{th}$ can be formed as long as $T_c < T_{th}$. As an example consider the Higgs model (2.1.10). The critical temperature is given by (2.3.20) and, for $\lambda \ll e^2$, we have $T_c \approx 2 \left(\lambda/e^2\right)^{1/2} \eta \ll \eta$.

The critical temperature cannot be arbitrarily small without violating the 'naturalness' condition. In the Higgs model, the values of $\lambda \ll e^4$ are unnatural in the sense that radiative corrections to λ are $\sim e^4$, and $\lambda \ll e^4$ would require fine-tuning. Hence, for natural values of the couplings, $T_c \gtrsim e\eta$. If e is the usual gauge coupling, $e^2 \sim 10^{-2}$, then T_c will not be below $\sim 0.1\,\eta$. However, one can consider models with a gauge singlet field φ, replacing the gauge coupling e by a Higgs coupling to another scalar or by a Yukawa coupling to a fermion. With an appropriate choice of the coupling constant, one can reduce T_c to the desired level. Another possibility is to consider a supersymmetric theory in which certain radiative corrections vanish because of supersymmetry. Lazarides, Panagiotakopoulos & Shafi [1987a,b] have argued that in some supersymmetric models T_c can be smaller than η by many orders of magnitude.

16.2 Phase transitions during inflation

A phase transition can occur during inflation if the symmetry breaking field φ is coupled to the inflaton field χ [Shafi & Vilenkin, 1984a,b]. To illustrate the idea, consider the simple model ,

$$\mathcal{L} = |\partial_\mu \varphi|^2 - m_0^2 |\varphi|^2 - \tfrac{1}{4}\lambda |\varphi|^4 , \qquad (16.2.1)$$

with an additional interaction term,

$$\mathcal{L}_{int} = g\chi^2 |\varphi|^2 . \qquad (16.2.2)$$

---

* A potential problem with making T_{th} very high is that we want magnetic monopoles to be inflated away, and thus the monopole-generating phase transition must have $T_c > T_{th}$.

We assume that m_0^2, λ, $g > 0$ and that the inflaton potential $V(\chi)$ is of the form shown in fig. 2.5(*a*). The effective mass-squared of the field φ is

$$m_\varphi^2 = m_0^2 - g\chi^2 \,. \tag{16.2.3}$$

The metric of the inflating universe is that of de Sitter space,

$$ds^2 = dt^2 - e^{2Ht}d\mathbf{x}^2 \,, \tag{16.2.4}$$

with the Hubble parameter H given by

$$H^2 \approx \frac{8\pi G}{3}V(\chi) \,. \tag{16.2.5}$$

During the course of inflation, the field χ rolls slowly down the potential from $\chi \approx 0$ to some large value η_χ. If $\eta_\chi > m_0/\sqrt{g} \equiv \chi_c$, then m_φ^2 becomes negative at $\chi > \chi_c$, and the field φ develops a vacuum expectation value. In our example φ is a complex scalar field, and this phase transition results in the formation of strings. As χ continues to roll towards η_χ, the characteristic length-scale of the string network grows exponentially. However, if the phase transition occurs sufficiently close to the end of inflation, this length does not become larger than the comoving size of the presently observable universe, and the strings are not completely inflated away. Domain walls, monopoles and textures can be formed in a similar way. Examples of specific models can be found in Shafi & Vilenkin [1984b], Kofman & Linde [1987], Vishniac, Olive & Seckel [1987] and Hodges & Primack [1991].

When χ is near χ_c, so that

$$\left|m_\varphi^2\right| \lesssim H^2 \,, \tag{16.2.6}$$

quantum fluctuations of the field φ are important. In the next section we shall see that these fluctuations can strongly affect the spatial distribution of defects if the range

$$\Delta\chi \sim H^2/m_0\sqrt{g} \,, \tag{16.2.7}$$

corresponding to (16.2.6), is traversed in more than a few Hubble times. In the opposite case when $\Delta\chi$ is traversed in a time $\Delta t \lesssim H^{-1}$, the phase transition is fast and the pattern of created defects is similar to that in a thermal phase transition. The correlation length in this case is $\xi \lesssim H^{-1}$ (at the time of the transition). The condition for a fast phase transition can be written using $\Delta t \sim \Delta\chi/\dot\chi$ with $\dot\chi$ from (2.4.6),

$$H\Delta t \sim \frac{H^4}{m_0\sqrt{g}\,|V'(\chi_c)|} \lesssim 1 \,. \tag{16.2.8}$$

In the 'chaotic inflation' scenario, the inflaton potential $V(\chi)$ is of the form shown in fig. 2.5(*b*), and the field χ rolls from some large value

χ_0 towards $\chi \approx 0$. The model (16.2.1–2) can be adapted to this case by choosing m_0^2, $g < 0$ and $\lambda > 0$. Defects will be formed if $|\chi_0| > |g^{-1/2}m_0|$.

Yokoyama [1988, 1989] has pointed out that the formation of defects during inflation can also be triggered by a coupling of the field φ to the curvature. As an example consider the model (16.2.1) with $m_0^2 < 0$ and with the inflaton coupling (16.2.2) replaced by

$$\mathcal{L}_\xi = -\xi R |\varphi|^2, \tag{16.2.9}$$

where R is the spacetime curvature. The effective mass-squared of the field φ is then

$$m_\varphi^2 = m_0^2 + \xi R, \tag{16.2.10}$$

where R can be found from the Einstein equations,

$$R \approx 32\pi G V(\chi). \tag{16.2.11}$$

As χ rolls down the potential, the curvature decreases and, with an appropriate choice of parameters, m_φ^2 can change sign at some point during inflation.

16.3 Formation of defects by quantum fluctuations

In cases when the symmetry is broken before or during the early stages of inflation, defects can still be formed by quantum fluctuations of the field φ [Vishniac, Olive & Seckel, 1987; Linde & Lyth, 1990; Hodges & Primack, 1991; Nagasawa & Yokoyama, 1992]. These fluctuations are particularly important when the effective mass-squared of φ is small,

$$m_\varphi^2 \ll H^2. \tag{16.3.1}$$

Before discussing the physics of defect formation, it will be useful to review some basic properties of a quantum scalar field in an inflationary universe.

For a massless, real field φ in de Sitter space, quantum fluctuations can be pictured as Brownian motion: the magnitude of φ on the scale of the de Sitter horizon (H^{-1}) fluctuates by an amount $\pm H/2\pi$ per Hubble time (H^{-1}). The quantum nature of the fluctuations becomes unimportant when the expansion of the universe stretches their wavelength beyond the horizon. At that point the amplitude of the fluctuations 'freezes', and can be subsequently treated as a part of the classical scalar field. The history of the field φ in each horizon-size region describes a random walk (on the φ-axis) of step-size $\sim H/2\pi$ at time intervals $\sim H^{-1}$. Different regions are described by different walks, and thus the value of φ varies in space. The average value of φ^2 grows linearly with time,

$$\langle \varphi^2 \rangle = \left(\frac{H}{2\pi}\right)^2 \frac{t}{H^{-1}} = \frac{H^3 t}{4\pi^2}, \tag{16.3.2}$$

where the averaging is over many horizon-size regions [Vilenkin & Ford, 1982; Linde, 1982b; Starobinsky, 1982]. The mean-square variation in the values of φ for two regions separated by a distance ℓ at a time t is

$$(\delta\varphi_\ell)^2 \approx \left(\frac{H^3}{4\pi^2}\right)(t - t_\ell) \approx \left(\frac{H}{2\pi}\right)^2 \ln(H\ell).\qquad (16.3.3)$$

Here, $t_\ell = t - H^{-1}\ln(H\ell)$ is the time when the comoving scale ℓ became greater than the horizon.

If φ has a small mass, $m_\varphi^2 \ll H^2$, then the growth of $\langle\varphi^2\rangle$ saturates after a time interval $\Delta t \sim H/m_\varphi^2$ at the value

$$\langle\varphi^2\rangle \approx \frac{3H^4}{8\pi^2 m_\varphi^2}.\qquad (16.3.4)$$

Equation (16.3.3) for $\delta\varphi_\ell$ is still valid on scales $\ell \ll \ell_*$, where ℓ_* is defined by $(\delta\varphi_{\ell_*})^2 \approx \langle\varphi^2\rangle$. In regions of size $\sim \ell_*$, the average values of φ are uncorrelated; their distribution is gaussian with a zero mean and rms value $\varphi_{\mathrm{rms}}^2 \approx \langle\varphi^2\rangle$.

Returning now to defect formation, we shall first consider the model

$$\mathcal{L} = \tfrac{1}{2}\left(\partial_\mu\varphi\right)^2 - \tfrac{1}{4}\lambda\left(\varphi^2 - \eta^2\right)^2,\qquad (16.3.5)$$

assuming for simplicity that φ is a real scalar field with no coupling to the inflaton or to curvature. If φ is treated classically, then one expects it to settle into one of the minima at $\varphi = \pm\eta$ and to be very uniform throughout the universe after inflation. However, quantum fluctuations can take φ over the barrier separating the two minima. This would lead to formation of domain walls.

The effective mass of φ at the minima is $m_\varphi^2 = 2\lambda\eta^2$, and for $\lambda\eta^2 \ll H^2$ we can use the Brownian picture of quantum fluctuations. From (16.3.4), the amplitude of fluctuations about the minimum is approximately given by $\langle(\delta\varphi)^2\rangle \approx 3H^4/16\pi^2\lambda\eta^2$. If $\langle(\delta\varphi)^2\rangle \ll \eta^2$, then domain wall formation is strongly suppressed. In the opposite limit, when $\langle(\delta\varphi)^2\rangle \gg \eta^2$, quantum fluctuations spread the field φ over a range far beyond the minima of the potential. To estimate this range, we can no longer use the value of m_φ^2 at $\varphi = \pm\eta$. For $|\varphi| \gg \eta$, $m_\varphi^2 = V''(\varphi) \approx 3\lambda\varphi^2$. Substituting this into (16.3.4) and setting $\varphi^2 \approx \langle\varphi^2\rangle$ we find

$$\varphi_{\mathrm{rms}} \approx \left(\langle\varphi^2\rangle\right)^{1/2} \approx (8\pi^2\lambda)^{-1/4} H.\qquad (16.3.6)$$

This range should now be compared with the expected variation of φ on scales within the observable universe. If ℓ_0 is the comoving scale of the present horizon at the end of inflation, then for reasonable values of

H, $\ln(H\ell_0) \lesssim 100$ and (16.3.3) gives

$$\delta\varphi_{\ell_0} \lesssim H.$$ (16.3.7)

The typical wall separation is determined by the scale ℓ_* on which $\delta\varphi_{\ell_*} \sim \varphi_{\rm rms}$. To ensure that this scale is smaller than the present horizon, we must require that $\delta\varphi_{\ell_0} \gtrsim \varphi_{\rm rms} \gtrsim \eta$, or

$$1 \lesssim 8\pi^2\lambda \lesssim \left(\frac{H}{\eta}\right)^4.$$ (16.3.8)

For $\ell_* \ll \ell_0$ the visible universe contains many domain walls and has an average value of φ very close to zero, $\bar{\varphi}_{\ell_0} \approx 0$. In the opposite limit, $\ell_* \gg \ell_0$, regions of size ℓ_0 have average values of φ anywhere in the range $-\varphi_{\rm rms} \lesssim \bar{\varphi}_{\ell_0} \lesssim \varphi_{\rm rms}$. Domain walls are present only in regions with $|\bar{\varphi}_{\ell_0}| \lesssim \delta\varphi_{\ell_0}$. The probability that the visible universe contains domain walls can then be estimated as [Hodges & Primack, 1991]

$$\mathcal{P}_{\rm w} \sim \delta\varphi_{\ell_0}/\varphi_{\rm rms} \sim (8\pi^2\lambda)^{1/4}.$$ (16.3.9)

The spatial distribution of walls inside individual regions depends on the ratio $r = |\bar{\varphi}_{\ell_0}|/\delta\varphi_{\ell_0}$. To illustrate this effect, Hodges & Primack [1991] performed a numerical simulation of a scalar field in a two-dimensional de Sitter space (which is equivalent to looking at a slice of a three-dimensional simulation). The simulation was performed on a square lattice representing a comoving region of initial size H^{-1}. The initial value of φ was set to a constant, $\varphi = \bar{\varphi}$, at all points of the lattice. After each two-folding of expansion the regions which had size H^{-1} at the previous step were divided into four pieces and a fluctuation, drawn from a gaussian distribution of width $(H/2\pi)(\ln 2)^{1/2}$, was added to φ in each of these pieces.[*] After N steps, the typical variation of φ in the whole comoving region is $\delta\varphi \simeq (H/2\pi)(N\ln 2)^{1/2}$. Domain walls are formed where $\varphi = 0$ and in fig. 16.1 the domain wall pattern is shown for two values of the ratio $r = \bar{\varphi}/\delta\varphi$. For $r \lesssim 1$, domain walls are abundant and there are many large-scale walls but, as r is increased above unity, the walls become sparse and only relatively small closed walls are present.

Strings, monopoles and textures can be formed by the same mechanism. For example, consider the model (16.3.5) where φ now is an n-component field and $\varphi^2 \equiv \varphi_1^2 + \ldots + \varphi_n^2$. The model has $O(n)$ symmetry and gives rise to strings, monopoles and textures for $n = 2, 3$ and 4, respectively. The requirement that the typical distance between the defects should be

---

[*] This method of simulating the quantum fluctuations of a scalar field in de Sitter space was introduced by Aryal & Vilenkin [1987].

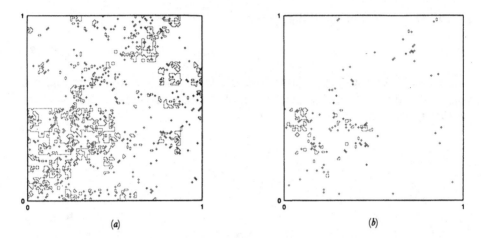

Fig. 16.1.   Domain walls formed by quantum fluctuations in regions of de Sitter space with different values of the ratio $r = \bar{\varphi}/\delta\varphi$: (a) $r = 1$, (b) $r = 2$ (from Hodges & Primack [1991]).

smaller than the present horizon leads to the same conditions (16.3.8). However, (16.3.9) for the probability of defect formation when $\lambda < 10^{-2}$ needs to be modified. Defects are formed when all the components φ_i with $i = 1, \ldots, n$ have average values smaller than $\delta\varphi_{\ell_0}$. The probabilities for strings, monopoles and textures can therefore be estimated as

$$\mathcal{P}_s \sim (8\pi^2\lambda)^{1/2}\,,$$
$$\mathcal{P}_m \sim (8\pi^2\lambda)^{3/4}\,, \qquad\qquad (16.3.10)$$
$$\mathcal{P}_t \sim 8\pi^2\lambda\,.$$

The spatial distributions of walls, strings and monopoles in this regime are related to one another. A realization of a string or monopole distribution can be obtained by intersecting two or three independent domain wall distributions, respectively. These 'domain walls' are, of course, just the surfaces where different components of φ are equal to zero. In regions where all components of $\bar{\varphi}_{\ell_0}$ are much smaller than $\delta\varphi_{\ell_0}$, domain walls are abundant, and we expect many large-scale strings and a large number of monopoles. However, if at least one component of $\bar{\varphi}_{\ell_0}$ is substantially greater than $\delta\varphi_{\ell_0}$, only small loops of string or nearby monopole pairs can be expected.

Can the mechanism described in this section be responsible for the production of superheavy defects required in galaxy formation scenarios? In the case when $\ell_* < \ell_0$, it follows from (16.3.8) and (2.4.14) that $\eta \lesssim H \lesssim 10^{15}\text{GeV}$. This is too low for gauge strings, global monopoles and textures, which require $\eta \sim 10^{16}\text{GeV}$, but could still work for global strings, in which case only $\eta \sim 10^{15}$GeV is required. Defects with $\eta \sim 10^{16}$GeV $> H$ can be formed when $\ell_* > \ell_0$ with probabilities given by (16.3.10). Note, however, that the second condition in (16.3.8) still has to be satisfied and the resulting probabilities are small: $\mathcal{P}_s \lesssim 10^{-2}$, $\mathcal{P}_m \lesssim 10^{-3}$, $\mathcal{P}_t \lesssim 10^{-4}$.

16.4 Defect solutions in de Sitter space

The rapid expansion of the universe during inflation can significantly affect the internal structure of topological defects. Consider, for example, the double-well model (16.3.5) which has domain wall solutions. In flat space, the thickness of the domain walls is

$$\delta_0 = \left(\sqrt{\lambda}\,\eta\right)^{-1}. \tag{16.4.1}$$

The corresponding solutions in an expanding universe will, in general, be time-dependent, so that for a wall lying in the xy-plane, $\varphi = \varphi(z,t)$ with

$$\varphi(z=0) = 0, \qquad \varphi(z \to \pm\infty) = \pm\eta. \tag{16.4.2}$$

A special feature of de Sitter space (16.2.4) describing an inflating universe is that it has a constant expansion rate H. This suggests that our model may have stationary solutions,

$$\varphi = \varphi(\zeta), \qquad \zeta = Hze^{Ht}, \tag{16.4.3}$$

in which the scalar field depends only on the proper distance from the xy-plane. And indeed, with this ansatz, the field equation for φ in de Sitter space becomes

$$(\zeta^2 - 1)\varphi'' + 4\zeta\varphi' + \lambda H^{-2}\varphi(\varphi^2 - \eta^2) = 0. \tag{16.4.4}$$

For a thin wall with $\delta_0 \ll H^{-1}$, most of the variation of φ between $-\eta$ and $+\eta$ occurs in a region where $|\zeta| \ll 1$ and (16.4.4) reduces to its flat-space form. In this case $\varphi(\zeta)$ is well approximated by the flat-space solution (13.1.8). The effect of de Sitter expansion is significant when $\delta_0 \sim H^{-1}$ and becomes dramatic when $\delta_0 \to (2H)^{-1}$ [Basu & Vilenkin, 1993]. In this limit, the wall thickness grows without bound,

$$\delta \sim H^{-1}[1 - 2(H\delta_0)^2]^{-1/2}, \tag{16.4.5}$$

and no regular solutions to (16.4.4) and (16.4.2) exist for $2H\delta_0 \geq 1$. Similar results have been obtained for strings and monopoles. In all three cases, stationary solutions do not exist when the flat-space size of the defect core, δ_0, exceeds some critical value, $\delta_{\max} \sim H^{-1}$. Thicker defects cannot survive in de Sitter space as coherent objects. They are smeared by the expansion of the universe, so that their thickness grows as $\delta \sim e^{Ht}$.

The formation of defects by quantum fluctuations, which was discussed in the previous section, assumes that $m_\varphi << H$, so that $\delta_0 >> H^{-1}$. Hence, in this case defects cannot be considered to be 'formed' until the end of inflation, and instead of 'defects' it is more appropriate to describe them as 'zeros of the field φ'. However, since defect formation eventually occurs where there are zeros of φ, the conclusions of §16.3 remain unchanged.

16.5 Quantum nucleation of defects

16.5.1 Physics of nucleation

As we observed at the beginning of this chapter, topological defects produced in phase transitions before or during early stages of inflation are diluted by an enormous factor and are never seen again. However, it can be shown that spherical domain walls, circular loops of string, and monopole–antimonopole pairs can be continuously created during inflation by quantum-mechanical tunnelling [Basu, Guth & Vilenkin, 1991]. As a result, topological defects corresponding to pre-inflationary phase transitions can still be present after inflation with non-negligible densities.

A semiclassical description of the nucleation process can be obtained using the instanton approach (for a review of instantons see, for example, Coleman [1985a]). Euclideanized de Sitter space is a four-sphere of radius H^{-1}. The instanton describing the nucleation of a domain wall is a maximal three-sphere (see fig. 16.2). The three-sphere represents the world surface of the wall, and the word 'maximal' means that its radius is equal to the radius H^{-1} of the four-sphere in which it is embedded. It is clear that this world surface is an extremum of the area, and thus the corresponding world history is a solution of the Euclidean equations of motion for the wall.

If the four-sphere is described by embedding it in a five-dimensional Euclidean space,

$$\vec{\xi}^2 + w^2 + \tau^2 = H^{-2}, \qquad (16.5.1)$$

where $\vec{\xi}$ is a three-vector, the instanton three-sphere can be described by

$$\vec{\xi}^2 + \tau^2 = H^{-2}, \qquad (16.5.2)$$
$$. \; w = 0 \, .$$

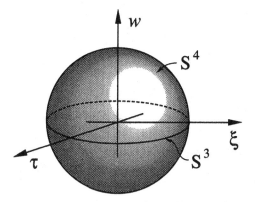

Fig. 16.2. The instanton describing the nucleation of a thin domain wall in de Sitter space is a maximal three-sphere embedded in a four-sphere of radius H^{-1}. This diagram shows two fewer dimensions than the actual instanton.

The Euclidean action of the instanton is the wall tension σ times the volume of the three-sphere (see (13.2.6)),

$$S_{\rm E} = \sigma \cdot 2\pi^2 H^{-3} \,. \tag{16.5.3}$$

For the double-well model (16.3.5), we have $\sigma = 4\sqrt{\lambda}\,\eta^3/3$ and the action is $S_{\rm E} = 8\pi^2\sqrt{\lambda}\,\eta^3/3H^3$. The tunnelling probability is proportional to $\exp(-S_{\rm E})$.

The use of the semiclassical approximation is justified if $S_{\rm E} \gg 1$. In deriving (16.5.3) we have also assumed implicitly that the wall is adequately described by the Nambu action (13.2.6). This is justified only if the wall thickness δ is much smaller than the wall radius, $\delta \ll H^{-1}$. In the general case, the instanton describing wall nucleation can be found by solving the corresponding field equations on a four-sphere. For the model (16.3.5), one has to look for a solution with $\varphi = 0$ at the equatorial three-sphere and $\nabla\varphi = 0$ at the poles. As in §16.4, one finds that non-trivial solutions exist only when $\delta_0 < \delta_{\rm max} = 1/2H$, where δ_0 is the flat-space wall thickness defined in (16.4.1) [Basu & Vilenkin, 1992]. Note that the value of $\delta_{\rm max}$ is the same as in §16.4. This is not surprising, since the instanton solutions can be obtained from the Lorentzian solutions by an appropriate analytic continuation.

As δ_0 approaches $\delta_{\rm max}$, the instanton solution approaches the homogeneous instanton,

$$\varphi = 0\,, \tag{16.5.4}$$

and for $\delta_0 > \delta_{\rm max}$ this is the only solution of the field equation that satisfies the boundary conditions. The instanton (16.5.4) describes a quantum fluctuation that brings φ to the top of the potential ($\varphi = 0$)

in a region of size $\sim H^{-1}$ [Hawking & Moss, 1983; Linde, 1992]. The corresponding action is given by the product of $V(0) = \frac{1}{4}\lambda\eta^4$ and the volume of S^4, $\Omega = (8\pi^2/3)H^{-4}$, that is

$$S_{\rm E} = \frac{2\pi^2\lambda\eta^4}{3H^4}.$$ (16.5.5)

Note that for $\delta_0 \sim \delta_{\rm max}$ this action has the same order of magnitude as that in the thin-wall approximation (16.5.3), indicating that (16.5.3) can be used for estimates in the whole range $0 < \delta_0 < \delta_{\rm max}$. For $\delta_0 >> H^{-1}$ the wall formation is more adequately described by the 'random walk' picture discussed in §16.3.

The evolution of the domain wall after nucleation can be described by analytically continuing (16.5.1–2) to a spacetime with Minkowski signature. One finds that de Sitter space is described by the usual hyperboloid

$$\vec{\xi}^2 + w^2 - \tau^2 = H^{-2},$$ (16.5.6)

and the world surface of the wall is given by

$$\vec{\xi}^2 = H^{-2} + \tau^2,$$
$$w = 0.$$ (16.5.7)

To understand how the evolving wall would be viewed by an observer in an inflationary universe, it is useful to transform to flat FRW coordinates $(t, \mathbf{x})$, with metric

$$ds^2 = dt^2 - e^{2Ht}d\mathbf{x}^2.$$ (16.5.8)

These coordinates are related to the hyperboloid coordinates by

$$\tau = H^{-1}\sinh(Ht) + \tfrac{1}{2}H\mathbf{x}^2\,e^{Ht}$$
$$w = H^{-1}\cosh(Ht) - \tfrac{1}{2}H\mathbf{x}^2\,e^{Ht}$$ (16.5.9)
$$\vec{\xi} = \mathbf{x}\,e^{Ht},$$

and the wall evolution equation (16.5.7) takes the form

$$\mathbf{x}^2 = H^{-2}\left(1 + e^{-2Ht}\right).$$ (16.5.10)

Thus, to an observer using coordinates (16.5.8), the wall appears to be a sphere of physical radius

$$R = H^{-1}\left(e^{2Ht} + 1\right)^{1/2}.$$ (16.5.11)

In the limit $t \to -\infty$, the radius of the wall approaches H^{-1}, suggesting that the wall nucleates with $R = H^{-1}$, but there seems to be no natural time at which one can say that the nucleation occurs. This appears puzzling, but we must remember that nucleation is a quantum-mechanical process, so the behaviour of the wall near the moment of nucleation cannot be described in purely classical terms. The wall radius at which the evolution becomes classical can be estimated by finding when the accumulated phase of the wave function becomes large,

$$\int_{H^{-1}}^{R} p_R \, dR \gtrsim 1, \tag{16.5.12}$$

where p_R is the momentum conjugate to R. This gives $R - H^{-1} \gtrsim H^5/\sigma^2$ [Basu *et al.*, 1991].

The instanton (16.5.2) and the resulting equations (16.5.7, 10) describe a nucleating wall in a particular configuration. In the general case, a simple analysis shows that (16.5.10) is replaced by

$$(\mathbf{x} - \mathbf{x}_0)^2 = H^{-2} \left(1 + \exp\left[-2H(t - t_0) \right] \right), \tag{16.5.13}$$

where the constants $\mathbf{x}_0$ and t_0 depend on the orientation of the three-sphere inside the four-sphere. The general wall trajectory (16.5.13) can be obtained from the particular trajectory (16.5.10) by space and time translations.

The nucleation of cosmic string loops can be described in a very similar way. The instanton in this case is a maximal two-sphere, and the Euclidean action is

$$S_{\mathrm{E}} = \mu \cdot 4\pi H^{-2}, \tag{16.5.14}$$

where μ is the string tension. The loops nucleate as circles, with a radius given by (16.5.11).

For monopole–antimonopole pairs the instanton is a circle of radius H^{-1} and

$$S_{\mathrm{E}} = m \cdot 2\pi H^{-1}, \tag{16.5.15}$$

where m is the monopole mass. We note that $S_{\mathrm{E}} = m/T_{\mathrm{GH}}$, where $T_{\mathrm{GH}} = H/2\pi$ is the Gibbons–Hawking [1977] temperature of de Sitter space. This process can be pictured, therefore, as 'thermal' monopole production.

The rate at which the defects are produced (per unit volume per unit time) can be written as

$$\Gamma = AH^4 \exp(-S_{\mathrm{E}}), \tag{16.5.16}$$

where A is a dimensionless coefficient. Since the only dimensionless combination of parameters in thin-defect models is the one appearing in the instanton action, A can be expressed as a function of S_{E}. To determine this function, one has to study perturbations about the instanton solutions, but in the most interesting regime, when S_{E} is not too large and the nucleation rate is non-negligible, we expect that $A \sim 1$.

16.5.2 Cosmological implications of nucleating defects

After nucleation, strings and walls are stretched by the exponential expansion of the universe, and by the end of inflation they have a wide spectrum of sizes. To find the resulting size distribution, we note that the number of defects formed per unit physical volume at time t_i in the interval dt_i is

$$dn(t_i) = \Gamma \, dt_i \,, \tag{16.5.17}$$

where Γ is the constant nucleation rate (16.5.16). The initial radii of strings or walls are $R_i \approx H^{-1}$. At later times, these radii are stretched, $R(t) \approx H^{-1} \exp[H(t - t_i)]$, and the density of defects is decreased by a factor $\exp[-3H(t - t_i)]$. Combining this with (16.5.17) and using $dR = HR|dt_i|$ we obtain [Basu *et al.*, 1991]

$$dn(t) \approx \frac{\Gamma}{H^4} \frac{dR}{R^4} \,. \tag{16.5.18}$$

At the end of inflation, this distribution spans the range of scales from $R \sim H^{-1}$ up to comoving scales beyond the present Hubble radius. Note that (16.5.18) has the same scale-invariant form as the distribution of string loops at a phase transition, (9.1.10).

After inflation, defects with $R \gg t$ are conformally stretched by the expansion and their size distribution continues to be described by (16.5.18). For strings, when loops come within the horizon, they begin to oscillate, and disregarding the slow decay due to gravitational radiation, the mass of loops remains constant. Hence, for $R < t$ we can write

$$dn(t) \approx \frac{\Gamma}{H^4} \left(\frac{a(R)}{a(t)} \right)^3 \frac{dR}{R^4} \,, \tag{16.5.19}$$

where R should now be understood as $R = M/2\pi\mu$ with M being the loop's mass. This distribution has the same form as that in the standard scenario of string evolution, where loops are chopped off an infinite string network (refer to (9.3.26) and (9.3.28)). The main difference is that the overall factor in (16.5.19) depends on the nucleation probability and is very model-dependent. With an appropriate choice of parameters the loops can serve as seeds for galaxies and clusters. In this model there are no infinite strings, the loops have zero velocities, and the corresponding structure formation scenario is similar to what we called 'the old string scenario' in chapter 11.

Since nucleating loops have the shape of a circle, one could expect that, instead of oscillating, they will all collapse to form black holes. However, there will be some deviations from a perfect circular shape due to quantum fluctuations. Analysis of small perturbations on a collapsing circular loop shows that the amplitude of the fluctuations remains roughly constant during the course of collapse [Vilenkin, 1981a; Garriga & Vilenkin, 1993]. For a loop of initial radius R, the Schwarzschild radius is $R_S = 4\pi G\mu R$, and for a black hole to form, we would require that $\delta R/R \lesssim 4\pi G\mu$. Garriga & Vilenkin [1992, 1993] developed a quantum theory of perturbations on strings in de Sitter space and used it to evaluate the probability of black hole formation and to derive observational bounds on the parameters $G\mu$ and S_E. It can be shown, however, that these bounds are always satisfied, provided that H satisfies the gravitational radiation constraint (2.4.14).

For domain walls, the size distribution of wall bubbles is described by the same equation (16.5.19), where R should be understood as the bubble radius at the time of horizon crossing. An important difference from strings is that walls of radius $R > (8\pi G\sigma)^{-1}$ inevitably collapse to form black holes. The walls also have a qualitatively different mass distribution. It is easily shown using (16.5.19) that the dominant contribution to the total mass of walls is given by the largest bubble within the present horizon.

If monopole–antimonopole pairs are produced, the monopole density at the end of inflation is of the order

$$n_M \sim H^3 \exp\left(-\frac{2\pi m}{H}\right).$$ (16.5.20)

Depending on the values of H and m, this density can be too high (so that the corresponding model is ruled out), it can be negligible, or it can lie in the 'interesting' range with potentially observable monopole densities.

16.6 String-driven inflation

Nucleating loops of string contribute to the vacuum energy density ρ_v and modify the rate of inflationary expansion. From (16.5.16) and (16.5.18), the string energy density is

$$\rho_s \approx \tfrac{1}{2} A H^2 \mu \exp(-S_E),$$ (16.6.1)

and the expansion rate is governed by the equation

$$H^2 = \frac{8\pi G}{3}(\rho_v + \rho_s).$$ (16.6.2)

For $G\mu \ll 1$, ρ_s is only a small correction to ρ_v,

$$\rho_s/\rho_v \approx \frac{4\pi}{3} G\mu\, A \exp(-S_E),$$ (16.6.3)

where $S_E = 4\pi\mu/H^2$ and we assume that $S_E \gtrsim 1$, $A \sim 1$. However, when the energy scales of both strings and inflation become comparable to the Planck scale, $G\mu \sim GH^2 \sim 1$, there is the interesting possibility that

ρ_v may vanish and inflation may be driven entirely by the string energy. Substituting (16.6.1) into (16.6.2) and setting $\rho_v = 0$ we obtain

$$\frac{4\pi}{3}G\mu\,A\,\exp\left(-\frac{4\pi\mu}{H^2}\right) = 1\,. \qquad (16.6.4)$$

This self-consistent inflationary model is similar to the Starobinsky [1980] model in which inflation is driven by the curvature-induced vacuum energy of the quantum fields.*

The possibility of string-driven inflation was suggested by Turok [1988a] who studied the behaviour of quantum fluctuations on strings in de Sitter space. He found that, as for a massless scalar field, the amplitude of each fluctuation mode 'freezes' when its wavelength becomes greater than H^{-1}. As fluctuations of different wavelengths accumulate, the string becomes 'wigglier' and its energy grows. Turok has argued that this energy gain can compensate for the energy lost by a string network due to the expansion of the universe and can provide a mechanism for string-driven inflation. An estimate of the corresponding string density and expansion rate gives

$$\rho_s \sim \mu m_{\mathrm{pl}}^2\,, \qquad H \sim \sqrt{\mu}\,. \qquad (16.6.5)$$

The characteristic scale of the string network can be found from $\rho_s \sim \mu/\xi^2$; this gives $\xi \sim m_{\mathrm{pl}}^{-1}$. Since ξ cannot be smaller than the string thickness, $\xi \gtrsim \delta \gtrsim \eta^{-1}$, we must have $\eta \sim m_{\mathrm{pl}}$ and $G\mu \sim 1$. (We cannot have $\eta \gg m_{\mathrm{pl}}$ because the deficit angle (7.1.31) of the strings would exceed 2π.) A similar picture has been discussed by Aharonov, Englert & Orloff [1987] who considered a situation in which the Gibbons–Hawking temperature of the de Sitter space, $T_{\mathrm{GH}} = H/2\pi$, is equal to the Hagedorn temperature of quantized strings, $T_{\mathrm{max}} \sim \sqrt{\mu}$. The possibility of inflation driven by nucleating loops was pointed out by Basu *et al.* [1991].

Although string-driven inflation is an intriguing idea, the quantitative descriptions attempted to date can only be considered as rough order-of-magnitude estimates. The reason for this is the fact that Planck-scale strings will have a large deficit angle and will cause strong deviations from homogeneity and isotropy on the horizon scale. The Friedmann evolution equation (16.6.2) will not be applicable in this case. Moreover, quantum gravity effects become important at the Planck density, so the classical Einstein equations will no longer be valid.

Planck-scale strings are naturally predicted in superstring theories (for a review see, for example, Green, Schwarz & Witten [1987]). It appears,

---

* Strictly speaking, ρ_s in (16.6.1) should be understood as the instanton contribution to the expectation value of T_0^0 and may be somewhat different from (16.6.1). The order of magnitude, however, is expected to be the same.

however, that superstrings cannot be used to drive inflation. The effective Lagrangian of superstring theories has the form

$$\mathcal{L} = e^{-2\varphi}[R + 4(\partial\varphi)^2] + \mathcal{L}_m, \qquad (16.6.6)$$

where the dilaton field φ determines the strength of the gravitational coupling and $\mathcal{L}_m$ is the 'matter' field Lagrangian. Einstein gravity is recovered by setting $\varphi = $ const., in which case a constant vacuum energy, $\mathcal{L}_m = -\rho_v = $ const., leads to inflationary expansion. However, for dilatonic gravity (16.6.6), the situation is quite different. It can be shown that, instead of accelerated expansion, a vacuum-dominated universe slows down and recontracts [Tseytlin & Vafa, 1992]. It appears, therefore, that superstring-driven inflation is impossible, even if quantum fluctuations could keep the string energy density at a constant level. For further discussion of superstring cosmology the reader is referred to papers by Brandenberger & Vafa [1988] and Tseytlin & Vafa [1992].

References

Abbott, L.F., & Sikivie, P. [1983], 'A cosmological bound on the invisible axion', *Phys. Lett.* **120B**, 133.

Abbott, L.F., & Wise, M.B. [1984], 'Constraints on generalized inflationary cosmologies', *Nucl. Phys.* **B244**, 541.

Abers, E.S., & Lee, B.W. [1973], 'Gauge theories', *Phys. Rep.* **9C**, 1.

Abrikosov, A.A. [1957], 'On the magnetic properties of superconductors of the second group', *Sov. Phys. JETP* **5**, 1174.

Accetta, F.S. [1988], 'Current quenching in superconducting cosmic strings', in *Cosmic Strings: The Current Status*, Accetta, F.S., & Krauss, L.M., eds. (World Scientific, Singapore).

Achucarro, A., Kuijken, K., Perivolaropoulos, L., & Vachaspati, T. [1992], 'Dynamical simulations of semilocal strings', *Nucl. Phys.* **B388**, 435.

Aharonov, A., Englert, F., & Orloff, J. [1987], 'Macroscopic fundamental strings in cosmology', *Phys. Lett.* **199B**, 366.

Aharonov, Y., & Bohm, D. [1959], 'Significance of electromagnetic potentials in quantum theory', *Phys. Rev.* **119**, 485.

Albrecht, A., Brandenberger, R.H., & Turok, N. [1987], 'Cosmic strings and cosmic structure', *New Scientist* **114**, No. 1556, 40.

Albrecht, A., & Stebbins, A. [1992a], 'Perturbations from cosmic strings in cold dark matter', *Phys. Rev. Lett.* **68**, 2121.

Albrecht, A., & Stebbins, A. [1992b], 'Cosmic string with a light massive neutrino', *Phys. Rev. Lett.* **69**, 2615.

Albrecht, A., & Steinhardt, P.J. [1982], 'Cosmology for grand unified theories with radiatively induced symmetry breaking', *Phys. Rev. Lett.* **48**, 1220.

Albrecht, A., & Turok, N. [1985], 'Evolution of cosmic strings', *Phys. Rev. Lett.* **54**, 1868.

Albrecht, A., & Turok, N. [1989], 'Evolution of cosmic string networks', *Phys. Rev.* **D40**, 973.

Albrecht, A., & York, T. [1988], 'A topological picture of cosmic string self-intersection', *Phys. Rev.* **D38**, 2958.

Alford, M.G., Benson, K., Coleman, S., March-Russell, J., & Wilczek, F. [1990], 'Interactions and excitations of nonabelian vortices', *Phys. Rev. Lett.* **64**, 1632. Erratum: *Phys. Rev. Lett.* **65**, 668.

Alford, M.G., Benson, K., Coleman, S., March-Russell, J., & Wilczek, F. [1991], 'Zero modes of nonabelian vortices', *Nucl. Phys.* **B349**, 414.

Alford, M.G., March-Russell, J., & Wilczek, F. [1989], 'Enhanced baryon number violation due to cosmic strings', *Nucl. Phys.* **B328**, 140.

Alford, M.G., & Wilczek, F. [1989], 'Aharonov–Bohm interaction of cosmic strings with matter', *Phys. Rev. Lett.* **62**, 1071.

Alfven, H., & Fälthammar, C.-G. [1963], *Cosmical Electrodynamics*, (Clarendon, Oxford).

Allen, B., & Caldwell, R.R. [1991a], 'Small scale structure on a cosmic string network', *Phys. Rev.* **D43**, 3173.

Allen, B., & Caldwell, R.R. [1991b], 'Generation of structure on a cosmic string network', *Phys. Rev. Lett.* **65**, 1705.

Allen, B., & Shellard, E.P.S. [1990], 'Cosmic string evolution—a numerical simulation', *Phys. Rev. Lett.* **64**, 119.

Allen, B., & Shellard, E.P.S. [1992], 'Gravitational radiation from a cosmic string network', *Phys. Rev.* **D45**, 1898.

Allen, B., & Shellard, E.P.S. [1993], 'Small-scale structure on cosmic strings', DAMTP-GR (in preparation).

Ambjørn, J., & Olesen, P. [1988], 'Superconducting cosmic strings and W-condensation', *Nucl. Phys.* **B310**, 625.

Ambjørn, J., & Olesen, P. [1989a], 'A magnetic condensate solution of the classical electroweak theory', *Phys. Lett.* **218B**, 67. Erratum: *Phys. Lett.* **220B**, 659.

Ambjørn, J., & Olesen, P. [1989b], 'On electroweak magnetism', *Nucl. Phys.* **B315**, 606.

Ambjørn, J., & Olesen, P. [1990a], 'A condensate solution of the electroweak theory which interpolates between the broken and symmetric phase', *Nucl. Phys.* **B330**, 193.

Ambjørn, J., & Olesen, P. [1990b], 'Electroweak magnetism: theory and application', *Int. J. Mod. Phys.* **A5**, 4525.

Ambjørn, J., & Olesen, P. [1990c], 'W-condensation near cosmic strings', in *Formation and Evolution of Cosmic Strings*, Gibbons, G.W., Hawking, S.W., & Vachaspati, T., eds. (Cambridge University Press).

Amsterdamski, P. [1988], 'Evolution of superconducting cosmic loops', *Phys. Rev.* **D39**, 1524.

Amsterdamski, P. & Laguna, P. [1988], 'Internal structure and the space-time of superconducting bosonic strings', *Phys. Rev.* **D37**, 877.

Amsterdamski, P. & O'Connor, D. [1988], 'Nonlinear electrodynamics near a superconducting string', *Nucl. Phys.* **B298**, 429.

Ansourian, M. [1977], 'Index theory and the axial current anomaly in two dimensions', *Phys. Lett.* **70B**, 301.

Aryal, M., & Everett, A.E. [1986], 'Gravitational effects of global strings', *Phys. Rev.* **D33**, 333.

Aryal, M., & Everett, A.E. [1987], 'Properties of Z_2 strings', *Phys. Rev.* **D35**, 3105.

Aryal, M., Everett, A.E., Vilenkin, A. & Vachaspati, T. [1986], 'Cosmic string networks', *Phys. Rev.* **D34**, 434.

Aryal, M., Ford, L.H. & Vilenkin, A. [1986], 'Cosmic strings and black holes', *Phys. Rev.* **D34**, 2263.

Aryal, M., Vachaspati, T., & Vilenkin, A. [1987], 'Notes on superconducting cosmic strings', *Phys. Lett.* **194B**, 25.

Aryal, M., & Vilenkin, A. [1987], 'The fractal dimension of an inflationary universe', *Phys. Lett.* **199B**, 351.

Atiyah, M.F., & Hitchin, N.J. [1985], 'Low-energy scattering of nonabelian monopoles', *Phys. Lett.* **107A**, 21.

Austen, D., Copeland, E., & Kibble, T.W.B. [1993], 'Evolution of cosmic string configurations', Imperial-TP-92-93-42.

Avelino, P.P., & Shellard, E.P.S. [1993], 'Cosmic-string-seeded structure formation', DAMTP preprint.

Babul, A., Paczynski, B., & Spergel, D.N. [1987], 'Gamma-ray bursts from superconducting cosmic strings at large redshifts', *Ap. J. Lett.* **316**, L49.

Babul, A., Piran, T., & Spergel, D.N. [1988a], 'Bosonic superconducting cosmic strings, 1. Classical field theory solutions', *Phys. Lett.* **202B**, 307.

Babul, A., Piran, T., & Spergel, D.N. [1988b], 'Superconducting cosmic strings, 2. Space-time curvature', *Phys. Lett.* **209B**, 477.

Bach, R., & Weyl, H. [1922], 'Nene Lösungen der Einsteinschen Gravitationsgleichungen', *Math. Z.* **13**, 134.

Bagger, J.A., Callan, C.G. & Harvey, J.A. [1986], 'Cosmic strings as orbifolds', *Nucl. Phys.* **B278**, 550.

Bahcall, N.A., & Soneira, R.M. [1983], 'The spatial correlation-function of rich clusters of galaxies', *Ap. J.* **270**, 20.

Barr, S.M., & Matheson, A.M. [1987a], 'Weak resistance in superconducting cosmic strings', *Phys. Rev.* **D36**, 2905.

Barr, S.M., & Matheson, A.M. [1987b], 'Limiting currents in fermionic superconducting strings', *Phys. Lett.* **198B**, 146.

Barrabes, C., & Israel, W. [1987], 'Relativistic kinetic theory of a system of cosmic strings: general formalism', *Ann. Phys.* **175**, 213.

Barriola, M. [1993], 'Numerical study of global texture collapse', Tufts preprint.

Barriola, M., & Vachaspati, T. [1991a], 'Strong gravity of a self-similar global texture', *Phys. Rev.* **D43**, 1056.

Barriola, M., & Vachaspati, T. [1991b], 'Analytic approximations to the self-similar global texture metric', *Phys. Rev.* **D43**, 2726.

Barriola, M., & Vachaspati, T. [1992], 'A new class of defects', *Phys. Rev. Lett.* **63**, 341.

Barriola, M., & Vilenkin, A. [1989], 'Gravitational field of a global monopole', *Phys. Rev. Lett.* **63**, 341.

Basu, R., Guth, A.H. & Vilenkin, A. [1991], 'Quantum creation of topological defects during inflation', *Phys. Rev.* **D44**, 340.

Basu, R., & Vilenkin, A., [1989], unpublished.

Basu, R. & Vilenkin, A. [1992], 'Nucleation of thick topological defects during inflation', *Phys. Rev.* **D46**, 2345.

Basu, R., & Vilenkin, A. [1993], 'Topological defects in de Sitter space', in preparation.

Battye, R.A., & Shellard, E.P.S. [1993], 'Global string radiation', to appear in *Nucl. Phys.* **B**, (1994).

Bennett, D.P. [1986a], 'The evolution of cosmic strings', *Phys. Rev.* **D33**, 872. Erratum: *Phys. Rev.* **D34**, 3932.

Bennett, D.P. [1986b], 'The evolution of cosmic strings: 2', *Phys. Rev.* **D34**, 3592.

Bennett, D.P. [1990], 'High resolution simulations of cosmic string evolution: numerics and long string evolution', in *Formation and Evolution of Cosmic Strings*, Gibbons, G.W., Hawking, S.W., & Vachaspati, T., eds. (Cambridge University Press).

Bennett, D.P., & Bouchet, F.R. [1988], 'Evidence for a scaling solution in cosmic string evolution', *Phys. Rev. Lett.* **60**, 257.

Bennett, D.P., & Bouchet, F.R. [1989], 'Cosmic string evolution', *Phys. Rev. Lett.* **63**, 1334.

Bennett, D.P., & Bouchet, F.R. [1990], 'High resolution simulations of cosmic string evolution: network evolution', *Phys. Rev.* **D41**, 2408.

Bennett, D.P., & Bouchet, F.R. [1991], 'Constraints on the gravity wave background generated by cosmic strings', *Phys. Rev.* **D43**, 2733.

Bennett, D.P., & Rhie, S.H. [1990], 'Cosmological evolution of global monopoles and the origin of large scale structure', *Phys. Rev. Lett.* **65**, 1709.

Bennett, D.P., & Rhie, S.H. [1993], 'COBE's constraints on the global monopole and texture theories of cosmic structure formation', *Ap. J.* **406**, 7.

Bennett, D.P., Rhie, S.H. & Weinberg, D.H., [1993], in preparation

Bennett, D.P., Stebbins, A., & Bouchet, F.R. [1992], 'The implications of the COBE DMR results for cosmic strings', *Ap. J.* **399**, 5.

Berezinsky, V., & Rubinstein, H.R. [1989], 'Evolution and radiation from superconducting cosmic strings', *Nucl. Phys.* **B323**, 95.

Bertschinger, E. [1985], 'Self-similar secondary infall and accretion in an Einstein–de Sitter universe', *Ap. J. Suppl.* **58**, 39.

Bertschinger, E. [1987a], 'Cosmological accretion wakes', *Ap. J.* **316**, 489.

Bertschinger, E. [1987b], 'Can cosmic strings generate large-scale streaming velocities?', *Ap. J.* **324**, 5.

Bertschinger, E. [1988], 'Resurrecting hot dark matter with cosmic strings', in *Cosmic Strings: The Current Status*, Accetta, F.S., & Krauss, L.M., eds. (World Scientific, Singapore).

Bertschinger, E. [1990], 'String-Seeded Galaxy Formation: Linear theory and N-body simulations', in *Formation and Evolution of Cosmic Strings*, Gibbons, G.W., Hawking, S.W., & Vachaspati, T., eds. (Cambridge University Press).

Bertschinger, E., & Watts, P.N. [1987], 'Galaxy formation with cosmic strings and massive neutrinos', *Ap. J.* **316**, 489.

Bhattacharjee, P. [1982], Ph. D. Thesis, (Imperial College).

Bhattacharjee, P. [1989], 'Cosmic strings and ultra-high energy cosmic rays', *Phys. Rev.* **D40**, 3968.

Bhattacharjee, P., Kibble, T.W.B., & Turok, N. [1982], 'Baryon number from collapsing cosmic strings', *Phys. Lett.* **119B**, 95.

Blau, S.K., & Guth, A.H. [1987], 'Inflationary cosmology', in *300 Years of Gravitation*, Hawking, S.W., & Israel, W., eds. (Cambridge University Press).

Bogomol'nyi, E.B. [1976], 'The stability of classical solutions', *Sov. J. Nucl. Phys.* **24**, 449.

Bogomol'nyi, E.B., & Vainstein, A.I. [1976], 'Stability of strings in gauge abelian theory', *Sov. J. Nucl. Phys.* **23**, 588.

Borrill, J., Copeland, E.J., & Liddle, A.R. [1991a], 'Initial conditions for global texture', *Phys. Lett.* **258B**, 310.

Borrill, J., Copeland, E.J., & Liddle, A.R. [1991b], 'The collapse of spherically symmetric textures', *Phys. Rev.* **D46**, 524.

Bouchet, F.R. [1990], 'High resolution simulations of cosmic string evolution: small scale structure', in *Formation and Evolution of Cosmic Strings*, Gibbons, G.W., Hawking, S.W., & Vachaspati, T., eds. (Cambridge University Press).

Bouchet, F.R., & Bennett, D.P. [1988], 'Properties of interacting cosmic string networks', in *Cosmic Strings: The Current Status*, Accetta, F.S., & Krauss, L.M., eds. (World Scientific, Singapore).

Bouchet, F.R., & Bennett, D.P. [1990], 'Millisecond pulsar constraints on cosmic strings', *Phys. Rev.* **D41**, 720.

Bouchet, F.R., Bennett, D.P. & Stebbins, A. [1988], 'Microwave anisotropy patterns from evolving string networks', *Nature* **335**, 410.

Bowick, M.J., Chandar, L., Schiff, E.A., & Srivastava, A.M. [1994], 'The cosmological Kibble mechanism in the laboratory: string formation in liquid crystals', *Science* **263**, 943.

Brandenberger, R.H. [1985], 'Quantum field theory methods and inflationary universe models', *Rev. Mod. Phys.* **57**, 1.

Brandenberger, R.H. [1987a], 'On the decay of cosmic string loops', *Nucl. Phys.* **B293**, 812.

Brandenberger, R.H. [1987b], 'Inflation and cosmic strings: two mechanisms for producing structure in the universe', *Int. J. Mod. Phys.* **A2**, 77.

Brandenberger, R.H. [1987c], 'Amplification of correlation functions by gravity in the cosmic string model', *Phys. Lett.* **191B**, 257.

Brandenberger, R.H., Albrecht, A., & Turok, N. [1986], 'Gravitational radiation from cosmic strings and the microwave background', *Nucl. Phys.* **B277**, 605.

Brandenberger, R.H., Davis, A.-C., & Hindmarsh, M.B. [1991], 'Baryogenesis from collapsing topological defects', *Phys. Lett.* **263B**, 239.

Brandenberger, R.H., Davis, A.-C. & Matheson, A.M. [1988a], 'Callan–Rubakov effect for strings', *Nucl. Phys.* **B307**, 909.

Brandenberger, R.H., Davis, A.-C. & Matheson, A.M. [1988b], 'Cosmic strings and baryogenesis', *Phys. Lett.* **218B**, 308.

Brandenberger, R.H., Kaiser, N., Schramm, D.N., & Turok, N. [1987], 'Galaxy and structure formation with hot dark matter and cosmic strings', *Phys. Rev. Lett.* **59**, 2371.

Brandenberger, R.H., Kaiser, N., Shellard, E.P.S., & Turok, N. [1987], 'Peculiar velocities from cosmic strings', *Phys. Rev.* **D36**, 335.

Brandenberger, R.H., Kaiser, N., & Turok, N. [1987], 'Dissipationless clustering of neutrinos around a cosmic string loop', *Phys. Rev.* **D36**, 2242.

Brandenberger, R.H., & Matheson, A.M. [1987], 'Cosmic string decay', *Mod. Phys. Lett.* **A2**, 461.

Brandenberger, R.H., & Perivolaropoulos, L. [1988], 'Proton decay catalyzed by superconducting cosmic strings', *Phys. Lett.* **208B**, 396.

Brandenberger, R.H., & Shellard, E.P.S. [1989], 'Angular momentum and the mass function of galaxies seeded by cosmic strings', *Phys. Rev.* **D40**, 2542.

Brandenberger, R.H., & Turok, N. [1986], 'Fluctuations from cosmic strings and the microwave background', *Phys. Rev.* **D33**, 82.

Brandenberger, R.H., & Vafa, C. [1988], 'Superstrings in the early universe', *Nucl. Phys.* **B316**, 391.

Brown, R.W., & DeLaney, D.B. [1989], 'A product representation for the harmonic series of a unit vector: a string application', *Phys. Rev. Lett.* **63**, 474.

Bucher, M. [1991], 'The Aharonov–Bohm effect and exotic statistics for nonabelian vortices', *Nucl. Phys.* **B350**, 163.

Bucher, M., [1992], unpublished.

Bucher, M., Lo, H.K., & Preskill, J. [1992], 'Topological approach to Alice electrodynamics', *Nucl. Phys.* **B386**, 3.

Burden, C.J. [1985], 'Gravitational radiation from a particular class of cosmic strings', *Phys. Lett.* **164B**, 277.

Burden, C. & Tassie, L.J. [1984], 'Additional rigidly rotating solutions in the string model of hadrons', *Aust. J. Phys.* **37**, 1.

Caldwell, R.R., & Allen, B. [1992], 'Cosmological constraints on cosmic string gravitational radiation', *Phys. Rev.* **D45**, 3447.

Callan, C.G. [1982a], 'Disappearing dyons', *Phys. Rev.* **D25**, 2141.

Callan, C.G. [1982b], 'Dyon-fermion dynamics', *Phys. Rev.* **D26**, 2058.

Callan, C.G., & Coleman, S. [1977], 'Fate of the false vacuum II: First quantum corrections', *Phys. Rev.* **D16**, 1662.

Callan, C.G., & Harvey, J.A. [1985], 'Anomalies and fermion zero modes on strings and domain walls', *Nucl. Phys.* **B250**, 427.

Campbell, B.A., Ellis, J., Nanopoulos, D.V., Olive, K.A., & Rudaz, S. [1986], 'Cosmic strings from hidden sector flux tubes', *Phys. Lett.* **179B**, 37.

Carlitz, R. [1972], 'Hadronic matter at high density', *Phys. Rev.* **D5**, 3231.

Caroli, C., de Gennes, P.G., & Matricon, J. [1964], 'Bound fermion states on a vortex line in a Type II superconductor', *Phys. Lett.* **9**, 307.

Carroll, S.M., Farhi, E., & Guth, A.H. [1992], 'An obstacle to building a time machine', *Phys. Rev. Lett.* **68**, 263. Erratum: *Phys. Rev. Lett.* **68**, 3368.

Carter, B. [1989a], 'Duality relation between charged elastic strings and superconducting cosmic strings', *Phys. Lett.* **224B**, 61.

Carter, B. [1989b], 'Stability and characteristic propagation speeds in superconducting cosmic and other string models', *Phys. Lett.* **228B**, 466.

Carter, B. [1990a], 'Integrable equation of state for noisy cosmic string', *Phys. Rev.* **D41**, 3869.

Carter, B. [1990b], 'Mechanics of cosmic rings', *Phys. Lett.* **238B**, 166.

Carter, B. [1990c], 'Covariant mechanics of simple and conducting strings and membranes', in *Formation and Evolution of Cosmic Strings*, Gibbons, G.W., Hawking, S.W., & Vachaspati, T., eds. (Cambridge University Press).

Carter, B. [1992], 'Basic brane theory', *Class. Quant. Grav.* **9**, 19.

Carter, B., & Frolov, V.P. [1989], 'Separability of string equilibrium equations in a generalized Kerr–de Sitter background', *Class. Quant. Grav.* **6**, 569.

Carter, B., Frolov, V.P., & Heinrich, O. [1991], 'Mechanics of stationary strings: separability of non-dispersive models in a black hole background', *Class. Quant. Grav.* **8**, 135.

Carter, B., & Martin, X. [1993], 'Stability conditions for circular string loops', CNRS Meudon preprint.

Cen, R.Y., Ostriker, J.P., Spergel, D.N., & Turok, N. [1991], 'A hydrodynamic approach to cosmology: texture seeded CDM and HDM cosmogonies', *Ap. J.* **383**, 1.

Chandrasekhar, S. [1943], 'Dynamical friction: I. General considerations: the coefficient of dynamical friction', *Ap. J.* **97**, 255.

Charlton, J.C. [1987], 'Cosmic string wakes and large-scale structure', *Ap. J.* **325**, 521.

Chase, S.T. [1986], 'Cosmic string, hydrodynamics and microanisotropies in the cosmic background radiation', *Nature* **323**, 42.

Chen, A.L., DiCarlo, D.A., & Hotes, S.A. [1988], 'Self-intersections in a three-parameter space of cosmic strings', *Phys. Rev.* **D37**, 863.

Chuang, I., Durrer, R., Turok, N., & Yurke, B. [1990], 'Cosmology in the laboratory: defect dynamics in liquid crystals', *Science* **251**, 1336.

Chudnovsky, E.M., Field, G.B., Spergel, D.N., & Vilenkin, A. [1986], 'Superconducting cosmic strings', *Phys. Rev.* **D34**, 944.

Chudnovsky, E.M., & Vilenkin, A. [1988], 'Strings in the sun?', *Phys. Rev. Lett.* **61**, 1043.

Clarke, C.J.S., Ellis, G.F.R., & Vickers, J.A.G. [1990], 'The large-scale bending of cosmic strings', *Class. Quant. Grav.* **7**, 1.

Cohen, A.G., & Kaplan, D.B. [1988], 'The exact metric about global cosmic strings', *Phys. Lett.* **215B**, 67.

Coleman, S. [1976], 'Quantum lumps and their quantum descendants', in *New Phenomena in Subnuclear Physics*, Zichichi, N., ed. (Plenum, New York).

Coleman, S. [1977], 'Fate of the false vacuum: semiclassical theory', *Phys. Rev.* **D15**, 2929.

Coleman, S. [1979], 'The uses of instantons', in *The Whys of Subnuclear Physics*, Zichichi, A., ed. (Plenum, New York).

Coleman, S. [1983], 'The magnetic monopole 50 years later', in *The Unity of Fundamental Interactions*, Zichichi, A., ed. (Plenum, New York).

Coleman, S. [1985a], *Aspects of Symmetry*, (Cambridge University Press).

Coleman, S. [1985b], 'Q-balls', *Nucl. Phys.* **B262**, 263.

Coleman, S., & de Luccia, F. [1980], 'Gravitational effects on and of vacuum decay', *Phys. Rev.* **D21**, 3305.

Coleman, S., & Weinberg, E.J. [1973], 'Radiative corrections as the origin of symmetry breaking', *Phys. Rev.* **D7**, 1887.

Comtet, A., & Gibbons, G.W. [1988], 'Bogomol'nyi bounds for cosmic strings', *Nucl. Phys.* **B299**, 719.

Copeland, E.J., Haws, D. & Hindmarsh, M.B. [1990], 'Classical theory of radiating strings', *Phys. Rev.* **D42**, 726.

Copeland, E.J., Haws, D., Hindmarsh, M.B., & Turok, N. [1988], 'Dynamics of and radiation from superconducting strings and springs', *Nucl. Phys.* **B306**, 908.

Copeland, E.J., Haws, D., Kibble, T.W.B., Mitchell, D., & Turok, N. [1986], 'Monopoles connected by strings and the monopole problem', *Nucl. Phys.* **B298**, 445.

Copeland, E.J., Haws, D., & Rivers, R. [1989], 'The effect of topological defects on phase transitions in the early universe', *Nucl. Phys.* **B319**, 687.

Copeland, E.J., Kibble, T.W.B., & Austen, D. [1992], 'Scaling solutions in cosmic string networks', *Phys. Rev.* **D45**, R1000.

Copeland, E.J., Kolb, E.W., & Liddle, A. [1990], 'Topological defects in extended inflation', *Phys. Rev.* **D42**, 2911.

Copeland, E.J., & Turok, N. [1986a], 'The stability of cosmic string loops', *Phys. Lett.* **173B**, 129.

Copeland, E.J., & Turok, N. [1986b], 'Intercommuting', Fermilab preprint (unpublished).

Copeland, E.J., Turok, N., & Hindmarsh, M.B. [1987], 'Dynamics of superconducting cosmic strings', *Phys. Rev. Lett.* **58**, 1910.

Cowie, L.L., & Hu, E.M. [1987], 'The formation of families of twin galaxies by string loops', *Ap. J.* **318**, L33.

Cragie, N.S. [1986], 'Monopoles and their quantum fields', in *Theory and Detection of Monopoles in Gauge Theories*, Cragie, N.S., ed. (World Scientific, Singapore).

Dabholkar, A., & Quashnock, J.M. [1990], 'Pinning down the axion', *Nucl. Phys.* **B333**, 815.

Daniel, N., & Vayonakis, C.E. [1981], 'The phase transition in the $SU(5)$ model at high temperatures', *Nucl. Phys.* **B180**, 301.

Davies, P.C.W., & Sahni, V. [1988], 'Quantum gravitational effects near cosmic strings', *Class. Quant. Grav.* **5**, 1.

Davis, R.L. [1985a], 'Goldstone bosons in string models of galaxy formation', *Phys. Rev.* **D32**, 3172.

Davis, R.L. [1985b], 'Nucleosynthesis problems for string models of galaxy formation', *Phys. Lett.* **161B**, 285.

Davis, R.L. [1986], 'Cosmic axions from cosmic strings', *Phys. Lett.* **180B**, 225.

Davis, R.L. [1987a], 'Texture: a cosmological topological defect', *Phys. Rev.* **D35**, 3705.

Davis, R.L. [1987b], 'Fermion masses on the vortex world sheet', *Phys. Rev.* **D36**, 2267.

Davis, R.L. [1987c], 'Twisted vortices, string solitons, and world sheet instantons', *Nucl. Phys.* **B294**, 867.

Davis, R.L. [1988], 'Semitopological solitons', *Phys. Rev.* **D38**, 3722.

Davis, R.L. [1989], 'Relativistic superfluids and vortex rings', *Phys. Rev.* **D40**, 4033.

Davis, R.L. [1990], 'Spinning vortices in Type II superconductors', *Mod. Phys. Lett.* **A5**, 955.

Davis, R.L., & Shellard, E.P.S. [1988a], 'The physics of vortex superconductivity I: Currents and quenching', *Phys. Lett.* **207B**, 404.

Davis, R.L., & Shellard, E.P.S. [1988b], 'The physics of vortex superconductivity II: Charge, angular momentum and the vorton', *Phys. Lett.* **209B**, 485.

Davis, R.L., & Shellard, E.P.S. [1988c], 'Antisymmetric tensors and spontaneous symmetry breaking', *Phys. Lett.* **214B**, 219.

Davis, R.L., & Shellard, E.P.S. [1988d], 'Radiation from axion strings', in *Cosmic Strings: The Current Status*, Accetta, F.S., & Krauss, L.M., eds. (World Scientific, Singapore).

Davis, R.L., & Shellard, E.P.S. [1989a], 'Cosmic vortons', *Nucl. Phys.* **B323**, 209.

Davis, R.L., & Shellard, E.P.S. [1989b], 'Global strings and superfluid vortices', *Phys. Rev. Lett.* **63**, 2021.

Davis, R.L., & Shellard, E.P.S. [1989c], 'Do axions need inflation?', *Nucl. Phys.* **B324**, 167.

DeLaney, D.B., Engle, K., & Scheick, X. [1989], 'The general two harmonic cosmic string', *Phys. Rev.* **D41**, 1775.

Derrick, G.H. [1964], 'Comments on nonlinear wave equations as models for elementary particles', *J. Math. Phys.* **5**, 1252.

Deser, S., & Jackiw, R. [1988], 'Classical and quantum scattering on a cone', *Comm. Math. Phys.* **118**, 495.

Deser, S., Jackiw, R., & 't Hooft, G. [1984], 'Three-dimensional Einstein gravity—dynamics of flat space', *Ann. Phys.* **152**, 220.

Deser, S., Jackiw, R., & 't Hooft, G. [1991], 'Physical cosmic strings do not generate closed time-like curves', *Phys. Rev. Lett.* **68**, 267.

Dicus, D.A., Page, D., & Teplitz, V.L. [1982], 'Two and three body contributions to monopole annihilation', *Phys. Rev.* **D26**, 1306.

Dimopoulos, S., Preskill, J., & Wilczek, F. [1982], 'Catalyzed nucleon decay in neutron stars', *Phys. Lett.* **119B**, 320.

Dine, M., & Fischler, W. [1983], 'The not-so-harmless axion', *Phys. Lett.* **120B**, 137.

Dine, M., Fischler, W., & Srednicki, M. [1981], 'A simple solution to the strong CP-problem with a harmless axion', *Phys. Lett.* **104B**, 199.

Dirac, P.A.M. [1931], 'Quantized singularities in the electromagnetic field', *Proc. Roy. Soc.* **A133**, 60.

Dixit, V.V., & Sher, M. [1992], 'Monopole annihilation and baryogenesis at the electroweak scale', *Phys. Rev. Lett.* **68**, 560.

Dobyns, Y.H. [1988], 'Gravitational accretion of hot dark matter', *Ap. J.* **329**, L5.

Dolan, L., & Jackiw, R. [1974], 'Symmetry breaking at finite temperature', *Phys. Rev.* **D9**, 3320.

Dolgov, A.D., & Zel'dovich, Ya.B. [1981], 'Cosmology and elementary particles', *Rev. Mod. Phys.* **53**, 1.

Dowker, J.S. [1977], 'Quantum field theory on a cone', *J. Phys.* **A10**, 115.

Dowker, J.S. [1987a], 'Field theory around a cosmic string', *Class. Quant. Grav.* **L157**, 4.

Dowker, J.S. [1987b], 'Casimir effect around a cone', *Phys. Rev.* **D36**, 3095.

Dowker, J.S. [1987c], 'Vacuum averages for arbitrary spin around a cosmic string', *Phys. Rev.* **D36**, 3742.

Durrer, R. [1989], 'Gravitational angular momentum radiation of cosmic strings', *Nucl. Phys.* **B328**, 238.

Durrer, R. [1990], 'Gauge invariant cosmological perturbation theory with seeds', *Phys. Rev.* **D42**, 2533.

Efstathiou, G., & Silk, J., [1983], *Fund. Cosm. Phys.*, **9**, 1.

Einhorn, M.B., & Sato, K. [1981], 'Monopole production in the very early universe in a first-order phase transition', *Nucl. Phys.* **B180**, 385.

Einhorn, M.B., Stein, D.L., & Toussaint, D. [1980], 'Are grand unified theories compatible with the standard cosmology', *Phys. Rev.* **D21**, 3295.

Embacher, F. [1993a], 'Cosmic string evolution by exact methods: I. Model for loop-production in flat space', *Nucl. Phys.* **B387**, 129.

Embacher, F. [1993b], 'Cosmic string evolution by exact methods: II. Gaussian model', *Nucl. Phys.* **B387**, 163.

Englert, F., Orloff, J., & Piran, T. [1988], 'Fundamental strings and large scale structure formation', *Phys. Lett.* **212B**, 423.

Everett, A.E. [1974], 'Observational consequences of a 'domain' structure of the universe', *Phys. Rev.* **D10**, 3161.

Everett, A.E. [1981], 'Cosmic strings in unified gauge theories', *Phys. Rev.* **D24**, 858.

Everett, A.E. [1988], 'New mechanism for superconductivity in cosmic strings', *Phys. Rev. Lett.* **61**, 1807.

Everett, A.E., & Aryal, M. [1986], 'Comment on 'Monopoles on strings'', *Phys. Rev. Lett.* **57**, 646.

Everett, A.E., Vachaspati, T., & Vilenkin, A. [1985], 'Monopole annihilation and causality', *Phys. Rev.* **D31**, 1925.

Everett, A.E., & Vilenkin, A. [1982], 'Left–right symmetric theories and vacuum domain walls and strings', *Nucl. Phys.* **B207**, 43.

Farhi, E., Guth, A.H., & Guven, J. [1990], 'Is it possible to create a universe in the laboratory?', *Nucl. Phys.* **B339**, 417.

Farris, T.H., Klephart, T.W., Weiler, T.J., & Yuan, T.C. [1992], 'The minimal electroweak model for monopole annihilation', *Phys. Rev. Lett.* **68**, 564.

Fewster, C., & Kay, B.S. [1993], 'Model dependence of baryon decay enhancement by cosmic strings', *Nucl. Phys.* **B399**, 89.

Fillmore, J.A., & Goldreich, P. [1984], 'Self-similar spherical voids in an expanding universe', *Ap. J.* **281**, 1.

Ford, L.H. & Vilenkin, A. [1981], 'A gravitational analog of the Aharonov–Bohm effect', *J. Phys.* **A14**, 2353.

Forgacs, P., Horvath, Z., & Palla, L. [1981a], 'Nonlinear superposition of monopoles', *Nucl. Phys.* **B192**, 141.

Forgacs, P., Horvath, Z., & Palla, L. [1981b], 'Exact multi-monopole solutions in the Bogomol'nyi–Prasad–Sommerfield limit', *Phys. Lett.* **99B**, 232. Erratum: *Phys. Lett.* **101B**, 457.

Förster, D. [1974], 'Dynamics of relativistic vortex lines and their relation to dual theory', *Nucl. Phys.* **B81**, 84.

Forte, S. [1988], 'Anomalous particle creation, spectral asymmetry and superconducting strings', *Phys. Rev.* **D38**, 1108.

Frautschi, S. [1971], 'Statistical bootstrap model of hadrons', *Phys. Rev.* **D3**, 2821.

Freese, K. [1984], 'Do monopoles keep white-dwarfs hot?', *Ap. J.* **286**, 216.

Freese, K., Turner, M.S., & Schramm, D.N. [1983], 'Monopole catalysis of nucleon decay in old pulsars', *Phys. Rev. Lett.* **51**, 1625.

Frenk, C.S., White, S.D.M., Efstathiou, G., & Davis, M. [1985], 'Cold dark matter, the structure of galactic halos and the origin of the Hubble sequence', *Nature* **317**, 595.

Friedberg, R., Lee, T.D., & Sirlin, A. [1976], 'Class of scalar field soliton solutions in three space dimensions', *Phys. Rev.* **D13**, 2739.

Frieman, J.A., Freese, K., & Turner, M.S. [1988], 'Superheavy magnetic monopoles and main sequence stars', *Ap. J.* **335**, 844.

Frieman, J.A., & Lynn, B.W. [1990], 'A new class of nontopological solitons', *Nucl. Phys.* **B329**, 1.

Frolov, V.P., Israel, W. and Unruh, W.G. [1989], 'Gravitational fields of straight and circular cosmic strings: relation between gravitational mass, angular deficit and internal structure', *Phys. Rev.* **D39**, 1084.

Frolov, V.P., & Serebrianii, E.M. [1987], 'Vacuum polarization in the gravitational field of a cosmic string', *Phys. Rev.* **D35**, 3779.

Frolov, V.P., Skarzhinsky, V.D., Zelnikov, A.I., & Heinrich, O. [1989], 'Equilibrium configurations of a cosmic string near a rotating black hole', *Phys. Lett.* **224B**, 255.

Fry, J. [1981], 'Gravitational correlations and the monopole problem', *Ap. J.* **246**, L93.

Futamase, T., & Garfinkle, D. [1988], 'What is the relation between $\delta\phi$ and μ for a cosmic string?', *Phys. Rev.* **D37**, 2086.

Garfinkle, D. [1985], 'General relativistic strings', *Phys. Rev.* **D32**, 1323.

Garfinkle, D. [1989], 'Tidal force from cosmic strings', *Phys. Rev.* **D40**, 1801.

Garfinkle, D., & Gregory, R. [1990], 'Corrections to the thin wall approximation in general relativity', *Phys. Rev.* **D41**, 1889.

Garfinkle, D., & Laguna, P. [1989], 'Contribution of gravitational self-interaction to $\Delta\phi$ and μ for a cosmic string', *Phys. Rev.* **D39**, 1552.

Garfinkle, D., & Vachaspati, T. [1987], 'Radiation from kinky, cuspless, cosmic loops', *Phys. Rev.* **D36**, 2229.

Garfinkle, D., & Vachaspati, T. [1988], 'Fields due to kinky, cuspless, cosmic loops', *Phys. Rev.* **D37**, 257.

Garfinkle, D., & Vachaspati, T. [1990], 'Cosmic string travelling waves', *Phys. Rev.* **D42**, 1960.

Garfinkle, D., & Will, C.M. [1987], 'The effect of dynamical friction on the motion of cosmic strings', *Phys. Rev.* **D35**, 1124.

Garriga, J., & Verdaguer, E. [1987], 'Cosmic strings and Einstein–Rosen soliton waves', *Phys. Rev.* **D36**, 2250.

Garriga, J., & Vilenkin, A. [1991], 'Perturbations on domain walls and strings: a covariant theory', *Phys. Rev.* **D44**, 1007.

Garriga, J., & Vilenkin, A. [1992], 'Quantum fluctuations on domain walls, strings and vacuum bubbles', *Phys. Rev.* **D45**, 3469.

Garriga, J., & Vilenkin, A. [1993], 'Black holes from nucleating strings', *Phys. Rev.* **D47**, 3265.

Gates, E., Krauss, L.M., & Terning, J. [1992], 'Monopole annihilation at the electroweak scale – not!', *Phys. Lett.* **284B**, 309.

Gelmini, G.B., Gleiser, M., & Kolb, E.W. [1989], 'Cosmology of biased discrete symmetry breaking', *Phys. Rev.* **D39**, 1558.

Geroch, R., & Traschen, J. [1987], 'Strings and other distributional sources in general relativity', *Phys. Rev.* **D36**, 1017.

Gibbons, G.W. [1990], 'Self-gravitating magnetic monopoles, global monopoles and black holes', in *The Physical Universe: The interface between cosmology, astrophysics and particle physics*, Barrow, J.D., Henriques, A.B., Lago, M.T.V.T, & Longair, M.S., eds. (Springer–Verlag).

Gibbons, G.W., & Hawking, S.W. [1977], 'Cosmological event horizons, thermodynamics, and particle creation', *Phys. Rev.* **D15**, 2738.

Gibbons, G.W., Ortiz, M.E., & Ruiz Ruiz, F.R. [1989], 'Existence of global strings coupled to gravity', *Phys. Rev.* **D39**, 1546.

Gibbons, G.W., Ortiz, M.E., Ruiz Ruiz, F.R., & Samols, T.M. [1992], 'Semilocal strings and monopoles', *Nucl. Phys.* **385**, 127.

Gibbons, G.W., Ruiz Ruiz, F.R., & Vachaspati, T. [1990], 'The nonrelativistic coulomb problem on a cone', *Comm. Math. Phys.* **127**, 295.

Ginzburg, V.L. [1960], 'Some remarks on phase transitions of the second kind and the microscopic theory of ferroelectric materials', *Sov. Phys. Solid State* **2**, 1824.

Goddard, P., Goldstone, J., Rebbi, C., & Thorn, C.B. [1973], 'Quantum dynamics of a massless relativistic string', *Nucl. Phys.* **B56**, 109.

Goldhaber, A.S. [1989], 'Collapse of a 'global monopole'', *Phys. Rev. Lett.* **67**, 2158.

Goldman, T., Kolb, E.W., & Toussaint, D. [1981], 'Gravitational clumping and the annihilation of monopoles', *Phys. Rev.* **D23**, 867.

Goldstone, J. [1961], 'Field theories with superconductor solutions', *Nuovo Cim.* **19**, 154.

Gooding, A.K., Spergel, D.N., & Turok, N. [1991], 'The formation of galaxies and quasars in a texture seeded CDM cosmogony', *Ap. J. Lett.* **372**, L5.

Goto, T. [1971], 'Relativistic quantum mechanics of a one-dimensional mechanical continuum and subsidiary condition of dual resonance model', *Prog. Theor. Phys.* **46**, 1560.

Gott, J.R. [1975], 'On the formation of elliptical galaxies', *Ap. J.* **201**, 296.

Gott, J.R. [1985], 'Gravitational lensing effects of vacuum string: exact results', *Ap. J.* **288**, 422.

Gott, J.R. [1991], 'Closed timelike curves produced by pairs of moving cosmic strings—exact solutions', *Phys. Rev. Lett.* **66**, 1126.

Gott, J.R., & Alpert, M. [1984], 'General relativity in a (2+1)-dimensional spacetime', *Gen. Rel. Grav.* **16**, 243.

Gott, J.R., & Rees, M.J. [1975], 'A theory of galaxy formation and clustering', *Astron. Ap.* **45**, 365.

Gradwohl, B.-A., Kalbermann, G., Piran, T., & Bertschinger, E. [1990], 'Global strings and superfluid vortices: analogies and differences', *Nucl. Phys.* **B338**, 371.

Green, M.B., Schwarz, J.H., & Witten, E. [1987], *Superstring theory: I & II*, (Cambridge University Press).

Gregory, R. [1987], 'Gravitational stability of local strings', *Phys. Rev. Lett.* **59**, 740.

Gregory, R. [1988a], 'Effective action for a cosmic string', *Phys. Lett.* **206B**, 199.

Gregory, R. [1988b], 'Global string singularities', *Phys. Lett.* **215B**, 663.

Gregory, R., Haws, D., & Garfinkle, D. [1990], 'The dynamics of domain walls and strings', *Phys. Rev.* **D42**, 343.

Gregory, R., Perkins, W.B., Davis, A.-C., & Brandenberger, R.H. [1990], 'Cosmic strings and baryon decay catalysis', in *Formation and Evolution of Cosmic Strings*, Gibbons, G.W., Hawking, S.W., & Vachaspati, T., eds. (Cambridge University Press).

Gunn, J.E. [1977], 'Massive galactic halos: I. Formation and evolution', *Ap. J.* **218**, 592.

Guth, A.H. [1981], 'The inflationary universe: a possible solution to the horizon and flatness problems', *Phys. Rev.* **D23**, 347.

Guth, A.H., [1991], 'Can a man-made universe be created by quantum tunneling without an initial singularity?', *Phys. Scripta*, 237.

Guth, A.H., & Tye, S.-H. [1980], 'Phase transitions and magnetic monopole production in the very early universe', *Phys. Rev. Lett.* **44**, 631. Erratum: *Phys. Rev. Lett.* **44**, 963.

Guth, A.H., & Weinberg, E.J. [1980], 'A cosmological lower bound on the Higgs boson mass', *Phys. Rev. Lett.* **45**, 1131.

Guth, A.H., & Weinberg, E.J. [1981], 'Cosmological consequences of a first-order phase transition in the $SU(5)$ grand unified model', *Phys. Rev.* **D23**, 876.

Guth, A.H., & Weinberg, E.J. [1983], 'Could the universe have recovered from a slow first-order phase transition?', *Nucl. Phys.* **B212**, 321.

Hagedorn, R. [1965], 'Statistical thermodynamics of strong interactions at high temperature', *Nuovo Cim. Suppl.* **3**, 147.

Hagedorn, R. [1968], 'Hadronic matter near the boiling point', *Nuovo Cim.* **56A**, 1027.

Hagmann, C., & Sikivie, P. [1991], 'Computer simulations of the motion and decay of global strings', *Nucl. Phys.* **B363**, 247.

Hara, T., & Miyoshi, S. [1987a], 'Formation of the first systems in the wakes of moving cosmic strings', *Prog. Theor. Phys.* **77**, 1152.

Hara, T., & Miyoshi, S. [1987b], 'Flareup of the universe after $z \sim 10^2$ for cosmic string model', *Prog. Theor. Phys.* **78**, 1081.

Hara, T., & Miyoshi, S. [1989], 'Large-scale structures and streaming velocities due to open cosmic strings', *Prog. Theor. Phys.* **81**, 1187.

Hara, T., & Miyoshi, S. [1990], 'Large scale structures due to wakes of open cosmic strings', *Prog. Theor. Phys.* **84**, 867.

Harari, D., & Lousto, C. [1990], 'Repulsive gravitational effects of global monopoles', *Phys. Rev.* **D42**, 2626.

Harari, D., & Polychronakos, A.P. [1990], 'A global string with an event horizon', *Phys. Lett.* **240B**, 55.

Harari, D., & Sikivie, P. [1987], 'On the evolution of global strings in the early universe', *Phys. Lett.* **195B**, 361.

Harari, D., & Sikivie, P. [1988], 'The gravitational field of a global string', *Phys. Rev.* **D37**, 3488.

Harrison, E.R. [1970], 'Generation of magnetic fields in the radiation era', *M. N. R. A. S.* **147**, 279.

Hartle, J.B., & Hawking, S.W. [1983], 'The wave function of the Universe', *Phys. Rev.* **D28**, 2960.

Harvey, J.A., Kolb, E.W., Reiss, D.B., & Wolfram, S. [1982], 'Calculation of cosmological baryon asymmetry in grand unified gauge models', *Nucl. Phys.* **B201**, 16.

Hawking, S.W., [1982a], The boundary conditions of the Universe, *Pontificia Academae Scientarium Scripta Varia*, 48.

Hawking, S.W. [1982b], 'The development of irregularities in a single bubble inflationary universe', *Phys. Lett.* **115B**, 295.

Hawking, S.W. [1989], 'Black holes from cosmic strings', *Phys. Lett.* **231B**, 237.

Hawking, S.W. [1990], 'Gravitational radiation from collapsing cosmic string loops', *Phys. Lett.* **246B**, 36.

Hawking, S.W. [1992], 'Chronology protection conjecture', *Phys. Rev.* **D46**, 603.

Hawking, S.W., & Ellis, G.F.R. [1973], *The Large Scale Structure of the Universe*, (Cambridge University Press).

Hawking, S.W., & Moss, I.G. [1983], 'Fluctuations in the inflationary universe', *Nucl. Phys.* **B224**, 180.

Hawking, S.W., Moss, I.G., & Stewart, J.M. [1982], 'Bubble collisions in the early universe', *Phys. Rev.* **D26**, 2681.

Haws, D., Hindmarsh, M.B., & Turok, N. [1988], 'Superconducting strings or springs?', *Phys. Lett.* **209B**, 255.

Hayashi, Y. [1988], 'Superconducting cosmic string loops as cosmic accelerators of cosmic rays', *Nuovo Cim.* **102B**, 81.

Hecht, M.W., & de Grand T.A. [1990], 'Radiation patterns from vortex–antivortex annihilation', *Phys. Rev.* **D42**, 519.

Helliwell, T.M., & Konkowski, D.A. [1986], 'Vacuum fluctuations outside cosmic strings', *Phys. Rev.* **D34**, 1918.

Higgs, P.W. [1964], 'Broken symmetries, massless particles and gauge fields', *Phys. Lett.* **12**, 132.

Hill, C.T., Hodges, H.M., & Turner, M.S. [1987], 'Variational study of ordinary and superconducting cosmic strings', *Phys. Rev. Lett.* **59**, 2493.

Hill, C.T., Hodges, H.M., & Turner, M.S. [1988], 'Bosonic superconducting cosmic strings', *Phys. Rev.* **D37**, 263.

Hill, C.T., Kagan, A.L. & Widrow, L.M..[1988], 'Are cosmic strings frustrated?', *Phys. Rev.* **D38**, 1100.

Hill, C.T., & Lee, K. [1988], 'Anomalies and effective Lagrangians of superconducting cosmic strings', *Nucl. Phys.* **B297**, 765.

Hill, C.T., Schramm, D.N., & Fry, J.N.,, [1989], *Comm. Nucl. Part. Sci.*, **19**, 25.

Hill, C.T., Schramm, D.N., & Walker, T.P. [1987], 'Ultrahigh-energy cosmic rays from superconducting cosmic strings', *Phys..Rev.* **D36**, 1007.

Hill, C.T., & Widrow, L.M. [1987], 'Superconducting cosmic strings with massive fermions', *Phys. Lett.* **189B**, 17.

Hilton, P.J. [1953], *Introduction to Homotopy Theory*, (Cambridge University Press).

Hindmarsh, M.B. [1988], 'Superconducting cosmic strings with coupled zero modes', *Phys. Lett.* **200B**, 429.

Hindmarsh, M.B. [1990a], 'Gravitational radiation from kinky infinite strings', *Phys. Lett.* **251B**, 28.

Hindmarsh, M.B. [1990b], 'Searching for cosmic strings', in *Formation and Evolution of Cosmic Strings*, Gibbons, G.W., Hawking, S.W., & Vachaspati, T., eds. (Cambridge University Press).

Hindmarsh, M.B., [1991], 'Massless modes on cosmic strings', Lecture presented at the Korber Symposium on Superfluid He-3 in Rotation, Helsinki, Finland, Jun 10–14, 1990.

Hindmarsh, M.B. [1992], 'Semilocal topological defects', *Phys. Rev. Lett.* **68**, 1263.

Hindmarsh, M.B., & Kibble, T.W.B. [1985], 'Beads on strings', *Phys. Rev. Lett.* **55**, 2398.

Hindmarsh, M.B., & Kibble, T.W.B. [1986], 'Hindmarsh and Kibble respond', *Phys. Rev. Lett.* **57**, 647.

Hindmarsh, M.B., & Wray, A. [1990], 'Gravitational effects of line sources and the zero width limit', *Phys. Lett.* **251B**, 498.

Hiscock, W.A. [1985], 'Exact gravitational field of a string', *Phys. Rev.* **D31**, 3288.

Hiscock, W.A. [1987], 'Semiclassical gravitational effects near cosmic strings', *Phys. Lett.* **188B**, 317.

Hiscock, W.A., & Lail, J.B. [1988], 'Relativistic fluid flows around cosmic strings', *Phys. Rev.* **D37**, 869.

Hodges, H.M. [1989], 'The formation of topological defects in phase transitions', *Phys. Rev.* **D39**, 3557.

Hodges, H.M., & Primack, J.R. [1991], 'String, texture and inflation', *Phys. Rev.* **D43**, 3155.

Hoffman, Y., & Shakem, J. [1985], 'Local density maxima – progenitors of structure', *Ap. J.* **297**, 16.

Hoffman, Y., & Zurek, W.H. [1988], 'Great attractor: is it a loop of a cosmic string?', *Nature* **333**, 46.

Hogan, C.J. [1984a], 'Cosmic strings and galaxies', *Nature* **310**, 365.

Hogan, C.J. [1984b], 'Massive black holes generated by cosmic strings', *Phys. Lett.* **143B**, 87.

Hogan, C.J. [1987a], 'Superconducting cosmic strings', *Nature* **326**, 742.

Hogan, C.J. [1987b], 'Cosmology – a new twist for cosmic strings', *Nature* **325**, 300.

Hogan, C.J., & Narayan, R. [1984], 'Gravitational lensing by cosmic strings', *M. N. R. A. S.* **211**, 575.

Hogan, C.J., & Rees, M.J. [1984], 'Gravitational interactions of cosmic strings', *Nature* **311**, 109.

Holman, R., Kibble, T.W.B., & Rey, S.-J. [1992], 'How efficient is the Langacker–Pi mechanism of monopole annihilation?', *Phys. Rev. Lett.* **69**, 241.

Huang, K., & Weinberg, S. [1970], 'Ultimate temperature and the early universe', *Phys. Rev. Lett.* **25**, 895.

Ipser, J., & Sikivie, P. [1984], 'Gravitationally repulsive domain wall', *Phys. Rev.* **D30**, 712.

Ishihara, H., Kubotani, H., & Nambu, Y. [1992], 'Cell structure formation of domain walls', *Phys. Lett. B*, in press.

Israel, W. [1966], 'Singular hypersurfaces and thin shells in general relativity', *Nuovo Cim.* **44B**, 1. Erratum: *Nuovo Cim.* **49B**, 463.

Israel, W. [1977], 'Line sources in general relativity', *Phys. Rev.* **D15**, 935.

Iyanaga, S., & Kawada, Y. [1977], *Encyclopedic Dictionary of Mathematics*, (MIT Press, Cambridge, Massachusetts).

Izawa, M., & Sato, H. [1987], 'Galaxy formation by cosmic strings and cooling of baryonic matter', *Prog. Theor. Phys.* **78**, 1219.

Jackiw, R. [1977], 'Quantum meaning of classical field theory', *Rev. Mod. Phys.* **49**, 681.

Jackiw, R. [1984], 'Lower dimensional gravity', *Nucl. Phys.* **B252**, 343.

Jackiw, R., & Rossi, P. [1981], 'Zero modes of the vortex–fermion system', *Nucl. Phys.* **B190**, 681.

Jacobs, L., & Rebbi, C. [1979], 'Interaction energy of superconducting vortices', *Phys. Rev.* **B19**, 4486.

Jaffe, A. & Taubes, C.M. [1980], *Vortices and Monopoles*, (Birkhäuser, Boston).

James, M., Perivolaropoulos, L., & Vachaspati, T. [1993], 'Detailed stability analysis of electroweak strings', *Nucl. Phys.* **B395**, 534.

Jetzer, P. [1992], 'Boson stars', *Phys. Rep.* **220**, 164.

Kaiser, N. [1984], 'On the spatial correlation of Abell clusters', *Ap. J. Lett.* **284**, L9.

Kaiser, N., & Stebbins, A. [1984], 'Microwave anisotropy due to cosmic strings', *Nature* **310**, 391.

Kalb, M. [1978], 'Magnetic monopoles in a gauge field theory from vortex-strings', *Phys. Rev.* **D17**, 2713.

Kalb, M. & Ramond, P. [1974], 'Classical direct interstring action', *Phys. Rev.* **D9**, 2273.

Kasner, E. [1921], 'Geometric theorems on Einstein's cosmological equations', *Amer. J. Math.* **43**, 217.

Kawano, L. [1990], 'Evolution of domain walls in the early universe', *Phys. Rev.* **D41**, 1013.

Kawasaki, M. & Maeda, K. [1988], 'Baryon number generation from cosmic string loops', *Phys. Lett.* **209B**, 271.

Kibble, T.W.B. [1976], 'Topology of cosmic domains and strings', *J. Phys.* **A9**, 1387.

Kibble, T.W.B. [1980], 'Some implications of a cosmological phase transition', *Phys. Rep.* **67**, 183.

Kibble, T.W.B. [1982], 'Phase transitions in the early universe', *Acta Phys. Polon.* **B13**, 723.

Kibble, T.W.B. [1985], 'Evolution of a system of cosmic strings', *Nucl. Phys.* **B252**, 227. Erratum: *Nucl. Phys.* **B261**, 750.

Kibble, T.W.B. [1986a], 'Configuration of Z_2 strings', *Phys. Lett.* **166B**, 311.

Kibble, T.W.B. [1986b], 'String dominated universe', *Phys. Rev.* **D33**, 328.

Kibble, T.W.B. [1990], 'Cosmic strings: an overview', in *Formation and Evolution of Cosmic Strings*, Gibbons, G.W., Hawking, S.W., & Vachaspati, T., eds. (Cambridge University Press).

Kibble, T.W.B., & Copeland, E. [1991], 'Evolution of small scale structure on cosmic strings', *Phys. Scripta* **T36**, 153.

Kibble, T.W.B., Lazarides, G., & Shafi, Q. [1982a], 'Strings in $SO(10)$', *Phys. Lett.* **113B**, 237.

Kibble, T.W.B., Lazarides, G., & Shafi, Q. [1982b], 'Walls bounded by strings', *Phys. Rev.* **D26**, 435.

Kibble, T.W.B., & Turok, N.G. [1982], 'Selfintersection of cosmic strings', *Phys. Lett.* **116B**, 141.

Kibble, T.W.B., & Weinberg, E.J. [1991], 'When does causality constrain the monopole abundance?', *Phys. Rev.* **D43**, 3188.

Kim, J.-E. [1987], 'Light pseudoscalars, particle physics and cosmology', *Phys. Rep.* **150**, 1.

Kirzhnits, D.A. [1972], 'Weinberg model and the 'hot' universe', *JETP Lett.* **15**, 745.

Kirzhnits, D.A., & Linde, A.D. [1972], 'Macroscopic consequences of the Weinberg model', *Phys. Lett.* **42B**, 471.

Kirzhnits, D.A., & Linde, A.D. [1974], 'A relativistic phase transition', *Sov. Phys. JETP* **40**, 628.

Kirzhnits, D.A., & Linde, A.D. [1976], 'Symmetry behaviour in gauge theories', *Ann. Phys.* **101**, 195.

Kiskis, J. [1977], 'Fermions in a pseudoparticle field', *Phys. Rev.* **D15**, 2329.

Kobayakawa, K., & Samura, T. [1992], 'High-energy particles by shock acceleration around superconducting cosmic strings', in *High Energy Neutrino Astrophysics*, Stenger *et al.*, eds. (World Scientific).

Kobzarev, I.Yu., Okun, L.B., & Zel'dovich, Ya.B., [1974], *Phys. Lett.* **50B**, 340.

Kofman, L., & Linde, A.D. [1987], 'Generation of density perturbations in the inflationary cosmology', *Nucl. Phys.* **B282**, 555.

Kolb, E.W., Colgate, S.A., & Harvey, J.A. [1982], 'Monopole catalysis of nucleon decay in neutron stars', *Phys. Rev. Lett.* **79**, 137.

Kolb, E.W., & Turner, M.S. [1990], *The Early Universe*, (Addison-Wesley, Redwood City, California).

Krauss, L.M., & Accetta, F.S. [1989], 'The stochastic gravitational wave spectrum resulting from cosmic string evolution', *Nucl. Phys.* **B319**, 747.

Kubotani, H. [1992], 'The domain wall network of explicitly broken $O(N)$ model', *Prog. Theor. Phys.* **87**, 387.

Kurki-Suonio, H., Matzner, R., Olive, K., & Schramm, D.N. [1990], 'Big bang nucleosynthesis and the quark–hadron transition', *Ap. J.* **353**, 406.

La, D., & Steinhardt, P.J. [1989], 'Extended inflation cosmology', *Phys. Rev. Lett.* **62**, 376. Erratum: *Phys. Rev. Lett.* **62**, 1066.

Laguna, P., & Garfinkle, D. [1989], 'Spacetime of supermassive $U(1)$ gauge cosmic strings', *Phys. Rev.* **D40**, 1011.

Laguna, P., & Matzner, R.A. [1987a], 'Discontinuity cylinder model of gravitating $U(1)$ cosmic strings', *Phys. Rev.* **D35**, 2933.

Laguna, P., & Matzner, R.A. [1987b], 'Coupled field solutions for $U(1)$ gauge cosmic strings', *Phys. Rev.* **D36**, 3663.

Laguna, P., & Matzner, R.A. [1989], 'Peeling $U(1)$ gauge cosmic strings', *Phys. Rev. Lett.* **62**, 1948.

Laguna, P., & Matzner, R.A. [1990], 'Numerical simulation of bosonic superconducting string interactions', *Phys. Rev.* **D41**, 1751.

Landau, L.D., & Lifshitz, E.M. [1975], *The Classical Theory of Fields*, (Pergamon Press, Oxford).

Landau, L.D., & Lifshitz, E.M. [1977], *Quantum Mechanics (Non-relativistic theory)*, (Pergamon Press, Oxford).

Landau, L.D., & Lifshitz, E.M. [1987], *Fluid Mechanics*, (Pergamon Press, Oxford).

Landau, L.D., Lifshitz, E.M., & Pitaevskii, L.P. [1984], *Electrodynomics of Continuous Media*, (Pergamon Press, Oxford).

Langacker, P., & Pi, S.-Y. [1980], 'Magnetic monopoles in grand unified theories', *Phys. Rev. Lett.* **45**, 1.

de Lapparent, V., Geller, M.J., & Huchra, J.P. [1986], 'The galaxy distribution and the large-scale structure of the universe', *Ap. J. Lett.* **302**, L1.

Lazarides, G., Panagiotakopoulos, C., & Shafi, Q. [1987a], 'Phenomenology and cosmology with superstrings', *Phys. Rev. Lett.* **56**, 432.

Lazarides, G., Panagiotakopoulos, C., & Shafi, Q. [1987b], 'Superheavy superconducting cosmic strings from superstring models', *Phys. Lett.* **183B**, 289.

Lazarides, G., Panagiotakopoulos, C., & Shafi, Q. [1988], 'Cosmic superconducting strings and colliders', *Nucl. Phys.* **B296**, 657.

Lazarides, G., & Shafi, Q. [1980], 'The fate of primordial magnetic monopoles', *Phys. Lett.* **94B**, 149.

Lazarides, G., & Shafi, Q. [1985], 'Superconducting strings in axion models', *Phys. Lett.* **151B**, 123.

Lazarides, G., Shafi, Q., & Trower, W.P. [1982], 'Consequences of a monopole with Dirac magnetic charge', *Phys. Rev. Lett.* **49**, 1756.

Lazarides, G., Shafi, Q., & Walsh, T.F. [1982], 'Cosmic strings and domains in unified theories', *Nucl. Phys.* **B195**, 157.

Lee, K., & Weinberg, E.J. [1984], 'Density fluctuations and particle–antiparticle annihilation', *Nucl. Phys.* **B246**, 354.

Lee, S.-H., & Weinberg, E.J. [1990], 'Causality bounds and monopoles connected by strings', *Phys. Rev.* **D42**, 3422.

Lee, T.D. [1976], 'Examples of four-dimensional soliton solutions and abnormal nuclear states', *Phys. Rep.* **23C**, 254.

Lee, T.D., & Wick, G.C. [1974], 'Vacuum stability and vacuum excitations in a spin-0 field theory', *Phys. Rev.* **D9**, 2291.

Leese, R.A., & Prokopec, T. [1992], 'Monte-Carlo simulation of texture formation', *Phys. Rev.* **D44**, 3749.

Leese, R.A., & Samols, T.M. [1992], 'Interaction of semilocal vortices', DAMTP-92-40.

Letelier, P.S. [1978], 'Conservation laws for a system of particles, strings, and membranes', *Phys. Rev.* **D18**, 359.

Letelier, P.S. [1983], 'String cosmologies', *Phys. Rev.* **D28**, 2414.

Letelier, P.S. [1987], 'Multiple cosmic strings', *Class. Quant. Grav.* **4**, L75.

Letelier, P.S., & Gal'tsov, D.V., [1993], 'Multiple moving crossed cosmic strings', *Class. Quant. Grav.*, in press.

Letelier, P.S., & Verdaguer, E. [1988], 'Cosmic strings in extra dimensions', *Phys. Rev.* **D37**, 2333.

Linde, A. [1977], 'On the vacuum instability and the Higgs meson mass', *Phys. Lett.* **70B**, 306.

Linde, A. [1981], 'Grand Bang', *Phys. Lett.* **99B**, 391.

Linde, A. [1982a], 'A new inflationary universe scenario', *Phys. Lett.* **108B**, 389.

Linde, A. [1982b], 'Scalar field fluctuations in the expanding universe and the new inflationary universe scenario', *Phys. Lett.* **116B**, 335.

Linde, A. [1983], 'Chaotic inflation', *Phys. Lett.* **B129**, 177.

Linde, A. [1984a], 'The inflationary universe', *Rep. Prog. Phys* **47**, 925.

Linde, A., [1984b], *Lett. Nuovo Cim.* **39**, 401

Linde, A. [1986], 'Eternally existing self-reproducing chaotic inflationary universe', *Phys. Lett.* **175B**, 395.

Linde, A. [1990], *Particle Physics and Inflationary Cosmology*, (Harwood, Chur, Switzerland).

Linde, A. [1992], 'Stochastic approach to tunneling and baby universe formation', *Nucl. Phys.* **B372**, 421.

Linde, A., & Lyth, D.H. [1990], 'Axionic domain wall production during inflation', *Phys. Lett.* **246B**, 353.

Linet, B. [1985], 'The static metrics with cylindrical symmetry describing a model of cosmic strings', *Gen. Rel. Grav.* **17**, 1109.

Linet, B. [1986a], 'Force on a charge in the space-time of a cosmic string', *Phys. Rev.* **D33**, 1833.

Linet, B. [1986b], 'On the wave equations in the space-time of a cosmic string', *Ann. Inst. Henri Poincaré* **A45**, 249.

Linet, B. [1987a], 'Quantum field theory in the space-time of a cosmic string', *Phys. Rev.* **D35**, 536.

Linet, B. [1987b], 'A vortex line model for infinite straight cosmic strings', *Phys. Lett.* **124A**, 240.

Linet, B. [1990], 'On the supermassive $U(1)$ gauge cosmic strings', *Class. Quant. Grav.* **7**, L79.

Lonsdale, S. & Moss, I. [1988], 'The motion of cosmic strings under gravity', *Nucl. Phys.* **B298**, 693.

Lozano, G., Maniasm, M.V., & Schaposnik, F.A. [1988], 'The charged vortex solution to spontaneously broken gauge theories with Chern–Simons term', *Phys. Rev.* **D38**, 601.

Lund, F., & Regge, T. [1976], 'Unified approach to strings & vortices with soliton solutions', *Phys. Rev.* **D14**, 1525.

Lyth, D.H. [1983], 'Measuring the curvature of space with stretched strings', *Eur. J. Phys.* **4**, 144.

Lyth, D.H. [1987], 'Cosmic strings and an improved upper bound on the energy density during inflation', *Phys. Lett.* **196B**, 126.

Ma, S. [1976], *Modern Theory of Critical Phenomena*, (W.A. Benjamin, New York).

MacDowell, S.W., & Törnkvist, O., [1991], in *Symposia Gaussiana*, Series A, vol. 1, 383, R.G. Lintz, ed., (Institutum Gaussianum, Toronto, Canada).

MacDowell, S.W., & Törnkvist, O. [1992], 'Structure of the ground state of the electroweak gauge theory in a strong magnetic field', *Phys. Rev.* **D45**, 3833.

MacGibbon, J.H., & Brandenberger, R.H. [1990], 'High energy neutrino flux from ordinary cosmic strings', *Nucl. Phys.* **B331**, 153.

MacKenzie, R. [1987], 'Energetics of bosonic superconducting strings', *Phys. Lett.* **197B**, 59. Erratum: *Phys. Lett.* **199B**, 596.

McLerran, L., Shaposhnikov, M., Turok, N., & Voloshin, N. [1991], 'Why the baryon asymmetry of the universe is approximately 10^{-10}', *Phys. Lett.* **256B**, 451.

Maeda, K., & Turok, N. [1988], 'Finite width corrections to the Nambu action for the Nielsen–Olesen string', *Phys. Lett.* **202B**, 376.

Magueijo, J.C.R. [1992], 'Inborn metric of cosmic strings', *Phys. Rev.* **D46**, 1368.

Manias, M.V., Naon, C.M., Schaposnik, F.A., & Trobo, M. [1986], 'Nonabelian charged vortices as cosmic strings', *Phys. Lett.* **171B**, 199.

Mankiewicz, L., & Zembowicz, R. [1988], 'Screening of current in superconducting cosmic strings by vacuum polarization effects', *Phys. Lett.* **202B**, 493.

Manohar, A. [1988], 'Anomalous vortices and electromagnetism 2', *Phys. Lett.* **206B**, 276.

Manton, N.S. [1982], 'A remark on the scattering of BPS monopoles', *Phys. Lett.* **110B**, 54.

Marder, L. [1959], 'Flat spacetimes with gravitational fields', *Proc. Roy. Soc.* **A252**, 45.

Massarotti, A. [1991a], 'Evolution of light domain walls interacting with dark matter', *Phys. Rev.* **D43**, 346.

Massarotti, A. [1991b], 'Light domain walls, massive neutrinos and the large scale structure', Fermilab-Pub-91-104-A.

Mather, J.C., *et. al.* [1990], 'A preliminary measurement of the cosmic background spectrum by the COBE satellite', *Ap. J.* **354**, L37.

Matzner, R.A. [1988], 'Interactions of $U(1)$ cosmic strings: Numerical intercommutation', *Computers in Physics* Sep/Oct, 51.

Matzner, R.A., & McCracken, J. [1988], 'Probability of intercommutation of cosmic strings', in *Cosmic Strings: The Current Status*, Accetta, F.S., & Krauss, L.M., eds. (World Scientific, Singapore).

Maunder, C.R.F. [1980], *Algebraic Topology*, (Cambridge University Press).

Mazur, P.O. [1986a], 'Induced angular momentum on superconducting cosmic strings', *Phys. Rev.* **D34**, 1925.

Mazur, P.O. [1986b], 'Spinning cosmic strings and quantization of energy', *Phys. Rev. Lett.* **57**, 929.

Melchiorri, F., Melchiorri, B., Ceccarelli, C., & Pietranera, L. [1981], 'Fluctuations in the microwave background at intermediate angular scales', *Ap. J. Lett.* **250**, L1.

Melott, A.L., & Scherrer, R.J. [1987], 'Formation of large scale structure from cosmic string loops and cold dark matter', *Nature* **328**, 691.

Melott, A.L., & Scherrer, R.J. [1988], 'Large-scale structure from cosmic string loops in a baryon-dominated universe', *Ap. J.* **331**, 38.

Mermin, N.D. [1979], 'Topological theory of defects', *Rev. Mod. Phys.* **51**, 625.

Mijic, M.B. [1988], 'Turning on a superconducting cosmic string', *Phys. Rev.* **D39**, 2864.

Mishustin, I.N., & Ruzmaikin, A.A. [1971], 'Occurrence of 'priming' magnetic fields during the formation of protogalaxies', *Sov. Phys. JETP* **34**, 233.

Mitchell, D., & Turok, N. [1987a], 'Statistical mechanics of cosmic strings', *Phys. Rev. Lett.* **58**, 1577.

Mitchell, D., & Turok, N. [1987b], 'Statistical properties of cosmic strings', *Nucl. Phys.* **B294**, 1138.

Mohapatra & Senjanović [1976], 'Broken symmetries at high temperature', *Phys. Rev.* **D20**, 3390.

Moriarty, K., Myers, E., & Rebbi, C. [1988a], 'Dynamical interactions of cosmic strings and flux vortices in superconductors', *Phys. Lett.* **207B**, 411.

Moriarty, K., Myers, E., & Rebbi, C., [1988b], 'Evolution and interaction of cosmic strings and flux vortices in superconductors', Presented at *Frontiers of Nonperturbative Field Theory*, Eger, Hungary, Aug 18-23, 1988.

Moriarty, K., Myers, E., & Rebbi, C. [1988c], 'Interactions of cosmic strings', in *Cosmic Strings: The Current Status*, Accetta, F.S., & Krauss, L.M., eds. (World Scientific, Singapore).

Moss, I., & Poletti, S. [1987], 'The gravitational field of a superconducting cosmic string', *Phys. Lett.* **199B**, 34.

Müller-Hartmann, E. [1966], 'λ-transition at the lower critical field in Type-II superconductors?', *Phys. Lett.* **23**, 521.

Myers, E., Rebbi, C., & Strilka, R. [1992], 'A study of the interaction and scattering of vortices in the abelian Higgs (or Ginzburg–Landau) model', *Phys. Rev.* **D45**, 1355.

Nagasawa, M., & Yokoyama, J. [1992], 'Phase transitions triggered by quantum fluctuations in the inflationary universe', *Nucl. Phys.* **B370**, 472.

Nambu, Y., [1966], in Proceedings of Int. Conf. on *Elementary Particles*, Y. Tanikawa, ed., (Publications Office, Progress in Theoretical Physics, Kyoto).

Nambu, Y., [1969], in Proceedings of Int. Conf. on *Symmetries & Quark Models* (Wayne State University); Lectures at the Copenhagen Summer Symposium, 1970.

Nambu, Y. [1977], 'String-like configurations in the Weinberg–Salam theory', *Nucl. Phys.* **B130**, 505.

Nambu, Y., Ishihara, H., Gouda, N., & Sugiyama, N. [1991], 'Anisotropies in the cosmic background radiation by domain networks', *Ap. J. Lett.* **375**, L35.

Newton, I. [1713], *Principia: Vol. II, The System of the World*, (republished by University of California Press, Berkeley).

Nielsen, N.K. [1980], 'Dimensional reduction and classical strings', *Nucl. Phys.* **B167**, 249.

Nielsen, H.B., & Olesen, P. [1973], 'Vortex-line models for dual strings', *Nucl. Phys.* **B61**, 45.

Nielsen, N.K., & Olesen, P. [1978], 'An unstable Yang–Mills field mode', *Nucl. Phys.* **B144**, 376.

Nielsen, N.K., & Olesen, P. [1987a], 'Dynamical properties of superconducting cosmic strings', *Nucl. Phys.* **B291**, 829.

Nielsen, N.K., & Olesen, P. [1987b], 'Dynamical properties of superconducting cosmic strings: 2. Heavy closed strings in a magnetic field', *Nucl. Phys.* **B298**, 776.

Nielsen, N.K., & Schroer, B. [1977], 'Topological fluctuations and breaking of chiral symmetry in gauge theories involving massless fermions', *Nucl. Phys.* **B120**, 62.

Notzold, D. [1991], 'Gravitational effects of global textures', *Phys. Rev.* **D43**, R961.

Olive, D., & Turok, N. [1982], 'Z_2 vortex strings in grand unified theories', *Phys. Lett.* **117B**, 193.

Olive, K.A. [1990], 'Inflation', *Phys. Rep.* **190**, 307.

Olive, K.A., Schramm, D.N., Steigman, G., & Walker, T. [1990], 'Big Bang nucleosynthesis revisited', *Phys. Lett.* **236B**, 454.

Ortiz, M.E. [1991], 'A new look at supermassive cosmic strings', *Phys. Rev.* **D43**, 2521.

Ostriker, J.P., & Thompson, C. [1987], 'Distortion of the cosmic background radiation by superconducting strings', *Ap. J.* **323**, L97.

Ostriker, J.P., Thompson, C., & Witten, E. [1986], 'Cosmological effects of superconducting strings', *Phys. Lett.* **180B**, 231.

Paczynski, B. [1986], 'Will cosmic strings be discovered using the Space Telescope?', *Nature* **319**, 567.

Paczynski, B., & Spergel, D.N. [1987], 'Death throes of a superconducting loop', Princeton preprint (unpublished).

Page, D.N., & Hawking, S.W. [1976], 'Gamma rays from primordial black holes', *Ap. J.* **206**, 1.

Pagels, H.R. [1987], 'Fractal geometry of cosmic strings and correlations among galaxies and Abell clusters', *Phys. Rev.* **D35**, 1141.

Panek, M., & Rudak, B. [1987], 'Superconducting cosmic strings—energy loss by plasma dissipation', *Phys. Lett.* **195B**, 357.

Park, C., Spergel, D.N., & Turok, N. [1991], 'Large scale structure in a texture seeded CDM cosmogony', *Ap. J.* **372**, L53.

Parker, E.N. [1970], 'The origin of magnetic fields', *Ap. J.* **160**, 383.

Parker, L. [1987], 'Gravitational particle production in the formation of cosmic strings', *Phys. Rev. Lett.* **59**, 1369.

Peccei, R.D., & Quinn, H.R. [1977], 'CP conservation and the presence of pseudoparticles', *Phys. Rev. Lett.* **38**, 1440; *Phys. Rev.* **D16**, 1791.

Peebles, P.J.E. [1971], *Physical Cosmology*, (Princeton University Press).

Peebles, P.J.E. [1980], *The Large-Scale Structure of the Universe*, (Princeton University Press).

Peebles, P.J.E. [1993], *Principles of Physical Cosmology*, (Princeton University Press).

Pen, U.L., Spergel, D.N., & Turok, N. [1993], 'Cosmic structure formation and microwave anisotropies from global field ordering', PUP-TH-1375.

Penrose, R. [1969], 'Gravitational collapse: the role of general relativity', *Rivista del Nuovo Cim.* **1**, 252.

Perivolaropoulos, L. [1992], 'Instabilities and interactions of global topological defects', *Nucl. Phys.* **B375**, 665.

Perivolaropoulos, L. [1993], 'Asymptotics of Nielsen–Olesen vortices', CfA preprint no. 3707.

Perivolaropoulos, L., Brandenberger, R.H., & Stebbins, A. [1990], 'Dissipationless clustering of neutrinos in cosmic string induced wakes', *Phys. Rev.* **D41**, 1764.

Perkins, W.B., [1991], 'Baryon decay catalysis and cosmic strings', Ph. D. Thesis, (University of Cambridge).

Perkins, W.B., Perivolaropoulos, L., Davis, A.-C., Brandenberger, R.H., & Matheson, A. [1991], 'Scattering of fermions from a cosmic string', *Nucl. Phys.* **B353**, 237.

Peter, P. [1992], 'Low mass current-carrying cosmic strings', *Phys. Rev.* **D46**, 3322.

Peter, P. [1993], 'The no cosmic-spring conjecture', *Phys. Rev.* **D47**, 3169.

Pollock, M. [1987], 'Is inflation compatible with string-induced galaxy formation?', *Phys. Lett.* **185B**, 34.

Polnarev, A., & Zembowicz, R. [1991], 'Formation of primordial black holes by cosmic strings', *Phys. Rev.* **D43**, 1106.

Polyakov, A.M. [1974], 'Particle spectrum in quantum field theory', *JETP Lett.* **20**, 194.

Polyakov, A.M. [1975], 'Isometric states of quantum fields', *JETP Lett.* **41**, 988.

Prasad, M.K., & Sommerfield, C.H. [1975], 'Exact classical solution for the 't Hooft monopole and the Julia–Zee dyon', *Phys. Rev. Lett.* **35**, 760.

Preskill, J. [1979], 'Cosmological production of superheavy monopoles', *Phys. Rev. Lett.* **43**, 1365.

Preskill, J. [1983], 'Monopoles in the early universe', in *The Very Early Universe*, Gibbons, G.W., & Hawking, S.W., eds. (Cambridge University Press).

Preskill, J. [1984], 'Magnetic monopoles', *Ann. Rev. Nucl. Part. Sci.* **34**, 461.

Preskill, J., [1985], 'Vortices and monopoles', Lectures presented at the 1985 Les Houches Summer School, Les Houches, France.

Preskill, J. [1992], 'Semilocal defects', CALT-68-1787.

Preskill, J., & Krauss, L.M. [1990], 'Local discrete symmetry and quantum mechanical hair', *Nucl. Phys.* **B341**, 50.

Preskill, J. & Vilenkin, A. [1993], 'Decay of metastable topological defects', *Phys. Rev.* **D47**, 2324.

Preskill, J., & Wise, M.B., [1983], unpublished.

Preskill, J., Wise, M.B., & Wilczek, F. [1983], 'Cosmology of the invisible axion', *Phys. Lett.* **120B**, 127.

Press, W.H. [1980], 'Spontaneous production of the Zel'dovich spectrum of cosmological fluctuations', *Phys. Scripta* **21**, 702.

Press, W.H., Ryden, B.S., & Spergel, D.N. [1989], 'Dynamical evolution of domain walls in an expanding universe', *Ap. J.* **347**, 590.

Press, W.H., & Vishniac, E.T. [1980], 'Tenacious myths about cosmological perturbations larger than the horizon size', *Ap. J.* **239**, 1.

Prokopec, T., Sornborger, A., & Brandenberger, R.H. [1992], 'Texture collapse', *Phys. Rev.* **D45**, 1971.

Quashnock, J.M. & Spergel, D.N. [1990], 'Gravitational self-interactions of cosmic strings', *Phys. Rev.* **D42**, 2505.

Rajaraman, R. [1982], *Solitons and Instantons*, (North-Holland, Amsterdam).

Rees, M.J. [1985], 'Mechanisms for biased galaxy formation', *M. N. R. A. S.* **213**, 75P.

Rees, M.J. [1986], 'Baryon concentration in string wakes at $z \gtrsim 200$—implications for galaxy formation', *M. N. R. A. S.* **222**, 3.

Rees, M.J. [1987], 'The origin and cosmogonic implications of seed magnetic fields', *Quar. J. Roy. Astron. Soc.* **28**, 197.

Rees, M.J., [1990], private communication.

Rhie, S.H., & Bennett, D.P. [1991], 'Global monopoles do not 'collapse'', *Phys. Lett.* **67B**, 1173.

Rohm, R., [1985], Ph. D. Thesis, (Princeton University).

Romani, R.W. [1988], 'Low frequency gravity wave spectra generated by cosmic strings', *Phys. Lett.* **215B**, 477.

Rosen, G. [1968], 'Charged particle solutions to nonlinear complex scalar field theories', *J. Math. Phys.* **9**, 999.

Ruback, P.J. [1988], 'Vortex string motion in the Abelian-Higgs model', *Nucl. Phys.* **B296**, 669.

Rubakov, V.A. [1981], 'Superheavy magnetic monopoles and decay of the proton', *JETP Lett.* **33**, 644.

Rubakov, V.A. [1982], 'ADJ anomaly and fermion number breaking in the presence of a magnetic monopole', *Nucl. Phys.* **B203**, 311.

Rudak, B., & Panek, M. [1987], 'Superconducting cosmic strings and the spectrum of microwave background radiation', *Phys. Lett.* **199B**, 346.

Ryden, B.S., Press, W.H., & Spergel, D.N. [1990], 'The evolution of networks of domain walls and cosmic strings', *Ap. J.* **357**, 293.

Sakellariadou, M. [1990], 'Gravitational waves emitted from infinite strings', *Phys. Rev.* **D42**, 354.

Sakellariadou, M. [1991], 'Radiation of Nambu–Goldstone bosons from infinitely long cosmic strings', *Phys. Rev.* **D44**, 3767.

Sakellariadou, M., & Vilenkin, A. [1988], 'Numerical experiments with cosmic strings in flat space-time', *Phys. Rev.* **D37**, 885.

Sakellariadou, M., & Vilenkin, A. [1990], 'Cosmic string evolution in flat space-time', *Phys. Rev.* **D42**, 349.

Sakharov, A.D. [1967], 'Violation of *CP* invariance, *C* asymmetry, and baryon asymmetry of the universe', *JETP Lett.* **5**, 24.

Salomonson, P. [1978], 'Stability of the Nielsen–Olesen vortex', *Phys. Rev.* **D18**, 2184.

Samols, T. [1990], 'Hermiticity of the metric on the vortex moduli space', *Phys. Lett.* **244B**, 285.

Samols, T. [1992], 'Vortex scattering', *Comm. Math. Phys.* **145**, 149.

Sanchez, N., & Signore, M. [1988], 'String theories and millisecond pulsars', *Phys. Lett.* **214B**, 14.

Sato, H. [1986a], 'Galaxy formation by cosmic strings', *Prog. Theor. Phys.* **75**, 1342.

Sato, H. [1986b], 'Cosmic strings and rotation velocity of spiral galaxies', *Mod. Phys. Lett.* **A1**, 9.

Sato, K. [1981a], 'Cosmological baryon-number domain structure and first-order phase transition of a vacuum', *Phys. Lett.* **99B**, 66.

Sato, K. [1981b], 'First order phase transition and the expansion of the Universe', *M.N.R.A.S.* **195**, 467.

Sato, K., Sasaki, M., Kodama, H., & Maeda, K. [1981], 'Creation of wormholes by first-order phase transition of a vacuum in the early universe', *Prog. Theor. Phys.* **65**, 1443.

Scherrer, R.J. [1987], 'Large-scale structure from cosmic string loops. I. Formation & linear evolution of perturbations', *Ap. J.* **130**, 1.

Scherrer, R.J. [1990], 'Cosmic string angular momentum', in *Formation and Evolution of Cosmic Strings*, Gibbons, G.W., Hawking, S.W., & Vachaspati, T., eds. (Cambridge University Press).

Scherrer, R.J., & Frieman, J.A. [1986], 'Cosmic strings as random walks', *Phys. Rev.* **D33**, 3556.

Scherrer, R.J., Melott, A.L., & Bertschinger, E. [1988], 'The formation of large scale structure from cosmic strings and massive neutrinos', *Phys. Rev. Lett.* **62**, 379.

Scherrer, R.J., & Press, W.H. [1989], 'Cosmic string loop fragmentation', *Phys. Rev.* **D39**, 371.

Scherrer, R.J., Quashnock, J.M. Spergel, D.N. & Press. W.H. [1991], 'Properties of realistic cosmic string loops', *Phys. Rev.* **D42**, 1908.

Schramm, D.N., & Vittorio, N. [1985], 'Galaxy formation, nonbaryonic matter and cosmic strings', *Comm. Nucl. Part. Phys.* **15**, 1.

Schwarz, A.S. [1982], 'Field theories with no local conservation of the electric charge', *Nucl. Phys.* **B208**, 141.

Schwinger, J. [1951], 'On gauge invariance and vacuum polarization', *Phys. Rev.* **82**, 664.

Shafi, Q., & Vilenkin, A. [1984a], 'Inflation with $SU(5)$', *Phys. Rev. Lett.* **52**, 691.

Shafi, Q., & Vilenkin, A. [1984b], 'Spontaneously broken global symmetries and cosmology', *Phys. Rev.* **D29**, 1870.

Shafi, Q., & Wetterich, C. [1983], 'Cosmology from higher dimensional gravity', *Phys. Lett.* **129B**, 387.

Shaposhnikov, M. [1987], 'Baryon asymmetry of the universe in standard electroweak theory', *Nucl. Phys.* **B287**, 757.

Shaposhnikov, M. [1988], 'Structure of the high temperature gauge ground state and electroweak production of the baryon asymmetry', *Nucl. Phys.* **B299**, 797.

Shellard, E.P.S. [1986a], 'Axionic domain walls & cosmology', in Proceedings of the 26th Liege International Astrophysical Colloquium, *The Origin and Early History of the Universe*, Demaret, J., ed. (University de Liege).

Shellard, E.P.S., [1986b], 'Quantum effects in the early universe', Ph.D. Thesis, (University of Cambridge).

Shellard, E.P.S. [1987], 'Cosmic string interactions', *Nucl. Phys.* **B283**, 624.

Shellard, E.P.S. [1988], 'Understanding intercommuting', in *Cosmic Strings: The Current Status*, Accetta, F.S., & Krauss, L.M., eds. (World Scientific, Singapore).

Shellard, E.P.S., [1989], unpublished.

Shellard, E.P.S. [1990], 'Axion strings and domain walls', in *Formation and Evolution of Cosmic Strings*, Gibbons, G.W., Hawking, S.W., & Vachaspati, V., eds. (Cambridge University Press).

Shellard, E.P.S. [1992], 'The numerical study of topological defects', in *Approaches to Numerical Relativity*, d'Inverno, R., ed. (Cambridge University Press).

Shellard, E.P.S. [1993], 'Reconnection', DAMTP-GR (in preparation).

Shellard, E.P.S., & Allen, B. [1990], 'On the evolution of cosmic strings', in *Formation and Evolution of Cosmic Strings*, Gibbons, G.W., Hawking, S.W., & Vachaspati, T., eds. (Cambridge University Press).

Shellard, E.P.S., & Brandenberger, R.H. [1988], 'Clusters of galaxies from cosmic strings', *Phys. Rev.* **D38**, 3610.

Shellard, E.P.S., Brandenberger, R.H., Kaiser, N., & Turok, N. [1987], 'Large scale peculiar velocities from cosmic strings', *Nature* **326**, 672.

Shellard, E.P.S. & Davis, R.L [1988], 'Charge, current and the vorton', in *Cosmic Strings: The Current Status*, Accetta, F.S., & Krauss, L.M., eds. (World Scientific, Singapore).

Shellard, E.P.S. & Davis, R.L [1989], 'Modelling superfluid vortex rings', MIT-CTP-1760 (unpublished).

Shellard, E.P.S., & Ruback, P.J. [1988], 'Vortex scattering in two dimensions', *Phys. Lett.* **209B**, 262.

Signore, M., & Sanchez, N. [1989], 'Cosmological gravitational wave background and pulsar timings', *Mod. Phys. Lett.* **4**, 799.

Signore, M., & Sanchez, N. [1990], 'The absence of distortion in the cosmic microwave background spectrum and superconducting strings', *Phys. Lett.* **241B**, 332.

Sikivie, P. [1982], 'Axions, domain walls and the early universe', *Phys. Rev. Lett.* **48**, 1156.

Sikivie, P., [1983], in Proceedings of Summer School on *Particle Physics*, (Gif-sur-Yvette).

Sikivie, P., [1992], 'Dark matter axions', Proceedings of First I.F.T. Workshop on *Dark Matter*, University of Florida, Gainseville, Feb. 14–16, 1992.

Silk, J., & Vilenkin, A. [1984], 'Cosmic strings and galaxy formation', *Phys. Rev. Lett.* **53**, 1700.

Simpson, J.A., [1983], in *Composition and Origin of Cosmic Rays*, Shapiro, M.M., ed. (Reidel, Dordrecht).

Skyrme, T.H.R. [1961], 'Particle states of a quantized meson field', *Proc. Roy. Soc.* **262**, 233.

Smith, A.G. [1990], 'Gravitational effects of infinite straight cosmic strings on classical and quantum fields', in *Formation and Evolution of Cosmic Strings*, Gibbons, G.W., Hawking, S.W., & Vachaspati, T., eds. (Cambridge University Press).

Smith, A.G., & Vilenkin, A. [1987a], 'Fragmentation of cosmic string loops', *Phys. Rev.* **D36**, 987.

Smith, A.G., & Vilenkin, A. [1987b], 'Numerical simulation of cosmic string evolution in flat space-time', *Phys. Rev.* **D36**, 990.

Smoot, G.F., Bennett, C.L., Kogut, A., Wright, E.L., *et al.* [1992], 'Structure in the COBE DMR first year maps', *Ap. J.* **396**, L1.

Soffel, M., Muller, B., & Greiner, W. [1982], 'Stability and decay of the Dirac vacuum in external gauge fields', *Phys. Rep.* **85**, 51.

Sokolov, D.D., & Starobinsky, A.A. [1977], 'The structure of the curvature tensor at conical singularities', *Sov. Phys. Dokl.* **22**, 312.

de Sousa Gerbert, P., & Jackiw, R. [1988], 'Classical and quantum scattering on a spinning cone', *Comm. Math. Phys.* **124**, 229.

de Sousa Gerbert, P. [1989], 'Aharonov–Bohm interaction of cosmic strings with fermionic matter', *Comm. Math. Phys.* **124**, 229.

Spergel, D.N., Piran, T., & Goodman, J. [1987], 'Dynamics of superconducting cosmic strings', *Nucl. Phys.* **B291**, 847.

Spergel, D.N., Press, W.H., & Scherrer, R.J. [1988], 'Self-excited cosmic string dynamos', *Nature* **334**, 682.

Spergel, D.N., Press, W.H., & Scherrer, R.J. [1989], 'Electromagnetic self-interaction of superconducting cosmic strings', *Phys. Rev.* **D39**, 379.

Spergel, D.N., Turok, N., Press, W.H., & Ryden, B.S. [1991], 'Global texture as the origin of large scale structure: numerical simulations of evolution', *Phys. Rev.* **D43**, 1038.

Srednicki, M.K., & Thiesen, S. [1987], 'Nongravitational decay of cosmic strings', *Phys. Lett.* **189B**, 397.

Srivastava, A.M. [1991], 'Importance of boundary conditions for topological production of textures and Skyrmions', *Phys. Rev.* **D43**, 1047.

Starobinsky, A.A. [1980], 'A new type of isotropic cosmological model without singularity', *Phys. Lett.* **91B**, 99.

Starobinsky, A.A. [1982], 'Dynamics of phase transitions in the new inflationary universe scenario and the generation of perturbations', *Phys. Lett.* **117B**, 175.

Staruszkiewicz, A. [1963], 'Gravitation theory in three-dimensional space', *Acta Phys. Pol.* **24**, 735.

Stauffer, D. [1979], 'Scaling theory of percolation clusters', *Phys. Rep.* **54**, 1.

Stebbins, A. [1986], 'Cosmic strings and cold matter', *Ap. J.* **303**, L21.

Stebbins, A. [1988], 'Cosmic strings and the microwave sky. 1: Anisotropy from moving strings', *Ap. J.* **327**, 584.

Stebbins, A., & Turner, M.S. [1989], 'Is the Great Attractor really a great wall?', *Ap. J. Lett.* **339**, L13.

Stebbins, A. & Veeraraghavan, S. [1992], 'MBR anisotropy from scalar field gradients', Fermilab-Pub-92-188-A.

Stebbins, A., Veeraraghavan, A., Brandenberger, R.H., Silk, J., & Turok, N. [1987], 'Cosmic string wakes', *Ap. J.* **322**, 1.

Stecker, F., & Shafi, Q. [1983], 'Axions & the evolution of structure in the universe', *Phys. Rev. Lett.* **50**, 928.

Stein-Schabes, J.A. [1986], 'Nonstatic vacuum strings: exterior and interior solutions', *Phys. Rev.* **D33**, 3545.

Stein-Schabes, J.A., & Burd, A.B. [1988], 'Cosmic strings in an expanding space-time', *Phys. Rev.* **D37**, 1401.

Stern, A. [1984], '$SO(10)$ vortices, zero modes and family cloning', *Phys. Rev. Lett.* **52**, 2118.

Stern, A., & Yajnik, U.A. [1986], '$SO(10)$ vortices and the electroweak phase transition', *Nucl. Phys.* **B267**, 158.

Stinebring, D.R., Ryba, M.F., Taylor, J.H., & Romani, R.W. [1990], 'The cosmic gravitational wave background: Limits from millisecond pulsar timing', *Phys. Rev. Lett.* **65**, 285.

Sunyaev, R.A., & Zel'dovich, Ya.B., [1970], *Astrophys. Space Sci.*, **7**, 3.

Sunyaev, R.A., & Zel'dovich, Ya.B. [1980], 'Microwave background radiation as a probe of contemporary structure and the history of the universe', *Ann. Rev. Astron. Astrophys.* **18**, 53.

Tassie, L.J. [1973], 'Magnetic flux lines as relativistic strings', *Phys. Lett.* **46B**, 397.

Tassie, L.J. [1974], 'Lines of quantized magnetic flux and the relativistic string of the dual resonance model of hadrons', *Int. J. Theor. Phys.* **9**, 167.

Taub, A.H. [1956], 'Plane symmetric spacetimes', *Phys. Rev.* **103**, 454.

Taubes, C.M. [1980], 'Arbitrary N-vortex solutions to the first order Ginzburg–Landau equations', *Comm. Math. Phys.* **72**, 277.

Taylor, J.C. [1976], *Gauge Theories of Weak Interactions*, (Cambridge University Press).

Thompson, C. [1988a], 'Dynamics of cosmic string', *Phys. Rev.* **D37**, 283.

Thompson, C., [1988b], Ph. D. Thesis, (Princeton University).

Thompson, C., [1988c], Private communication (unpublished).

Thompson, C. [1990], 'Cosmic magnetic fields and superconducting strings', *IAU Symposia* **140**, 507.

't Hooft, G. [1974], 'Magnetic monopoles in unified gauge models', *Nucl. Phys.* **B79**, 276.

't Hooft, G. [1976a], 'Magnetic charge quantization and fractionally charged quarks', *Nucl. Phys.* **B105**, 538.

't Hooft, G. [1976b], 'Symmetry breaking through Adler–Bell anomalies', *Phys. Rev. Lett.* **37**, 8.

't Hooft, G. [1988], 'Non-perturbative two particle scattering amplitudes in (2+1)-dimensional quantum gravity', *Comm. Math. Phys.* **117**, 685.

Tipler, F. [1976], 'Causality violation in asymptotically flat spacetimes', *Phys. Rev. Lett.* **37**, 879.

Traschen, J., Turok, N., & Brandenberger, R.H. [1986], 'Microwave anisotropies from cosmic strings', *Phys. Rev.* **D34**, 919.

Tremaine, S., & Gunn, J.E. [1979], 'Dynamical role of light neutral leptons in cosmology', *Phys. Rev. Lett.* **42**, 407.

Tseytlin, A.A., & Vafa, C. [1992], 'Elements of string cosmology', *Nucl. Phys.* **B372**, 443.

Turner, M.S. [1982], 'Thermal production of superheavy magnetic monopoles in the early universe', *Phys. Lett.* **115B**, 95.

Turner, M.S. [1985a], 'Comment on 'String dominated universe (SDU)"', *Phys. Rev. Lett.* **54**, 252.

Turner, M.S., [1985b], in Proceedings of the Cargese School on *Fundamental Physics and Cosmology*, Audouze, J., & Tran Tranh Van, J., eds., (Editions Frontiers: Gif-sur-Yvette).

Turner, M.S., Parker, E.N., & Bogdan, T. [1982], 'Magnetic monopoles and the survival of galactic magnetic fields', *Phys. Rev.* **D26**, 239.

Turner, M.S., Watkins, R., & Widrow, L.M. [1991], 'Microwave distortions from collapsing domain wall bubbles', *Ap. J. Lett.* **367**, L43.

Turok, N. [1983a], 'The production of string loops in an expanding universe', *Phys. Lett.* **123B**, 387.

Turok, N. [1983b], 'The evolution of cosmic density perturbations around grand unified strings', *Phys. Lett.* **126B**, 437.

Turok, N. [1984], 'Grand unified strings and galaxy formation', *Nucl. Phys.* **B242**, 520.

Turok, N. [1985], 'Cosmic string and the correlations of Abell clusters', *Phys. Rev. Lett.* **55**, 1801.

Turok, N. [1988a], 'String driven inflation', *Phys. Rev. Lett.* **60**, 549.

Turok, N., [1988b], 'Phase transitions as the origin of large-scale structure', presented at the 27th Int. Universitatswochen fur Kernphysik, Schladming, Austria, Feb. 22–Mar. 3, 1988.

Turok, N. [1989a], 'String statistical mechanics', *Physica* **158**, 516.

Turok, N. [1989b], 'Global texture as the origin of cosmic structure', *Phys. Rev. Lett.* **63**, 2625.

Turok, N. [1991], 'Global texture as the origin of cosmic structure', *Physica Scripta.* **T36**, 135.

Turok, N., & Bhattacharjee, P. [1984], 'Stretching cosmic strings', *Phys. Rev.* **D29**, 1557.

Turok, N., & Brandenberger, R.H. [1986], 'Cosmic strings and the formation of galaxies and clusters of galaxies', *Phys. Rev.* **D33**, 2175.

Turok, N., & Schramm, D.N. [1984], 'Strings and the origins of galaxies', *Nature* **312**, 598.

Turok, N., & Spergel, D.N. [1990], 'Global texture and the microwave background', *Phys. Rev. Lett.* **64**, 2736.

Turok, N., & Spergel, D.N. [1991], 'Scaling solution for cosmological sigma models at large N', *Phys. Rev. Lett.* **66**, 3093.

Turok, N., & Zadrozny, J. [1990], 'Dynamical generation of baryons at the electroweak transition', *Phys. Rev. Lett.* **65**, 2331.

Tyupkin, Y.S., Fateev, V.A., & Schwarz, A.S., [1975], *Sov. Phys. JETP Lett.* **21**, 42.

Unruh, W.G., Hayward, G., Israel, W., & McManus, D. [1989], 'Cosmic string loops are straight', *Phys. Rev. Lett.* **62**, 2897.

Vachaspati & Vachaspati, T. [1990], 'Travelling waves on domain walls and cosmic strings', *Phys. Lett.* **238B**, 41.

Vachaspati, T. [1986], 'Gravitational effects of cosmic strings', *Nucl. Phys.* **B277**, 593.

Vachaspati, T. [1987a], 'Gravity of cosmic loops', *Phys. Rev.* **D35**, 1767.

Vachaspati, T. [1987b], 'The gravity of cosmic loops', *Gen. Rel. Grav.* **19**, 1053.

Vachaspati, T. [1988a], 'Cosmic strings & the large-scale structure of the universe', *Phys. Rev. Lett.* **57**, 1655.

Vachaspati, T. [1988b], 'Cosmic string loop self-intersections and intercommuting', *Phys. Rev.* **D39**, 1768.

Vachaspati, T., [1988c], unpublished.

Vachaspati, T. [1991], 'The formation of topological defects', *Phys. Rev.* **D44**, 3723.

Vachaspati, T. [1992], 'The structure of wiggly cosmic string wakes', *Phys. Rev.* **D45**, 3487.

Vachaspati, T. [1993], 'Electroweak strings', *Nucl. Phys.* **B397**, 648.

Vachaspati, T. & Achucarro, A. [1991], 'Semi-local cosmic strings', *Phys. Rev.* **D44**, 3067.

Vachaspati, T., Everett, A.E., & Vilenkin, A. [1984], 'Radiation from vacuum strings and domain walls', *Phys. Rev.* **D30**, 2046.

Vachaspati, T., & Vilenkin, A. [1984], 'Formation and evolution of cosmic strings', *Phys. Rev.* **D30**, 2036.

Vachaspati, T., & Vilenkin, A. [1985], 'Gravitational radiation from cosmic strings', *Phys. Rev.* **D31**, 3052.

Vachaspati, T., & Vilenkin, A. [1987], 'Evolution of cosmic networks', *Phys. Rev.* **D35**, 1131.

Vachaspati, T., & Vilenkin, A. [1991], 'Large scale structure from wiggly cosmic strings', *Phys. Rev. Lett.* **67**, 1057.

Van Dalen, A., & Schramm, D.N. [1988], 'Cosmic string induced peculiar velocities', *Ap. J.* **326**, 70.

Veeraraghavan, S., & Stebbins, A. [1990], 'Causal compensated perturbations in cosmology', *Ap. J.* **365**, 37.

Veeraraghavan, S., & Stebbins, A. [1992], 'Large scale microwave anisotropy from gravitating seeds', *Ap. J.* **395**, 55.

de Vega, H.J., & Schaposnik, F.A. [1976], 'A classical vortex solution of the Abelian Higgs model', *Phys. Rev.* **D14**, 1100.

de Vega, H.J., & Schaposnik, F.A. [1986], 'Electrically charged vortices in nonabelian gauge theories', *Phys. Rev. Lett.* **56**, 2564.

Vickers, J.A.G. [1987], 'Generalized cosmic strings', *Class. Quant. Grav.* **4**, 1.

Vilenkin, A. [1981a], 'Cosmological density fluctuations produced by vacuum strings', *Phys. Rev. Lett.* **46**, 1169. Erratum: *Phys. Rev. Lett.* **46**, 1496.

Vilenkin, A. [1981b], 'Gravitational field of vacuum domain walls and strings', *Phys. Rev.* **D23**, 852.

Vilenkin, A. [1981c], 'Cosmic strings', *Phys. Rev.* **D24**, 2082.

Vilenkin, A. [1981d], 'Gravitational radiation from cosmic strings', *Phys. Lett.* **107B**, 47.

Vilenkin, A. [1982a], 'Creation of universes from nothing', *Phys. Lett.* **117B**, 25.

Vilenkin, A. [1982b], 'Cosmological density fluctuations produced by a Goldstone field', *Phys. Rev. Lett.* **48**, 59.

Vilenkin, A. [1982c], 'Cosmological evolution of monopoles connected by strings', *Nucl. Phys.* **B196**, 240.

Vilenkin, A. [1983a], 'Quantum fluctuations in the new inflationary universe', *Nucl. Phys.* **B226**, 527.

Vilenkin, A. [1983b], 'Birth of inflationary universes', *Phys. Rev.* **D27**, 2848.

Vilenkin, A. [1983c], 'Gravitational field of vacuum domain walls', *Phys. Lett.* **133B**, 177.

Vilenkin, A. [1984a], 'Cosmic strings as gravitational lenses', *Ap. J.* **L51**, 282.

Vilenkin, A. [1984b], 'String dominated universe', *Phys. Rev. Lett.* **53**, 1016.

Vilenkin, A. [1984c], 'Causality gives no constraints on monopole annihilation', *Phys. Lett.* **136B**, 47.

Vilenkin, A. [1985], 'Cosmic strings and domain walls', *Phys. Rep.* **121**, 263.

Vilenkin, A. [1986a], 'Boundary conditions in quantum cosmology', *Phys. Rev.* **D33**, 3560.

Vilenkin, A. [1986b], 'Looking for cosmic strings', *Nature* **322**, 613.

Vilenkin, A. [1987a], 'Gravitational interactions of cosmic strings', in *300 Years of Gravitation*, Hawking, S.W., & Israel, W., eds. (Cambridge University Press).

Vilenkin, A. [1988], 'Gamma rays from superconducting cosmic strings', *Nature* **332**, 610.

Vilenkin, A. [1990], 'Effect of small-scale structure on the dynamics of cosmic strings', *Phys. Rev.* **D41**, 3038.

Vilenkin, A. [1991], 'Cosmic string dynamics with friction', *Phys. Rev.* **D43**, 1060.

Vilenkin, A., & Everett, A.E. [1982], 'Cosmic strings and domain walls in models with Goldstone and pseudo-Goldstone bosons', *Phys. Rev. Lett.* **48**, 1867.

Vilenkin, A., & Field, G.B. [1987], 'Quasars and superconducting cosmic strings', *Nature* **326**, 772.

Vilenkin, A., & Ford, L.H. [1982], 'Gravitational effects upon cosmological phase transitions', *Phys. Rev.* **D26**, 1231.

Vilenkin, A., & Shafi, Q. [1983], 'Density fluctuations from strings and galaxy formation', *Phys. Rev. Lett.* **51**, 1716.

Vilenkin, A., & Vachaspati, T. [1987a], 'Radiation of Goldstone bosons from cosmic strings', *Phys. Rev.* **D35**, 1138.

Vilenkin, A., & Vachaspati, T. [1987b], 'Electromagnetic radiation from superconducting cosmic strings', *Phys. Rev. Lett.* **58**, 1041.

Vishniac, E.T., Olive, K.A., & Seckel, D. [1987], 'Cosmic strings and inflation', *Nucl. Phys.* **B289**, 717.

Vollick, D.N. [1992a], 'Small-scale structure on cosmic strings and galaxy formation', *Phys. Rev.* **D45**, 1884.

Vollick, D.N. [1992b], 'String induced density perturbations in hot dark matter', *Ap. J.* **397**, 14.

Voloshin, N.B., Kobzarev, I.Yu., & Okun', L.B. [1975], 'Bubbles in a metastable vacuum', *Sov. J. Nucl. Phys.* **20**, 644.

Wang, J. [1990], 'Cosmic strings in the Sun', *Mod. Phys. Lett.* **6**, 2101.

Weinberg, D.H., & Gunn, J.E. [1990], 'Large-scale structure and the adhesion approximation', *Ap. J.* **247**, 260.

Weinberg, E.J. [1979], 'Multi-vortex solutions of the Ginzburg–Landau equations', *Phys. Rev.* **D19**, 3008.

Weinberg, E.J. [1981], 'Index calculations for the fermion–vortex system', *Phys. Rev.* **D24**, 2669.

Weinberg, E.J. [1983], 'Constraints on the annihilation of primordial magnetic monopoles', *Phys. Lett.* **126B**, 441.

Weinberg, S. [1972], *Gravitation and Cosmology*, (Wiley, New York).

Weinberg, S. [1973], 'Perturbative calculations of symmetry breaking', *Phys. Rev.* **D7**, 2887.

Weinberg, S. [1974], 'Gauge and global symmetries at high temperature', *Phys. Rev.* **D9**, 3357.

Weinberg, S. [1978], 'A new light boson?', *Phys. Rev. Lett.* **40**, 223.

Weinberg, S. [1979], 'Cosmological production of baryons', *Phys. Rev. Lett.* **42**, 850.

Widrow, L.M. [1988], 'Zero modes and anomalies in superconducting strings', *Phys. Rev.* **D38**, 1684.

Widrow, L.M. [1989a], 'Dynamics of thick domain walls', *Phys. Rev.* **D40**, 1002.

Widrow, L.M. [1989b], 'The collapse of nearly spherical domain walls', *Phys. Rev.* **D39**, 3576.

Widrow, L.M. [1989c], 'General relativistic domain walls', *Phys. Rev.* **D39**, 3571.

Wilczek, F. [1978], 'Problem of strong P and T invariance in the presence of instantons', *Phys. Rev. Lett.* **40**, 279.

Wilson, K.G., & Kogut, I. [1974], 'The renormalization group and the ϵ expansion', *Phys. Rep.* **12C**, 75.

Witten, E. [1981], 'Cosmological consequences of a light Higgs boson', *Nucl. Phys.* **B177**, 477.

Witten, E. [1984], 'Cosmic separation of phases', *Phys. Rev.* **D30**, 272.

Witten, E. [1985a], 'Superconducting strings', *Nucl. Phys.* **B249**, 557.

Witten, E. [1985b], 'Cosmic superstrings', *Phys. Lett.* **153B**, 243.

Witten, E. [1985c], 'Symmetry breaking in superstring models', *Nucl. Phys.* **B258**, 75.

Xanthopoulos, B.C. [1986a], 'A rotating cosmic string', *Phys. Lett.* **178B**, 163.

Xanthopoulos, B.C. [1986b], 'Cylindrical waves and cosmic strings of Petrov type D', *Phys. Rev.* **D34**, 3608.

Xanthopoulos, B.C. [1987], 'Cosmic strings coupled with gravitational and electromagnetic waves', *Phys. Rev.* **D35**, 3713.

Yajnik, U.A. [1986], 'Phase transition induced by cosmic strings', *Phys. Rev.* **D34**, 1237.

Yajnik, U.A. [1987], 'Exotic configurations for gauge theory strings', *Phys. Lett.* **184B**, 229.

Yajnik, U.A., & Padmanabhan, T. [1987], 'Analytical approach to string induced phase transition', *Phys. Rev.* **D35**, 3100.

Yokoyama, J. [1988], 'Natural way out of the conflict between cosmic strings and inflation', *Phys. Lett.* **212B**, 273.

Yokoyama, J. [1989], 'Inflation can save cosmic strings', *Phys. Rev. Lett.* **63**, 712.

York, T. [1989], 'Fragmentation of cosmic string loops', Fermilab-Pub-89/32-A (unpublished).

Zel'dovich, Ya.B. [1970], 'Gravitational instability: an approximate theory for large density perturbations', *Astron. Ap.* **5**, 84.

Zel'dovich, Ya.B. [1980], 'Cosmological fluctuations produced near a singularity', *M. N. R. A. S.* **192**, 663.

Zel'dovich, Ya.B., & Khlopov, M.Yu. [1978], 'On the concentration of relic magnetic monopoles in the Universe', *Phys. Lett.* **79B**, 239.

Zel'dovich, Ya.B., Kobzarev, I.Yu., & Okun, L.B. [1974], 'Cosmological consequences of the spontaneous breakdown of a discrete symmetry', *Sov. Phys. JETP* **40**, 1.

Zel'dovich, Ya.B., & Novikov, I.D. [1983], *Relativistic Astrophysics Vol. 2*, (University of Chicago Press).

Zel'dovich, Ya.B., Ruzmaikin, A.A., & Sokoloff, D.D. [1983], *Magnetic Fields in Astrophysics*, (Gordon and Breach, New York).

Zhang, S. [1987], 'Quantum fluctuations of the superconducting cosmic string', ITP-SB-87-33 (unpublished).

Zhou, Z. [1988], 'The accretion of moving cosmic string', *Chin. Phys. Lett.* **5**, 93.

Zurek, W.H. [1985a], 'Cosmological experiments in superfluid helium?', *Nature* **317**, 505.

Zurek, W.H. [1985b], 'String-seeded spirals: origin of rotation & the density of the halo', *Phys. Rev. Lett.* **57**, 2326.

Zurek, W.H. [1988], 'Richness distribution of string-seeded Abell clusters: accretion by a loop vs. hierarchical clustering', *Ap. J.* **324**, 19.

Zurek, W.H., & Quinn, P.J. [1987], 'Rotation profile of a string-seeded halo', Los Alamos preprint (unpublished).

Index